FRONTIERS IN STATISTICS

FRONTIERS IN STATISTICS

Dedicated to Peter John Bickel
in Honor of His 65th Birthday

Editors

Jianqing Fan
Princeton University, USA

Hira L Koul
Michigan State University, USA

Imperial College Press

Published by

Imperial College Press
57 Shelton Street
Covent Garden
London WC2H 9HE

Distributed by

World Scientific Publishing Co. Pte. Ltd.
5 Toh Tuck Link, Singapore 596224
USA office: 27 Warren Street, Suite 401-402, Hackensack, NJ 07601
UK office: 57 Shelton Street, Covent Garden, London WC2H 9HE

British Library Cataloguing-in-Publication Data
A catalogue record for this book is available from the British Library.

ISBN   1-86094-670-4
ISBN   1-86094-698-4 (pbk)

Printed by Mainland Press Pte Ltd

To Peter J. Bickel

Our Teacher and Friend

Preface

The innovations in information and technology have revolutionized scientific research and knowledge discovery. They allow us to collect massive amount of data at relatively low cost. Observations with one-dimensional curves, two-dimensional images, and three-dimensional movies are commonly seen at the frontiers of scientific research. Statistical challenges arise from diverse fields of sciences, engineering, and humanities, ranging from genomic research, biomedical studies, and natural resource discovery to machine learning, financial engineering, and risk managements. High-dimensionality and massive amounts of data are the common features at the frontiers of statistical studies. These have profound impacts on statistical thinking, methodological development, and theoretical studies.

The 1980s and 1990s saw a phenomenal growth in numerous areas of statistical modeling and inferences. These areas include semiparametric models, data-analytic nonparametric methods, statistical learning, network tomography, analysis of longitudinal data, financial econometrics, time series, bootstrap and other resampling methodologies, statistical computing, mixed effects models, and robust multivariate analysis. This volume consists of 16 review articles and 6 research papers in these and some other related areas contributed by eminent leaders in statistics, who are the friends and former students of Professor Peter J. Bickel. It gives an overview of new developments and a contemporary outlook on the various frontiers of statistics. It is dedicated to Professor Peter John Bickel in honor of his 65th birthday. The monograph features an article by Kjell Doksum and Ya'acov Ritov that summarizes some of the Bickel's distinguished contributions. Most of the articles in the book will be presented at the "Workshop on Frontiers of Statistics", May 18 – 20, 2006, Princeton, New Jersey, co-sponsored by the Institute of Mathematical Statistics and International Indian Statistical Association, chaired by Jianqing Fan.

Ever since the publication of the monograph by Bickel, Klaassen, Ritov and Wellner (1993), the development in the asymptotically efficient inference in semiparametric models has seen an exponential growth as is indicated in the chapter by Wellner, Klaassen and Ritov in this book. The contents of the chapter by Schick and Wefelmeyer review the developments of the asymptotically efficient inference in semiparametric time series models that have been developed only in the last 12 years and are on the cutting edge of this area of research. The chapter by Xia and Tong reviews numerous estimation methods in generalized linear regression models with an unknown link function, the so-called single index models, an area of research presently at the forefront. Van der Laan and Rubin discuss some innovative procedures based on cross validation for selecting estimating functions for estimation of parameters of interest. An oracle property is established under a general condition.

Closely related to the semiparametric modeling is the nonparametric methods for function estimation and their related applications. The area has been very active over the last twenty years, thanks to the exponential increase of computing power and availability of large data sets. Doksum and Schafer propose bandwidth selection procedures in local linear regression by maximizing the limiting power when testing the hypothesis of constant regression against local nonparametric Pitman alternatives. Among other things they show that the power optimal bandwidth can give much higher power than the bandwidth chosen by minimizing the mean squared error. The chapter by Hall and Kay has bearing on the classification of a given observation. They provide a new estimator of an index of atypicality and assess some of its optimality properties. Wang reviews wavelet statistical methodologies developed in the past fifteen years with emphasis on estimating regression functions, detecting and estimating change-points, solving statistical inverse problems, and studying self-similarity of a stochastic process. A classical problem in statistics is that of model diagnostics. In particular the most investigated model diagnostic problems in the literature are those of the lack-of-fit testing of a parametric regression model and the goodness-of-fit hypothesis of an error distribution in a given regression model. The chapter by Koul demonstrates that nonparametric techniques play an important role for lack-of-fit testing. It gives a brief review of asymptotically distribution free tests for these hypotheses based on certain marked empirical processes and residual empirical processes, respectively. The underlying theme is that all of these tests can be based on certain martingale transforms of these processes.

Related to nonparametric methods are boosting and bootstrap. An effective method of nonparametric classification and regression is boosting. The paper of Bühlmann and Lutz provides a review of boosting and proposes a new bootstrap method for high-multivariate, linear time series. Lahiri reviews the literature on bootstrap methodology under independence and for different classes of dependent processes including Markov processes, long range dependent time series and spatial processes. An important tool for theoretical justification of bootstrap methods is asymptotic expansions. Such expansions are well understood when underlying distributions are either absolutely continuous or pure lattice distributions. Götze and van Zwet begin an investigation of discrete non-lattice distributions and discuss some new challenges in this interesting area of research.

The last two decades have seen a considerable amount of literature on the analysis of longitudinal and functional data. The chapter by Fan and Li presents an overview on recent developments in non- and semi-parametric inference for longitudinal data. Müller and Yao give a review of functional regression models and of the principal component analysis through a conditional expectation algorithm when having sparse longitudinal data.

Many exciting statistical developments over the last two decades lie at the interface between statistics and other subject disciplines. The chapter by Servidea and Meng demonstrates the fruitfulness of cross-fertilization between statistical physics and statistical computation, by focusing on the celebrated Swendsen-Wang algorithm for the Ising model and its recent perfect sampling implementation by Huber. Furthermore, it outlines some important results and open problems in these areas. Lawrence, Michailidis, Nair and Xi provide a review of the statistical issues and developments in network tomography with an emphasis on active tomography and illustrate the results with an application to internet telephony. A comprehensive overview on likelihood inferences for diffusion processes is given by Aït-Sahalia, based on discretely sampled data. This is an important subject in financial econmetrics where statistics plays an important role. Aït-Sahalia gives extensive treatments on the expansions of transition densities and their applications. The chapter by Park and Jeong gives a review of the recently developed theory for several promising nonparametric estimators of frontiers or boundaries in productivity analysis.

Many contemporary statistical problems borrow the ideas from traditional statistics. Ghosh and Tokdar presents some of the recent development on the convergence and consistency proof of Newton's algorithm for estimating a mixing distribution. Jiang and Ge give an overview of linear, generalized linear and nonlinear mixed effects models with emphases on recent developments and challenges. It is well known that the sample mean vector and the sample covariance matrix are optimal when the underlying data are normal but extremely sensitive to outliers and heavy tailed data. The chapter by Zuo surveys numerous robust alternatives of these classical location and scatter estimators and discusses their applications to the multivariate data analysis. Wong proposes an innovative idea for estimating the loss of an estimate.

All the chapters in the book have been reviewed by world experts. We take this opportunity to thank all the referees for rendering their invaluable help: Sarat Dass, Irene Gijbels, Jayanta Ghosh, Heng Peng, Marian Hristache, Jinhua Huang, Ildar Ibragimov, Gareth James, Jiming Jiang, Soumendra Lahiri, Linyuan Li, Yin Li, Shiqing Ling, Neal Madras, Moushen Pourhamadi, Anton Schick and Chunming Zhang. We are most grateful to the enthusiastic support of all the people who have helped us. We are particularly indebted to Dr. Heng Peng and Dr. Chongqi Zhang for invaluable help and editorial assistance, turning individual contributions into such a wonderful book. Assistance from graduate students at Princeton, including Yingying Fan, Clifford Lam, Yue Niu and Jinchi Lv are gratefully acknowledged. The generous financial supports from National Science Foundation, Minerva Foundation, Bendheim Center for Finance, and Department of Operations Research and Financial Engineering make the workshop and this volume possible.

Professor Peter J. Bickel has made numerous significant contributions to many of the fields discussed in this volume as is evidenced by references to his work. We present this volume to him in celebration of his 65th birthday. We look forward to his many more innovative contributions to statistical science.

Jianqing Fan, Princeton
Hira L. Koul, East Lansing
December 22, 2005

Contents

Part I. Semiparametric Modeling

---

3. Efficient Estimator for Time Series

4. On the Efficiency of Estimation for a Single-index Model

5. Estimating Function Based Cross-Validation

Part II. Nonparametric Methods

6. Powerful Choices: Tuning Parameter Selection Based on Power

Part III. Statistical Learning and Bootstrap

Part IV. Longitudinal Data Analysis

Part V. Statistics in Science and Technology

15. Statistical Physics and Statistical Computing: A Critical Link

16. Network Tomography: A Review and Recent Developments

Part VI. Financial Econometrics

17. Likelihood Inference for Diffusions: A Survey

Part VII. Parametric Techniques and Inferences

Peter Bickel, age 3, with his mother Madeleine Moscovici Bickel and father Eliezer Bickel, September 1943, Bucharest, Rumania.

Peter with his uncle and aunt Shlomo and Yetta Bickel and cousin Alexander Bickel, New York City in the early 1950s.

Left: Peter and Nancy Kramer Bickel, June 1964, shortly after their marriage, Berkeley, California. Right: Peter and Nancy Bickel, expecting their first child, November 1966, Berkeley.

Peter and Nancy with son, Stephen, and daughter, Amanda, Winter 1968, Berkeley.

Yahav and Bickel families in Berkeley in August 1975, from left to right, back row, Yossi (Joseph), Gili and Aviva Yahav, Peter Bickel, front row, Amanda Bickel, Amit Yahav, Steve Bickel.

Nancy and Peter with their newly married children, from top, Amanda Bickel, Peter Mayer, Stephen Bickel, Eliyana Adler, August 1993.

Jack Kiefer (1924–81) introduces Peter and Nancy to mushroom hunting and they find 6 kilos of chanterelles at Point Reyes National Seashore.

Peter Bickel and Hira Koul at the French Statistical Society meeting in Brussels, 2002.

David Blackwell, Peter Bickel and Erich Lehmann at the party celebrating Peter's election to the National Academy of Sciences in May 1986.

Peter Bickel's mother, Madeleine Korb, left, chats with Peter and Gene Hammel at the same party.

Dean Len Kuhi and Joe Hodges celebrate with Peter at the party in May 1986.

Peter Bickel and Ya'acov Ritov chat over lunch.

Peter Bickel, Juliet Shaffer and her husband Erich Lehmann, Kjell Doksum and David Freedman at Erich Lehmann's 65th birthday celebration,
Berkeley 1982.

Kjell Doksum, Peter Bickel and Willem Van Zwet on the same occasion.

Peter Bickel and Ya'acov Ritov at the Multivariate Analysis Meeting in Hong Kong, May 1997.

Mary Lou Breiman, Susan Olshen, Leo Breiman and Richard Olshen at the party for Nancy and Peter's 40th anniversary, Berkeley, August 2004.

Leo Goodman and David Siegmund at Nancy and Peter's party, August 2004.

Nancy and Peter celebrating 40 years together, Berkeley, August 2004.

Peter Bickel with Jianqing Fan (left) and Ping Zhang (right), 2005, Princeton University.

Peter Bickel with Yonghua Wang, Princeton University, 2005.

CHAPTER 1

Our Steps on the Bickel Way

Kjell Doksum, Ya'acov Ritov

Department of Statistics
University of Wisconsin
Madison, WI 53706
doksum@stat.wisc.edu

Department of Statistics
Hebrew University of Jerusalem
Mt. Scopus, Jerusalem 91905, Israel
yaacov@mscc.huji.ac.il

This chapter reviews some of Peter Bickel's research contributions briefly.

1.1. Introduction

Bickel has been a leading figure in the field of statistics in the forty-three years since he received his Ph.D. in Statistics at the age of 22 from UC Berkeley in 1963 under the guidance of Erich Lehmann. His contributions are in education, research, and service to the profession and society. In education we find more than fifty students who have received their Ph.D. in statistics under Bickel's guidance and a number of these are contributing to this volume. In addition we have found that doing joint work with Bickel is a real educational experience.

Figure 1.1.   Peter J. Bickel

Bickel's many collaborators have benefited from his statistical insights and wisdom. Bickel has also contributed to the statistical education of a large number of statistics Ph.D. students who took courses using his

1

joint text book "Mathematical Statistics: Basic Ideas and Selected Topics", Bickel and Doksum (2000).

His professional service includes twice being chair of the Berkeley statistics department, twice being dean of the Physical Sciences, being director of the Statistical Computing Facility and being director of the Statistical Laboratory, all at UC Berkeley. It includes also being a member and chair of a number of national and international committees and commissions including the National Research Council, the National Institute of Statistical Sciences, the National Academy of Science, the American Association for the Advancement of Science, and Council of Scientific Advisors, EURANDOM. He has been associate editor of the Annals of Statistics, Bernoulli, Statistica Sinica and the Proceedings of the National Academy of Sciences. He has received most of the honors in our profession including Guggenheim, NATO, Miller and the MacArthur Fellowships. He has received the COPSS prize, been Wald lecturer and has been president of the Institute of Mathematical Statistics and the Bernoulli Society. In addition he is a member of the National Academy of Science, the National Academy of Arts and Sciences and the Royal Netherlands Accademie of the Arts and Sciences.

In the rest of this piece we will focus on Bickel's research contribution to statistics. We are not going to discuss all of Bickel's contribution to statistics — it is too great, in too many fields, to be included in one survey. We are not going even to discuss all of his important work — we do not know them well enough. The following is just some of Bickel's work that *we* found interesting and influential. The subject of this summary is what *we* learnt from Bickel's, not on what can be learnt from him. Even so, it is shorter than his work deserves.

1.2. Doing Well at a Point and Beyond

The paper Bickel (1983) is not known very well. The toy problem is simple. We wish to estimate μ, the mean of a normal variable with variance one. We look for an estimator which behaves reasonably well whatever is the true value of the parameter, but we want it to behave really well at a given point, say 0. Bickel translates this to the minimization of the loss at 0, subject to a bound on the loss else. A reasonable approximation to the estimator is given by $\hat{\theta}(X) = (|X| - c)_+ \text{sgn}(x)$. Bickel generalizes the model to the multinormal distribution with an estimator that does well on a subspace, but this is still a toy model.

This however is an example of some of the Bickel's research. It deals with a very big philosophical problem by leaving out non-essential elements and contributing important theoretical consideration.

The "standard" way to deal with the above mentioned problem is pre-testing: test whether $\mu = 0$. If the null is accepted, estimate μ by 0, otherwise by x. This method lacks a rigorous justification, and the subject deserved a rigorous treatment. The claim of the paper is a soft-Bayesian. You should not ignore your assumptions. If you believe that the parameter belongs to a subset, you should use this fact. However, you should not ignore the possibility that you are wrong, and therefore you should use an estimator, that although does well on the specific subset, behaves reasonably well everywhere else.

The resulting estimator resembles the pre-testing estimate. However, it is different. If the observation is greater than the threshold, then $X - c$ is used and not just X. Moreover, the level of the test is not arbitrary, but is a result of the imposed natural constraint.

Bickel extends the ideas to practical frameworks in the paper Bickel (1984). Here he considers procedures that are "optimal" with respect to a given risk function over a submodel M_0 of models subject to doing "well" over a larger class M_1 of models. One of the problems considered is the familiar linear model problem where we are to choose between using a model M_0 with r parameters, or a model M_1 with s parameters, where $r < s$. He first solves the problem for known covariance matrices, then shows that the results hold asymptotically when these matrices are replaced by estimates for models where the parameter is of the Pitman form $\theta_n = \theta_0 + an^{-1/2}$ with $\theta_0 \in R^r$ and $a \in R^s$. In the case where the risk is mean square error, he finds that for a certain class of estimates, the asymptotic solution to the "optimal" at M_0 and good over M_1, formulation is a linear combination of the MLE's under the models M_0 and M_1, with the weights in the combination determined by a function of the Wald statistic for testing M_0 vs. M_1. Bickel's idea of using parameters in the Pitman form is what makes it possible to derive asymptotic procedures and results for this difficult problem where inference and model selection problems are tackled together from a robustness point of view. More recently, Clasekens and Hjorth (2003) have used parameters in Pitman form to deal with estimation and model selection problems from the point of view of minimizing asymptotic mean squared error under the general model M_1. A comparison of these procedures with those of Bickel is of interest.

1.3. Robustness, Transformations, Oracle-free Inference, and Stable Parameters

Much of Bickel's work is concerned with robustness, that is, the performance of statistical procedures over a wide class of models that typically are semiparametric models that describe neighborhoods of ideal models. The frameworks considered include one and two sample experiments, regression experiments, and time series as well as general frameworks. In the case of Box and Cox (1964) transformation regression models where a transformation $h(Y; \lambda)$ of the response Y follow a linear regression model with coefficient vector β, error variance σ^2, transformation parameter λ and error distribution F, Bickel considered in the joint work Bickel and Doksum (1981) robust estimation of the Euclidean parameters for semiparametric models where F is general. In particular, they established the extra variability of the estimate of β due to the unknown λ. This work was controversial, cf. Box and Cox (1982) and Hinkley and Runger (1984), for a brief period, because it seemed that the coefficient vector β depends on λ and results that depend on a model that assumes that lambda is unknown were claimed to be "scientifically irrelevant". Fortunately the coefficient vector has an intuitive interpretation independent of λ and there is no need to depend on an oracle to provide the true λ: Brillinger (1983) shows how to do inference when a transformation is unknown, by showing that the parameter $\alpha = \beta/|\beta|$, where $|\ |$ is the Euclidean norm, is identifiable and independent of the transformation. It has the interpretation giving the relative importance of the covariates provided they have been standardized to have SD's equal to one. He also provided $\sqrt{n}$ inference for α and gave the inflation in variability due to the unknown transformation. These ideas and results were developed further by Stoker (1986) and lead to the field of index models in statistics and econometrics as well as the concept of stable parameters, e.g. Cox and Reid (1987), Doksum and Johnson (2002). In Bickel and Ritov (1997), Bickel developed asymptotically efficient rank based estimates of alpha for general transformation regression models with increasing transformations, a project that had been approximated by Doksum (1987).

1.4. Distribution Free Tests, Higher Order Expansions, and Challenging Projects

Bickel is not one to shy away from challenging problems. One of the most challenging projects ever carried out in statistics was Bickel's joint work

Bickel and van Zwet (1978) on higher order expansions of the power of distribution free tests in the two sample case and the joint work Alber, Bickel and van Zwet (1976) in the one sample case. They considered situations where the $n^{-1/2}$ term in the asymptotic expansion of the power is zero and n^{-1} term is required. In a classic understatement they wrote "the proofs are a highly technical matter". They used their results to derive the asymptotic Hodges-Lehmann deficiency d of Pitman efficient distribution free tests with respect to optimal parametric tests for given parametric models. Here d is defined as the limit of the difference between the sample size required by a Pitman efficient distribution free test to reach a given power to the sample size required by the optimal parametric test to reach the same power. They found that in normal models, the optimal permutation tests have deficiency zero, while this is not true for locally optimal rank (normal scores) tests.

1.5. From Adaptive Estimation to Semiparametric Models

In his 1980 Wald lecture, Bickel (1982), Bickel discussed the ideas of Stein (1957) on adaptive estimation. It was already well established (Beran (1974), Stone (1975), Sacks (1975), Fabian (1978)) that adaptive estimation is possible for the estimation of the center θ of symmetric distribution on the line. That is, asymptotically θ can be estimated as well when the density f of $x - \theta$ is unknown as when f is known. In his paper Bickel discussed the conditions needed to ensure that adaptive estimation is generally possible. The meeting with Jon Wellner resulted in extending the scope to the general semiparametric model. A project of almost 10 years started in which the estimation in the presence of non-Euclidean nuisance parameters was discussed and analyzed to the fine details. It included a general analysis, mainly in Bickel, Klaassen, Ritov and Wellner (1993), and discussion of specific models.

The statistical models considered in the book and the relevant papers were interesting themselves, but almost all of them presented a more general issue. They were chosen not only because they had an interesting application, but because they were fitted to present a new theoretical aspect.

Thus the title of Bickel and Ritov (1988) is "Estimating integrated squared density derivatives". Although the estimation of the integral of the square of the density can be motivated, and was done in the past, this was not the reason the paper was written. Information bounds in regular parametric model are achievable. That is, there are estimators which achieve these bounds. There was a conjecture that the same is true also for semiparametric models. Before this paper, different estimators for different parameters and models were presented which achieved the information bound, but always more conditions were needed in the construction of the estimator section, than were needed to establish the bounds. An example was needed to clarify this point, and in Bickel and Ritov (1988) it was shown that if not enough smoothness is assumed, there may be rate bounds which are not even of the $\sqrt{n}$ rate. The title of Bickel and Ritov (1990) was less modest and presented a more general claim (although the paper just gave some more examples): "Achieving information bounds in semi and non parametric models."

Similarly, Bickel and Ritov (1993) dealt with a relatively minor situation (although, it generalizes nicely to deal with many censoring situations). The title is "Efficient estimation using both direct and indirect observations." The interest in this problem was in effect different. The situation is thus that the information bound for the parameter of interest using only the indirect observation is 0. The question asked was whether a sub-sample which seemingly carries no information is important when it can be combined with an informative sample. The surprising answer was yes.

The book by Bickel, Klaasen, Ritov and Wellner (1993) on efficient and adaptive estimation was a real project, which ended in 1993. It needed careful and tedious work. The effort was to cover all relevant aspects of semi-parametric models, from information bounds to efficient estimation of both the Euclidean and non-Euclidean parameters.

The book dealt only with i.i.d. observations. Bickel and Kwon (2001) extended the ideas beyond the i.i.d. model. In this paper the authors formulate a "calculus" similar to that of the i.i.d. case that enables them to analyze the efficiency of procedures in general semiparametric models when a nonparametric model has been defined. The extension includes regression models (in their functional form), counting process models in survival analysis and submodels of Markov chains, which traditionally require elaborate special arguments.

1.6. Hidden Markov Models

The hidden Markov model (HMM) is a simple extension of the parametric Markov model. The only difference is that we have noisy observations of the states of the chain. Formally, let $X_1, X_2, \ldots, X_n$ be an unobserved Markov chain with a finite state space. Let $Y_1, Y_2, \ldots, Y_n$ be independent given the Markov chain, where the conditional distribution of Y_i given $X_1, X_2, \ldots, X_n$

depends only on X_i and maybe on an unknown parameter $\theta \in \Theta \subseteq R^d$.

After the long time spent on semi-parametric models, this seemed to be just a simple parametric model. The theory of regular parametric models for Markov chains is a natural extension of that of the i.i.d. case. HMMs looked to be the next natural step, and developing their theory seemed to be an interlude in the drama of semi-parametric research. It was not. The model needed serious work. The first paper, Bickel and Ritov (1996) was a long tedious analysis with more than 20 lemmas, which resulted at the end with a too weak result (the LAN condition is satisfied for the HMM models). What was needed was the help of Tobias Rydén, who could write a one page formula without a mistake, Bickel, Ritov and Rydén (1998). The third paper, Bickel, Ritov and Rydén (2002), was elegant, if its complicated notation is deciphered.

1.7. Non- and Semi-parametric Testing

The paper Bickel and Rosenblatt (1973) is seemingly about density estimation. But it set the stage to our further work on testing. Limit theorems are obtained for the maximum of the normalized deviation of the density estimate from its expected value, and for quadratic norms of the same quantity. This sets the stage for chi-square type of tests and for a consideration of local large deviation in some unknown location from the null hypotheses. Aït-Sahalia, Bickel and Stoker (2001), applied these ideas to regression models and econometrical data.

There is a real theoretical problem with testing. It is not clear what can be really achieved in non-parametric testing: should the researcher be greedy and look for (almost) everything? That is, all deviations are considered equally likely. The price for such an omnipotent test, is having an impotent test — no real power in any direction. In particular deviations on the $\sqrt{n}$ level cannot be detected. The alternative is looking for tests which concentrate their power in a restricted set of directions. The latter test would typically be able to detect deviation from the null in all directions on the $\sqrt{n}$ scale, but their power would be mostly minor. However, in a few directions they would be powerful. Thus we meet again the theme of doing well at a point. Bickel, Ritov and Stoker (2005a,b) consider this problem, argue that there is no notion of optimal or efficient test, and the test to use should be Taylor made to the problem at hand.

In a different direction, Bickel and Bühlmann (1997) argue that given even an infinitely long data sequence, it is impossible (with any test statistic) to distinguish perfectly between linear and nonlinear processes. Their approach was to consider the set of moving-average (linear) processes and study its closure. The closure is, surprisingly very large and contains non-ergodic processes.

1.8. The Road to Real Life

Much of Bickel's work is theoretical. There are real world examples, but they are just that, examples. In recent years, however, Bickel has been devoting much of his time to real world projects, where his main interest has been in the subject matter per se, and not as a experimental lab for statistical ideas.

The first field considered was traffic analysis. The first problem was travel time estimation using loop detector micro-data, Petty, Bickel, Kwon, Ostland, Rice, Ritov and Schoenberg (1998). In Kwon, Coifman, and Bickel (2000), the estimation of future travel time was considered. The proposed estimation method uses as input flow and occupancy data on one hand and historical travel-time information on the other hand.

The second field which really fascinates Bickel is molecular biology. The practical work of Bickel as a statistician in the field is in topics like motif discoveries, Kechris, van Zwet, Bickel and Eisen (2004), important sites in protein sequences, Bickel, Kechris, Spector, Wedemayer, and Glazer (2003), or finding critical structural features of HIV proteins as targets for therapeutic intervention, Bickel, Cosman, Olshen, Spector and Rodrigo (1996).

References

Aït-Sahalia, Y., Bickel, P. J., and Stoker, T. M. (2001). Goodness-of-fit tests for kernel regression with an application to option implied volatilities. *J. Econometrics*, **105**, 363–412.

Albers, W., Bickel, P. J., and van Zwet, W. R. (1976). Asymptotic expansions for the power of distribution free tests in the one-sample problem. *Ann. Statist.*, **4**, 108–156.

Beran, R. (1974). Asymptotically efficient adaptive rank estimates in location models. *Ann. Statist.*, **2**, 63–74.

Bickel, P. J. (1982). On adaptive estimation. *Ann. Statist.*, **10**, 641–671.

Bickel, P. J. (1983). Minimax estimation of the mean of a normal distribution subject to doing well at a point. In *Recent advances in statistics*. Academic Press, New York.

Bickel, P. J. (1984). Parametric robustness: small biases can be worthwhile. *Ann. Statist.*, **12**, 864–879.

Bickel, P. J. and Bühlmann, P. (1997). Closure of linear processes. *J. Theoret. Probab.*, **10**, 445–479.

Bickel, P. J., Klaassen, C. A. J., Ritov, Y., and Wellner, J. A. (1993). *Efficient and Adaptive Estimation for Semiparametric Models*, Vol. Reprinted 1998. Johns Hopkins University Press, Baltimore.

Bickel, P. J., Cosman, P. C., Olshen, R. A., Spector, P., Rodrigo, A. G., and Mullins, J. I. (1996). Covariability of v3 loop amino acids. *AIDS Res Hum Retroviruses*, **12**, 1401–1411.

Bickel, P. J. and Doksum, K. A. (1981). An analysis of transformations revisited. *J. Amer. Statist. Assoc.*, **76**, 296–311.

Bickel, P. J. and Doksum, K. A. (2000). *Mathematical Statistics: Basic Ideas and Selected Topics*, volume I. Prentice Hall, Upper Saddle River.

Bickel, P. J., Kechris, K. J., Spector, P., Wedemayer, G. J., and Glazer, A, N. (2003). Finding important sites in protein sequences. *Proceedings of The National Academy of Sciences*, **99**, 14764-14771.

Bickel, P. J. and Kwon, J. (2001). Inference for semiparametric models: some questions and an answer. *Statist. Sinica*, **11**, 863–960.

Bickel, P. J. and Ritov, Y. (1988). Estimating integrated squared density derivatives: sharp best order of convergence estimates. *Sankhya*, **A50**, 391–393.

Bickel, P. J. and Ritov, Y. (1990). Achieving information bounds in non- and semi-parametric models. *Ann. Statist.*, **18**, 925–938.

Bickel, P. J. and Ritov, Y. (1993). Efficient estimation using both direct and indirect observations. (russian); English version in theory probab. appl. 38 (1993), no. 2, 194–213. *Teor. Veroyatnost. i Primenen.*, **38**, 233–238.

Bickel, P. J. and Ritov, Y. (1996). Inference in hidden markov models. i. local asymptotic normality in the stationary case. *Bernoulli*, **2**, 199–228.

Bickel, P. J. and Ritov, Y. (1997). Local asymptotic normality of ranks and covariates in transformation models. In D. Pollard, E. Torgerson, and G. Yang (Eds.), *Festschrift for Lucien Le Cam*. Springer, New York.

Bickel, P. J., Ritov, Y., and Rydén, T. (1998). Asymptotic normality of the maximum-likelihood estimator for general hidden markov models. *Ann. Statist.*, **26**, 1614–1635.

Bickel, P. J., Ritov, Y., and Rydén, T. (2002). Hidden markov model likelihoods and their derivatives behave like i.i.d. ones. *Ann. Inst. H. Poincaré Probab. Statist.*, **38**, 825–846.

Bickel, P. J., Ritov, Y., and Stoker, T. M. (2005a). Nonparametric testing of an index model. In D. W. K. Andrews and J. H. Stock (Eds.), *Identification and Inference for Econometric Models:A Festschrift in Honor of Thomas J.Rothenberg*. Cambridge University Press, Cambridge.

Bickel, P. J., Ritov, Y., and Stoker, T. M. (2006). Tailor-made tests for goodness-of-fit to semiparametric hypotheses. *Ann. Statist.*, to appear.

Bickel, P. J. and Rosenblatt, M. (1973). On some global measures of the deviations of density function estimates. *Ann. Statist.*, **1**, 1071–1095.

Bickel, P. J. and van Zwet, W. R. (1978). Asymptotic expansions for the power of distribution free tests in the two-sample problem. *Ann. Statist.*, **6**, 937–1004.

Box, G. E. P. and Cox, D. R. (1964). An analysis of transformations. (with discussion). *J. Roy. Statist. Soc. Ser. B*, **26**, 211–252.

Box, G. E. P. and Cox, D. R. (1982). Comment on: "an analysis of transformations revisited". *J. Amer. Statist. Assoc.*, **77**, 209–211.

Brillinger, D. R. (1983). A generalized linear model with "gaussian" regressor variables. In P. J. Bickel, J. L. Doksum, and J. Hodges (Eds.), *A Festschrift for Erich L. Lehmann*. Wadsworth, Belmont.

Claeskens, G. and Hjorth, N. L. (2003). The focused information criterion. *J. Amer. Statist. Assoc.*, **98**, 900–916.

Cox, D. R. and Reid, N. (1987). Parameter orthogonality and approximate conditional inference (with discussion). *J. Roy. Statist. Soc. Ser. B*, **49**, 1–39.

Doksum, K. A. (1987). An extension of partial likelihood methods for proportional hazard models to general transformation models. *Ann. Statist.*, **15**, 325–345.

Doksum, K. A. and Johnson, R. (2002). Comments on "box-cox transformations in linear models: large sample theory and tests of normality" by Chen, G. and lockhart, R. A. and Stephens, M. A. *Canad. J. Statist.*, **30**, 177–234.

Fabian, V. (1978). On asymptotically efficient recursive estimation. *Ann. Statist.* **6**, 854–866.

Hinkley, D. V. and Runger, G. (1984). The analysis of transformed data. *J. Amer. Statist. Assoc.*, **79**, 302–320.

Kechris, K. J., van Zwet E., Bickel, P. J., and Eisen, M. B. (2004). Detecting DNA regulatory motifs by incorporating positional trends in information content. *Genome Biology*, **5**, R50.

Kwon, J., Coifman, B., and Bickel, P. J. (2000). Day-to-day travel-time trends and travel-time prediction from loop-detector data. *Transportation Research Record*, **1717**, 120–129.

Petty, K. F., Bickel, P. J., Kwon, J., Ostland, M., Rice, J. A., Ritov, Y., and Schoenberg, F. (1998). Accurate estimation of travel times from single-loop detectors. *Transportation Research: Part A—Policy and practice*, **32**, 1–17.

Sacks, J. (1975). An asymptotically efficient sequence of estimators of a location parameter. *Ann. Statist.*, **3**, 285–298.

Stein, C. (1957). Efficient nonparametric testing and estimation. In *Proceedings of the Third Berkeley Symposium on Mathematical Statistics and Probability, 1954–1955, Vol. I, pp. 187–195*, Berkeley and Los Angeles, University of California Press.

Stoker, T. M. (1986). Consistent estimation of scaled coefficients. *Econometrica*, **54**, 1461–1481.

Stone, C. J. (1975). Adaptive maximum likelihood estimators of a location parameter. *Ann. Statist.*, **3**, 267–284.

Bickel's Publication

Books

1. Bickel, P. J. (1971). *Mathematical Statistics, Part I, Basic ideas and selected topics.* Holden-Day Inc., San Francisco, CA.
2. Bickel, P. J., Andrews, D., Hampel, F., Huber, P. J., Rogers, W. H. and Tukey, J. W. (1972). *Robust Estimation of Location: Survey and Advances.* Princeton University Press, Princeton, NJ.
3. Bickel, P. J. and Doksum, K. (1976). *Mathematical statistics: Basic ideas and selected topics.* (Textbook), Holden-Day, Inc., San Francisco, CA.
4. Bickel, P. J. and Doksum, K. A. editors. (1983). *A Festschrift for Erich L. Lehmann.* In honor of his sixty-fifth birthday. Wadsworth Statistics/Probability Series. Wadsworth Advanced Books and Software, Belmont, CA.
5. Bickel, P. J., Klaassen, C. A. J., Ritov, Y. and Wellner, J. A. (1993). *Efficient and adaptive estimation in semiparametric models.* Johns Hopkins Series in the Mathematical Sciences. Johns Hopkins University Press, Baltimore, MD.
6. Bickel, P. J., Klaassen, C. A. J., Ritov, Y. and Wellner, J. A. (1998). *Efficient and adaptive estimation for semiparametric models.* Springer-Verlag, NY. Reprint of the 1993 original.
7. Bickel, P. J. and Doksum, K. A. (2001). *Mathematical Statistics, Vol 1: Basic ideas and selected topics.* Second Edition. Prentice Hall, New Jersey.

Articles

1964 − 1969

8. Bickel, P. J. (1964). On some alternative estimates for shift in the p-variate one sample problem. *Ann. Math. Statist.*, 35:1079–1090.
9. Bickel, P. J. (1965). On some asymptotically nonparametric competitors of hotelling's T^2. *Ann. Math. Statist.*, 36:160–173.
10. Bickel, P. J. (1965). On some robust estimates of location. *Ann. Math. Statist.*, 36:847–859.
11. Bickel, P. J. and Yahav, J. A. (1965). Renewal theory in the plane. *Ann. Math. Statist.*, 36:946–955.
12. Bickel, P. J. and Yahav, J. A. (1965). The number of visits of vector walks to bounded regions. *Israel Journal of Mathematics*, 3:181–186.
13. Bickel, P. J. and Yahav, J. A. (1966). Asymptotically pointwise optimal procedures in sequential analysis. In *Proc. Fifth Berkeley Sympos. Math. Statist. and Probability (Berkeley, Calif., 1965/66), Vol. I: Statistics*, 401–415. Univ. California Press, Berkeley, CA.
14. Bickel, P. J. (1966). Some contributions to the theory of order statistics. In *Proc. Fifth Berkeley Sympos. Math. Statist. and Probability (Berkeley,*

Calif., 1965/66), Vol. I: Statistics, 575–593. Univ. California Press, Berkeley, CA.

15. Bickel, P. J. and Hodges Jr, J. L. (1967). The asymptotic theory of Galton's test and a related simple estimate of location. *Ann. Math. Statist.*, 38:73–89.

16. Bickel, P. J. and Blackwell, D. (1967). A note on Bayes estimates. *Ann. Math. Statist.*, 38:1907–1912.

17. Bickel, P. J. and Yahav, J. A. (1968). Asymptotically optimal Bayes and minimax procedures in sequential estimation. *Ann. Math. Statist.*, 39:442–456.

18. Bickel, P. J. and Bahadur, R. R. (1968). Substitution in conditional expectation. *Ann. Math. Statist.*, 39:377–378.

19. Bickel, P. J. and Yahav, J. A. (1969). Some contributions to the asymptotic theory of Bayes solutions. *Z. Wahrscheinlich-keitstheorie und verw. Geb.*, 11:257–276.

20. Bickel, P. J. (1969). A distribution free version of the Smirnov two sample test in the p-variate case. *Ann. Math. Statist.*, 40:1–23.

21. Bickel, P. J. and Yahav, J. A. (1969). On an a.p.o. rule in sequential estimation with quadratic loss. *Ann. Math. Statist.*, 40:417–426.

22. Bickel, P. J. and Berk, R. A. (1968). On invariance with almost invariance. *Ann. Math. Statist.*, 39:1573–1577.

23. Bickel, P. J. and Doksum, K. A. (1969). Tests for monotone failure rate based on normalized sample spacings. *Ann. Math. Statist.*, 40:1216–1235.

24. Bickel, P. J. (1969). A remark on the Kolmogorov-Petrovskii criterion. *Ann. Math.Statist.*, 40:1086–1090.

25. Bickel, P. J. (1969). Test for monotone failure rate ii. *Ann. Math. Statist.*, 40:1250–1260.

26. Bickel, P. J. (1969). Review of "theory of rank tests" by J. Hajek. J. Amer. Statist. Assoc., Vol. 64 (1969), pp. 397-399.

27. Bickel, P. J. and Lehmann, E. (1969). Unbiased estimation in convex families. *Ann. Math. Statist.*, 40:1523–1535.

28. Bickel, P. J. (1969). Une generalisation de type hajek-renyi d'une inegalite de M. P. Levy. *Resume of No. 18*, 269:713–714.

1970 – 1974

29. Bickel, P. J. (1970). A Hajek-Renyi extension of levy's inequality and its applications. *Acta. Math. Acad. Sci. Hungar*, 21:199–206.

30. Bickel, P. J. and Wichura, M. (1971). Convergence criteria for multiparameter stochastic processes and some applications. *Ann. Math. Statist.*, 42:1656–1669.

31. Bickel, P. J. (1971). On some analogues to linear combinations of order statistics in the linear model. *Statist.*, 1:207–216.

32. Bickel, P. J. and Bahadur, R. R. (1971). On conditional test levels in large samples. In *Essays in Probability and Statistics*, 25–34. Univ. of North Carolina Press, Chapel Hill, N.C.

33. Bickel, P. J. and Yahav, J. A. (1972). On the Wiener process approximation

to Bayesian sequential testing problems. In *Proceedings of the Sixth Berkeley Symposium on Mathematical Statistics and Probability (Univ. California, Berkeley, Calif., 1970/1971), Vol. I: Theory of statistics*, 57–83, Berkeley, CA, 1972. Univ. California Press.

34. Bickel, P. J. (1973). On some analogues to linear combinations of order statistics in the linear model. *Ann. Statist.*, 1:597–616.

35. Bickel, P. J. (1973). On the asymptotic shape of Bayesian sequential tests of $\leq o$ versus $\geq o$ for exponential families. *Ann. Statist.*, 1:231–240.

36. Bickel, P. J. and Rosenblatt, M. (1973). On some global measures of the deviations of density function estimates. *Ann. Statist.*, 1:1071–1095.

37. Bickel, P. J. (1974). Edgeworth expansions in nonparametric statistics. *Ann.Statist.*, 2:1–20.

38. Bickel, P. J. and Rosenblatt, M. (1973). Two-dimensional random fields. In *Multivariate analysis, III (Proc. Third Internat. Sympos., Wright State Univ., Dayton, Ohio, 1972)*, 3–15. Academic Press, New York.

39. Bickel, P. J. and Lehmann, E. L. (1974). Measures of location and scale. In *Proceedings of the Prague Symposium on Asymptotic Statistics (Charles Univ., Prague, 1973), Vol. I*, 25–36, Prague, Charles Univ.

1975 − 1979

40. Bickel, P. J. (1975). One-step Huber estimates in the linear model. *J. Amer. Statist. Assoc.*, 70:428–434.

41. Bickel, P. J., Hammel, E. and O'Connell, J. (1975). Sex bias in graduate admissions: Data from berkeley. Science, Vol. 187, pp. 398–404.

42. Bickel, P. J. and Lehmann, E. L. (1975). Descriptive statistics for nonparametric models. I. introduction. *Ann. Statist.*, 3(5):1038–1044.

43. Bickel, P. J. and Lehmann, E. L. (1975). Descriptive statistics for nonparametric models. II. location. *Ann. Statist.*, 3(5):1045–1069.

44. Albers, W., Bickel, P. J. and van Zwet, W. R. (1976). Asymptotic expansions for the power of distribution-free tests in the one-sample problem. *Annals of Statistics*, 4(1):108–156.

45. Bickel, P. J. and Lehmann, E. L. (1976). Descriptive statistics for nonparametric models. III. dispersion. *Ann. Statist.*, 6(6):1139–1158.

46. Bickel, P. J. (1976). Another look at robustness: a review of reviews and some new developments. *Scand. J. Statist.*, 3(4):145–168. With discussion by Sture Holm, Bengt Rosén, Emil Spjøtvoll, Steffen Lauritzen, Søren Johansen, Ole Barndorff-Nielsen and a reply by the author.

47. Bickel, P. J. and Yahav, J. A. (1977). On selecting a set of good populations. In *Statistical decision theory and related topics. II (Proc. Sympos., Purdue Univ., Lafayette, Ind., 1976)*, 37–55. Academic Press, NY.

48. Bickel, P. J. (1978). Using residuals robustly I: Testing for heteroscedastisity, nonlinearity, nonadditivity. *Ann. Statist.*, 6(2):266–291.

49. Bickel, P. J. and van Zwet, W. R. (1978). Asymptotic expansions for the power of distribution free tests in the two-sample problems. *Ann. Statist.*, 6(5):937–1004.

50. Bickel, P. J. and Herzberg, A. (1979). Robustness of design against auto-correlation in time 1: Asymptotic theory, optimality for location and linear regression. *Ann. Statist.*, 7(1):77–95.

51. Bickel, P. J. and Lehmann, E. L. (1979). Descriptive statistics for non-parametric models IV spread. In *Contributions to statistics*, 33–40. Reidel, Dordrecht.

1980 – 1984

52. Bickel, P. J. and van Zwet, W. R. (1980). On a theorem of Hoeffding. In *Asymptotic theory of statistical tests and estimation (Proc. Adv. Internat. Sympos., Univ. North Carolina, Chapel Hill, N.C., 1979)*, 307–324. Academic Press, NY.

53. Bickel, P. J. and Doksum, K. A. (1981). An analysis of transformations revisited. 76(374):296–311.

54. Bickel, P. J. (1981). Minimax estimation of the mean of a normal distribution when the parameter space is restricted. *Ann. Statist.*, 9(6):1301–1309.

55. Bickel, P. J., Chibisov, D. M. and van Zwet, W. R. (1981). On efficiency of first and second order. *Int. Stat. Review*, 49(2):169–175.

56. Bickel, P. J. (1981). Quelques aspects de la statistique robuste. In *Ninth Saint Flour Probability Summer School—1979 (Saint Flour, 1979)*, volume 876 of *Lecture Notes in Math.*, 1–72. Springer, Berlin.

57. Bickel, P. J. and Lehmann, E. L. (1981). A minimax property of the sample mean in finite populations. *Ann. Statist.*, 9(5):1119–1122.

58. Bickel, P. J.,Herzberg, A. and Schilling, M. F. (1981). Robustness of design against autocorrelation in time II: optimality, theoretical and numerical results for the first-order autoregressive process. *J. Amer. Statist. Assoc.*, 76(376):870–877.

59. Bickel, P. J. and D.A. Freedman, D. A. (1981). Some asymptotic theory for the bootstrap. *Ann. Statist.*, 9(6):1218–1228.

60. Bickel, P. J. and Yahav, J. A. (1982). Asymptotic theory of selection procedures and optimality of Gupta's rules. In *Statistics and Probablity: Essays in Honor of C.R. Rao*, (J.K. Ghosh, G. Kallianpur, P.R. Krishnaiah, eds.), 109–124. North-Holland, Amsterdam.

61. Bickel, P. J. and Robinson, J. (1982). Edgeworth expansions and smooth-ness. *Ann. Prob.*, 10(2):500–503.

62. Bickel, P. J. (1983). Minimax estimation of the mean of normal distribution subject to doing well at a point. In *Recent advances in statistics*, 511–528. Academic Press, NY.

63. Bickel, P. J. (1982). On adaptive estimation. *Ann. Statist.*, 10(3):647–671.

64. Bickel, P. J. and Breiman, L. (1983). Sums of functions of nearest neighbor distances, moment bounds, limit theorems and a goodness of fit test. *Ann. Prob.*, 11(1):647–671.

65. Bickel, P. J. and Collins, J. (1983). Minimizing Fisher information over mixtures of distributions. *Sankhya*, 45(1):185–214.

66. Bickel, P. J. and Hodges Jr, J. L. (1983). Shifting integer valued random vari-

ables. In *A Festschrift for Erich L. Lehmann*, Wadsworth Statist./Probab. Ser., 49–61. Wadsworth, Belmont, CA.

67. Bickel, P. J. and Freedman, D. (1983). Bootstrapping regression models with many parameters. In *A Festschrift for Erich L. Lehmann*, Wadsworth Statist./Probab. Ser., 28–48. Wadsworth, Belmont, CA.

68. Bickel, P. J. and Freedman, D. (1984). Asymptotic normality and the bootstrap in stratified sampling. *Ann. Statist.*, 12(2):470–482.

69. Bickel, P. J. (1984). Review: Contributions to a general asymptotic statistical theory by J. Pfanzagl. *Ann. Statist.*, 12:786-791.

70. Bickel, P. J. (1984). Review: Theory of point estimation by E. L. Lehmann. *Metrika*, pp. 256.

71. Bickel, P. J. (1984). Robust regression based on infinitesimal neighbourhoods. *Ann. Statist.*, 12(4):1349–1368.

72. Bickel, P. J. (1984). Parametric robustness: small biases can be worthwhile. *Ann. Statist.*, 12(3):864–879.

1985 – 1989

73. Bickel, P. J., Götze, F. and van Zwet, W. R. (1985). A simple analysis of third-order efficiency of estimates. In *Proceedings of the Berkeley conference in honor of Jerzy Neyman and Jack Kiefer, Vol. II (Berkeley, Calif., 1983)*, Wadsworth Statist./Probab. Ser., 749–768, Belmont, CA, Wadsworth.

74. Bickel, P. J. and Ritov, J. (1987). Efficient estimation in the errors in variables model. *Ann. Statist.* , 15(2):513–541.

75. Bickel, P. J., Götze, F. and van Zwet, W. R. (1987). The edgeworth expansion for U statistics of degree 2. *Ann. Statist.*, 15(4):1463–1484.

76. Bickel, P. J. and Klaassen, C. A. J. (1987). Empirical Bayes estimation in functional and structural models, and uniformly adaptive estimation of location. *Adv. in Appl. Math.*, 7(1):55–69.

77. Bickel, P. J. (1986). Efficient testing in a class of transformation models. In *Proceedings of the 1st World Congress of the Bernoulli Society, Vol. 2 (Tashkent, 1986)*, 3–11, VNU Sci. Press. Utrecht.

78. Bickel, P. J. and Yahav, J. A. (1987). On estimating the total probability of the unobserved outcomes of an experiment. In J. van Ryzin Ed., *Adaptive statistical procedures and related topics (Upton, N.Y., 1985)*, volume 8 of *IMS Lecture Notes Monogr. Ser.*, 332–337. Inst. Math. Statist., Hayward, CA.

79. Bickel, P. J. (1987). Robust estimation. In Johnson and Eds. S. Kotz, editors, *Encyclopedia of Statistical Sciences*. J. Wiley, NY.

80. Bickel, P. J. and Yahav, J. A. (1988). Richardson extrapolation and the bootstrap. *J. Amer. Statist. Assoc.*, 83(402):387–393.

81. Bickel, P. J. and Yahav, J. A. (1988). On estimating the number of unseen species and system reliability. In *Statistical decision theory and related topics, IV, Vol. 2 (West Lafayette, Ind., 1986)*, 265–271. Springer, NY.

82. Bickel, P. J. and Mallows, C. L. (1988). A note on unbiased Bayes estimates. *American Statistician*, 42(2):132–134.

83. Bickel, P. J. and Ritov, Y. (1988). Estimating integrated squared density derivatives: sharp best order of convergence estimates. *Sankhyā Ser. A*, 50(3):381–393.

84. Bickel, P. J. (1988). Estimating the size of a population. Proc. IVth Purdue Symposium on Decision Theory and Related Topics.

85. Bickel, P. J. and Krieger, A. (1989). Confidence bands for a distribution function using the bootstrap. *J. Amer. Statist. Assoc.*, 84(405):95–100,

86. Bickel, P. J., Olshen, R. and Bai, C. (1989). The bootstrap for prediction. Proceedings of an Oberwolfach Conference, Springer-Verlag.

1990 – 1994

87. Bickel, P. J. and Ghosh, J. K. (1990). A decomposition for the likelihood ratio statistic and the Bartlett correction–a Bayesian argument. *Ann. Statist.*, 18(3):1070–1090.

88. Bickel, P. J. and Ritov, Y. (1990). Achieving information bounds in non and semiparametric models. *Ann. Statist.*, 18(2):925–938.

89. Bai, C., Bickel, P. J. and Olshen, R. A. (1990). Hyperaccuracy of bootstrap based prediction. In *Probability in Banach spaces, 7 (Oberwolfach, 1988)*, volume 21 of *Progr. Probab.*, 31–42. Birkhäuser Boston, MA.

90. Bickel, P. J. and Ritov, J. (1991) Large sample theory of estimation in biased sampling regression models. I. *Ann. Statist.*, 19(2):797–816.

91. Bickel, P. J., Ritov, Y. and Wellner, J. A. (1991). Efficient estimation of linear functionals of a probability measure P with known marginal distributions. *Ann. Statist.*, 19(3):1316–1346.

92. Bickel, P. J. and Ritov, Y. (1992). Testing for goodness of fit: a new approach. In A.K. Md.E. Saleh, editor, *Nonparametric statistics and related topics (Ottawa, ON, 1991)*, 51–57. North-Holland, Amsterdam.

93. Bickel, P. J. and Zhang, P. (1992). Variable selection in non-parametric regression with categorical covariates. *J. Amer. Statist. Assoc.*, 87(417):90–98.

94. Bickel, P. J. (1992). Inference and auditing: The stringer bound. *International Statistical Review*, 60(2):197–209

95. Bickel, P. J. (1992). Theoretical comparison of bootstrap t confidence bounds. In L. Billard and R. Le Page, editors, *Exploring the Limits of the Bootstrap*, Wiley Ser. Probab. Math. Statist., 65–76. Wiley, NY.

96. Bickel, P. J. and Millar, P. W. (1992). Uniform convergence of probability measures on classes of functions. *Statist. Sinica*, 2(1):1–15.

97. Bickel, P. J., Nair, V. N.and Wang, Paul C. C. (1992). Nonparametric inference under biased sampling from a finite population. *Ann. Statist.*, 20(2):853–878.

98. Bickel, P. J. (1993). Estimation in semiparametric models. In C.R. Rao, editor, *Chapter 3 in Multivariate Analysis: Future Directions*, volume 5 of *North-Holland Ser. Statist. Probab.*, 55–73. North-Holland, Amsterdam.

99. Bickel, P. J. and Krieger, A. M. (1993). Extensions of Chebychev's inequality with applications. *Probability and Mathematical Statistics*, 13(2):293–310.

100. Bickel, P. J. (1993). Efficient estimation using both direct and indirect observations. *Teor. Veroyatnost. i Primenen.*, 38(2):233–258, 1993. Translated into English in Theory Probab. Appl. 38, No.2, 194-213.

101. Bickel, P. J. and Ritov, Y. (1994). Ibragimov Hasminskii models. In Shanti S. et al. Gupta, editor, *Statistical decision theory and related topics V*, 51–60. Springer, NY. Proceedings of the fifth Purdue international symposium on statistical decision theory and related topics held at Purdue University, West Lafayette, IN, June 14-19, 1992.

102. Bickel, P. J. (1994). Asymptotic distribution of the likelihood ratio statistic in a prototypical non regular problem. In S.K. Mitra, editor, *Bahadur Volume*. Wiley Eastern, New Dehli.

103. Bickel, P. J. (1994). Discussion of papers by Feigelson and Nousek. In E. Feigelson and G.J. Babu, editors, *Statistical Challenges in Modern Astronomy*. Springer, NY.

1995 – 1999

104. Bickel, P. J. and Ritov, Y. (1995). Estimating linear functionals of a pet image. *IEEE Tr. of Medical Imaging*, 14:81-87.

105. Bickel, P. J., Hengartner, N., Talbot, L. and Shepherd, I. (1995). Estimating the probability density of the scattering cross section from Rayleigh scattering experiments. *J. Opt. Soc. Am. A*, 12:1316-1323.

106. Bickel, P. J. and Fan, J. (1996). Some problems on estimation of unimodal densities. *Statist. Sinica*, 6(1):23–45.

107. Bickel, P. J. and Ren, J. (1996). The m out of n bootstrap and goodness of fit tests with double censored data. In Helmut Rieder, Ed., *Robust statistics, data analysis, and computer intensive methods (Schloss Thurnau, 1994)*, volume 109 of *Lecture Notes in Statist.*, 35–47. Springer, NY.

108. Bickel, P. J. and Nair, V. N. (1995). Asymptotic theory of linear statistics in sampling proportional to size without replacement. *Probab. Math. Statist.*, 15:85–99. Dedicated to the memory of Jerzy Neyman.

109. Bickel, P. J. and Ritov, Y. (1995). An exponential inequality for U-statistics with applications to testing. *Probability in the Engineering and Informational Sciences*, 9(1):39–52.

110. Bickel, P. J., Cosman, P. C., Olshen, R. A., Spector, P. C., Rodrigo, A. G., and Mullins. J. I. (1996). Covariability of v3 loop amino acids.

111. Bickel, P. J. (1996). What is a linear process? *Proceedings of National Academy of Sciences*, 93:12128–12131.

112. Bickel, P. J. and Ritov, Y. (1996). Inference in hidden Markov models. I. Local asymptotic normality in the stationary case. *Bernoulli*, 2(3):199–228.

113. Bickel, P. J., Götze, F. and van Zwet, W. R. (1997). Resampling fewer than n observations: gains, losses, and remedies for losses. *Statist. Sinica*, 1(1):1–31.

114. Bickel, P. J. and Ritov, Y. (1997). Local asymptotic normality of ranks and covariates in transformation models. In *Festschrift for Lucien Le Cam*, 43–54. Springer, NY.

115. Bickel, P. J. (1997). Discussion of statistical aspects of hipparcos photometric data (F. van Leeuwen *et al.*). In G.J. Babu and Fr. Fergelson, Eds, *Statistical Challenges in Modern Astronomy II*. Springer-Verlag.

116. Bickel, P. J. (1997). An overview of scma ii in statistical challenges in modern astronomy ii. In G.J. Babu and Fr. Fergelson, Eds, *Statistical Challenges in Modern Astronomy II*. Springer-Verlag.

117. Bickel, P. J. (1997). Lan for ranks and covariates. In E. Torgersen D. Pollard and G. Yang, Eds, *Festschrift for Lucien Le Cam*. Springer Verlag.

118. Bickel, P. J. and Bühlmann, P. (1997). Closure for linear processes. *J. Theoret. Probab.*, 10(2):445–479. Dedicated to Murray Rosenblatt.

119. Bickel, P. J., Petty, K. F., Jiang, J., Ostland, M., Rice, J., Ritov, Y. and Schoenberg, R. (1997). Accurate estimation of travel times from single loop detectors. *Transportation Research A*.

120. Nielsen, J. P., Linton, O. and Bickel, P. J. (1998). On a semiparametric survival model with flexible covariate effect. *Ann. Statist.*, 26(1):215–241.

121. van der Laan, M. J., Bickel, P. J. and Jewell, N. P. (1998). Singly and doubly censored current status data: estimation, asymptotics and regression. *Scand. J. Statist.*, 24(3):289–307.

122. Bickel, P. J., Ritov, Y. and Rydén, T. (1998). Asymptotic normality of the maximum-likelihood estimator for general hidden Markov models. *Ann. Statist.*, 26(4):1614–1635.

123. Bickel, P. J. and Bühlmann, P. (1999). A new mixing notion and functional central limit theorem for a sieve bootstrap in time series. *Bernoulli*, 5(3):413–446.

2000 – 2004

124. Sakov, A. and Bickel, P. J. (2000). An Edgeworth expansion for the m out of n bootstrapped median. *Statistics and Probability Letters*, 49(3):217–223.

125. Bickel, P. J. Coifman, B. and Kwon, J. (2000). Day to day travel time trends and travel time prediction from loop dectector data. *Transportation Research Board Record*. 1717:120-129.

126. Bickel, P. J. and Ritov, Y. (2000). Non- and semiparametric statistics: compared and contrasted. *J. Statist. Plann. Inference*, 91(2):209–228.

127. Bickel, P. J. (2000). Statistics as the information science. *Opportunities for the Mathematical Sciences*, 9–11.

128. Bickel, P. J. and Levina, E. (2001). The earth mover's distance is the Mallows distance: some insights from statistics. In *Proceedings of ICCV '01*, 251–256, Vancouver.

129. Ait-Sahalia, Y., Bickel, P. J. and Stoker, T. M. (2001). Goodness-of-fit tests for kernel regression with an application to option implied volatilities. *J. Econometrics*, 105(2):363–412.

130. Bickel, P. J., Chen, C., Kwon, J., Rice, J.,Varaiya, P. and van Zwet, E. (2002). Traffic flow on a freeway network. In *Nonlinear estimation and classification (Berkeley, CA, 2001)*, Vol. 171 of *Lecture Notes in Statist.*, 63–81. Springer, NY.

131. Bickel, P. J. and Ren, J. (2001). The bootstrap in hypothesis testing. In *State of the art in probability and statistics (Leiden, 1999)*, Vol. 36 of *IMS Lecture Notes Monogr. Ser.*, 91–112. Inst. Math. Statist., Beachwood, OH. Festschrift for W. R. van Zwet.

132. Bickel, P. J., Buyske, S., Chang, H., and Ying, Z. (2001). On maximizing item information and matching difficulty with ability. *Psychometrika*, 66(1):69–77.

133. Bickel, P. J. and Lehmann, E. L. (2001). Frequentist interpretation of probability. In *International Encyclopedia ot the Social and Behavioral Sciences*, 5796–5798. Elsevier Science Ltd., Oxford.

134. Bickel, P. J. and Kwon, J. (2001). Inference for semiparametric models: some questions and an answer (with discussion). *Statist. Sinica*, 11(4):863–960.

135. Bickel, P. J. and Sakov, A. (2002). Equality of types for the distribution of the maximum for two values of n implies extreme value type. *Extremes*, 5(1):545–553.

136. Bickel, P. J.,Petty, K., Ostland, M., Kwon, J., and Rice, J. (2002). A new methodology for evaluating incident detection algorithms. *Transportation Research*, C10:189.

137. Kwon, J., Min, K., Bickel, P. J. and Renne, P. R. (2002). Statistical methods for jointly estimating the decay constant of ^{40}K and the age of a dating standard. *Math. Geology*, 34(4):457–474.

138. Bickel, P. J. and Sakov, A. (2002). Extrapolation and the bootstrap. *Sankhyā Ser. A*, 64(3, part 1):640–652. Special issue in memory of D. Basu.

139. Bickel, P. J., Kechris, K. J., Spector, P. C., Wedemayer, G. J. and Glazer, A. N. (2002). Finding important sites in protein sequences. *Proc. Natl. Acad. Sci. USA*, 99(23):14764–14771.

140. Bickel, P. J., Ritov, Y. and Rydén, T. (2002). Hidden Markov model likelihoods and their derivatives behave like i.i.d. ones. *Ann. Inst. H. Poincaré Probab. Statist.*, 38(6):825–846. En l'honneur de J. Bretagnolle, D. Dacunha-Castelle, I. Ibragimov.

141. Berk, R. A., Bickel, P. J., Campbell, K., Fovell, R., Keller-McNulty, S., Kelly, E., Linn, R., Park, B., Perelson, A., Rouphail, N., Sacks, J. and Schoenberg, F. (2002). Workshop on statistical approaches for the evaluation of complex computer models. *Statist. Sci.*, 17(2):173–192.

142. Bickel, P. J. and Ritov, Y. (2003). Inference in hidden markov models. In *Proc. Intl. Congress of Mathematicians*, Vol. 2. World Publishers, Hong Kong.

143. Kim, N. and Bickel, P. J. (2003). The limit distribution of a test statistic for bivariate normality. *Statist. Sinica*, 13(2):327–349.

144. Bickel, P. J. and Ritov, Y. (2003). Nonparametric estimators which can be "plugged-in". *Ann. Statist.*, 31(4):1033–1053.

145. Bickel, P. J., Eisen, M. B., Kechris, K. and van Zwet, E. (2004). Detecting dna regulatory motifs by incorporating positional trends in information content. *Genome Biology* Vol. 5 Issue 7.

146. Ge, Z., Bickel, P. J. and Rice, J. A. (2004). An approximate likelihood approach to nonlinear mixed effects models via spline approximation. *Comput.*

Statist. Data Anal., 46(4):747–776.

147. Bickel, P. J. (2004). Unorthodox bootstraps. *Journal of Korean Statistical Society*, 32:213–224.

148. Bickel, P. J. (2004). Robustness of prewhitening of heavy tailed sources. Springer Lecture Notes in Computer Science.

Comments and Discussion

149. Bickel, P. J. and Rosenblatt, M. (1975). Corrections to: "On some global measures of the deviations of density function estimates" (Ann. Statist. **1** (1973), 1071–1095). *Ann. Statist.*, 3(6):1370.

150. Albers, W., Bickel, P. J. and van Zwet, W. R. (1978). Correction to: "Asymptotic expansions for the power of distribution free tests in the one-sample problem" (*Ann. Statist.* **4** (1976), no. 1, 108–156). *Ann. Statist.*, 6(5):1170–1171.

151. Bickel, P. J., Ritov, Y. and Ryden, T. (2002). Hidden Markov and state space models asymptotic analysis of exact and approximate methods for prediction, filtering, smoothing and statistical inference. In *Proceedings of the International Congress of Mathematicians, Vol. I (Beijing, 2002)*, 555–556, Higher Ed. Press, Beijing.

152. Bickel, P. J. (1993). Correction to: Inference and auditing: The stringer bound. *Int. Stat. Rev.*, 61(3):487.

153. Bickel, P. J., Chibisov, D. M. and van Zwet, W. R. (1981). On efficiency of first and second order. *International Statistical Review*, 49:169-175.

154. Bickel, P. J. (1979). Comment on "Conditional independence in statistical theory" by Dawid, A. P. *Journal of the Royal Statistical Society B*, 41(1):21.

155. Bickel, P. J. (1979). Comment on "Edgeworth saddle point approximations with statistical applications" by Barndorff-neilsen, O. and Cox, D. R. *Journal of Royal Statistical Society B*, 41(3):307.

156. Bickel, P. J. (1980). Comment on Box, G.: "Sampling and bayes" inference in scientific modelling and robustness. *Journal Royal Statistical Society A*, 143:383–431.

157. Bickel, P. J. (1983). Comment on "Bounded influence regression" by Huber, P. J. *J. Amer. Statist. Assoc.*, 78:75–77.

158. Bickel, P. J. (1984). Comment on: Hinkley D. and Runger S. "Analysis of transformed data". *J. Amer. Statist. Assoc.*, 79:309.

159. Bickel, P. J. (1984). Comment on "Adaptive estimation of nonlinear regression models" by Manski, C. F. *Econometric Reviews*, 3(2):145–210.

160. Bickel, P. J. (1987). Comment on "Better bootstrap confidence intervals" by Efron, B. *J. Amer. Statist. Assoc.*, 82:191.

161. Bickel, P. J. (1988). Mathematical Sciences: Some research trends (Section on Statistics). National Academy Press, Washington.

162. Bickel, P. J. (1988). Comment on "Theoretical comparison of bootstrap confidence intervals" by Hall, P. *Ann. Statist.*, 959–960.

163. Bickel, P. J. (1988). Comment on "Rank based robust analysis of models" by Draper, D. *Stat. Science.*

164. Bickel, P. J. and Ritov, Y. (1990). Comment on: Silverman, B. W., Jones, M. C., Wilson, J. D. and Nytchka, D. "A smoothed em approach etc.". *Journal of Royal Statistical Society B*, 271–325. (with discussion).

165. Bickel, P. J. (1990). Comment on: "Fisherian inference in likelihood and prequential frames of inference" by Dawid, A. P. *Journal of Royal Statistical Society B*.

166. Bickel, P. J. and Le Cam, L. (1990). A conversation with Ildar Ibragimov. *Statist. Sci.*, 5(3):347–355.

167. Bickel, P. J. (1990). Renewing us mathematics: A plan for the 1990's. Report of NRC Committee. N.A.S. Press.

168. Bickel, P. J. (1994). What academia needs. *The American Statistician*, 49:1–6.

169. Bickel, P. J., Ostland, M., Petty, K., Jiang, J., Rice, J. and Schoenberg, R. (1997). Simple travel time estimation from single trap loop detectors. *Intellimotion*, 6(1).

170. Bickel, P. J. (1997). Discussion of evaluation of forensic dna evidence. *Proc. Natl. Acad. Sci. USA*, 94:5497.

171. Bickel, P. J. (2000). Comment on "Hybrid resampling intervals for confidence intervals" by Chuang and Lai. *Statistica Sinica*, 10:1–50.

172. Bickel, P. J. and Ritov, Y. (2000). Comment on "On profile likelihood" by Murphy, S. and van der Vaart, A. *J. Amer. Statist. Assoc.*, 95(450):449–485.

173. Bickel, P. J. (2000). Statistis as the information scinence *Opportunities for the mathematical sciences, NSF Workshop*.

174. Bickel, P. J. (2002). Comment on "What is a statistical model" by Mccullagh, P. *Ann. Statist.*, 30(2):1225–1310.

175. Bickel, P. J. (2002). Comment on Chen, Lockhart and Stephens "Boxcox transformations in linear models". *Canadian Journal of Statistics*, 30(2):177–234.

176. Bickel, P. J. and Lehmann, E. L. (2001). Frequentist inference. In *International Encyclopedia ot the Social and Behavioral Sciences*, 5789–5796. Elsevier Science Ltd., Oxford.

177. Bickel, P. J. (2002). The board on mathematical sciences has evolved. *IMS Bulletin*.

Part I

Semiparametric Modeling

CHAPTER 2

Semiparametric Models: A Review of Progress since BKRW (1993)

Jon A. Wellner, Chris A. J. Klaassen, Ya'acov Ritov

Department of Statistics
University of Washington
Seattle, Washington 98195-4322
jaw@stat.washington.edu

Korteweg-de Vries Institute for Mathematics
University of Amsterdam
Plantage Muidergracht 24
1018 TV Amsterdam, The Netherlands
chrisk@science.uva.nl

Department of Statistics
The Hebrew University of Jerusalem
Mt. Scopus, Jerusalem 91905, Israel
yaacov@mscc.huji.ac.il

This chapter sketches a review of the developments in semiparametric statistics since the publication in 1993 of the monograph by Bickel, Klaassen, Ritov, and Wellner.

2.1. Introduction

This chapter gives a brief review of some of the major theoretical developments in the theory of semiparametric models since publication of our jointly authored book, Bickel, Klaassen, Ritov, and Wellner (1993), henceforth referred to as BKRW (1993). It is, for reasons of space, a very selective and somewhat personal review. We apologize in advance to all of those whose works we have not covered for any reason.

A major focus in semiparametric theory is on asymptotic efficiency. A special case of this semiparametric efficiency occurs when the least favorable parametric submodel of the semiparametric model is a natural parametric model. Typically this natural parametric model is the model from which the semiparametric model is built by relaxing distributional assumptions. For example in the semiparametric symmetric location model the least favorable parametric submodel is the symmetric location model with the symmetric density known. In the semiparametric linear regression model with the error distribution unknown with mean zero, the least favorable parametric submodel is the linear regression model with the error distribution known. The first semiparametrically efficient procedures were developed in these cases and they were called adaptive because they adapted themselves to the unknown underlying density. Stone (1975) and Beran (1974) were the first to construct efficient estimators in the semiparametric symmetric location model, preceded by van Eeden (1970) who proved efficiency of her estimator with the symmetric densities assumed to be strongly unimodal. A milestone in this line of research was the paper by Bickel (1982). Actually, existence of adaptive procedures was suggested already in the fifties by Stein (1956) and the first adaptive test was presented by Hajek (1962).

Meanwhile Cox (1972) introduced the proportional hazards model and his estimators for the parametric part of the model, and Cox (1975) introduced his notion of "partial likelihood". There was a flurry of work in the late 1970s in an effort to understand the efficiency properties of Cox's "partial likelihood" estimators: e.g. Efron (1977), Oakes (1977), and Kay (1979). The work of the first author of the present paper (culminating in Begun, Hall, Huang and Wellner (1983), began with an effort to rework and generalize the calculations of Efron (1977) along the lines of some of the modern information bound theory in papers of Beran (1977a) and Beran (1977b).

The first appearances of the term "semiparametric" in the literature (of which we are aware) occur in a *Biometrics* paper by Gail, Santner and Brown (1980) and in a paper in *Demography* by Finnas and Hoem (1980). Within a year the term was also used by Oakes (1981) in his influential review of work on the Cox (1972) model, and by Turnbull (1981) in his *Mathematical Reviews* review of Kalbfleisch and Prentice (1980). Subsequently the term was applied by Epstein (1982) and Epstein (1983) in reviews of Whitehead (1980) (who used the terminology "partially parametric") and Oakes (1981), by Louis, Mosteller and McPeek (1982), page 95, and by Andersen (1982), page 67. Unfortunately, the first author of the present paper in writing Begun, Hall, Huang and Wellner (1983) used the terminology "parametric - nonparametric" and "mixed model". The terminology "semiparametric" became accepted, however, as can be seen from Figure 2.1 which shows appearance of the term "semiparametric" in title, keywords, or abstract in three major indexes of statistical literature: MathSciNet, the Current Index of Statistics, and the ISI Web of Science.

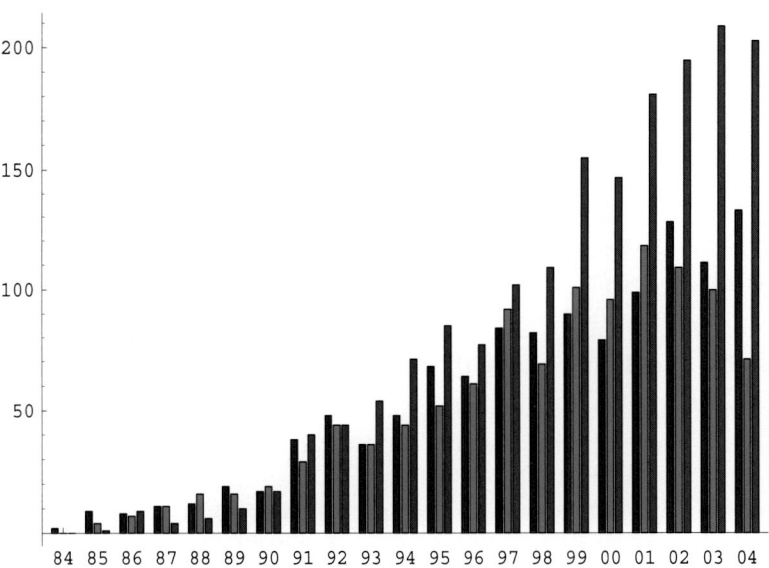

Figure 2.1.   Numbers of papers with "semiparametric" in title, keywords, or abstract, by year, 1984 - 2004. Red = MathSciNet; Green = Current Index of Statistics (CIS); Blue = ISI Web of Science.

The theory of estimation and testing for semiparametric models has been developing rapidly since the publication of BKRW in 1993. Here we briefly survey some of the most important of these developments from our perspective and pose some questions and challenges for the future. We will not attempt to review the many applications of semiparametric models since they have become too numerous to review in the limited space available here. [A search on MathSciNet in early May 2005 for "semiparametric" gave 1185 hits.] Our short review will be broken down according to the following (somewhat arbitrary and overlapping) categories, which are depicted in the following sections.

2.2. Missing Data Models

[A search on MathSciNet in early May 2005 for "semiparametric" and "missing data" gave 15 hits.] A major development in this area was the systematic development of information bounds for semiparametric regression models with covariates missing at random by Robins, Rotnitzky and Zhao (1994), Robins, Rotnitzky and Zhao (1995), and Robins, Hsieh, and Newey (1995); see also Robins and Rotnitzky (1992), Robins and Rotnitzky (1995), and Nan, Emond, and Wellenr (2004). For another recent treatment of the information bound calculations under Coarsening At Random (CAR) and Missing At Random (MAR), see van der Vaart (1998), pages 379–383.

The information bounds for missing data models have been shown to be achievable in some special cases: for examples involving two-stage sampling, see Breslow, McNeney and Wellner (2003), and see Chen (2002), Chen (2004), and Wang, Linton and Härdle (2004) for further examples and recent developments. Much more work is needed in this area.

2.3. Testing and Profile Likelihood Theory

[A search on MathSciNet in early May 2005 for "semiparametric" and "profile likelihood" gave 13 hits; a search for "semiparametric" and "testing" gave 233 hits.] Although BKRW (1993) did not manage to treat the theory of tests for semiparametric models, the literature developed rapidly in this area during the mid and late 1990s, with contributions by Choi, Hall and Schick (1996), Murphy and van der Vaart (1997), and Murphy and van der Vaart (2000). In particular, Su and Wei (1991) initiated the study of profile likelihood methods in semiparametric settings, and their study was developed further by Murphy and van der Vaart (1997) and Murphy and van der Vaart (2000). Murphy and van der Vaart (1997, 2000) show that semiparametric profile likelihoods have quadratic expansions in the efficient scores under appropriate Donsker type hypotheses on the scores corresponding to a least favorable sub-model and a certain "no-bias" condition. This important development opens the door to likelihood ratio type tests and confidence intervals in many semiparametric models for which the least favorable sub-models can be constructed. The main difficulty in applying the results of Murphy and van der Vaart (2000) seems often to be in construction of least favorable submodels with the right properties. Severini and Staniswalis (1994) develop methods based on combining techniques from profile likelihood and quasi-likelihood considerations, while Lee, Kosorok and Fine (2005) propose Bayesian MCMC methods applied to the semiparametric profile likelihood. Banerjee (2005) has studied the power behavior of likelihood ratio tests under contiguous alternatives, and shows that the limiting distributions under local alternatives are non-central chi-square with shift parameter involving a quadratic form in the efficient information matrix. Murphy and van der Vaart (1999) study the use of "observed information" in semiparametric models and applications thereof to testing.

Testing a parametric fixed link single-index regression model against a semiparametric alternative single-index model was considered by Horowitz and Härdle (1994), but with the parameter θ_0 involved in the single index under both null and alternative. The case of index parameter allowed to differ under the alternative was considered by Härdel, Spokoiny and Sperlich (1997). Kauermann and Tutz (2001) consider testing certain classes of parametric and semiparametric regression models against general smooth alternatives. On the other hand, Bickel, Ritov, and Stoker (2005a) argue that there is no real notion of optimality in testing of semiparametric composite hypotheses. Any test would have negligible power against departures in most directions. Their recommendation is to use tailor-made tests, which concentrate power in directions which are important to the investigator. These general ideas were applied to index models in Bickel, Ritov, and Stoker (2005b).

2.4. Semiparametric Mixture Model Theory

[A search on MathSciNet in early May 2005 for "semiparametric" and "mixture model" gave 37 hits.] In a classical paper proposing models alternative to those considered by Neyman and Scott (1948), Kiefer and Wolfowitz (1956) showed that maximum likelihood estimators are consistent in a large class of semiparametric mixture models before the term "semiparametric" was in existence. Although other less satisfactory estimators had been constructed for many models of this type during the 1970s and 1980s (see e.g. van der Vaart (1988)), efficiency and asymptotic normality of maximum likelihood estimators were completely unknown through the mid-1990s. But van der Vaart (1996) succeeded in using empirical process theory methods together with methods and results of Pafanzagl (1988) and Pafanzagl (1990) to establish asymptotic normality and efficiency of the maximum likelihood estimators for several important examples of this type of model (including an exponential frailty model, a normal theory errors in variables model, and a model involving scale mixtures of Gaussians). It seems difficult to formulate a completely satisfactory general theorem, but it also seems clear that the methods of van der Vaart (1996) will apply to a wide range of semiparametric mixture models.

2.5. Rates of Convergence via Empirical Process Methods

[A search on MathSciNet in early may 2005 for "semiparametric" and "convergence rate" gave 27 hits. Searching for "nonparametric" and "convergence rate" gave 214 hits.] Rates of convergence of minimum contrast estimators, maximum likelihood estimators, and variants of maximum likelihood involving sieves and penalization, mostly aimed at nonparametric settings, were a topic of considerable research during the 1990s, beginning with Wong and Severini (1991), Birgé and Massart (1993), Shen and Wong (1994) and Wong and Shen (1995). The results of these authors rely on sharp bounds for local oscillations of empirical processes indexed by the classes of functions involved in the maximization, and hence are closely related to the available bounds for suprema of empirical processes. See van deer Vaart and Wellner (1996) Sections 3.2 and 3.4 for a recasting of those results.

This initial progress continued with van de Geer (1996), Birgé, and Massart (1998), Shen (1997), and Shen and He (1997).

The results in these works have important consequences for maximum likelihood estimators as well as sieved and penalized maximum likelihood estimators in semiparametric models. For example, Huang (1996) used the methods of Birgé and Massart (1993) and Wong and Shen (1995) to obtain rates of convergence of maximum likelihood estimators for Cox's proportional hazards model with current status data. See van der Vaart (2002), section 8, pages 424 - 432, for a summary of the methods and an alternative treatment of Huang's results. van der Vaart (2002) also gives a number of other useful applications of empirical process theory to problems in semiparametric models.

2.6. Bayes Methods and Theory

[A search on MathSciNet in early May 2005 for "semiparametric" and "Bayesian" gave 121 hits.] Bayes estimators and procedures have been proposed for a wide range of semiparametric models: see Lenk (1999) for a Bayesian approach to semiparametric regression, Müller and Roeder (1997) for a Bayesian model for case-control studies, Vidakovic (1998) for Bayes methods in connnection with wavelet based nonparametric estimation, Newton, Czado and Chappell (1996) for Bayes inference for semiparametric binary regression, and Ghosal and van der Vaart (2000) for Bayes estimation with mixtures of normal densities. Lazar (2003) gives an interesting Bayes approach to empirical likelihood (see below), while Sinha, IBrahim, and Chen (2003) give an interesting Bayesian justification of Cox's partial likelihood. Much of the popularity of Bayes methods is due to the new computational tools available; see e.g. Gilks, Richardson, and Spiegelhalter (1999), Carlin and Louis (2000), and Robert and Casella (2004). Considerable progress has been made in understanding consistency issues and rates of convergence of Bayes procedures (see e.g. Ghosal, Ghosh and Samanta (1995), Ghosal, Ghosh and van der Vaart (2000), Shen and Wasserman (2001), and Huang (2004)), but major gaps remain in the theory. For example, a suitably general Bernstein - von Mises theorem is still lacking despite several initial efforts in this direction. Freedman (1999) gives negative results for non-smooth functionals in the context of nonparametric regression (which are not surprising from the perspective of semiparametric information bounds), while Kim and Lee (2004) give more positive results in a right censoring model with smooth functionals, and Shen (2002) gives some preliminary general results. Kleijin and van der Vaart (2002) give a treatment of Bayes estimators in the case of miss-specified models. From the examples, it is clear that Bayes procedures need not even be consistent in general and that care is needed with respect to the choice of priors. The growing number of examples and special cases point to the need for a more complete theoretical understanding.

2.7. Model Selection Methods

[A search on MathSciNet in early May 2005 for "semiparametric" and "model selection" gave 28 hits.] Theoretical understanding of model selection methods in nonparametric estimation problems has progressed rapidly during the last 10 years with major contributions by Birgé, and Massart (1997), Birgé, and Massart (1998), Barron, Birgé, and Massart (1999), and Birgé, and Massart (2001). Also see Massart (2000) and Massart's forthcoming St. Flour Lecture notes from 2003. These developments have begun to have some impact on semiparametric estimation as well: Raftery, Madigan and Volinsky (1996) and Raftery, Madigan, Volinsky and Kronmal (1997) used Bayesian methods for covariate selection in Cox models. Tibshirani (1997) introduces "lasso" methods for proportional hazards models (which involve an L_1 penalty term), while Fan and Li (2002) propose an alternative approach based on a non-concave penalty, and extend these methods to a class of frailty models (see below). Bunea (2004) studies the effect of covariate selection methods on inference in a partly linear regression model, and this is carried over to non-proportional hazards models in survival analysis by Bunea and McKeague (2005). It would be of some interest to extend these developments to the models considered by Huang (1999). We suspect that much more remains to be done to fully understand the advantages and disadvantages of various model-selection strategies.

2.8. Empirical Likelihood

[A search on MathSciNet in early May 2005 for "semiparametric" and "empirical likelihood" gave 18 hits; searching for "empirical likelihood" alone gave 192 hits.] Owen (1988) and Owen (1990) introduced the notion of empirical likelihood and showed how a reasonable facsimile of the standard theorem for the likelihood ratio statistic in regular parametric models continues to hold for finite-dimensional smooth functionals $\nu(P)$, $\nu : \mathcal{P} \to \mathbb{R}^d$. The basic notion in Owen's theory involves estimation of P in the restricted model $\mathcal{P}_0 = \{P \in \mathcal{M} : \nu(P) = t_0\}$ where $t_0 \in \mathbb{R}^d$ is fixed. The resulting model can be viewed as a semiparametric model with a tangent space having finite co-dimension in $L_2^0(P)$, one of the topics treated in BKRW (see section 6.2, pages 222- 229) and much earlier by Koshevnik and Levit (1976). This led to a considerable development of "empirical likelihood" based methods in connection with estimating equation approaches to a wide variety of semiparametric models: see e.g. Owen (1991), Qin (1993), Qin and Lawless (1994), Qin and Lawless (1995), and Lazar and Mykland (1999). Qin (1998) and Zou, Fine and Yandell (2002) give applications to mixture models.

2.9. Transformation and Frailty Models

[A search on MathSciNet in early May 2005 for "semiparametric" and "frailty" gave 28 hits searching for "semiparametric" and "transformation" gave 35 hits.] Cox (1972) introduced the proportional hazards model, resulting in one of the most cited papers in statistics ever. The Cox proportional hazards model for survival data is a transformation model with the baseline cumulative hazard function as unknown transformation. It is one of the prime examples of a semiparametric model in BKRW. Clayton and Cuzick (1985) generalized the Cox model by introducing frailty as an unobservable random factor in the hazard function. The results of Clayton and Cuzick inspired the key theoretical development in Bickel (1986); also see Klaassen (2005) for further discussion of the differential equations determining the efficient score functions in transformation models . These developments started a stream of papers that propose pragmatic methods in ever more complicated frailty models incorporating e.g. cluster-frailty, censoring etc. Efficiency of the proposed inference procedures is not necessarily a goal here; e.g. Li and Ryan (2002). Murphy (1995) provided key asymptotic theory for maximum likelihood estimation methods in a basic gamma-frailty model, and her methods were extended to more complicated frailty models by Parner (1998).

Another fundamental paper is Bickel and Ritov (1997) which discusses the issue of loss of efficiency when going from core to transformation model. The methods advocated by Bickel and Ritov have been implemented in the case of the binormal ROC model by Zou and Hall (2000), Zou and Hall (2002); see Hjellvik and Tjøstheim (1996) for background material concerning ROC curve estimation. Cai and Moskowitz (2004) give a profile likelihood approach to estimation in ROC models that deserves further study and evaluation.

Copula models for joint distributions have started receiving increasing interest in econometrics, finance, and other application areas. Klaassen and Wellner (1997) identified normal location-scale families as the least favorable sub-models for the class of bivariate normal copula models and constructed efficient estimators. Progress on efficient estimation for other copula models remains as a challenging open problem (with potential recent progress by Chen, Fan and Tsyrennikov (2004)).

A study of transformation and other semiparametric models with a focus on applications in econometrics is given in Horowitz (1998). A Bayesian approach is presented by Mallick and Walker (2003).

2.10. Semiparametric Regression Models

[A search on MathSciNet in early May 2005 for "semiparametric" and "regression" gave 652 hits.] A motivating class of models for BKRW were the "partially linear semiparametric regression models" studied in section 4.3, pages 107 - 112, BKRW (1993). It was already shown in Ritov and Bickel (1990) that efficient estimators cannot be constructed in these semiparametric models without smoothness assumptions on the class of functions allowed (also see Section 2.11 below). On the other hand, Schick (1993) gave a treatment of information bounds and construction of efficient estimators for the parametric component for a general class of models encompassing those treated in BKRW that improved on the results of Cuzick (1992). His approach was via estimation of the efficient influence function.

In the meantime information bounds and efficient estimators have now been constructed for many generalizations and variants of these models. For example, Sasieni (1992a) and Sasieni (1992b) calculated information bounds for a partially linear version of the Cox proportional hazards model. Estimates achieving the bounds were constructed by Huang (1999). Efficient estimation in a different but related class of models was studied by Neilsen, Linton, Bickel (1998).

Interesting classes of semilinear semiparametric regression models for applications to micro-array data have recently been introduced by Fan, Peng, and Huang (2005) and Huang, Wang, and Zhang (2005). There seem to be a large number of open questions connected with these models and their applications.

For a Bayesian approach to semiparametric regression models, see Seifu, Severini, a 1 Tanner (1999).

2.11. Extensions to Non-i.i.d. Data

[A search on MathSciNet in early May 2005 for "semiparametric" and "time series" gave 139 hits.] Semiparametric theory for times series models was already well underway by the time BKRW was published in (1993). For example, Kreiss (1987) had considered adaptive estimation (in the sense of Stein (1956) and BKRW) for stationary (causal) ARMA processes, while Gassiat (1990) and Gassiat (1993) showed that adaptation is not possible in the case of non-causal ARMA models. Drost, Klaassen, and Werker (1997) and Drost and Klaassen (1997) generalized the results of Kreiss and others (notably Koul and Schick (1997)) to classes of non-linear time series models. Koul and Schick (1996) considered an interesting random coefficient autoregressive model. The current state of the art is summarized in Greenwood, Müller, and Wefemeyer (2004).

Consideration of information bounds and efficient estimation more generally for a wide variety of Markov chains and other Markov processes finally emerged in the mid- and late 1990s: see e.g. Greenwood and Wefemeyer (1995), Schick and Wefelmeyer (1999), and Kessler, Schick, and Wefelmeyer (2001). Bickel and Kwon (2001) (following up on Bickel (1993)) reformulated much of this work and provided considerable unification. Also see the discussion piece by Greenwood, Schick, and Wefelmeyer following Bickel and Kwon (2001).

Another interesting direction of generalization concerns relaxing from "structural" (or i.i.d.) modeling to "functional" (non - i.i.d.) modeling. For an interesting study of the classical normal theory "errors in variables" model under "functional" or "incidental nuisance parameters", see Murphy and van der Vaart (1996). McNeney and Wellner (2000) give a review of information bounds for functional models.

2.12. Critiques and Possible Alternative Theories

In two key papers Bickel and Ritov (1988) and Ritov and Bickel (1990) pointed out that attainment of information bounds in semiparametric and nonparametric situations requires additional assumptions on the dimensionality of the parameter space. They gave several explicit examples of differentiable functionals, for which the information bounds are finite and yet are not attained in general by any estimator. One of these examples is the functional $\nu(P) = \int_0^1 p^2(x)dx$ for probability measures P on $[0,1]$ with density p with respect to Lebesgue measure. Another example involves estimation of θ in the partly linear regression model $Y = \theta^T Z + r(X) + \epsilon$ based on observation of (Y, Z, X). In both examples, the standard semiparametric bounds are attained when the parameters p and r are assumed to be smooth enough. However, for other smoothness classes, one can show that the (attained) minimax rate is much slower than $n^{-1/2}$. Moreover, in general, there is not even a consistent estimator. Thus there exists a "gap" between the semiparametric information bounds based on Hellinger differentiability, and the "real" information bounds that consider the amount of smoothness assumed. Birgé and Massart (1995) develop theory for other nonlinear functionals of the form $\nu(P) = \int_0^1 \phi(p(x), p'(x), \ldots, p^{(k)}(x))dx$ for densities p of smoothness s, and for $s \geq 2k + 1/4$ they construct estimators converging at rate $n^{-1/2}$. Moreover they show that $\nu(P)$ cannot be estimated faster than $n^{-\gamma}$ with $\gamma = 4(s - k)/(4s + 1)$ when $s < 2k + 1/4$.

These examples and others have been developed further by Robins and Ritov (1997) who make some steps toward development of a "Curse of Dimensionality Appropriate" (or CODA) asymptotic theory of semiparametric models. Robins and Ritov argue via a class of models involving missing data that the existing theory is inadequate and should be altered to incorporate more uniformity in convergence to the limiting distributions. Bickel and Ritov (2000) make a somewhat different suggestion involving regularization of parameters. These ideas deserve further exploration and development.

References

Section 2.1: Introduction.

Andersen, P. K. (1982). Testing goodness of fit of Cox's regression and life model. *Biometrics* **38**, 67–77.

Begun, J. M., Hall, W. J., Huang, W. M., and Wellner, J. A. (1983). Information and asymptotic efficiency in parametric-nonparametric models. *Ann. Statist.* **11**, 432–452.

Beran, R. (1974). Asymptotically efficient adaptive rank estimates in location models. *Ann. Statist.* **2**, 63–74.

Beran, R. (1977a). Estimating a distribution function. *Ann. Statist.* **5**, 400–404.

Beran, R. (1977b). Minimum Hellinger distance estimates for parametric models. *Ann. Statist.* **5**, 445–463.

Bickel, P. J. (1982). On adaptive estimation. *Ann. Statist.* **10**, 647–671.

Bickel, P. J., Klaassen, C. A. J., Ritov, Y., and Wellner, J. A. (1993). *Efficient and Adaptive Estimation for Semiparametric Models.* Johns Hopkins University Press, Baltimore. Reprinted (1998), Springer, New York.

Cox, D. R. (1972). Regression models and life tables (with discussion). *J. R. Stat. Soc. Ser. B Stat. Methodol.*, **34**, 187–220.

Cox, D. R. (1975). Partial likelihood. *Biometrika* **62**, 269–276.

Efron, B. (1977). The efficiency of Cox's likelihood function for censored data. *J. Amer. Statist. Assoc.* **72**, 557–565.

Epstein, B. (1982). Review of *White, J.; Fitting Cox's regression model to survival data using GLIM. Math. Rev.* **82c:62139**.

Epstein, B. (1983). Review of *Oakes, D.; Survival times: aspects of partial likelihood. Math. Rev.* **83f:62145**.

Finnas, F. and Hoem, J. M. (1980). Starting age and subsequent birth intervals in cohabitational unions in current Danish cohorts, 1975. *Demography* **17**, 275–295.

Gail, M. H., Santner, T. J., and Brown, C. C. (1980). An analysis of comparative carcinogenesis experiments based on multiple times to tumor. *Biometrics* **36**, 255–266.

Hájek, J. (1962). Asymptotically most powerful rank-order tests. *Ann. Math. Statist.* **33**, 1124–1147.

Kalbfleisch, J. D. and Prentice, R. L. (1980). *The Statistical Analysis of Failure Time Data.* Wiley, New York.

Kay, R. (1979). Some further asymptotic efficiency calculations for survival data regression models. *Biometrika* **66**, 91–96.

Louis, T. A., Mosteller, F., and McPeek, B. (1982). Timely topics in statistical methods for clinical trials. *Ann. Rev. Biophys. Bioeng.* **11**, 81–104.

Oakes, D. (1977). The asymptotic information in censored survival data. *Biometrika* **64**, 441–448.

Oakes, D. (1981). Survival times: aspects of partial likelihood. With a discussion and reply by the author. *Internat. Statist. Rev.* **49**, 235–264.

Stein, C. (1956). Efficient nonparametric testing and estimation. *Proc. Third Berkeley Symp. Math. Statist. Probab.* **1**, 187–195, University of California Press, Berkeley.

Stone, C. J. (1975). Adaptive maximum likelihood estimators of a location parameter. *Ann. Statist.* **3**, 267–284.

Turnbull, B. W. (1981). Review of *Kalbfleisch, J. D. and Prentice, R. L.; The Statistical Analysis of Failure Time Data. Math. Rev.* **81i:62186**.

Whitehead, J. (1980). Fitting Cox's regression model to survival data using GLIM. *Appl. Statist.* **29**, 268–275.

van Eeden, C. (1970). Efficiency-robust estimation of location. *Ann. Math. Statist.* **41**, 172–181.

Section 2.2: Missing data models.

Breslow, N., McNeney, B., and Wellner, J. A. (2003). Large sample theory for semiparametric regression models with two-phase, outcome dependent sampling. *Ann. Statist.* **31**, 1110–1139.

Chen, H. Y. (2002). Double-semiparametric method for missing covariates in Cox regression models. *J. Amer. Statist. Assoc.* **97**, 565–576.

Chen, H. Y. (2004). Nonparametric and semiparametric models for missing covariates in parametric regression. *J. Amer. Statist. Assoc.* **99**, 1176–1189.

Nan, B., Emond, M. J., and Wellner, J. A. (2004). Information bounds for Cox regression models with missing data. *Ann. Statist.* **32**, 723–753.

Robins, J. M., Hsieh, F., and Newey, W. (1995). Semiparametric efficient estimation of a conditional density with missing or mismeasured covariates. *J. R. Stat. Soc. Ser. B Stat. Methodol.* **57**, 409–424.

Robins, J. M. and Rotnitzky, A. (1992). Recovery of information and adjustment for dependent censoring using surrogate markers. *AIDS Epidemiology: Methodological Issues*, 297–331. eds. N. P. Jewell, K. Dietz, and V. T. Farewell. Birkhäuser, New York.

Robins, J. M. and Rotnitzky, A. (1995). Semiparametric efficiency in multivariate regression models with missing data . *J. Amer. Statist. Assoc.* **90**, 122–129.

Robins, J. M., Rotnitzky, A., and Zhao, L. P. (1994). Estimation of regression coefficients when some regressors are not always observed. *J. Amer. Statist. Assoc.* **89**, 846–866.

Robins, J. M., Rotnitzky, A., and Zhao, L. P. (1995). Analysis of semiparametric regression models for repeated outcomes in the presence of missing data . *J. Amer. Statist. Assoc.* **90**, 106–121.

van der Vaart, A. W. (1998). *Asymptotic Statistics.* Cambridge Univ. Press, Cambridge.

Wang, Q., Linton, O. B., and Härdle, W. (2004). Semiparametric regression analysis with missing response at random. *J. Amer. Statist. Assoc.* **99**, 334–345.

Section 2.3: Testing and profile likelihood theory.

Banerjee, M. (2005). Likelihood ratio tests under local alternatives in regular semiparametric models. *Statist. Sinica* **15**, 635–644.

Bickel, P. J., Ritov, Y., and Stoker, T. M. (2005a). Tailor-made tests for goodness-of-fit to semiparametric hypotheses. *Ann. Statist.*, to appear.

Bickel, P. J., Ritov, Y., and Stoker, T. M. (2005b). Nonparametric testing of an index model. *Identification and Inference for Econometric Models: A Festschrift in Honor of Thomas J. Rothenberg*, ed. by D. W. K. Andrews and J. H. Stock. Cambridge University Press, Cambridge (2005b).

Choi, S., Hall, W. J., and Schick, A. (1996). Asymptotically uniformly most powerful tests in parametric and semiparametric models. *Ann. Statist.* **24**, 841–861.

Härdle, W., Spokoiny, V., and Sperlich, S. (1997). Semiparametric single index versus fixed link function modelling. *Ann. Statist.* **25**, 212–243.

Horowitz, J. L. and Härdle, W. (1994). Testing a parametric model against a semiparametric alternative. *Econometric Theory* **10**, 821–848.

Kauermann, G. and Tutz, G. (2001). Testing generalized linear and semiparametric models against smooth alternatives. *J. R. Stat. Soc. Ser. B Stat. Methodol.* **63**, 147–166.

Lee, B. L., Kosorok, M. R., and Fine, J. P. (2005). The profile sampler. *J. Amer. Statist. Assoc.* **100**, 960–969.

Murphy, S. A. and van der Vaart, A. W. (1997). Semiparametric likelihood ratio inference. *Ann. Statist.* **25**, 1471–1509.

Murphy, S. A. and van der Vaart, A. W. (1999). Observed information in semiparametric models. *Bernoulli* **5**, 381–412.

Murphy, S. A. and van der Vaart, A. W. (2000). On profile likelihood. *J. Amer. Statist. Assoc.* **95**, 449–485.

Severini, T. A. and Staniswalis, J. G. (1994). Quasi-likelihood estimation in semiparametric models. *J. Amer. Statist. Assoc* **89**, 501–511.

Severini, T. A. and Wong, W. H. (1991). Profile likelihood and conditionally parametric models. *Ann. Statist.* **20**, 1768–1802.

Section 2.4: Semiparametric mixture model theory.

Kiefer, J. and Wolfowitz, J. (1956). Consistency of the maximum likelihood estimator in the presence of infinitely many nuisance parameters. *Ann. Math. Statist.* **27**, 887–906.

Neyman, J. and Scott, E. (1948). Consistent estimates based on partially consistent observations. *Econometrica* **16**, 1–32.

Pfanzagl, J. (1988). Consistency of maximum likelihood estimators for certain nonparametric familes, in particular: mixtures. *J. Statist. Plann. Inference* **19**, 137–158.

Pfanzagl, J. (1990). *Estimation in Semiparametric Models. Lecture Notes in Statistics* **63**. Springer, New York.

van der Vaart, A. W. (1988). Estimating a real parameter in a class of semiparametric models. *Ann. Statist.* **16**, 1450–1474.

van der Vaart, A. W. (1996). Efficient maximum likelihood estimation in semiparametric mixture models. *Ann. Statist.* **24**, 862–878.

Section 2.5: Rates of convergence via empirical process methods.

Birgé, L. and Massart, P. (1993). Rates of convergence for minimum contrast estimators. *Probab. Theory Related Fields* **97**, 113–150.

Birgé, L. and Massart, P. (1998). Minimum contrast estimators on sieves: exponential bounds and rates of convergence. *Bernoulli* **4**, 329–375.

Huang, J. (1996). Efficient estimation for the proportional hazards model with interval censoring. *Ann. Statist.* **24**, 540–568.

Shen, X. (1997). On methods of sieves and penalization. *Ann. Statist.* **25**, 2555–2591.

Shen, X. and He, K. (1997). The rate of convergence of PMLEs. *Probab. Math. Statist.* **17**, 21–28.

Shen, X. and Wong, W. H. (1994). Convergence rate of sieve estimates. *Ann. Statist.* **22**, 580–615.

van de Geer, S. (1996). Rates of convergence for the maximum likelihood estimator in mixture models. *J. Nonparametr. Stat.* **6**, 293–310.

van der Vaart, A. W. (2002). Semiparametric Statistics. *Lectures on Probability Theory and Statistics, Ecole d'Eté de Probabilités de Saint-Flour XXIX - 1999, Lecture Notes in Mathematics* **1781**, 330–457.

van der Vaart, A. W. and Wellner, J. A. (1996). *Weak Convergence and Empirical Processes.* Springer, New York.

Wong, W. H. and Severini, T. A. (1991). On maximum likelihood estimation in infinite dimensional parameter spaces. *Ann. Statist.* **19**, 603–632.

Wong, W. H. and Shen, X. (1995). Probability inequalities for likelihood ratios and convergence rates of sieve MLE's. *Ann. Statist.* **23**, 339–362.

Section 2.6: Bayes methods and theory.

Carlin, B. P. and Louis, T. A. (2000). *Bayes and Empirical Bayes methods for Data Analysis, 2nd Ed.* Chapman and Hall / CRC Texts, Boca Raton.

Freedman, D. (1999). On the Bernstein-Von Mises theorem with infinite-dimensional parameters. *Ann. Statist.* **27**, 1119–1140.

Ghosal, S., Ghosh, J. K., and Samanta, T. (1995). On convergence of posterior distributions. *Ann. Statist.* **23**, 2145–2152.

Ghosal, S., Ghosh, J. K., and van der Vaart, A. W. (2000). Convergence rates of posterior distributions. *Ann. Statist.* **28**, 500–531.

Ghosal, S. and van der Vaart, A. W. (2001). Entropies and rates of convergence for maximum likelihood and Bayes estimation for mixtures of normal densities. *Ann. Statist.* **29**, 1233–1263.

Gilks, W. R., Richardson, S., and Spiegelhalter, D. J. (1999). *Markov Chain MonteCarlo in Practice.* Chapman and Hall, Boca Raton.

Huang, T-M. (2004). Convergence rates for posterior distributions and adaptive estimation. *Ann. Statist.* **32**, 1556–1593.

Kim, Y. and Lee, J. (2004). A Bernstein-von Mises theorem in the nonparametric right-censoring model. *Ann. Statist.* **32**, 1492–1512.

Kleijn, B. J. K. and van der Vaart, A. W. (2002). Misspecification in Infinite -Dimensional Bayesian statistics. *Technical Report* **2002-16**, Free University, Amsterdam.

Lazar, N. A. (2003). Bayesian empirical likelihood. *Biometrika* **90**, 319–326.

Lenk, P. J. (1999). Bayesian inference for semiparametric regression using a

Fourier representation. *J. R. Stat. Soc. Ser. B Stat. Methodol.* **61**, 863–879.

Müller, P. and Roeder, K. A. (1997). Bayesian semiparametric model for case-control studies with errors in variables. *Biometrika* **84**, 523 - 537.

Newton, M. A., Czado, C., and Chappell, R. (1996). Bayesian inference for semiparametric binary regression. *J. Amer. Statist. Assoc.* **91**, 142–153.

Robert, C. P. and Casella, G. (2004). *Monte Carlo Statistical Methods* Springer, New York.

Shen, X. (2002). Asymptotic normality of semiparametric and nonparametric posterior distributions. *J. Amer. Statist. Assoc.* **97**, 222–235.

Shen, X. and Wasserman, L. (2001). Rates of convergence of posterior distributions. *Ann. Statist.* **29**, 687–714.
sic:03

Sinha, D., Ibrahim, J. G., and Chen, M-H. (2003). A Bayesian justification of Cox's partial likelihood. *Biometrika* **90**, 629–641.

Vidakovic, B. (1998). Wavelet-based nonparametric Bayes methods. *Practical Nonparametric and Semiparametric Bayesian Statistics*, 133–155, *Lecture Notes in Statistics* **133**, Springer, New York.

Section 2.7: Model selection methods.

Barron, A., Birgé, L., and Massart, P. (1999). Risk bounds for model selection via penalization. *Probab. Theory Related Fields* **113**, 301–413.

Birgé, L. and Massart, P. (1997). From model selection to adaptive estimation. *Festschrift for Lucien Le Cam*, 55–87, Springer, New York.

Birgé, L. and Massart, P. (1998). Minimum contrast estimators on sieves: exponential bounds and rates of convergence. *Bernoulli* **4**, 329–375.

Birgé, L. and Massart, P. (2001). Gaussian model selection. *J. Eur. Math. Soc.* **3**, 203–268.

Bunea, F. (2004). Consistent covariate selection and post model selection inference in semiparametric regression. *Ann. Statist.* **32**, 898–927.

Bunea, F. and McKeague, I. W. (2005). Covariate selection for semiparametric hazard function regression models. *J. Multivariate Anal.* **92**, 186–204.

Fan, J. and Li, R. (2002). Variable selection for Cox's proportional hazards model and frailty model. *Ann. Statist.* **30**, 74–99.

Huang, J. (1999). Efficient estimation of the partly linear additive Cox model. *Ann. Statist.* **27**, 1536–1563.

Massart, P. (2000). Some applications of concentration inequalities to statistics. *Ann. Fac. Sci. Toulouse Math.* **(6) 9**, 245–303.

Raftery, A. E., Madigan, D., and Volinsky, C. T. (1996). Accounting for model uncertainty in survival analysis improves predictive performance. *Bayesian Statistics 5 – Proceedings of the Fifth Valencia International Meeting, Valencia, Spain.* Oxford Science Publications, Oxford.

Raftery, A. E., Madigan, D., Volinsky, C. T., and Kronmal, R. A. (1997). Bayesian model averaging in proportional hazards models: predicting the risk of a stroke. *Appl. Statist.* **46**, 443–448.

Tibshirani, R. J. (1997). The lasso method for variable selection in the Cox model.

Statist. in Medicine **16**, 385–395.

Section 2.8: Empirical likelihood.

Koshevnik, Yu. A. and Levit, B. Y. (1976). On a non-parametric analogue of the information matrix. *Theory Probab. Appl.* **21**, 738–753.

Lazar, N. A. and Mykland, P. A. (1999). Empirical likelihood in the presence of nuisance parameters. *Biometrika* **86**, 203–211.

Owen, A. B. (1988). Empirical likelihood ratio confidence intervals for a single functional. *Biometrika* **75**, 237–249.

Owen, A. B. (1990). Empirical likelihood ratio confidence regions. *Ann. Statist.* **18**, 90–120.

Owen, A. (1991). Empirical likelihood for linear models. *Ann. Statist.* **19**, 1725–1747.

Qin, J. (1993). Empirical likelihood in biased sample problems. *Ann. Statist.* **21**, 1182–1196.

Qin, J. (1998). Semiparametric likelihood based method for goodness of fit tests and estimation in upgraded mixture models. *Scand. J. Statist.* **25**, 681–691.

Qin, J. and Lawless, J. (1994). Empirical likelihood and general estimating equations. *Ann. Statist.* **22**, 300–325.

Qin, J. and Lawless, J. (1995). Estimating equations, empirical likelihood and constraints on parameters. *Canad. J. Statist.* **23**, 145–159.

Zou, F., Fine, J. P., and Yandell, B. S. (2002). On empirical likelihood for a semi-parametric mixture model. *Biometrika* **89**, 61–75.

Section 2.9: Transformation and frailty models.

Bickel, P. J. (1986). Efficient testing in a class of transformation models . *Proceedings of the 45th Session of the International Statistical Institute* 23.3.63–23.3.81. ISI, Amsterdam.

Bickel, P. J. and Ritov, Y. (1997). Local asymptotic normality of ranks and co-variates in transformation models. *Festschrift for Lucien Le Cam*, 43–54, Springer, New York.

Cai, T. X. and Moskowitz, C. S. (2004). Semi-parametric estimation of the binormal ROC curve for a continuous diagnostic test. *Biostatistics* **5**, 573–586.

Chen, X., Fan, Y., and Tsyrennikov, V. (2004). Efficient estimation of semi-parametric multivariate copula models. *Working Paper Nol. 04-W20*, Department of Economics, Vanderbilt University. Available at: *http://www.vanderbilt.edu/Econ/wparchive/working04.html*

Clayton, D. G. and Cuzick, J. (1985). The semi-parametric Pareto model for regression analysis of survival times. *Bull. Int. Statist. Inst.* **51**, 23.3.175–23.3.180.

Cox, D. R. (1972). Regression models and life tables (with discussion). *J. R. Stat. Soc. Ser. B Stat. Methodol.* **34**, 187–220.

Horowitz, J. L. (1998). Semiparametric Methods in Econometrics. *Lecture Notes in Statistics* **131**, Springer-Verlag, New York.

Hsieh, F. and Turnbull, B. W. (1996). Nonparametric and semiparametric estimation of the receiver operating characteristic curve. *Ann. Statist.* **24**, 25–40.

Klaassen, C. A. J. (2005). A Sturm-Liouville problem in semiparametric transformation models . Festschrift volume for Kjell Doksum, to appear.

Klaassen, C. A. J. and Wellner, J. A. (1997). Efficient estimation in the bivariate normal copula model: normal margins are least favorable. *Bernoulli* **3**, 55–77.

Li, Y. and Ryan, L. M. (2002). Modeling spatial survival data using semiparametric frailty models . *Biometrics* **58**, 287–297.

Mallick, B. K. and Walker, S. (2003). A Bayesian semiparametric transformation model incorporating frailties. *Model selection, model diagnostics, empirical Bayes and hierarchial Bayes, Special Issue II, J. Statist. Plann. Inference* **112**, 159–174.

Murphy, S. A. (1995). Asymptotic theory for the frailty model. *Ann. Statist.* **23**, 182–198.

Parner, E. (1998). Asymptotic theory for the correlated gamma-frailty model. *Ann. Statist.* **26**, 183–214.

Zou, K. H. and Hall, W. J. (2000). Two transformation models for estimating an ROC curve derived from continuous data. *J. Appl. Stat.* **27**, 621–631.

Zou, K. H. and Hall, W. J. (2002). Semiparametric and parametric transformation models for comparing diagnostic markers with paired design. *J. Appl. Stat.* **29**, 803–816.

Section 2.10: Semiparametric regression models.

Cuzick, J. (1992). Efficient estimates in semiparametric additive regression models with unknown error distribution. *Ann. Statist.* **20**, 1129–1136.

Fan, J., Peng, H., and Huang, T. (2005). Semilinear high-dimensional model for normalization of microarray data: a theoretical analysis and partial consistency. With discussion. *J. Amer. Statist. Assoc.* **100**, 781–813.

Huang, J. (1999). Efficient estimation of the partly linear additive Cox model. *Ann. Statist.* **27**, 1536–1563.

Huang, J., Wang, D., and Zhang, C-H. (2005). A two-way semilinear model for normalization and analysis of cDNA microarray data. *J. Amer. Statist. Assoc.* **100**, 814–829.

Nielsen, J. P., Linton, O. B., and Bickel, P. J. (1998). On a semiparametric survival model with flexible covariate effect. *Ann. Statist.* **26**, 215–241.

Ritov, Y. and Bickel, P. J. (1990). Achieving information bounds in non and semiparametric models. *Ann. Statist.* **18**, 925 - 938.

Sasieni, P. (1992a). Information bounds for the conditional hazard ratio in a nested family of regression models. *J. R. Stat. Soc. Ser. B Stat. Methodol.* **54**, 617–635.

Sasieni, P. (1992b). Non-orthogonal projections and their application to calculating the information in a partly linear Cox model. *Scand. J. Statist.* **19**, 215–233.

Schick, A. (1993). On efficient estimation in regression models. *Ann. Statist.* **21**, 1486–1521. Correction and addendum. *Ann. Statist.* **23**, 1862–1863.

Seifu, Y., Severini, T. A., and Tanner, M. A. (1999). Semiparametric Bayesian inference for regression models. *Canad. J. Statist.* **27**, 719–734.

Section 2.11: Extensions to non-i.i.d. data.

Bickel, P. J. (1993). Estimation in semiparametric models. *Multivariate Analysis: Future Directions*, 55–73.

Bickel, P. J. and Kwon, J. (2001). Inference for semiparametric models: some questions and an answer. With comments and a rejoinder by the authors. *Statist. Sinica* **11**, 863–960.

Drost, F. C. and Klaassen, C. A. J. (1997). Efficient estimation in semiparametric GARCH models. *J. Econometrics* **81**, 193–221.

Drost, F. C., Klaassen, C. A. J., and Werker, B. J. M. (1997). Adaptive estimation in time-series models. *Ann. Statist.* **25**, 786–817.

Gassiat, E. (1990). Estimation semi-paramétrique d'un modèle autorégressif stationnaire multiindice non nécessairement causal. [Semiparametric estimation of a not necessarily causal multi-index stationary autoregressive model]. *Ann. Inst. H. Poincaré Probab. Statist.* **26**, 181–205.

Gassiat, E. (1993). Adaptive estimation in noncausal stationary AR processes. *Ann. Statist.* **21**, 2022–2042.

Greenwood, P. E., Müller, U. U., and Wefelmeyer, W. (2004). An introduction to efficient estimation for semiparametric time series. *Parametric and semiparametric models with applications to reliability, survival analysis, and quality of life*, 253–269, Stat. Ind. Technol., Birkhäuser, Boston.

Greenwood, P. E. and Wefelmeyer, W. (1995). Efficiency of empirical estimators for Markov chains. *Ann. Statist.* **23**, 132–143.

Kessler, M., Schick, A., and Wefelmeyer, W. (2001). The information in the marginal law of a Markov chain. *Bernoulli* **7**, 243–266.

Koul, H. L. and Schick, A. (1996). Adaptive estimation in a random coefficient autoregressive model. *Ann. Statist.* **24**, 1025–1052.

Koul, H. L. and Schick, A. (1997). Efficient estimation in nonlinear autoregressive time-series models. *Bernoulli* **3**, 247–277.

Kreiss, J-P. (1987). On adaptive estimation in stationary ARMA processes. *Ann. Statist.* **15**, 112–133.

McNeney, B. and Wellner, J. A. (2000). Application of convolution theorems in semiparametric models with non-i.i.d. data. *Prague Workshop on Perspectives in Modern Statistical Inference: Parametrics, Semiparametrics, Non-parametrics (1998). J. Statist. Plann. Inference* **91**, 441–480.

Murphy, S. A. and van der Vaart, A. W. (1996). Likelihood inference in the errors-in-variables model. *J. Multivariate Anal.* **59**, 81–108.

Schick, A. and Wefelmeyer, W. (1999). Efficient estimation of invariant distributions of some semiparametric Markov chain models. *Math. Methods Statist.* **8**, 426–440.

Stein, C. (1956). Efficient nonparametric testing and estimation. *Proc. Third Berkeley Symp. Math. Statist. Prob.* **1**, 187–195. University of California Press, Berkeley.

Section 2.12: Critiques and possible alternative theories.

Bickel, P. J. and Ritov, Y. (1988). Estimating integrated squared density derivatives: sharp best order of convergence estimates. *Sankhyā Ser. A* **50**, 381–393.

Bickel, P. J. and Ritov, Y. (2000). Non- and semiparametric statistics: compared and contrasted. *J. Statist. Plann. Inference* **91**, 209–228.

Birgé, L. and Massart, P. (1995). Estimation of integral functionals of a density. *Ann. Statist.* **23**, 11–29.

Ritov, Y. and Bickel, P. J. (1990). Achieving information bounds in non and semiparametric models. *Ann. Statist.* **18**, 925–938.

Robins, J.M. and Ritov, Y. (1997). Toward a curse of dimensionality appropriate (CODA) asymptotic theory for semi-parametric models. *Statistics in Medicine* **16**, 285–319.

CHAPTER 3

Efficient Estimator for Time Series

Anton Schick and Wolfgang Wefelmeyer

Department of Mathematical Sciences
Binghamtom University
Binghamton, NY 13902-6000
anton@math.binghamton.edu

Mathematisches Institut
Universität zu Köln
Weyertal 86–90, 50931 Köln, Germany
wefelm@math.uni-koeln.de

We illustrate several recent results on efficient estimation for semiparametric time series models with a simple class of models: first-order nonlinear autoregression with independent innovations. We consider in particular estimation of the autoregression parameter, the innovation distribution, conditional expectations, the stationary distribution, the stationary density, and higher-order transition densities.

3.1. Introduction

Inference for semiparametric parametric time series is well-studied. Two recent monographs are Taniguchi and Kakizawa (2000) and Fan and Yao (2003). The classical nonparametric estimators are however inefficient. In the last twenty years, efficient estimators for various functionals of such models have been constructed. The main effort was on estimators for the Euclidean parameters, but recently other functionals of time series have also been treated. We describe some of these results in a simple situation, observations $X_0, \ldots, X_n$ from a stationary nonlinear autoregressive model $X_i = r_\vartheta(X_{i-1}) + \varepsilon_i$ with independent innovations ε_i. For notational simplicity we restrict our attention to the first-order case and assume that ϑ is one-dimensional. The innovations are assumed to have mean zero, finite variance, and a positive density f. The model is semiparametric, with

45

"parameter of interest" ϑ and "nuisance parameter" f. We will also be interested in f, in which case ϑ would be the nuisance parameter. In the time series and regression literature, our model would be called "parametric" because the (auto-)regression function depends on a finite-dimensional parameter.

In Section 2 we recall a characterization of regular and efficient estimators in the context of our model. Section 3 describes an efficient estimator for ϑ as a one-step improvement of a $n^{1/2}$-consistent initial estimator. Section 4 shows that appropriately weighted residual-based empirical estimators are efficient for linear functionals $E[h(\varepsilon)]$ of the innovation distribution. Section 5 introduces similarly weighted residual-based kernel estimators $\hat{f}_w$ for the innovation density and shows that plug-in estimators $\int h(y)\hat{f}_w(y)\,dy$ are also efficient for linear functionals $E[h(\varepsilon)]$. Section 6 uses the representation $E(h(X_{n+1}) \mid X_n) = E[h(\varepsilon + r_\vartheta(x))]$ for a conditional expectation to construct $n^{1/2}$-consistent and efficient estimators for it. The results extend to higher-order lags. As m tends to infinity, the conditional expectation $E(h(X_{n+m} \mid X_n)$ of lag m converges to the expectation $E[h(X)]$ under the stationary law. This gives rise to efficient estimators for such expectations, as shown in Section 7. The stationary density g has the representation $g(y) = E[f(y - r_\vartheta(X))]$. In Section 8 we use this representation to construct $n^{1/2}$-consistent and efficient estimators for g. The two-step transition density q_2 has the representation $q_2(x, z) = E[f(z - r_\vartheta(\varepsilon + r_\vartheta(x)))]$. Section 9 suggests $n^{1/2}$-consistent and efficient estimators for q_2. This extends to higher-order lags.

Our estimators for the autoregression parameter, the innovation distribution and the stationary distribution in Sections 3, 4 and 7 have the same parametric convergence rate as the usual nonparametric estimators, but smaller asymptotic variances. On the other hand, our estimators for conditional expectations, the stationary density and higher-order transition densities in Sections 5, 8 and 9 have better, parametric, convergence rates than the nonparametric estimators. The parametric rates in Sections 5 and 9 require a parametric form of the autoregression function, because the representations of the functionals there are functions of the value of the autoregression function at a point. The parametric rates in Sections 4, 7 and 8 would extend to models with semiparametric or nonparametric autoregression functions, because the functionals considered there are smooth functionals of the autoregression function, and the plug-in principle works.

3.2. Characterization of Efficient Estimators

In this section we recall a characterization of efficient estimators in the context of our semiparametric time series model. The standard reference in the case of i.i.d. data is Bickel, Klaassen, Ritov and Wellner (1998). The theory for time series is similar, especially for processes driven by independent innovations, as in our model.

Let $X_0, \ldots, X_n$ be observations from the stationary nonlinear time series $X_i = r_\vartheta(X_{i-1}) + \varepsilon_i$. Assume that the innovations ε_i are i.i.d. with mean zero, finite variance σ^2, and density f. Conditions for geometric ergodicity in terms of the growth of r_ϑ are in Mokkadem (1987), Bhattacharya and Lee (1995a), Bhattacharya and Lee (1995b) and An and Huang (1996). The model is described by the transition density $f(y - r_\vartheta(x))$ from $X_0 = x$ to $X_1 = y$, and parametrized by ϑ and f. Write F for the distribution function of f. Let g denote the stationary density. We will write (X, ε) for (X_0, ε_1) and (X, Y) for (X_0, X_1). In order to characterize efficient estimators of smooth functionals of (ϑ, f), we show that the model is locally asymptotically normal. For this, fix ϑ and f and introduce perturbations $\vartheta_{nu} = \vartheta + n^{-1/2}u$ and $f_{nv}(x) \doteq f(x)(1 + n^{-1/2}v(x))$ (in the sense of Hellinger differentiability). Here the local parameter u runs through $\mathbb{R}$. For f_{nv} to be again a probability density with mean zero, the local parameter v must lie in the linear space

$$V = \{v \in L_2(F) : E[v(\varepsilon)] = E[\varepsilon v(\varepsilon)] = 0\}.$$

In other words: $v(\varepsilon)$ must be orthogonal to 1 and ε. If r_ϑ is appropriately differentiable in ϑ with derivative $\dot{r}_\vartheta$, then the transition density is perturbed as

$$f_{nv}(y - r_{\vartheta_{nu}}(x)) \doteq f(y - r_\vartheta(x))\big(1 + n^{-1/2}\big(v(y - r_\vartheta(x)) + u\dot{r}_\vartheta(x)\ell(y - r_\vartheta(x))\big)\big),$$

with $\ell = -f'/f$ the score function for location of the innovation distribution. The perturbation of the transition density is the *tangent*; it is convenient to write it as a random variable

$$t_{uv}(X, Y) = v(\varepsilon) + u\dot{r}_\vartheta(X)\ell(\varepsilon).$$

The *tangent space* of the model is

$$T = \{t_{uv}(X, Y) : u \in \mathbb{R}, \ v \in V\}.$$

Let P_{n+1} denote the joint law of $(X_0, \ldots, X_n)$, with density

$$g(X_0) \prod_{i=1}^{n} f(X_i - r_\vartheta(X_{i-1})),$$

and write $P_{n+1,uv}$ for the joint law under ϑ_{nu} and f_{nv}. Koul and Schick (1997) prove *local asymptotic normality*, i.e. a quadratic approximation of the local log-likelihood of the form

$$\log \frac{dP_{n+1,uv}}{dP_{n+1}}(X_0, \ldots, X_n)$$

$$= n^{-1/2} \sum_{i=1}^{n} t_{uv}(X_{i-1}, X_i) - \frac{1}{2} E[t_{uv}^2(X, Y)] + o_p(1),$$

where the linear term $n^{-1/2} \sum_{i=1}^{n} t_{uv}(X_{i-1}, X_i)$ is asymptotically normal with variance $E[t_{uv}^2(X, Y)]$ by a martingale central limit theorem.

The tangent space T is a subspace of

$$S = \{s(X, Y) \in L_2(P_2) : E(s(X, Y) \mid X) = 0\}.$$

Consider a real-valued functional κ of (ϑ, f). Call κ *differentiable* at (ϑ, f) with *gradient* $s \in S$ if

$$n^{1/2}(\kappa(\vartheta_{nu}, f_{nv}) - \kappa(\vartheta, f)) \to E[s(X, Y)t_{uv}(X, Y)], \quad (u, v) \in \mathbb{R} \times V.$$

The gradient is not uniquely determined, but its projection t_* onto T, the *canonical gradient*, is. Let $\hat{\kappa}$ be an estimator of κ. Call $\hat{\kappa}$ *regular* at (ϑ, f) with *limit* L if

$$n^{1/2}(\hat{\kappa} - \kappa(\vartheta_{nu}, f_{nv})) \Rightarrow L \quad \text{under } P_{n+1,uv}, \quad (u, v) \in \mathbb{R} \times V.$$

Call $\hat{\kappa}$ *asymptotically linear* at (ϑ, f) with *influence function* $s \in S$ if

$$n^{1/2}(\hat{\kappa} - \kappa(\vartheta, f)) = n^{-1/2} \sum_{i=1}^{n} s(X_{i-1}, X_i) + o_p(1).$$

By a martingale central limit theorem, such an estimator is asymptotically normal with variance $E[s^2(X, Y)]$. The convolution theorem of Hájek (1970) and Le Cam (1971) in the version of Bickel *et al.* (1998, Section 2.3) implies the following three results:

(1) The distribution of L is a convolution, $L = N + M$ in distribution, where N is normal with variance $E[t_*^2(X, Y)]$ and M is independent of N.

(2) A regular estimator has limit $L = N$ if and only if it is asymptotically linear with influence function t_*.

(3) An asymptotically linear estimator is regular if and only if its influence function is a gradient.

A regular estimator with limit $L = N$ is least dispersed among all regular estimators. Such an estimator is called *efficient*. It follows from (1)–(3) that $\hat{\kappa}$ is regular and efficient if and only if it is asymptotically linear with influence function equal to the canonical gradient,

$$n^{1/2}(\hat{\kappa} - \kappa(\vartheta, f)) = n^{-1/2} \sum_{i=1}^{n} t_*(X_{i-1}, X_i) + o_p(1). \qquad (3.1)$$

Remark 3.1: Note that S is the tangent space of the nonparametric model of *all* first-order Markov chains on the state space $\mathbb{R}$. Such a chain is described by its transition distribution $Q(x, dy)$, which is perturbed as $Q_{ns}(x, dy) \doteq Q(x, dy)(1 + n^{-1/2}s(x, y))$ with $s \in S$. The reason for embedding T into S, i.e. the autoregressive model into a nonparametric Markov chain model, is the following. Often the functional of interest has a natural extension to a larger model. In such a larger model it is typically easier to determine a gradient, for example as the influence function of some nonparametric estimator. The canonical gradient is then found by projecting the given gradient onto T. Also, in some cases an efficient estimator is found by correcting the given nonparametric estimator.

The choice of the larger space S determines how many (regular) asymptotically linear estimators exist. For the choice $S = T$, any functional would have a unique gradient, and all (regular) asymptotically linear estimators would be asymptotically equivalent. We could also pick a larger S than above, for example the tangent space of all Markov chains of *arbitrary* order, which would give more "asymptotically linear" estimators, but for our purposes the chosen S turns out to be large enough.

Remark 3.2: We have introduced gradients and influence functions as elements of the tangent space S of *transition* distributions $Q(x, dy)$. Bickel (1993) and Bickel and Kwon (2001) describe Markov chain models by the *joint* law $P_2(dx, dy)$ of two successive observations. This is particularly convenient when the model and the functional of interest are naturally described in terms of P_2. Results for Markov chains can then be obtained from results for bivariate i.i.d. models, and vice versa. See also the discussion in Greenwood, Schick and Wefelmeyer (2001).

To calculate canonical gradients, it is convenient to decompose the tangents t_{uv} into orthogonal components. We have $E[\varepsilon\ell(\varepsilon)] = 1$. Hence the projection of $\ell(\varepsilon)$ onto V is $\ell_V(\varepsilon) = \ell(\varepsilon) - \sigma^{-2}\varepsilon$. Write $\mu = E[\dot{r}_\vartheta(X)]$. Then

$$t_{uv}(X, Y) = v(\varepsilon) + u\mu\ell_V(\varepsilon) + us_0(X, Y)$$

with $s_0(X, Y) = \dot{r}_\vartheta(X)\ell(\varepsilon) - \mu\ell_V(\varepsilon)$ orthogonal to V. The variance of $s_0(X, Y)$ is

$$\Lambda = E[s_0^2(X, Y)] = RJ - \mu^2 J_V$$

with $R = E[\dot{r}_\vartheta^2(X)]$, $J = E[\ell^2(\varepsilon)]$ and $J_V = E[\ell_V^2(\varepsilon)] = J - \sigma^{-2}$.

In the following we describe the canonical gradients for some of the functionals considered in Sections 3–9 below.

Autoregression parameter. By Section 2 of Schick and Wefelmeyer (2002a), the canonical gradient of $\kappa(\vartheta, f) = \vartheta$ is

$$t_*(X, Y) = \Lambda^{-1} s_0(X, Y). \tag{3.2}$$

If f happens to be normal, then $\ell(\varepsilon) = \sigma^{-2}\varepsilon$, so $J = \sigma^{-2}$, $J_V = 0$, $\Lambda = R\sigma^{-2}$ and hence $s_0(X, Y) = \dot{r}_\vartheta(X)\sigma^{-2}\varepsilon$ and $t_*(X, Y) = R^{-1}\dot{r}_\vartheta(X)\varepsilon$.

Innovation distribution. By Section 2 of Schick and Wefelmeyer (2002a), the canonical gradient of a linear functional $\kappa(\vartheta, f) = E[h(\varepsilon)]$ of the innovation distribution has canonical gradient

$$t_*(X, Y) = h_V(\varepsilon) - \mu E[h_V(\varepsilon)\ell(\varepsilon)]\Lambda^{-1} s_0(X, Y), \tag{3.3}$$

where $h_V(\varepsilon) = h(\varepsilon) - E[h(\varepsilon)] - \sigma^{-2}E[\varepsilon h(\varepsilon)]\varepsilon$ is the projection of $h(\varepsilon)$ onto V. In the submodel with ϑ *known*, the canonical gradient of $E[h(\varepsilon)]$ is $h_V(\varepsilon)$.

Conditional expectation. The conditional expectation with lag one of a function h can be written

$$E(h(Y) \mid X = x) = \int h(y)f(y - r_\vartheta(x))\, dy = \int h(y + r_\vartheta(x))f(y)\, dy.$$

This is the (unconditional) expectation $E[h(\varepsilon, \vartheta)]$ of a function $h(y, \vartheta) = h(y + r_\vartheta(x))$ depending on ϑ, and the gradient is similar to (3.3), with additional terms from this dependence on ϑ,

$$t_*(X, Y) = h_V(\varepsilon, \vartheta) + (\dot{r}_\vartheta(x) - \mu)E[h_V(\varepsilon, \vartheta)\ell(\varepsilon)]\Lambda^{-1} s_0(X, Y). \tag{3.4}$$

3.3. Autoregression Parameter

A simple estimator of ϑ is the *least squares estimator*. It is defined as the minimizer in ϑ of $\sum_{i=1}^n (X_i - r_\vartheta(X_{i-1}))^2$ and is the solution of the martingale estimating equation

$$\sum_{i=1}^n \dot{r}_\vartheta(X_{i-1})(X_i - r_\vartheta(X_{i-1})) = 0.$$

By Taylor expansion, its influence function is seen to be $s(X,Y) = R^{-1}\dot{r}_\vartheta(X)\varepsilon$. We have seen in Section 2 that this equals the canonical gradient of ϑ only if the innovations happen to be normally distributed.

An efficient estimator of ϑ is obtained in Koul and Schick (1997) as a one-step improvement of an initial $n^{1/2}$-consistent estimator $\hat{\vartheta}$, for example the least squares estimator. Rewrite the canonical gradient of ϑ as

$$t_*(X,Y) = \Lambda^{-1}\big((\dot{r}_\vartheta(X) - \mu)\ell(\varepsilon) - \mu\sigma^{-2}\varepsilon\big).$$

Estimate $\ell = -f'/f$ by $\hat{\ell} = -\hat{f}'/\hat{f}$ with $\hat{f}$ an appropriate kernel estimator. Then estimate J by $\hat{J} = (1/n)\sum_{i=1}^n \hat{\ell}^2(\hat{\varepsilon}_i)$, and μ, R and σ^2 by empirical estimators $\hat{\mu} = (1/n)\sum_{i=1}^n \dot{r}_{\hat{\vartheta}}(X_i)$, $\hat{R} = (1/n)\sum_{i=1}^n \dot{r}_{\hat{\vartheta}}^2(X_i)$ and $\hat{\sigma}^2 = (1/n)\sum_{i=1}^n \hat{\varepsilon}_i^2$, where $\hat{\varepsilon} = X_i - r_{\hat{\vartheta}}(X_{i-1})$ are the residuals. For $\Lambda = RJ - \mu^2 J_V = (R - \mu^2)J - \mu^2\sigma^{-2}$ we obtain the estimator $\hat{\Lambda} = (\hat{R} - \hat{\mu}^2)\hat{J} - \hat{\mu}^2\hat{\sigma}^{-2}$, and for $t_*(X_{i-1}, X_i)$ we obtain the estimator

$$\hat{t}_*(X_{i-1}, X_i) = \hat{\Lambda}^{-1}\big((\dot{r}_{\hat{\vartheta}}(X_i) - \hat{\mu})\hat{\ell}(\hat{\varepsilon}_i) - \hat{\mu}\hat{\sigma}^{-2}\hat{\varepsilon}_i\big).$$

The efficient one-step improvement of $\hat{\vartheta}$ is then

$$\hat{\vartheta} + \frac{1}{n}\sum_{i=1}^n \hat{t}_*(X_{i-1}, X_i).$$

It does not require sample splitting. A related result for parameters of the moving average coefficients in invertible linear processes is in Schick and Wefelmeyer (2002b).

Remark 3.3: The *linear* autoregressive model $X_i = \vartheta X_{i-1} + \varepsilon_i$ is a degenerate case. Here $r_\vartheta(X) = \vartheta X$, $\dot{r}_\vartheta(X) = X$, and $\mu = 0$. Hence $s_0(X,Y) = X\ell(\varepsilon)$. Furthermore, $R = E[\dot{r}_\vartheta^2(X)] = E[X^2]$, which is the stationary variance,

$$R = \tau^2 = \sigma^2\sum_{j=0}^\infty \vartheta^{2j} = \frac{\sigma^2}{1 - \vartheta^2}.$$

Hence $\Lambda = \tau^2 J$, and the canonical gradient (3.2) reduces to $t_*(X,Y) = \tau^{-2}J^{-1}X\ell(\varepsilon)$. The least squares estimator is $\hat{\vartheta} = \sum_{i=1}^n X_{i-1}X_i / \sum_{i=1}^n X_{i-1}^2$. An efficient estimator for ϑ is the one-step improvement

$$\hat{\vartheta} + \hat{\tau}^{-2}\hat{J}^{-1}\frac{1}{n}\sum_{i=1}^n X_i\hat{\ell}(\hat{\varepsilon}_i)$$

with $\hat{\varepsilon} = X_i - \hat{\vartheta}X_{i-1}$ and $\hat{\tau}^2 = (1 - \hat{\vartheta}^2)^{-1}(1/n)\sum_{i=1}^n \hat{\varepsilon}_i^2$, and with $\hat{J}$ and $\hat{\ell}$ as before.

The canonical gradient equals the one in the submodel with f *known*. Hence ϑ can be estimated *adaptively* with respect to f. To prove efficiency in this situation, we need local asymptotic normality only for fixed f. Kreiss (1987a) and Kreiss (1987b) constructs adaptive estimators for parameters in ARMA models with symmetric innovation density and in AR models with mean zero innovation density. Jeganathan (1995) and Drost, Klaassen, and Werker (1997) generalize Kreiss (1987a) to nonlinear and heteroscedastic autoregression. See also Koul and Pflug (1990) and Koul and Schick (1996) for adaptive estimation in explosive linear autoregression and in random coefficient autoregression. General results on adaptive estimation in the i.i.d. case are in Putter and Klaassen (2005).

3.4. Innovation Distribution

In this section we consider estimation of a linear functional $E[h(\varepsilon)]$ of the innovation distribution. Suppose first that ϑ is *known*. Then we know the innovations $\varepsilon_i = X_i - r_\vartheta(X_{i-1})$ and can estimate $E[h(\varepsilon)]$ by the empirical estimator $(1/n)\sum_{i=1}^n h(\varepsilon_i)$. Its influence function is $h(\varepsilon) - E[h(\varepsilon)]$. The canonical gradient for ϑ known is $h_V(\varepsilon) = h(\varepsilon) - E[h(\varepsilon)] - \sigma^{-2}E[\varepsilon h(\varepsilon)]\varepsilon$, so the empirical estimator is not efficient. The reason is that it does not use the information that the innovations have mean zero. There are different ways of using this information.

1. Following Levit (1975) and Haberman (1984), an estimator with influence function $h_V(\varepsilon)$ is obtained by estimating $\sigma^{-2}E[\varepsilon h(\varepsilon)]$ empirically and using the *corrected empirical estimator*

$$\frac{1}{n}\sum_{i=1}^n h(\varepsilon_i) - \frac{\sum_{i=1}^n \varepsilon_i h(\varepsilon_i)}{\sum_{i=1}^n \varepsilon_i^2}\frac{1}{n}\sum_{i=1}^n \varepsilon_i.$$

2. Following Owen (1988) and Owen (2001), choose random weights w_i such that the weighted empirical distribution has mean zero, $\sum_{i=1}^n w_i\varepsilon_i = 0$, and use the *weighted empirical estimator*

$$\frac{1}{n}\sum_{i=1}^n w_i h(\varepsilon_i).$$

By the method of Lagrange multipliers, the weights are seen to be of the form $w_i = 1/(1 + \lambda\varepsilon_i)$. This implies $\lambda = \sigma^{-2}(1/n)\sum_{i=1}^n \varepsilon_i + o_p(n^{-1/2})$ and therefore

$$\frac{1}{n}\sum_{i=1}^n w_i h(\varepsilon_i) = \frac{1}{n}\sum_{i=1}^n h(\varepsilon_i) - \sigma^{-2}E[\varepsilon h(\varepsilon)]\frac{1}{n}\sum_{i=1}^n \varepsilon_i + o_p(n^{-1/2}).$$

Hence this estimator also has influence function $h_V(\varepsilon)$.

Now return to the autoregressive model of interest, with ϑ *unknown*. The *parametric plug-in principle* says that an efficient estimator for $E[h(\varepsilon)]$ is obtained by replacing ϑ by an efficient estimator $\hat{\vartheta}$. Then the true innovations are replaced by the residuals $\hat{\varepsilon}_i = X_i - r_{\hat{\vartheta}}(X_{i-1})$. Correspondingly, choose weights $\hat{w}_i$ such that $\sum_{i=1}^n \hat{w}_i \hat{\varepsilon}_i = 0$. Efficient estimators for $E[h(\varepsilon)]$ are then obtained as the *residual-based corrected empirical estimator*

$$\frac{1}{n} \sum_{i=1}^n h(\hat{\varepsilon}_i) - \frac{\sum_{i=1}^n \hat{\varepsilon}_i h(\hat{\varepsilon}_i)}{\sum_{i=1}^n \hat{\varepsilon}_i^2} \frac{1}{n} \sum_{i=1}^n \hat{\varepsilon}_i \tag{3.5}$$

and the *residual-based weighted empirical estimator*

$$\frac{1}{n} \sum_{i=1}^n \hat{w}_i h(\hat{\varepsilon}_i). \tag{3.6}$$

Depending on h, these improvements can lead to drastic variance reductions. As with the true innovations, we have the expansion

$$\frac{1}{n} \sum_{i=1}^n \hat{w}_i h(\hat{\varepsilon}_i) = \frac{1}{n} \sum_{i=1}^n h(\hat{\varepsilon}_i) - \sigma^{-2} E[\varepsilon h(\varepsilon)] \frac{1}{n} \sum_{i=1}^n \hat{\varepsilon}_i + o_p(n^{-1/2}). \tag{3.7}$$

The same expansion holds for the estimator (3.5). For any $n^{1/2}$-consistent estimator $\hat{\vartheta}$,

$$\frac{1}{n} \sum_{i=1}^n h(\hat{\varepsilon}_i) = \frac{1}{n} \sum_{i=1}^n h(\varepsilon_i) - \mu E[h'(\varepsilon)](\hat{\vartheta} - \vartheta) + o_p(n^{-1/2}).$$

By (3.2), an efficient estimator $\hat{\vartheta}$ has influence function $\Lambda^{-1} s_0(X, Y)$. With $E[h'(\varepsilon)] = E[h(\varepsilon)\ell(\varepsilon)]$, the estimators (3.5) and (3.6) are seen to have influence functions equal to the canonical gradient (3.3). Hence they are efficient.

Efficient estimators of the type (3.5) were obtained by Wefelmeyer (1994) for linear autoregression, by Schick and Wefelmeyer (2002a) for nonlinear and heteroscedastic autoregression $X_i = r_{\vartheta}(X_{i-1}) + s_{\vartheta}(X_{i-1})\varepsilon_i$, and by Schick and Wefelmeyer (2002b) for invertible linear processes with moving average coefficients depending on a finite-dimensional parameter ϑ.

Related results are possible for autoregression $X_i = r(X_{i-1}) + \varepsilon_i$ with a semiparametric or nonparametric model for the autoregression function r. Then $\hat{\varepsilon}_i = X_i - \hat{r}(X_{i-1})$ with $\hat{r}$ a nonparametric estimator of r. Here the (unweighted) residual-based empirical estimator $(1/n) \sum_{i=1}^n h(\hat{\varepsilon}_i)$ is already efficient. The reason is that $\hat{r}$ uses the information in $E[\varepsilon] = 0$. For the corresponding (heteroscesdastic) nonparametric *regression* model, Akritas and van Keilegom (2001) obtain a functional central limit theorem for the residual-based empirical distribution function. Different estimators for the regression function are used in Müller, Schick and Wefelmeyer (2004a), Schick and Wefelmeyer (2004c). See also Cheng (2002), Cheng (2004) and Cheng (2005).

Remark 3.4: The *linear* autoregressive model $X_i = \vartheta X_{i-1} + \varepsilon_i$ is a degenerate case. Then $r_\vartheta(X) = \vartheta X$, $\dot{r}_\vartheta(X) = X$, and $\mu = 0$. By Section 2, the canonical gradient of $E[h(\varepsilon)]$ is $h_V(\varepsilon)$ and equals the canonical gradient in the submodel with ϑ *known*. Hence $E[h(\varepsilon)]$ can be estimated *adaptively* with respect to ϑ. It follows in particular that the estimators (3.5) and (3.6) are efficient even if an inefficient estimator for ϑ is used. Also, to prove efficiency, we need local asymptotic normality only for fixed ϑ.

3.5. Innovation Density

In this section we describe weighted residual-based kernel estimators $\hat{f}_w$ for the innovation density f. They will be efficient in the (weak) sense that they lead to efficient plug-in estimators $\int h(x)\hat{f}_w(x)\,dx$ for linear functionals $E[h(\varepsilon)]$. This will be used in Sections 6–9 to construct efficient estimators for conditional expectations, the stationary distribution and density, and higher-order transition densities.

As in Section 4, let $\hat{\vartheta}$ be $n^{1/2}$-consistent, introduce residuals $\hat{\varepsilon}_i = X_i - r_{\hat{\vartheta}}(X_{i-1})$ and choose weights $\hat{w}_i$ such that $\sum_{i=1}^n \hat{w}_i \hat{\varepsilon}_i = 0$. The innovation density f can be estimated by unweighted and weighted residual-based kernel estimators

$$\hat{f}(y) = \frac{1}{n}\sum_{i=1}^n k_b(y - \hat{\varepsilon}_i) \quad \text{and} \quad \hat{f}_w(y) = \frac{1}{n}\sum_{i=1}^n \hat{w}_i k_b(y - \hat{\varepsilon}_i),$$

where $k_b(x) = k(x/b)/b$ with k a kernel and b a bandwidth. For an appropriate choice of bandwidth, a Taylor expansion gives

$$\hat{f}_w(y) = \hat{f}(y) - \sigma^{-2} y f(y) \frac{1}{n}\sum_{i=1}^n \hat{\varepsilon}_i + o_p(n^{-1/2}).$$

This is analogous to expansion (3.7) for the weighted residual-based empirical estimator $(1/n)\sum_{i=1}^n \hat{w}_i h(\hat{\varepsilon}_i)$. We see that $\hat{f}_w$ differs from $\hat{f}$ by a term of order $n^{-1/2}$. Since $\hat{f}_w$ and $\hat{f}$ converge to f at a slower rate, weighting has a negligible effect if we are interested in estimating f itself.

Now we compare the residual-based kernel estimator $\hat{f}$ with the kernel estimator f^* based on the true innovations, $f^*(y) = n^{-1}\sum_{i=1}^n k_b(y - \varepsilon_i)$.

We obtain

$$\hat{f}(y) = f^*(y) + f'(y)\mu(\hat{\vartheta} - \vartheta) + o_p(n^{-1/2}).$$

For the weighted residual-based kernel estimator we therefore have

$$\hat{f}_w(y) = f^*(y) + \sigma^{-2}yf(y)\frac{1}{n}\sum_{i=1}^{n}\varepsilon_i - (\sigma^{-2}yf(y) + f'(y))\mu(\hat{\vartheta} - \vartheta) + o_p(n^{-1/2}).$$

Müller, Schick, and Wefelmeyer (2005a) give conditions under which these stochastic expansions hold for some norms such as the supremum norm and the V-norm $\|f\|_V = \int |f(y)|V(y)\,dy$, where typically $V(y) = (1 + |y|)^m$ for some non-negative integer m. For $m = 0$ this is the L_1-norm with respect to Lebesgue measure.

The stronger norms, with $m > 0$, are useful when we want to estimate e.g. a moment $E[\varepsilon^m]$ with a plug-in estimator $\int y^m \hat{f}_w(y)\,dy$. This is an example of the *nonparametric plug-in principle*: Even though $\hat{f}_w$ converges to f at a rate slower than $n^{-1/2}$, the smooth functional $\int y^m \hat{f}_w(y)\,dy$ of $\hat{f}_w$ converges to $\int y^m f(y)\,dy = E[\varepsilon^m]$ at the parametric rate $n^{-1/2}$.

The estimator $\int y^m \hat{f}_w(y)\,dy$ is even efficient if an efficient estimator $\hat{\vartheta}$ is used. More generally, for all sufficiently regular h bounded by (a multiple of) V, the plug-in estimators $\int h(y)\hat{f}_w(y)\,dy$ are efficient for $\int h(y)f(y)\,dy = E[h(\varepsilon)]$. We may therefore call $\hat{f}_w$ *efficient for plug-in*. This (weak) efficiency concept for function estimators was introduced by Klaassen, Lee, and Ruymgaart (2001).

Weighting can lead to considerable variance reduction. For example, in the *linear* autoregression model $X_i = \vartheta X_{i-1} + \varepsilon_i$ we have $\mu = 0$ and

$$\int y^m \hat{f}(y)\,dy = \frac{1}{n}\sum_{i=1}^{n}\varepsilon_i^m + o_p(n^{-1/2}),$$

$$\int y^m \hat{f}_w(y)\,dy = \frac{1}{n}\sum_{i=1}^{n}\varepsilon_i^m - \sigma^{-2}E[\varepsilon^{m+1}]\frac{1}{n}\sum_{i=1}^{n}\varepsilon_i + o_p(n^{-1/2}).$$

The asymptotic variances are $E[\varepsilon^{2m}]$ and $E[\varepsilon^{2m}] - \sigma^{-2}(E[\varepsilon^{m+1}])^2$, respectively. For $m = 3$ and f normal these variances are $15\sigma^6$ and $6\sigma^6$, respectively, a variance reduction of nearly two thirds.

3.6. Conditional Expectation

The conditional expectation $E(h(X_{n+1}) \mid X_n = x)$ with lag one of a known function h can be stimated by a nonparametric estimator

$$\frac{\sum_{i=1}^{n} k_b(x - X_{i-1})h(X_i)}{\sum_{i=1}^{n} k_b(x - X_{i-1})},$$

where $k_b(x) = k(x/b)/b$ with k a kernel and b a bandwidth. If the time series is known to be first-order Markov with transition density $q(x, y)$ from x to y, then we have

$$E(h(X_{n+1}) \mid X_n = x) = \int h(y) q(x, y) \, dy,$$

and an estimator is obtained by plugging in a (kernel) estimator for q. In the nonlinear autoregressive model $X_i = r_\vartheta(X_{i-1}) + \varepsilon_i$, the transition density is $q(x, y) = f(y - r_\vartheta(x))$, and we can write

$$E(h(X_{n+1}) \mid X_n = x) = \int h(y + r_\vartheta(x)) f(y) \, dy = E[h(\varepsilon + r_\vartheta(x))].$$

This is an (unconditional) expectation under the innovation distribution as in Section 4, but now of a function $h(y, \vartheta) = h(y + r_\vartheta(x))$ depending on ϑ. This suggests estimating the conditional expectation by

$$\frac{1}{n} \sum_{i=1}^{n} h(\hat{\varepsilon}_i + r_{\hat{\vartheta}}(x)) = \int h(\cdot, \hat{\vartheta}) \, d\hat{\mathbb{F}}$$

with $\hat{\mathbb{F}}(y) = (1/n) \sum_{i=1}^{n} \mathbf{1}(\hat{\varepsilon}_i \leq y)$ the empirical distribution function of the residuals $\hat{\varepsilon}_i = X_i - r_{\hat{\vartheta}}(X_{i-1})$. If $\hat{\vartheta}$ is $n^{1/2}$-consistent, then $\int h(\cdot, \hat{\vartheta}) \, d\hat{\mathbb{F}}$ will be $n^{1/2}$-consistent. For efficiency we need to use an efficient estimator $\hat{\vartheta}$ and to replace $\hat{\mathbb{F}}$ by a version that uses the information $E[\varepsilon] = 0$. As seen in Section 4, one way of doing this is by taking a weighted version $\hat{\mathbb{F}}_w(y) = (1/n) \sum_{i=1}^{n} w_i \mathbf{1}(\hat{\varepsilon}_i \leq y)$. The resulting estimator for the conditional expectation is $\int h(\cdot, \hat{\vartheta}) \, d\hat{\mathbb{F}}_w$. Similar as in Section 5, a Taylor expansion gives

$$\int h(\cdot, \hat{\vartheta}) \, d\hat{\mathbb{F}}_w = \frac{1}{n} \sum_{i=1}^{n} h_V(\varepsilon_i, \vartheta) + (\dot{r}_\vartheta(x) - \mu) E[h'_V(\varepsilon, \vartheta)](\hat{\vartheta} - \vartheta) + o_p(n^{-1/2}).$$

We have $E[h'_V(\varepsilon, \vartheta)] = E[h_V(\varepsilon, \vartheta)\ell(\varepsilon)]$. By (3.2), an efficient estimator has influence function $\Lambda^{-1} s_0(X, Y)$; so $\int h(\cdot, \hat{\vartheta}) \, d\hat{\mathbb{F}}_w$ has influence function equal to the canonical gradient (3.4) and is therefore efficient.

These results extend to higher lags. For example, for lag two the conditional expecation $E(h(X_{n+2}) \mid X_n = x)$ becomes $E[h(\varepsilon_2 + r_\vartheta(\varepsilon_1 + r_\vartheta(x)))]$ and is estimated $n^{1/2}$-consistently by the residual-based von Mises statistic

$$\frac{1}{n^2} \sum_{i=1}^{n} \sum_{j=1}^{n} h(\hat{\varepsilon}_j + r_{\hat{\vartheta}}(\hat{\varepsilon}_i + r_{\hat{\vartheta}}(x))) = \iint h(z + r_{\hat{\vartheta}}(y + r_{\hat{\vartheta}}(x))) \, d\hat{\mathbb{F}}(y) d\hat{\mathbb{F}}(z).$$

An efficient estimator is obtained if we replace $\hat{\mathbb{F}}$ by the weighted version $\hat{\mathbb{F}}_w$ and use an efficient estimator for ϑ.

To treat lags higher than two, set $\varrho_{1\vartheta}(x, y) = y + r_\vartheta(x)$ and define recursively $\varrho_{m\vartheta}(x, y_1, \ldots, y_m) = y_m + r_\vartheta(\varrho_{m-1,\vartheta}(x, y_1, \ldots, y_{m-1}))$. Then an m-step conditional expectation can be written

$$\nu(h) = E(h(X_{n+m}) \mid X_n = x) = E[h(\varrho_{m\vartheta}(x, \varepsilon_1, \ldots, \varepsilon_m))].$$

With $\hat{\vartheta}$ efficient, an efficient estimator for $E(h(X_{n+m}) \mid X_n = x)$ is the weighted residual-based von Mises statistic

$$\int \cdots \int h(\varrho_{m\hat{\vartheta}}(x, y_1, \ldots, y_m)) \, d\hat{\mathbb{F}}_w(y_1) \ldots d\hat{\mathbb{F}}_w(y_m).$$

To prove $n^{1/2}$-consistency of such von Mises statistics, we need an appropriate balance of smoothness assumptions on h and on f. For discontinuous h we must assume that f is smooth. Then we can replace $\hat{\mathbb{F}}$ or $\hat{\mathbb{F}}_w$ by smoothed versions $d\hat{\mathbb{F}}(y) = \hat{f}(y) \, dy$ and $d\hat{\mathbb{F}}_w(y) = \hat{f}_w(y) \, dy$ with residual-based unweighted or weighted kernel estimators $\hat{f}$ and $\hat{f}_w$ as in Section 5. Write

$$\hat{\nu}(h) = \int \cdots \int h(\varrho_{m\hat{\vartheta}}(x, y_1, \ldots, y_m)) \hat{f}_w(y_1) \ldots \hat{f}_w(y_m) \, dy_1 \cdots dy_m.$$

Müller, Schick, and Wefelmeyer (2005b) prove functional central limit theorems for processes $\{n^{1/2}(\nu(h) - \hat{\nu}(h)) : h \in \mathcal{H}\}$ and appropriate function classes $\mathcal{H}$.

Simulations show that smoothing also improves the small-sample behavior of the von Mises statistics, especially for discontinuous h.

Remark 3.5: The parametric rates for estimators of conditional expectations do not extend to autoregression $X_i = r(X_{i-1}) + \varepsilon_i$ with semiparametric or nonparametric autoregression function r. A conditional expectation of lag one has representation $E(h(X_{n+1}) \mid X_n = x) = E[h(\varepsilon + r_\vartheta(x))]$. This is a (smooth) function of $r(x)$, and the convergence rate of an estimator for the conditional expectation is in general determined by the rate at which we can estimate $r(x)$. Nevertheless, estimators based on this representation may still be better than nonparametric estimators. For results in nonparametric (censored) *regression* we refer to van Keilegom, Akritas and Veraverbeke (2001), van Keilegom and Veraverbeke (2001) and van Keilegom and Veraverbeke (2002).

3.7. Stationary Distribution

A simple estimator for the expectation $E[h(X)]$ of a known function h under the stationary distribution is the empirical estimator $(1/n) \sum_{i=1}^n h(X_i)$.

The estimator does not make use of the autoregressive structure of our model. A better estimator can be obtained as follows. As m tends to infinity, the m-step conditional expectation $E(h(X_m) \mid X_0 = x)$ converges to $E[h(X)]$ at an exponential rate. By Section 6, for fixed m, an efficient estimator of $E(h(X_m) \mid X_0 = x)$ is the weighted residual-based von Mises statistic

$$\hat{\kappa}_m(x) = \int \cdots \int h(\varrho_{m\hat{\vartheta}}(x, y_1, \ldots, y_m)) \, d\hat{\mathbb{F}}_w(y_1) \ldots d\hat{\mathbb{F}}_w(y_m)$$

or its smoothed version. We expect that $\hat{\kappa}_{m(n)}(x)$ is efficient for $E[h(X)]$ if $m(n)$ increases with n at an appropriate (logarithmic) rate.

The bias induced by the choice of starting point x can be removed by averaging, i.e. by using instead of $\hat{\kappa}_{m(n)}(x)$ the estimator

$$\frac{1}{n} \sum_{i=1}^{n} \hat{\kappa}_{m(n)}(X_i).$$

For invertible linear processes with moving average coefficients depending on a finite-dimensional parameter, a corresponding result is proved in Schick and Wefelmeyer (2004b).

3.8. Stationary Density

The usual estimator for the stationary density g of a time series is the kernel estimator $(1/n) \sum_{i=1}^{n} k_b(x - X_i)$, where $k_b(x) = k(x/b)/b$ with kernel k and bandwidth b. For our nonlinear autoregressive model, the stationary density can be written

$$g(y) = \int f(y - r_\vartheta(x)) g(x) \, dx = E[f(y - r_\vartheta(X))].$$

This is an expectation under the stationary distribution as in Section 7, but now for a function $h(x, \vartheta, f) = f(y - r_\vartheta(x))$ depending on ϑ and f. By the plug-in principle mentioned in Section 5, we expect to obtain a $n^{1/2}$-consistent estimator if we plug appropriate kernel estimators $\hat{f}$ for f and $\hat{g}$ for g and a $n^{1/2}$-consistent estimator $\hat{\vartheta}$ for ϑ into this representation,

$$g^*(y) = \int \hat{f}(y - r_{\hat{\vartheta}}(x)) \hat{g}(x) \, dx.$$

Note however that g is the density of the convolution of ε and $r_\vartheta(X)$. Even if the density g of X is nice, the distribution of $r_\vartheta(X)$ may be unpleasant. A degenerate case would be a constant autoregression function, say $r_\vartheta =$

0. Then we observe independent $X_i = \varepsilon_i$ with density $g = f$, and $n^{1/2}$-consistent estimation of g is not possible. There may be a problem even if r_ϑ is smooth and strictly increasing but with derivative vanishing at some point. However, if r_ϑ has a derivative that is bounded away from zero, $g^*(y)$ will be $n^{1/2}$-consistent. For efficiency we need an efficient estimator $\hat{\vartheta}$ as in Section 3 and a weighted residual-based kernel estimator $\hat{f}_w$ for the innovation density as in Section 5, and we must replace the estimator $\hat{g}(x)\,dx$ by an efficient estimator as in Section 7.

In the i.i.d. case, $n^{1/2}$-consistent estimators for convolution densities are studied by Frees (1994), Saavedra and Cao (2000) and Schick and Wefelmeyer (2004c). A $n^{1/2}$-consistent estimator for the stationary density of a first-order moving average process is obtained in Saavedra and Cao (1999). Schick and Wefelmeyer (2004a) introduce an efficient version, and Schick and Wefelmeyer (2004d) prove functional central limit theorems for higher-order moving average processes and density estimators viewed as elements of function spaces. For general invertible linear processes, $n^{1/2}$-consistent estimators of the stationary density are constructed in Schick and Wefelmeyer (2005).

3.9. Transition Density

The one-step transition density $q(x, y)$ from $X_0 = x$ to $X_1 = y$ of a first-order Markov chain can be estimated by the Nadaraya–Watson estimator

$$\hat{q}(x, y) = \frac{\sum_{i=1}^n k_b(x - X_{i-1})k_b(y - X_i)}{\sum_{i=1}^n k_b(x - X_{i-1})},$$

where $k_b(x) = k(x/b)/b$ with kernel k and bandwidth b. The two-step transition density $q_2(x, z)$ from $X_0 = x$ to $X_2 = z$ has the representation $q_2(x, z) = \int q(x, y)q(y, z)\,dy$ and can be estimated by $\int \hat{q}(x, y)\hat{q}(y, z)\,dy$. For our nonlinear autoregressive model, the two-step transition density can be written

$$q_2(x, z) = \int f(z - r_\vartheta(y))f(y - r_\vartheta(x))\,dy = E[f(z - r_\vartheta(\varepsilon + r_\vartheta(x)))].$$

This is an expectation under the innovation distribution as in Section 4, but now for a function $h(y, \vartheta, f) = f(z - r_\vartheta(y + r_\vartheta(x)))$ depending on ϑ and f. There is some formal similarity with the representation of the stationary density in Section 6, but also an essential difference: There we had an expectation under the *stationary* distribution; here the expectation is taken with respect to the *innovation* distribution. This makes efficient estimation of the transition density easier, because it is easier to estimate the innovation distribution efficiently. As efficient estimator of $q_2(x, z)$ we suggest

$$\hat{q}_2(x, z) = \int \hat{f}_w(z - r_{\hat{\vartheta}}(y)) \hat{f}_w(y - r_{\hat{\vartheta}}(x)) \, dy$$

with $\hat{\vartheta}$ efficient. The result extends to higher-order lags.

Acknowledgments

Anton Schick was supported by NSF grant DMS 0405791.

References

Akritas, M. G. and van Keilegom, I. (2001). Non-parametric estimation of the residual distribution. *Scand. J. Statist.* **28**, 549–567.

An, H. Z. and Huang, F. C. (1996). The geometrical ergodicity of nonlinear autoregressive models. *Statist. Sinica* **6**, 943–956.

Bhattacharya, R. N. and Lee, C. (1995a). Ergodicity of nonlinear first order autoregressive models. *J. Theoret. Probab.* **8**, 207–219.

Bhattacharya, R. N. and Lee, C. (1995b). On geometric ergodicity of nonlinear autoregressive models. *Statist. Probab. Lett.* **22**, 311–315. Erratum: **41** (1999), 439–440.

Bickel, P. J. (1993). Estimation in semiparametric models. In: *Multivariate Analysis: Future Directions* (C. R. Rao, ed.), pp. 55–73, North-Holland, Amsterdam.

Bickel, P. J., Klaassen, C. A. J., Ritov, Y. and Wellner, J. A. (1998). *Efficient and Adaptive Estimation for Semiparametric Models*. Springer, New York.

Bickel, P. J. and Kwon, J. (2001). Inference for semiparametric models: Some questions and an answer (with discussion). *Statist. Sinica* **11**, 863–960.

Cheng, F. (2002). Consistency of error density and distribution function estimators in nonparametric regression. *Statist. Probab. Lett.* **59**, 257–270.

Cheng, F. (2004). Weak and strong uniform consistency of a kernel error density estimator in nonparametric regression. *J. Statist. Plann. Inference* **119**, 95–107.

Cheng, F. (2005). Asymptotic distributions of error density and distribution function estimators in nonparametric regression. *J. Statist. Plann. Inference* **128**, 327–349.

Drost, F. C., Klaassen, C. A. J. and Werker, B. J. M. (1997). Adaptive estimation in time-series models, *Ann. Statist.* **25**, 786–817.

Fan, J. and Yao, Q. (2003). *Nonlinear Time Series. Nonparametric and Parametric Methods*. Springer Series in Statistics, Springer, New York.

Frees, E. W. (1994). Estimating densities of functions of observations. *J. Amer. Statist. Assoc.* **89**, 517–525.

Greenwood, P. E., Schick, A. and Wefelmeyer, W. (2001). Comment [on Bickel and Kwon (2001)]. *Statist. Sinica* **11**, 892–906.

Haberman, S. J. (1984). Adjustment by minimum discriminant information. *Ann. Statist.* **12**, 971–988. Erratum: **14** (1986), 358.

Hájek, J. (1970). A characterization of limiting distributions of regular estimates. *Z. Wahrsch. Verw. Gebiete* **14**, 323–330.

Jeganathan, P. (1995). Some aspects of asymptotic theory with applications to time series models. *Econometric Theory* **11**, 818–887.

Klaassen, C. A. J., Lee, E.-J. and Ruymgaart, F. H. (2001). On efficiency of indirect estimation of nonparametric regression functions. In: *Algebraic Methods in Statistics and Probability* (M. A. G. Viana and D. S. P. Richards, eds.), 173–184, Contemporary Mathematics 287, American Mathematical Society, Providence, RI.

Koul, H. L. and Pflug, G. C. (1990). Weakly adaptive estimators in explosive autoregression. *Ann. Statist.* **18**, 939–960.

Koul, H. L. and Schick, A. (1996). Adaptive estimation in a random coefficient autoregressive model. *Ann. Statist.* **24**, 1025–1052.

Koul, H. L. and Schick, A. (1997). Efficient estimation in nonlinear autoregressive time-series models. *Bernoulli* **3**, 247–277.

Kreiss, J-P. (1987a). On adaptive estimation in stationary ARMA processes. *Ann. Statist.* **15**, 112–133.

Kreiss, J-P. (1987b). On adaptive estimation in autoregressive models when there are nuisance functions. *Statist. Decisions* **5**, 59–76.

Le Cam, L. (1971). Limits of experiments. *Proc. Sixth Berkeley Symp. Math. Statist. Probab.* **1**, 245–261.

Levit, B. Y. (1975). Conditional estimation of linear functionals. *Problems Inform. Transmission* **11**, 39–54.

Mokkadem, A. (1987). Sur un modèle autorégressif non linéaire: ergodicité et ergodicité géométrique. *J. Time Ser. Anal.* **8**, 195–204.

Müller, U. U., Schick, A. and Wefelmeyer, W. (2004a). Estimating linear functionals of the error distribution in nonparametric regression. *J. Statist. Plann. Inference* **119**, 75–93.

Müller, U. U., Schick, A. and Wefelmeyer, W. (2004b). Estimating functionals of the error distribution in parametric and nonparametric regression. *J. Nonparametr. Statist.* **16**, 525–548.

Müller, U. U., Schick, A. and Wefelmeyer, W. (2005a). Weighted residual-based density estimators for nonlinear autoregressive models. *Statist. Sinica* **15**, 177–195.

Müller, U. U., Schick, A. and Wefelmeyer, W. (2005b). Efficient prediction for linear and nonlinear autoregressive models. Technical Report, Department of Mathematical Sciences, Binghamton University.

Müller, U. U., Schick, A. and Wefelmeyer, W. (2005c). Estimating the error distribution function in nonparametric regression. Technical Report, Department of Mathematical Sciences, Binghamton University.

Owen, A. B. (1988). Empirical likelihood ratio confidence intervals for a single functional. *Biometrika* **75**, 237–249.

Owen, A. B. (2001). *Empirical Likelihood.* Monographs on Statistics and Applied Probability **92**, Chapman & Hall / CRC, London.

Putter, H. and Klaassen, C. A. J. (2005). Efficient estimation of Banach parameters in semiparametric models. *Ann. Statist.* **33**.

Saavedra, A. and Cao, R. (1999). Rate of convergence of a convolution-type estimator of the marginal density of an MA(1) process. *Stochastic Process. Appl.* **80**, 129–155.

Saavedra, A. and Cao, R. (2000). On the estimation of the marginal density of a moving average process. *Canad. J. Statist.* **28**, 799–815.

Schick, A. and Wefelmeyer, W. (2002a). Estimating the innovation distribution in nonlinear autoregressive models. *Ann. Inst. Statist. Math.* **54**, 245–260.

Schick, A. and Wefelmeyer, W. (2002b). Efficient estimation in invertible linear processes. *Math. Methods Statist.* **11**, 358–379.

Schick, A. and Wefelmeyer, W. (2004a). Root n consistent and optimal density estimators for moving average processes. *Scand. J. Statist.* **31**, 63–78.

Schick, A. and Wefelmeyer, W. (2004b). Estimating invariant laws of linear processes by U-statistics. *Ann. Statist.* **32**, 603-632.

Schick, A. and Wefelmeyer, W. (2004c). Root n consistent density estimators for sums of independent random variables. *J. Nonparametr. Statist.* **16**, 925–935.

Schick, A. and Wefelmeyer, W. (2004d). Functional convergence and optimality of plug-in estimators for stationary densities of moving average processes. *Bernoulli* **10**, 889–917.

Schick, A. and Wefelmeyer, W. (2005). Root n consistent density estimators for invertible linear processes. Technical Report, Department of Mathematical Sciences, Binghamton University.

Taniguchi, M. and Kakizawa, Y. (2000). *Asymptotic Theory of Statistical Inference for Time Series.* Springer Series in Statistics, Springer, New York.

van Keilegom, I., Akritas, M. G. and Veraverbeke, N. (2001). Estimation of the conditional distribution in regression with censored data: a comparative study. *Comput. Statist. Data Anal.* **35**, 487–500.

van Keilegom, I. and Veraverbeke, N. (2001). Hazard rate estimation in nonparametric regression with censored data. *Ann. Inst. Statist. Math.* **53**, 730–745.

van Keilegom, I. and Veraverbeke, N. (2002). Density and hazard estimation in censored regression models. *Bernoulli* **8**, 607–625.

Wefelmeyer, W. (1994). An efficient estimator for the expectation of a bounded function under the residual distribution of an autoregressive process. *Ann. Inst. Statist. Math.* **46**, 309–315.

CHAPTER 4

On the Efficiency of Estimation for a Single-index Model

Yingcun Xia and Howell Tong

Department of Statistics and Applied Probability
National University of Singapore, Singapore
staxyc@stat.nus.edu.sg

Department of Statistics
London School of Economics, U.K.
h.tong@lse.ac.uk

The single-index model is one of the more popular semi-parametric models in econometrics and statistics. Many methods, some of which have been specifically developed for this model, are available for the estimation of the index parameter. The more popular ones are the ADE method and its variations (Härdle and Stoker (1989), Hristache, Juditski, Polzehl and Spokoiny (2001)), the single-indexed kernel method (Härdle, Hall and Ichimura (1993), Xia, Tong, Li and Zhu (2002)) and sliced inverse regression methods (Li (1991)). In this chapter, we review these methods, propose alternative approaches for the estimation, obtain the asymptotic distributions of these estimators and compare their efficiencies based on asymptotic distributions. This chapter includes new results as well as survey material.

4.1. Introduction

Single index models (SIMs) are widely used in applied quantitative sciences. Although the context of application for SIMs rarely prescribes the distributional form of the involved statistical error, one approach is to specify the latter (by assumption and often via an explicit "link" function) up to a low-dimensional parameter. The focus is then on the estimation of the low-dimensional parameter, typically via its likelihood function. Both from a theoretical and a practical point of view, this approach has been criticized. A more recent approach is based on the semiparametric modelling, which allows the high (in fact infinite) dimensional parameter (namely the link function) to be flexible and unknown (except for some smoothness assumptions) while retaining the focus on the finite dimensional index parameter. Specifically, consider the following single-index model,

$$Y = g(\theta_0^\top X) + \varepsilon, \qquad (4.1)$$

where $E(\varepsilon|X) = 0$ almost surely, g is an unknown link function, and θ_0 is a single-index with $\|\theta_0\| = 1$ for identification. In this model there is a single linear combination of covariates X that is designed to capture maximal information about the relation between the response variable Y and the covariates X, thereby ameliorating the "curse of dimensionality". Estimation of the single-index model, focusing on the index parameter, is very attractive both in theory and in practice. In the last decade or so, a series of papers (e.g. Powell, Stock and Stoker (1989), Härdle and Stoker (1989), Ichimura (1993), Klein and Spady (1993), Härdle, Hall and Ichimura (1993), Horowitz and Härdle (1996), Hristache, Juditski and Spokoiny (2001), Xia, Tong, Li and Zhu (2002)) have considered the estimation of the parametric index and the nonparametric part (i.e. the function g), focusing on the root-n consistency of the former; efficiency issues have also been studied. Amongst the various methods of estimation, the more popular ones are the average derivative estimation (ADE) method investigated by Härdle and Stoker (1989), the sliced inverse regression (SIR) method proposed by Li (1991), and the simultaneous minimization method of Härdle, Hall and Ichimura (1993), referenced as HHI hereafter.

The ADE method is the first approach that uses the structure of the model directly. Because the index θ_0 is proportional to the derivatives $\partial E(Y|X = x)/\partial x = g'(\theta_0^\top x)\theta_0$, the index can be estimated by the average of the derivatives or gradients. The original version of ADE uses high dimensional kernel regression to estimate the derivatives at the observation points. The method is simple to implement with an easily understood algorithm. However, because a high dimensional kernel estimation method is used, the estimation still suffers from the "curse of dimensionality". Hristache, Juditski and Spokoiny (2001) (referenced as HJS hereafter), adopted the same idea as ADE and proposed a dynamic procedure to adapt to the structure of the model by lowering the dimension of the kernel smoothing. By simulations, they showed that their method has much better performance than the original ADE method. The idea can be extended easily to the multi-index model:

$$Y = g(\theta_1^\top X, \theta_2^\top X, \cdots, \theta_q^\top X) + \varepsilon. \qquad (4.2)$$

See Hristache, Juditski, Polzehl and Spokoiny (2001). They proved that root-n consistency for the index estimators can be achieved when $q \leq 3$. Because HJS did not give the asymptotic distribution of their estimator, it is hard to compare the efficiency of the method with other methods.

The sliced inverse regression (SIR) method proposed by Li (1991) uses a striking idea. The approach is originally proposed for a more general dimension reduction model,

$$Y = g(\theta_1^\top X, \theta_2^\top X, \cdots, \theta_q^\top X, \varepsilon). \qquad (4.3)$$

Here, ε is independent of X. Thus, the method does not require that the noise is additive as in Model (4.1) or Model (4.2). As a special case, Model (4.1) can be estimated using the SIR method. Especially, when the link function g is monotonic, the SIR method can produce a good estimator for the index θ_0. However, the approach needs a strong assumption on the design of X in order to obtain the asymptotics of the estimators theoretically. The asymptotic distribution was obtained by Zhu and Fang (1996). Because the idea is simple and very easy to implement, it is very popular for the analysis of huge data sets. However, because the limiting distribution depends on functions that do not appear in the model, it is again hard for us to compare the SIR with other estimation methods theoretically. Recently, methods based on the *joint* distribution of (X, y) have been developed. See for examples Delecroix, Härdle and Hristache (2003) and Yin and Cook (2005).

In simple terms, the model may be estimated as follows. Locally estimate the regression function, which depends on the index θ, and obtain the estimator of the index by minimizing the sum of squares of the residuals with respect to the index. HHI adopted the idea by taking the pilot variable h as a parameter and minimizing the residuals respect to both h and the index. Strikingly, they found that the commonly used under-smoothing approach is unnecessary and that the root-n consistency can be achieved for the index estimator. However, the procedure of HHI is very difficult to implement. Moreover, it is not easy to extend the idea to the multi-index Model (4.2) or the dimension reduction Model (4.3). Xia, Tong, Li and Zhu (2002) proposed the minimum average conditional variance method (MAVE) to estimate the link function and the index simultaneously and gave a simple algorithm for the estimation procedure. This idea can also be applied to the multi-index Model (4.2). More recent work by Xia (2005) suggests that the idea can also be applied to the dimension reduction Model (4.3). Simulations suggest that this approach has very good performance for finite samples. In the proof of HHI, the lower information bound cannot be achieved because of the trimming function. In fact, to our knowledge no theoretical comparison of the efficiency of estimation among the various estimation methods has been conducted systematically to-date.

In this chapter, we propose alternative but equivalent versions for the ADE method and the HJS method. The alternative versions are easy to implement and in each case the limiting distribution is obtained. The asymptotic normality of MAVE is proved. By choosing an appropriate trimming function, we further show that the MAVE can achieve the lower information bound under some assumptions on the distribution functions. We give asymptotic distribution for the SIR estimator of the single-index model. Finally, the efficiencies of these estimation methods are compared based on the limiting distributions.

4.2. Estimation via Outer Product of Gradients

Let $G(x) = E(Y|X = x)$. It is easy to see from model (4.1) that

$$\nabla G(x) = \frac{\partial}{\partial x} G(x) = \frac{\partial}{\partial x} g(\theta_0^\top x) = g'(\theta_0^\top x)\theta_0. \qquad (4.4)$$

Therefore, the derivative of the regression function has the same direction as the index θ_0.

4.2.1. *Average derivative estimation method*

Based on the above observation, Powell, Stock and Stoker (1989) and Härdle and Stoker (1989) proposed the Average Derivative Estimation (ADE) method, which estimates the θ_0 via $E \nabla G(x)$. Let $f(x)$ be the marginal density of X. Integration by parts gives

$$\delta \overset{def}{=} E\{\nabla G(X)\} = -E\{\nabla \log f(X)Y\}.$$

Instead of estimating $\partial G(x)/\partial x$, Powell, Stock and Stoker (1989) proposed that we estimate instead $l(x) = \partial \log f(x)/\partial x$, which is done by the higher order kernel smoother. The estimation can be implemented as follows. Let $H(v)$ be a p-variate kernel function, and $H_b(v) = H(v/b)/b^p$. The density function of X is then estimated by

$$\hat{f}(x) = n^{-1} \sum_{i=1}^{n} H_b(X_i - x).$$

Then $E\{\partial G(X)/\partial x\}$ can be estimated by

$$\hat{\delta} = n^{-1} \sum_{j=1}^{n} \rho_n(\hat{f}(X_j)) Y_j \hat{l}(X_j),$$

where $\hat{l}(x) = \partial \log \hat{f}(x)/\partial x$ and ρ_n is a trimming function. Finally, the estimator of θ_0 is

$$\hat{\theta}_{\mathrm{ADE}} = \hat{\delta}/|\hat{\delta}|.$$

Because the nonparametric estimation of a high dimensional density function is used, the ADE procedure suffers from the "curse of dimensionality". Another disadvantage of the methods is that when $E(g'(\theta_0^\top X)) = 0$, that is,

$$E \bigtriangledown G(x) = E(g'(\theta_0^\top X))\theta_0 = 0,$$

the method fails to estimate the direction.

4.2.2. *Single-index estimation via outer product of gradients and a refinement*

As one way to overcome the drawback in ADE when $Eg'(\theta_0^\top X) = 0$, Samarov (1993) proposed the use of the outer product of the gradients, namely $E\{\bigtriangledown G(X) \bigtriangledown^\top G(X)\}$. The index θ_0 is the first eigenvector of $E\{\bigtriangledown G(X) \bigtriangledown^\top G(X)\}$. This idea can be easily extended to the estimation of the multi-index Model (4.2).

To implement the estimation, we first estimate the gradients by local linear smoothing. Specifically, we consider the local linear fitting in the form of the following minimization problem

$$\min_{a_j, b_j} \sum_{i=1}^{n} \left[Y_i - a_j - b_j^\top X_{ij} \right]^2 w_{ij}, \tag{4.5}$$

where $X_{ij} = X_i - X_j$ and w_{ij} is a weight depending on the distance between X_i and X_j. Calculate

$$\hat{\Sigma} = \frac{1}{n} \sum_{j=1}^{n} \rho_j \hat{b}_j \hat{b}_j^\top,$$

where $\hat{b}_j$ is the minimizer from (4.5) and ρ_j is a trimming function employed to handle the boundary points; see the discussion below. Then the first eigenvector, $\hat{\theta}$, of $\hat{\Sigma}$ is an estimator of θ_0.

Note that in the above estimation, a multi-dimensional kernel smooth is used, which is well known to be very inefficient. To improve the efficiency, we can consider using the structure of the dependency of Y on X; following the idea of Xia, Tong, Li and Zhu (2002), we consider a single-index kernel weight

$$w_{ij} = K_h(\hat{\theta}^\top X_{ij}) \tag{4.6}$$

and repeat the above procedure. We call this method the refined method of outer product of gradients (rOPG) estimation. Hristache, Juditski and Spokoiny (2001) also discussed a similar idea. To implement the estimation procedure, we propose the following rOPG algorithm.

Let $\rho_n(\cdot)$ be any bounded function with bounded third order derivatives on $\mathbb{R}$ such that $\rho_n(v) = 1$ if $v > 2n^{-\epsilon}$; $\rho_n(v) = 0$ if $v \leq n^{-\epsilon}$ for some $\epsilon > 0$. Details will be given below. A simple algorithm can be stated as follows. Suppose θ is an initial estimate of θ_0.

Step 1: Calculate

$$\begin{pmatrix} a_j^\theta \\ b_j^\theta \end{pmatrix} = \left\{ \sum_{i=1}^n K_h(\theta^\top X_{ij}) \begin{pmatrix} 1 \\ X_{ij} \end{pmatrix} \begin{pmatrix} 1 \\ X_{ij} \end{pmatrix}^\top \right\}^{-1} \sum_{i=1}^n K_h(\theta^\top X_{ij}) \begin{pmatrix} 1 \\ X_{ij} \end{pmatrix} Y_i,$$

where h is a bandwidth (see the details below).
Step 2: Calculate

$$\hat{\Sigma} = n^{-2} \sum_{j=1}^n \hat{\rho}_j^\theta b_j^\theta (b_j^\theta)^\top,$$

where $\hat{\rho}_j^\theta = \rho_n(\hat{f}_\theta(\theta^\top X_j))$ and $\hat{f}_\theta(\theta^\top x) = n^{-1} \sum_{i=1}^n K_h(\theta^\top (X_i - x))$. Replace θ by the first eigenvector of $\hat{\Sigma}$ and repeat Steps 1 and 2 until convergence.

The limiting eigenvector is the estimator of θ_0. Denote it by $\hat{\theta}_{rOPG}$.

4.3. Global Minimization Estimation Methods

Because the link function is unknown, we can use local linear smoother to estimate it. On the other hand, because θ_0 is a global parameter, it should be estimated via minimizing a global function. Based on this idea, many estimation methods have been proposed. Klein and Spady (1993) proved that the estimator based on this idea can achieve the asymptotic informa-tion bound. HHI showed that the estimator of the single-index can achieve root-n consistency without under-smoothing the link function. Weisberg and Welsh (1994) proposed a Newton-Raphson-type algorithm. Delecroix, Hristache and Patilea (2005) allowed the errors to admit more general distributions by using a general loss function. Sherman (1994) investigated the theoretical techniques and gave some useful results.

4.3.1. *HHI estimation*

The idea can be stated in the N-W kernel smoothing framework as follows. We estimate θ_0 by minimizing

$$n^{-1} \sum_{j=1}^{n} \{y_j - \hat{g}_n^{\backslash j}(\theta^\top X_j)\}^2$$

with respect to θ (and h), where $\hat{g}_n^{\backslash j}(\theta^\top X_j)$ is the estimated function value of g at $\theta^\top X_j$ with observation (X_j, Y_j) deleted, i.e.

$$\hat{g}_n^{\backslash j}(\theta^\top X_j) = \sum_{\substack{i=1 \\ i \neq j}}^{n} K_h(\theta^\top X_{ij}) Y_i / \sum_{\substack{i=1 \\ i \neq j}}^{n} K_h(\theta^\top X_{ij}).$$

Denote the estimator of θ_0 by $\hat{\theta}_{\text{HHI}}$.

Note that the above minimization problem is very difficult to implement. Actually, HHI used the grid search algorithm for the calculation. The algorithm is feasible in practice only when the number of variates is small.

4.3.2. *The minimum average conditional variance estimation*

Xia, Tong, Li and Zhu (2002) simplified the procedure in their general MAVE. As a special case of their procedure, we can estimate Model (4.1) as follows. Note that under Model (4.1), all the gradients $\partial g(\theta_0^\top x)/\partial x$ at different x have the same direction θ_0 as shown in Equation (4.4). To use this observation, we can replace b_j in (4.5) by θd_j, where d_j is a scalar, and have the following local approximation

$$n^{-1} \sum_{i=1}^{n} \{Y_i - a_j - d_j \theta^\top X_{ij}\}^2 K_h(X_{ij}).$$

This is a measure of the local average departure of the estimated model from the observed Y_i when X_i is close to X_j. It is actually an estimator of local variance if a_j, b_j and θ are close to the true values; see Xia, Tong, Li and Zhu (2002). Obviously, the best approximation of θ should minimize the overall departure for all $X_j, j = 1, \cdots, n$. Thus, we propose to estimate θ_0 by

$$\hat{\theta} = \arg \min_{\theta:|\theta|=1} \sum_{j=1}^{n} \sum_{i=1}^{n} \{Y_i - a_j - d_j \theta X_{ij}\}^2 w_{ij}.$$

Similar to rOPG, we can make the weight w_{ij} adaptive to the model as in (4.6). We call the estimation method the refined MAVE (rMAVE).

The corresponding algorithm can be stated as follows. Suppose θ is an initial estimate of θ_0.

Step 1: Calculate

$$\begin{pmatrix} a_j^\theta \\ d_j^\theta h \end{pmatrix} = \{ \sum_{i=1}^{n} K_h(\theta^\top X_{ij}) \begin{pmatrix} 1 \\ X_{ij} \end{pmatrix} \begin{pmatrix} 1 \\ \theta^\top X_{ij} \end{pmatrix}^\top \}^{-1}$$

$$\times \sum_{i=1}^{n} K_h(\theta^\top X_{ij}) \begin{pmatrix} 1 \\ \theta^\top X_{ij} \end{pmatrix} Y_i,$$

where h is a bandwidth (see the details below).

Step 2: Calculate

$$\theta = \{ \sum_{i,j=1}^{n} K_h(\theta^\top X_{ij}) \hat{\rho}_j^\theta (d_j^\theta)^2 X_{ij} X_{ij}^\top / \hat{f}_\theta(\theta^\top X_j) \}^{-1}$$

$$\times \sum_{i,j=1}^{n} K_h(\theta^\top X_{ij}) \hat{\rho}_j^\theta d_j^\theta X_{ij}(y_i - a_j^\theta) / \hat{f}_\theta(\theta^\top X_j),$$

where $\hat{\rho}_j^\theta$ is defined in section 2.2. Repeat step 1 with $\theta := \theta/|\theta|$ and and step 2 until convergence.

The limiting value of θ is an estimator of θ_0. Denote it by $\hat{\theta}_{r\text{MAVE}}$.

4.4. Sliced Inverse Regression Method

If X is elliptically distributed and independent of ε, we can use the sliced inverse regression approach to estimate θ_0. For ease of exposition, we further assume that X is standardized. By the model, we have

$$E(X|Y) = E\{E(X|\theta_0^\top X, Y)|Y\} = E\{E(X|\theta_0^\top X)|Y\}$$
$$= E\{\theta_0\theta_0^\top X|Y\} = \theta_0 m(Y),$$

where $m(y) = E(\theta_0^\top X|Y = y)$ and $\theta_1, \cdots, \theta_{p-1}$ are such that $(\theta_0, \theta_1, \cdots, \theta_{p-1})^\top (\theta_0, \theta_1, \cdots, \theta_{p-1}) = I_p$, a $p \times p$ identity matrix. See Li (1991) for more details.

Based on the above fact, a simple algorithm is as follows. Estimate $E(X|Y = y)$ by the local linear kernel smoother

$$\hat{\xi}(y) = \sum_{i=1}^{n} w_h(Y_i, y) X_i \Big/ \sum_{i=1}^{n} w_h(Y_i, y),$$

where $w_h(Y_i, y) = \{s_{n2}(y) K_h(Y_i - y) - s_{n1}(y) K_h(Y_i - y)(Y_i - y)/h\}$ and $s_{nk}(y) = n^{-1} \sum_{i=1}^{n} K_h(Y_i - y)\{(Y_i - y)/h\}^k$ for $k = 0, 1, 2$. Then the estimator of θ_0 is the first eigenvector of

$$\sum_{j=1}^{n} \hat{\rho}_j \hat{\xi}(Y_j) \hat{\xi}(Y_j)^\top,$$

where $\hat{\rho}_j = \rho_n(\hat{f}_Y(Y_j))$ with $\hat{f}_Y(Y_j) = n^{-1} \sum_{i=1}^{n} K_h(Y_i - y)$. Denote the estimator by $\hat{\theta}_{\text{SIR}}$.

Theoretically, the SIR method has good *asymptotic* properties because it uses *univariate* nonparametric regression. The estimators can achieve root-n consistency even when the dimension, q in Model (4.3), is very high. However, in terms of efficiency, it is not so attractive. For finite samples, the performance can be far from satisfactory.

4.5. Asymptotic Distributions

We impose the following conditions in order to obtain some useful asymptotics.

(C1) [Design] The density function $f(x)$ of X has bounded second order derivatives on $\mathbb{R}^p$, $E|X|^r < \infty$ and $E|Y|^r < \infty$ for all $r > 0$.

(C2) [Link function] The link function $g(v)$ has bounded third order derivatives.

(C3) [Kernel function] $K(v)$ is a symmetric density function; there exists a constant $c > 0$ such that $e^{cv} K(v) \to 0$ as $v \to \infty$ and that the Fourier transform of $K(v)$ is absolutely integrable.

(C4) [Bandwidth and trimming parameter] $h \propto n^{-1/5}$ and $\epsilon < 1/10$.

Condition (C1) can be relaxed to the existence of moments up to a finite order. In HHI, they require that X is bounded. See also Härdle and Stoker (1989). The smoothness of the link function in (C2) can be weakened to the existence of the second derivative at the cost of more complicated proofs. Another approach is to use a different trimming function to trim off the boundary points and allow a more general link function; in this case the MAVE estimates will not necessarily be the most efficient in the semiparametric sense with respect to the limiting distribution. The kernel functions include the Gaussian kernel and the Epanechnikov kernel. The bandwidth (C4) is practicable because it is of the order of an optimal bandwidth in both the estimation of the link function and the estimation of the index; see HHI. Because our interest here is asymptotic efficiency, we assume in the following that the initial value θ is in a small neighbor of θ_0. Define $\Theta_n = \{\theta : |\theta - \theta_0| \le C_0 n^{-1/2+1/10}\}$ where C_0 is a constant. (In HHI, they even assume that the initial value is in a root-n neighbourhood $\{\theta : |\theta - \theta_0| \le C_0 n^{-1/2}\}$.) This assumption is feasible because such an initial value is obtainable using existing methods.

Let $\mu_\theta(x) = E(X|\theta^\top X = \theta^\top x)$, $\nu_\theta(x) = E(X|\theta^\top X = \theta^\top x) - x$, $w_\theta(x) = E(XX^\top|\theta^\top X = \theta^\top x)$, $W_0(x) = \nu_{\theta_0}(x)\nu_{\theta_0}^\top(x)$ and $W(x) = w_{\theta_0}(x) - \mu_{\theta_0}(X)\mu_{\theta_0}^\top(X)$. For any symmetric matrix A, A^+ denotes the Moore-Penrose inverse matrix.

For the ADE estimator, Powell, Stock and Stoker (1989) proved that under some regularity conditions, $\sqrt{n}(\hat\delta - \delta) \xrightarrow{D} N(0, \tilde\Sigma_{\mathrm{ADE}})$, where $\tilde\Sigma_{\mathrm{ADE}}$ is the covariance matrix of $l(X)\varepsilon + g'(\theta_0^\top X)\theta_0$. Based on this, we further have $\sqrt{n}(\hat\theta_{\mathrm{ADE}} - \theta_0) \xrightarrow{D} N(0, \Sigma_{\mathrm{ADE}})$, where

$$\Sigma_{\mathrm{ADE}}$$
$$= \{Eg'(\theta_0^\top X)\}^{-2}(I - \theta_0\theta_0^\top)E\{[\frac{\partial \log f(X)}{\partial x}][\frac{\partial \log f(X)}{\partial x}]^\top \varepsilon^2\}(I - \theta_0\theta_0^\top).$$

HHI proved that under appropriate regularity conditions,

$$\sqrt{n}(\hat\theta_{\mathrm{HHI}} - \theta_0) \xrightarrow{D} N(0, \Sigma_{\mathrm{HHI}}),$$

where

$$\Sigma_{\mathrm{HHI}}$$
$$= [E\{g'(\theta_0 X)^2 W(X)\}]^+ E\{g'(\theta_0 X)^2 W_0(X)\varepsilon^2\}[E\{g'(\theta_0 X)^2 W(X)\}]^+.$$

If we further assume that ε is normally distributed and independent of X, or more generally, the conditional distribution of Y given X belongs to a canonical exponential family

$$f_{Y|X}(y|x) = \exp\{y\eta(x) - \mathcal{B}(\eta(x)) + \mathcal{C}(y)\}$$

for some known functions $\mathcal{B}$, $\mathcal{C}$ and η, then Σ_{HHI} is the information lower bound in the semiparametric sense. See the proofs in Carroll, Fan, Gijbels and Wand (1997).

Theorem 4.1: [rOPG] Under the conditions (C1)-(C4), we have

$$\sqrt{n}(\hat\theta_{r\mathrm{OPG}} - \theta_0) \xrightarrow{D} N(0, \Sigma_{r\mathrm{OPG}}),$$

where

$$\Sigma_{r\text{OPG}} = E\{g'(\theta_0 X)^2 W(X)^+ W_0(X)\varepsilon^2 W(X)^+\}/\{Eg'(\theta_0^\top X)^2\}^2.$$

Theorem 4.2: [rMAVE] Under the conditions (C1)-(C4), we have

$$\sqrt{n}(\hat{\theta}_{r\text{MAVE}} - \theta_0) \xrightarrow{D} N(0, \Sigma_{r\text{MAVE}}),$$

where $\Sigma_{r\text{MAVE}} = \Sigma_{\text{HHI}}$.

Thus, the rMAVE estimator is the most efficient estimator in the semi-parametric sense if ε is normally distributed and independent of X.

Theorem 4.3: [SIR] If X is elliptically distributed, then

$$\sqrt{n}(\hat{\theta}_{\text{SIR}} - \theta_0) \xrightarrow{D} N\left(0, \Sigma_{\text{SIR}}\right),$$

where

$$\Sigma_{\text{SIR}} = \{Em(Y)^2\}^{-2} E\{m(Y)^2[X - \theta_0 m(Y)][X - \theta_0 m(Y)]^\top\}.$$

4.6. Comparisons in Some Special Cases

Since all the limiting distributions are normal, we can in principle compare the efficiencies by their covariance matrices. We can do this explicitly for the following special cases.

Corollary 4.1: The information matrix $\Sigma_{r\text{MAVE}} = \Sigma_{\text{HHI}}$. If ε is independent of X, then

$$\Sigma_{r\text{OPG}} \geq \Sigma_{r\text{MAVE}}.$$

By Corollary 4.1, $\hat{\theta}_{r\text{MAVE}}$ is more efficient than $\hat{\theta}_{r\text{OPG}}$ if X is independent of ε. Because the distribution of $\hat{\theta}_{\text{ADE}}$ depends on the derivative of the distribution function, it is not easy to compare its covariance matrix with others. However, under the normal assumption, we have the following conclusion.

Corollary 4.2: If $(X^\top, \varepsilon)$ is normal, then

$$\Sigma_{r\text{MAVE}} = \Sigma_{r\text{OPG}} \leq \Sigma_{\text{ADE}}.$$

The equality holds only when g is a linear function.

For SIR, its distribution depends on another unknown function; see Theorem 4.3. General comparison is difficult to make. However, for the following simple example, we can evaluate explicitly the different efficiencies for different methods.

Example 4.1: Suppose that $(X^\top, \varepsilon)$ is standard normal and that the true model is

$$Y = \theta_0^\top X + \sigma\varepsilon. \tag{4.7}$$

Then $\Sigma_{\text{ADE}} = \Sigma_{r\text{OPG}} = \Sigma_{r\text{MAVE}} = \sigma^2(I - \theta_0\theta_0^\top)$. It is interesting to find that *without knowing the link function g* the efficiency of each of the above three semi-parametric methods is the same as that of the MLE within the framework of a parametric model *with a known link function g*. Note that $E(\theta_0^\top X|Y) = Y/(1 + \sigma^2)$ in Model (4.7). Simple calculations lead to

$$\Sigma_{\text{SIR}} = (1 + \sigma^2)(I - \theta_0\theta_0^\top).$$

Note that $I - \theta_0\theta_0^\top$ is a positive semi-definite matrix; we have

$$\Sigma_{\text{ADE}} = \Sigma_{r\text{OPG}} = \Sigma_{r\text{MAVE}} < \Sigma_{\text{SIR}}.$$

It is interesting to see that θ_0 cannot be estimated efficiently by SIR even when there is no noise term ε in the model.

4.7. Proofs of the Theorems

Recall that K is a density function. Thus, $\int K = 1$. For ease of exposition, we further assume that $\mu_2 = \int v^2 K(v)dv = 1$. Otherwise, we can redefine $K(v) := \mu_2^{-1/2} K(\mu_2^{-1/2}v)$. Let $f_\theta(v)$ be the density function of $\theta^\top X$ and $\mathcal{D}_n = \{(\theta, x) : |x| < n^c, f_\theta(\theta^\top x) > n^{-\epsilon}, \theta \in \Theta_n\}$ where $c \geq 1$. Suppose A_n is a matrix. By $A_n = O(a_n)$, we mean all elements in A_n are $O(a_n)$ almost surely. Let $\delta_n = (nh/\log n)^{-1/2}$ and $\tau_n = h^2 + \delta_n$. Let $W_{g0} = E\{g'(\theta_0^\top X)^2 \nu_{\theta_0}(\theta_0^\top X)\nu_{\theta_0}^\top(\theta_0^\top X)\}$.

Lemma 4.1: *[Basic results for kernel smoothing] Suppose* $E(Z|\theta^\top X = \theta^\top x) = m(\theta^\top x)$ *has bounded derivatives up to third order and* $E|Z|^r < \infty$ *for all* $r > 0$. *Let* $(X_i, Y_i, Z_i), i = 1, 2, \cdots, n$ *be a sample from* (X, Y, Z). *Then*

$$n^{-1}\sum_{i=1}^{n} K_h(\theta^\top X_{ix})(\theta^\top X_{ix}/h)^d Z_i$$

$$= f(\theta^\top x)m(\theta^\top x)\mu_d + \{f(\theta^\top x)m(\theta^\top x)\}'\mu_{d+1}h + O(\tau_n), \tag{4.8}$$

uniformly for $(\theta, x) \in \mathcal{D}_n$, *where* $\mu_d = \int K(v)v^d dv$. *If* $E(Z|Y = y) = m(y)$ *has bounded third derivatives, then*

$$n^{-1} \sum_{i=1}^{n} K_h(Y_i - y)\{(Y_i - y)/h\}^d Z_i$$

$$= f_Y(y)m(y)\mu_d + \{f_Y(y)m(y)\}'\mu_{d+1}h + O(\tau_n), \qquad (4.9)$$

uniformly for $y \in \{y : |y| < n^c\}$ *for any positive c.*

Lemma 4.2: *[Kernel smoothers in rOPG] Let* $X_{ix} = X_i - x$,

$$S_n^\theta(x) = n^{-1} \sum_{i=1}^{n} K_h(\theta^\top X_{ix}) \begin{pmatrix} 1 \\ X_{ix} \end{pmatrix} \begin{pmatrix} 1 \\ X_{ix} \end{pmatrix}^\top$$

and

$$\begin{pmatrix} a_x^\theta \\ b_x^\theta \end{pmatrix} = \{S_n^\theta(x)\}^{-1} \sum_{i=1}^{n} K_h(\theta^\top X_{ix}) \begin{pmatrix} 1 \\ X_{ix} \end{pmatrix} Y_i.$$

Under assumptions (C1)–(C4), we have

$$b_x^\theta = \theta_0 g'(\theta_0^\top x) + \{nf(\theta_0^\top x)\bar{w}(x)\}^+ \sum_{i=1}^{n} K_h(\theta_0^\top X_{ix})\nu_{\theta_0}(X_i)\varepsilon_i$$

$$+o(n^{-1/2})(1 + |x|^3),$$

uniformly for $(\theta, x) \in \mathcal{D}_n$, *where* $\bar{w}(x) = w_{\theta_0}(x) - \mu_{\theta_0}(x)\mu_{\theta_0}^\top(x)$.

It is interesting to see that the bias term is negligible. Thus, the rOPG method does not need under-smoothing for the estimator of the single-index to achieve root-n consistency.

Lemma 4.3: *[Kernel smoothers in rMAVE] Let*

$$\Sigma_n^\theta(x) = n^{-1} \sum_{i=1}^{n} K_h(\theta^\top X_{ix}) \begin{pmatrix} 1 \\ \theta^\top X_{ix}/h \end{pmatrix} \begin{pmatrix} 1 \\ \theta^\top X_{ix}/h \end{pmatrix}^\top$$

and

$$\begin{pmatrix} a_x^\theta \\ d_x^\theta h \end{pmatrix} = \{\Sigma_n^\theta(x)\}^{-1} \sum_{i=1}^{n} K_h(\theta^\top X_{ix}) \begin{pmatrix} 1 \\ \theta^\top X_{ix}/h \end{pmatrix} Y_i.$$

Under assumptions (C1)–(C4), we have

$$a_x^\theta = g(\theta_0^\top x) + g'(\theta_0^\top x)(\theta_0 - \theta)^\top \nu_{\theta_0}(x) + \frac{1}{2}g''(\theta_0^\top x)h^2 + \mathcal{E}_{n,1}(x)$$

$$+o(n^{-1/2})(1 + |x|^3),$$

$$d_x^\theta h = g'(\theta_0^\top x)h + \mathcal{E}_{n,2}(x) + o(n^{-1/2})(1 + |x|^3),$$

uniformly for $(\theta, x) \in \mathcal{D}_n$, where $\mathcal{E}_{n,1}(x) = \{nf(\theta^\top x)\}^{-1} \sum_{i=1}^n K_h(\theta^\top X_{ix}) \varepsilon_i$, $\mathcal{E}_{n,2}(x) = \{nhf(\theta^\top x)\}^{-1} \sum_{i=1}^n K_h(\theta^\top X_{ix}) \theta^\top X_{ix} \varepsilon_i$.

Lemma 4.4: *[Kernel smoothers in SIR] Under assumptions (C1)–(C4),*

$$\xi(y) = \theta_0 m(y) + \frac{1}{2}\theta_0 m''(y)h^2 + \{nf_Y(y)\}^{-1} \sum_{i=1}^n K_h(Y_i - y)\{X_i - \theta_0 m(y)\}$$

$$+o(n^{-1/2})$$

uniformly for $y \in \{y : f(y) > n^{-\epsilon}\}$.

Lemma 4.5: *[Numerator in the MAVE] Under assumptions (C1)–(C4), we have*

$$n^{-2} \sum_{i,j=1}^n \hat{\rho}_j^\theta K_h(\theta^\top X_{ij}) X_{ij}\{Y_i - a_j^\theta - d_j^\theta \theta_0^\top X_{ij}\}$$

$$= W_{g0}(\theta - \theta_0) + \frac{1}{n} \sum_{i=1}^n g'(\theta_0^\top X_i)\nu_{\theta_0}(X_j)\varepsilon_i + o(n^{-1/2})$$

uniformly for $\theta \in \Theta_n$, *where* $W_{g0} = E\{(g'(\theta_0^\top X))^2 \nu_{\theta_0}(X)\nu_{\theta_0}^\top(X)\}$.

Lemma 4.6: *[Denominator in the MAVE] Under assumptions (C1)–(C4), we have*

$$\{n^{-1} \sum_{i,j=1}^n \hat{\rho}_j^\theta (d_j^\theta)^2 K_h(\theta^\top X_{ij}) X_{ij} X_{ij}^\top / \hat{f}_\theta(\theta^\top X_j)\}^{-1}$$

$$= \theta_0 \theta_0^\top \{g'(\theta_0^\top x)\}^2 h^{-2} + \frac{1}{2}W_{g0}^+ + O(\tau_n n^\epsilon / h)$$

uniformly for $\theta \in \Theta_n$.

Proof of ADE estimator. Härdle and Stoker (1989) showed that (A.5, p. 992)

$$\hat{\delta} = Eg'(\theta_0^\top X)\theta_0 + n^{-1} \sum_{j=1}^n \{l(X_j)\varepsilon_j + [g'(\theta_0^\top X_j) - Eg'(\theta_0^\top X_j)]\theta_0\}$$

$$+o_p(n^{-1/2}).$$

It follows that

$$|\hat{\delta}|^2 = \{Eg'(\theta_0^\top X)\}^2 \Big(1 + 2\{nEg'(\theta_0^\top X)\}^{-1} \sum_{j=1}^n \{\theta_0^\top l(X_j)\varepsilon_j$$

$$+[g'(\theta_0^\top X_j) - Eg'(\theta_0^\top X_j)]\} + o_p(n^{-1/2})\Big).$$

Thus

$$\hat{\delta}/|\hat{\delta}| = \theta_0 + \{nEg'(\theta_0^\top X)\}^{-1}(I - \theta_0\theta_0^\top)\sum_{j=1}^{n} l(X_j)\varepsilon_j + o(n^{-1/2}).$$

Therefore, the asymptotic distribution of ADE follows from the central limit theorem. □

Proof of Theorem 4.1. By assumption (C1), we have

$$\sum_{n=1}^{\infty} P(\bigcup_{i=1}^{n}\{X_i \notin \mathcal{D}_n\}) \leq \sum_{n=1}^{\infty} nP(X_i \notin \mathcal{D}_n) \leq \sum_{n=1}^{\infty} nP(|X_i| > n^c)$$

$$\leq \sum_{n=1}^{\infty} nn^{-4c}E|X|^4 < \infty.$$

It follows from the Borel-Cantelli Lemma that

$$P(\bigcup_{i=1}^{n}\{X_i \notin \mathcal{D}_n\}, \ i.o.) = 0. \tag{4.10}$$

Let $\tilde{\mathcal{D}}_n = \{x : f_\theta(\theta^\top x) > 2n^{-\epsilon}\}$. Similarly, we have

$$P(\bigcup_{i=1}^{n}\{X_i \notin \tilde{\mathcal{D}}_n\}, \ i.o.) = 0. \tag{4.11}$$

Thus, we can exchange summations over $\{X_j : j = 1, \cdots, n\}$, $\{X_j : X_j \in \mathcal{D}_n, j = 1, \cdots, n\}$ and $\{X_j : X_j \in \tilde{\mathcal{D}}_n, j = 1, \cdots, n\}$ in the sense of almost surely consistency. Let

$$M_n \stackrel{def}{=} n^{-1}\sum_{j=1}^{n} \hat{\rho}_j^\theta g'(\theta_0^\top X_j)\{nf_\theta(\theta^\top X_j)\bar{w}_\theta(X_j)\}^+ \sum_{i=1}^{n} K_h(\theta^\top X_{ij})\nu_\theta(X_i)\varepsilon_i.$$

By Lemma 4.1 and (C1), we have

$$\hat{f}_\theta(\theta^\top x) = f_\theta(\theta^\top x) + O(\tau_n).$$

Thus by the smoothness of $\rho_n(.)$ and (4.11), we have

$$\rho_n(\hat{f}_\theta(\theta^\top x)) = \rho_n(f_\theta(\theta^\top x)) + O(\tau_n) = 1 + O(\tau_n). \tag{4.12}$$

We introduce operator E_j as follows. For any function $G(X_i, Y_i, X_k, Y_k, X_\ell, Y_\ell)$ with $i \neq j \neq k$, we define

$$E_k G(X_i, Y_i, X_k, Y_k, X_\ell, Y_\ell) = E\{G(X_i, Y_i, X_k, Y_k, X_\ell, Y_\ell)|X_i, Y_i, X_\ell, Y_\ell\}.$$

Let $C_{ij} = K_h(\theta^\top X_{ij})\rho_n(f_\theta(\theta^\top X_j))g'(\theta_0^\top X_j)\{f_\theta(\theta^\top X_j)\bar{w}_\theta(X_j)\}^+$. We have

$$E_j C_{ij} = g'(\theta_0^\top X_i)\bar{w}_\theta(X_i)^+ + O(\delta_\theta + h^2).$$

Thus, by (C1) and Lemma (4.8),

$$M_n = n^{-2} \sum_{i=1}^{n} \sum_{j=1}^{n} \{C_{ij} - E_j C_{ij}\} \nu_\theta(X_i)\varepsilon_i + n^{-1} \sum_{i=1}^{n} E_j C_{ij} \nu_\theta(X_i)\varepsilon_i$$
$$+ o(n^{-1/2})$$

$$= n^{-1} \sum_{i=1}^{n} g'(\theta_0^\top X_i) W(X_i)^+ \nu_{\theta_0}(X_i)\varepsilon_i + o(n^{-1/2}).$$

By (4.10), (4.11), (4.12) and Lemma 4.2, we have almost surely

$$n^{-1} \sum_{j=1}^{n} \hat{\rho}_j^\theta b_j^\theta (b_j^\theta)^\top = n^{-1} \sum_{X_j \in D_n} \hat{\rho}_j^\theta b_j^\theta (b_j^\theta)^\top$$

$$= n^{-1} \sum_{X_j \in D_n} \{1 + O(\tau_n + \delta_\theta)\} g'(\theta_0^\top X_j)^2 \theta_0 \theta_0^\top + \theta_0 M_n^\top + M_n \theta_0^\top + o(n^{-1/2})$$

$$= n^{-1} \sum_{j=1}^{n} \{1 + O(\tau_n + \delta_\theta)\} g'(\theta_0^\top X_j)^2 (\theta_0 + \tilde{M}_n)(\theta_0 + \tilde{M}_n) + o(n^{-1/2}), \quad (4.13)$$

where

$$\tilde{M}_n = \{Eg'(\theta_0^\top X)^2\}^{-1} n^{-1} \sum_{i=1}^{n} g'(\theta_0^\top X_i) W(X_i)^+ \nu_{\theta_0}(X_i)\varepsilon_i + o(n^{-1/2}).$$

Thus Theorem 4.1 follows from the central limit theorem and (4.13).   □

Proof of Theorem 4.2 Suppose we start with $\theta \in \Theta_n$. After one iteration, we have a new θ as

$$\tilde{\theta} = \theta_0 + \{\sum_{i,j=1}^{n} K_h(\theta^\top X_{ij}) \hat{\rho}_j^\theta (d_j^\theta)^2 X_{ij} X_{ij}^\top / \hat{f}_\theta(\theta^\top X_j)\}^{-1}$$

$$\times \sum_{i,j=1}^{n} K_h(\theta^\top X_{ij}) \hat{\rho}_j^\theta d_j^\theta X_{ij}(y_i - a_j^\theta - d_j^\theta X_{ij}\theta_0) / \hat{f}_\theta(\theta^\top X_j).$$

By definition, we have $\Gamma_{\theta_0} = W_{g0}^+ W_{g0} = I - \theta_0\theta_0^\top$. By Lemmas 4.5 and 4.6 and Equations (4.10) and (4.11), we have

$$\tilde{\theta} = a_n\theta_0 + \frac{1}{2}\Gamma_{\theta_0}(\theta - \theta_0) + \frac{1}{2}n^{-1}W_{g0}^+ \sum_{i=1}^{n} g'(\theta_0^\top X_i)\nu_{\theta_0}(X_j)\varepsilon_i + o(n^{-1/2}),$$

where $a_n = 1 + o(1)$. Note that $\theta_0^\top(\theta - \theta_0) = O(\delta_\theta^2)$ and $\theta_0^\top W_{g0}^+ = 0$. Thus $|\tilde{\theta}| = a_n\{1 + o(n^{-1/2})\}$. It follows that

$$\tilde{\theta}/|\tilde{\theta}| = \theta_0 + \frac{1}{2}\Gamma_{\theta_0}(\theta - \theta_0) + \frac{1}{2}n^{-1}W_{g0}^+ \sum_{i=1}^{n} g'(\theta_0^\top X_i)\nu_{\theta_0}(X_i)\varepsilon_i + o(n^{-1/2}).$$

Let $\theta^{(k)}$ be the value of θ after k iteration. We have

$$\theta^{(k+1)} = \theta_0 + \frac{1}{2}\Gamma_{\theta_0}(\theta^{(k)} - \theta_0) + \frac{1}{2}n^{-1}W_{g0}^+ \sum_{i=1}^{n} g'(\theta_0^\top X_i)\nu_{\theta_0}(X_i)\varepsilon_i + o(n^{-1/2}).$$

Thus Theorem 4.2 follows immediately from the above recursive relation and the central limit theorem. □

Proof of Theorem 4.3. By Lemma 4.3 and the methods for the trimming function above, we have

$$n^{-1}\sum_{j=1}^{n}\hat{\rho}_j\xi(Y_j)\xi^\top(Y_j) = \theta_0\theta_0^\top n^{-1}\sum_{j=1}^{n}\{m(Y_j) + \frac{1}{2}m''(Y_j)h^2\}^2$$

$$+\theta_0 N_n^\top + N_n\theta_0^\top + o(n^{-1/2}),$$

where

$$N_n = n^{-2}\sum_{j=1}^{n}m(Y_j)\{hf(Y_j)\}^{-1}\sum_{i=1}^{n}K_h(Y_i - Y_j)\{X_i - \theta_0 m(Y_i)\}$$

$$= n^{-1}\sum_{i=1}^{n}m(Y_i)\{X_i - \theta_0 m(Y_i)\} + o(n^{-1/2}).$$

The rest of the proof is similar to that of Theorem 4.1. □

Proof of Corollary 4.1. If ε is independent of X, we have

$$\Sigma_{rOPG} = E\{g'(\theta_0^\top X)^2 W(X)^+\}\sigma^2/E\{g'(\theta_0^\top X)^2\}^2,$$

where $\sigma^2 = E(\varepsilon^2)$ and

$$\Sigma_{rMAVE} = [E\{g'(\theta_0^\top X)^2 W(X)\}]^+\sigma^2.$$

Let $A = \{g'(\theta_0^\top X)\}^{-2}W(X)^+$ and $C = -g'(\theta_0^\top X)^{-2}(EA^+)^+$. Then,

$$\Sigma_{rOPG} - \Sigma_{rMAVE} = \sigma^2 E\{g'(\theta_0^\top X)^4[A^{1/2}/Eg'(\theta_0^\top X)^2 + C(A^+)^{1/2}]$$

$$\times [A^{1/2}/Eg'(\theta_0^\top X)^2 + C(A^+)^{1/2}]^\top\}$$

$$\geq 0.$$

Thus, Corollary 4.1 follows.

Proof of Corollary 4.2. Without loss of generality, we assume X is standard normal $N(0, I)$. Then $x = \Gamma_{\theta_0}x + \theta_0(\theta_0^\top x)$. We have

$$\mu_{\theta_0}(x) = \theta_0(\theta_0^\top x), \quad \nu_{\theta_0}(x) = -\Gamma_{\theta_0}x, \quad w_{\theta_0}(x) = \Gamma_{\theta_0} + \theta_0\theta_0^\top(\theta_0^\top x)^2,$$

$$W_0(x) = \Gamma_{\theta_0}xx^\top\Gamma_{\theta_0}, \quad W(x) = \Gamma_{\theta_0}xx^\top\Gamma_{\theta_0} - \theta_0\theta_0^\top(\theta_0^\top x)^2. \quad (4.14)$$

Note that $\theta_0^\top X$ is independent of $\Gamma_{\theta_0}X$ and $EXX^\top = I$. We have

$$E\{g'(\theta_0 X)^2 W(X)\} = E\{g'(\theta_0 X)^2\}\Gamma_{\theta_0} - \theta_0\theta_0^\top E\{g'(\theta_0 X)^2(\theta_0 X)^2\}.$$

Because Γ_{θ_0} is projection matrix and $\theta_0^\top \Gamma_{\theta_0} = 0$, we have

$$[E\{g'(\theta_0 X)^2 W(X)\}]^+ = \Gamma_{\theta_0}/E\{g'(\theta_0 X)^2\} - \theta_0 \theta_0^\top/E\{g'(\theta_0 X)^2 (\theta_0 X)^2\}.$$

Similarly

$$E\{g'(\theta_0 X)^2 W_0(X)\varepsilon^2\} = E\{g'(\theta_0 X)^2\}\Gamma_{\theta_0}\sigma^2.$$

We have from the above equations that

$$\Sigma_{r\text{MAVE}} = \Gamma_{\theta_0}\sigma^2/E\{g'(\theta_0 X)^2\}.$$

By the definition of Moore-Penrose inverse and that $\theta_0^\top \Gamma_{\theta_0} = 0$, we have

$$\{W(x)\}^+ = \Gamma_{\theta_0} xx^\top \Gamma_{\theta_0}(x^\top \Gamma_{\theta_0} x)^{-1} - \theta_0 \theta_0^\top (\theta_0^\top x)^{-2}.$$

By $\Gamma_{\theta_0}^2 = \Gamma_{\theta_0}$ and $\theta_0^\top \Gamma_{\theta_0} = 0$, it follows that

$$\{W(x)\}^+ W_0(x)\{W(x)\}^+ = \Gamma_{\theta_0} xx^\top \Gamma_{\theta_0}.$$

Because $\theta_0^\top X$ is independent of $\Gamma_{\theta_0} X$, we have

$$\Sigma_{r\text{OPG}} = E\{g'(\theta_0^\top X)^2\}E\varepsilon^2 E\{\Gamma_{\theta_0} XX^\top \Gamma_{\theta_0}\}/E\{g'(\theta_0^\top X)^2\}^2$$
$$= \sigma^2 \Gamma_{\theta_0}/E\{g'(\theta_0^\top X)^2\}. \tag{4.15}$$

Note that $\partial \log f(x)/\partial x = x$. We have

$$E\{[\frac{\partial \log f(X)}{\partial x}][\frac{\partial \log f(X)}{\partial x}]^\top \varepsilon^2\} = E(XX^\top \varepsilon^2) = I\sigma^2.$$

Thus

$$\Sigma_{\text{ADE}} = \sigma^2 \Gamma_{\theta_0}/\{Eg'(\theta_0^\top X)\}^2. \tag{4.16}$$

Corollary 4.2 follows from (4.15), (4.16) and the fact that $\{Eg'(\theta_0^\top X)\}^2 \le E\{g'(\theta_0^\top X)^2\}$, in which the equality holds only when g' is constant. □

4.7.1. Proofs of the lemmas

We give detailed proofs of some of the lemmas mentioned above, leaving the other proofs to Xia, Tong and Li (2002). We first cite some further lemmas, the proofs of which can also be found in the above paper.

Lemma 4.7: *Suppose $m_n(\chi, Z), n = 1, 2, \cdots$, are measurable functions of Z with index $\chi \in \mathbb{R}^d$, where d is any integer number, such that (i) $|m_n(\chi, Z)| \le a_n M(Z)$ with $E(M(Z)^r) < \infty$ for some $r > 2$ and a_n increases with n such that $a_n < c_0 n^{1-1/(r-1)}/\log^3 n$ and $n^{-1}a_n \to 0$; (ii) $E(m_n(\chi, Z))^2 < a_n m_0^2(\chi)$ (without loss of generality, we assume $m_0(\chi) > 1$); and (iii) $|m_n(\chi, Z) - m_n(\chi', Z)| = |\chi - \chi'|^{\alpha_1} n^{\alpha_2} G(Z)$ with* some $\alpha_1, \alpha_2 > 0$ and $EG^2(Z)$ exists. Suppose $\{Z_i, i = 1, \cdots, n\}$ is a random sample from Z. Then, for any positive α_0, we have

$$\sup_{|\chi| \le n^{\alpha_0}} \left| \{nm_0(\chi)\}^{-1} \sum_{i=1}^{n} \{m_n(\chi, Z_i) - Em_n(\chi, Z_i)\} \right| = O\{(a_n \log n/n)^{1/2}\}.$$

Lemma 4.8: Let $\xi(\theta)$ is a measurable function of (X, y). Suppose $E\{\xi(\theta)|\theta^\top X\} = 0$ for all $\theta \in \Theta$ and $|\xi(\theta) - \xi(\vartheta)| \le |\theta - \vartheta|\tilde{\xi}$ with $E\tilde{\xi}^r < \infty$ for some $r > 2$. Let φ_i be defined in Lemma A.1. If (C1) and (C6) hold, then

$$\sup_{\theta \in \Theta} \left| \frac{1}{n^2} \sum_{i=1}^{n} \sum_{j=1}^{n} \left\{ K_h(\theta^\top X_{ij})\varphi_j(\theta) - E_j K_h(\theta^\top X_{ij})\varphi_j(\theta)) \right\} \xi_i(\theta) \right| = O(\delta_n^2).$$

Proof of Lemma 4.2. By Lemma 4.7, we have

$$S_n^\theta(x) = f_\theta(\theta^\top x) \begin{pmatrix} 1 & \nu_\theta^\top(x) \\ \nu_\theta(x) & \tilde{w}_\theta(x) \end{pmatrix} + O(\tau_n)(1 + |x|^2),$$

where $\tilde{w}_\theta(x) = w_\theta(x) - x\mu_\theta^\top(x) - \mu_\theta(x)x^\top + xx^\top$. If $f_\theta(\theta^\top x) > n^{-\epsilon}$, then

$$\{S_n^\theta(x)\}^{-1} = f_\theta^{-1}(\theta^\top x) \begin{pmatrix} c(x) & \gamma_\theta^\top(x) \\ \gamma_\theta(x) & \tilde{w}_\theta(x)^+ \end{pmatrix} + O(\tau_n n^\epsilon)(1 + |x|^2), \quad (4.17)$$

where $c(x) = \{1 - \nu_\theta^\top(x)\tilde{w}_\theta(x)^+\nu_\theta(x)\}^{-1}$ and $\gamma_\theta(x) = -\bar{w}_\theta(x)^+\nu_\theta(x)$. By the Taylor expansion of $g(\theta_0^\top X_i)$ at $\theta_0^\top x$ and $\theta_0 = \theta + (\theta_0 - \theta)$, we have

$$Y_i = g(\theta_0^\top x) + g'(\theta_0^\top x)\theta_0^\top X_{ix} + \frac{1}{2}g''(\theta_0^\top x)(\theta_0^\top X_{ix})^2 + \varepsilon_i + O(|\theta_0^\top X_{ix}|^3)$$

$$= g(\theta_0^\top x) + g'(\theta_0^\top x)\theta_0^\top X_{ix} + \frac{1}{2}g''(\theta_0^\top x)(\theta^\top X_{ix})^2 + \varepsilon_i + \Delta_n(x, X_i, \theta),$$
$$(4.18)$$

where $\Delta_n(x, X_i, \theta) = O(|\theta_0^\top X_{ix}|^3 + |\theta^\top X_{ix}| \cdot |X_{ix}|\delta_\theta + |X_{ix}|^2\delta_\theta^2)$ and $\delta_\theta = |\theta - \theta_0|$. It is easy to see that

$$\{nS_n^\theta(x)\}^{-1} \sum_{i=1}^{n} K_h(\theta^\top X_{ix}) \begin{pmatrix} 1 \\ X_{ix} \end{pmatrix} \{g(\theta_0^\top x) + g'(\theta_0^\top x)\theta_0^\top X_{ix}\}$$

$$= \begin{pmatrix} g(\theta_0^\top x) \\ \theta_0 g'(\theta_0^\top x) \end{pmatrix}. \quad (4.19)$$

It follows from Lemma 4.1 that

$$\{nS_n^\theta(x)\}^{-1} \sum_{i=1}^n K_h(\theta^\top X_{ix}) \begin{pmatrix} 1 \\ X_{ix} \end{pmatrix} g''(\theta_0^\top x)(\theta^\top X_{ix})^2$$

$$= \begin{pmatrix} c(x) & \gamma_\theta^\top(x) \\ \gamma_\theta(x) & \bar{w}_\theta(x)^+ \end{pmatrix} g''(\theta_0^\top x) \begin{pmatrix} 1 \\ \nu_\theta(x) \end{pmatrix} h^2 + O(h^2 \tau_n n^\epsilon)(1 + |x|^2)$$

$$= \begin{pmatrix} (c(x) + \gamma_\theta^\top \nu_\theta(x))g''(x)h^2 \\ 0 \end{pmatrix} + O(h^2 \tau_n n^\epsilon)(1 + |x|^2). \tag{4.20}$$

By Lemma 4.7, it follows that

$$n^{-1} \sum_{i=1}^n K_h(\theta^\top X_{ix}) \begin{pmatrix} 1 \\ X_{ix} \end{pmatrix} |\theta^\top X_{ix}|^k |X_{ix}|^\ell = O(h^k)(1 + |x|^{\ell+1}),$$

for $k, \ell = 0, 1, 2, 3$. Thus,

$$\{nS_n^\theta(x)\}^{-1} \sum_{i=1}^n K_h(\theta^\top X_{ix}) \begin{pmatrix} 1 \\ X_{ix} \end{pmatrix} \Delta_n(x, X_i, \theta)$$

$$= O\{(h^3 + \delta_\theta^2)n^\epsilon\}(1 + |x|^4). \tag{4.21}$$

For the stochastic term, we have from (4.17)

$$\{nS_n^\theta(x)\}^{-1} \sum_{i=1}^n K_h(\theta^\top X_{ix}) \begin{pmatrix} 1 \\ X_{ix} \end{pmatrix} \varepsilon_i$$

$$= \{nf_\theta(\theta^\top x)\}^{-1} \begin{pmatrix} c(x) & \gamma_\theta^\top(x) \\ \gamma_\theta(x) & \bar{w}_\theta(x)^+ \end{pmatrix} \sum_{i=1}^n K_h(\theta^\top X_{ix}) \begin{pmatrix} 1 \\ X_{ix} \end{pmatrix} \varepsilon_i + O(\tau_n \delta_n n^\epsilon)$$

$$= \{nf_\theta(\theta^\top x)\}^{-1} \sum_{i=1}^n K_h(\theta^\top X_{ix}) \begin{pmatrix} c(x) + \gamma^\top(x)X_{ix} \\ \bar{w}_\theta(x)^+ \{X_i - \mu_\theta(x)\} \end{pmatrix} \varepsilon_i$$

$$+ O(\tau_n \delta_n n^\epsilon)(1 + |x|^2). \tag{4.22}$$

Combining (4.18)–(4.22), we complete the proof of Lemma 4.2. $\qquad \square$

Proof of Lemma 4.3. It follows from Lemma 4.1 that

$$\Sigma_n^\theta(x) = \begin{pmatrix} f_\theta(\theta^\top x) & f_\theta'(\theta^\top x)h \\ f_\theta'(\theta^\top x)h & f_\theta(\theta^\top x) \end{pmatrix} + O(\tau_n).$$

Thus, if $f_\theta(\theta^\top x) > n^{-\epsilon}$ then

$$\{\Sigma_n^\theta(x)\}^{-1} = f_\theta^{-1}(\theta^\top x) \left\{ I - f_\theta^{-1}(\theta^\top x) \begin{pmatrix} 0 & f_\theta'(\theta^\top x) \\ f_\theta'(\theta^\top x) & 0 \end{pmatrix} h \right\} + O(\tau_n n^\epsilon). \tag{4.23}$$

Rewrite (4.18) as

$$Y_i = g(\theta_0^\top x) + g'(\theta_0^\top x)\theta^\top X_{ix} + \frac{1}{2}g''(\theta_0^\top x)(\theta^\top X_{ix})^2 + g'(\theta_0^\top x)(\theta_0 - \theta)^\top X_{ix}$$
$$+\varepsilon_i + \Delta_n(x, X_i, \theta).$$

It is easy to see that

$$\{n\Sigma_n^\theta(x)\}^{-1} \sum_{i=1}^n K_h(\theta^\top X_{ix}) \begin{pmatrix} 1 \\ \theta^\top X_{ix} \end{pmatrix} \{g(\theta_0^\top x) + g'(\theta_0^\top x)\theta^\top X_{ix}\}$$
$$= \begin{pmatrix} g(\theta_0^\top x) \\ g'(\theta_0^\top x)h \end{pmatrix}. \tag{4.24}$$

Now, by Lemma 4.1 we have

$$\{n\Sigma_n^\theta(x)\}^{-1} \sum_{i=1}^n K_h(\theta^\top X_{ix}) \begin{pmatrix} 1 \\ \theta^\top X_{ix}/h \end{pmatrix} X_{ix}^\top (\theta_0 - \theta)g'(\theta_0^\top x)$$
$$= \begin{pmatrix} g'(\theta_0^\top x)(\theta_0 - \theta)\nu_\theta(x) \\ O(h\delta_\theta)(1 + |x|) \end{pmatrix} + O(\tau_n\delta_\theta n^\epsilon)(1 + |x|). \tag{4.25}$$

Again, by Lemma 4.1 and (4.23), we have

$$\{n\Sigma_n^\theta(x)\}^{-1} \sum_{i=1}^n K_h(\theta^\top X_{ix}) \begin{pmatrix} 1 \\ \theta^\top X_{ix}/h \end{pmatrix} g''(\theta_0^\top x)(\theta^\top X_{ix})^2$$
$$= \begin{pmatrix} g''(\theta_0^\top x)h^2 \\ g''(\theta_0^\top x)f_\theta'(\theta^\top x)h^3 \end{pmatrix} + O(h\tau_n n^\epsilon). \tag{4.26}$$

Similar to (4.21), we have

$$\{n\Sigma_n^\theta(x)\}^{-1} \sum_{i=1}^n K_h(\theta^\top X_{ix}) \begin{pmatrix} 1 \\ \theta^\top X_{ix}/h \end{pmatrix} \Delta_n(x, X_i, \theta)$$
$$= O\{(h^3 + \delta_\theta^2)n^\epsilon\}(1 + |x|^3). \tag{4.27}$$

For the noise term, we have from (4.23)

$$\{n\Sigma_n^\theta(x)\}^{-1} \sum_{i=1}^n K_h(\theta^\top X_{ix}) \begin{pmatrix} 1 \\ \theta^\top X_{ix}/h \end{pmatrix} \varepsilon_i$$
$$= \{nf_\theta(\theta^\top x)\}^{-1} \sum_{i=1}^n K_h(\theta^\top X_{ix}) \begin{pmatrix} 1 \\ \theta^\top X_{ix}/h \end{pmatrix} \varepsilon_i + O(\tau_n\delta_n n^\epsilon). \tag{4.28}$$

Combining (4.23), (4.24), (4.25), (4.26), (4.27) and (4.28), we complete the proof of Lemma 4.3. □

Acknowledgments

The work is partially supported by NUS FRG R-155-000-048-112.

References

Carroll, R.J., Fan, J., Gijbels, I. and Wand, M. P. (1997). Generalized partially
    linear single-Index models. *J. Amer. Stat. Assoc.*, **92**, 477-489.

Delecroix, M., Hristache, M. and Patilea, V. (2005). On semiparametric M-
    estimation in single-index regression. *Journal of Statistical Planning and
    Inference*, to appear.

Delecroix, M., Härdle, W. and Hristache, M. (2003). Efficient estimation in con-
    ditional single-index regression. *Journal of Multivariate Analysis*, **86**,
    213-226.

Fan, J. and Gijbels, I. (1996) *Local Polynomial Modelling and Its Applications.*
    Chapman & Hall, London.

Härdle, W., Hall, P. and Ichimura, H. (1993) Optimal smoothing in single-index
    models. *Ann. Statist.*, **21**, 157-178.

Härdle, W. and Stoker, T. M. (1989) Investigating smooth multiple regression by
    method of average derivatives. *J. Amer. Stat. Ass.* **84** 986-995.

Horowitz, J. L. and Härdle, W. (1996) Direct semiparametric estimation of
    single-index models with discrete covariates. *Journal of the American
    Statistical Association* **91** 1632-1640.

Hristache, M., Juditski, A. and Spokoiny, V. (2001). Direct estimation of the index
    coefficients in a single-index model, *Annals of Statistics* **29** 595-623.

Hristache, M., Juditski, A, Polzehl, J., Spokoiny, V. (2001). Structur adaptive
    approach for dimension reduction. *Annals of Statistics* **29** 1537–1566.

Ichimura, H. (1993) Semiparametric least squares (SLS) and weighted SLS esti-
    mation of single-index models. *Journal of Econometrics* **58** 71-120.

Klein, R. W. and R. H. Spady (1993) An Efficient Semiparametric Estimator for
    Binary Response Models. *Econometrica*, **61**, 387–421.

Li, K. C. (1991) Sliced inverse regression for dimension reduction (with discus-
    sion). *Amer. Statist. Ass.*, **86**, 316-342.

Powell, J. L., Stock, J. H. and Stoker, T. M. (1989) Semiparametric estimation
    of index coefficients. *Econometrica* **57** 1403-1430.

Samarov, A. M. (1993) Exploring regression structure using nonparametric func-
    tional estimation. *Journal of the American Statistical Association*, **88**,
    836-847.

Sherman, R. P. (1994) U-processes in the analysis of a generalized semiparametric
    regression estimator. *Econometric Theory*, **10**, 372-395.

Weisberg, S. and Welsh, A. H. (1994) Adapting for the missing link. *Annals of
    Statistics*, **22**, 1674-1700.

Xia, Y., and Härdle, W. (2002) Semi-parametric estimation of generalized par-
    tially linear single-index models. Discussion paper n.2002-56, SFB373,
    Berlin.

Xia, Y., Tong, H. and Li, W. K. (2002). Single-index volatility models and estimation. Statistica Sinica, 12, 785-799.

Xia, Y., Tong, H., Li, W. K. and Zhu, L. (2002) An adaptive estimation of dimension reduction space (with discussions). *J. Roy. Statist. Soc. B.*, **64**, 363-410.

Xia, Y. (2005) A constructive approach to the estimation of dimension reduction directions. Manuscript.

CHAPTER 5

Estimating Function Based Cross-Validation

M.J. van der Laan and Dan Rubin

Division of Biostatistics
University of California, Berkeley
140 Warren Hall #7360, Berkeley, CA 94720-7360
laan@stat.berkeley.edu

Suppose we observe a sample of independent and identically distributed realizations of a random variable. Given a model for the data generating distribution, assume that the parameter of interest can be characterized as the parameter value which makes the population mean of a possibly infinite dimensional estimating function equal to zero. Given a collection of candidate estimators of this parameter, and specification of the vector estimating function, we propose a norm of the cross-validated estimating equation as criteria for selecting among these estimators. For example, if we use the Euclidean norm, then our criteria is defined as the Euclidean norm of the empirical mean over the validation sample of the estimating function at the candidate estimator based on the training sample. We establish a finite sample inequality of this method relative to an oracle selector, and illustrate it with some examples. This finite sample inequality provides us also with asymptotic equivalence of the selector with the oracle selector under general conditions. We also study the performance of this method in the case that the parameter of interest itself is pathwise differentiable (and thus, in principle, root-n estimable).

5.1. Introduction

Suppose that one observes n independent and identically distributed copies of a random variable (referred to as the experimental unit), and that one wishes to learn from this data set a parameter of the distribution of this random variable. It is often possible to generate a (potentially large) set of estimators of this parameter of interest indexed by one or more fine tuning parameters such as quantities measuring the degree in which the estimator is informed by the actual data (instead of by priors or modeling assumptions). These estimators differ in variance and bias. An important problem in statistics is the construction of a selector based on the data, which selects among these candidate estimators, so that the corresponding data adaptively selected estimator behaves asymptotically well relative to an optimally selected estimator based on actually knowing the truth. The latter type of selector is often referred to as an oracle selector.

Cross-validation is a particular selection method which has been extensively studied in the past in the context of density estimation (for example, bandwidth selection for kernel density estimators, and model selection for model-specific maximum likelihood estimators), and regression. As discussed by Breiman (1996a) in the context of dimensionality selection in regression, criteria such as Mallow's C_p, Akaike information's criterion (AIC), and the Bayesian information criterion (BIC), do not account for the data-driven selection of the sequence of models and thus provide biased assessment of prediction error in finite sample situations. Instead, risk estimation methods based on sample reuse have been favored. The main procedures include: leave-one-out cross-validation, V-fold cross-validation, Monte Carlo cross-validation, and the bootstrap (Chapter 3 in Breiman, Friedman, Olshen, and Stone (1984), Breiman and Spector (1992), Breiman (1996a), Breiman (1996b), Chapter 17 in Erfon and Tibshirani (1993), Chapters 7 and 8 in Györfi, Kohler, Krzyźak and Walk (2002), Chapter 7 in Hastie, Tibshirani and Friedman (2001), Chapter 3 in Ripley (1996), Stone (1974), and Stone (1977)). Thus, a variety of cross-validation procedures are available for estimating the risk of a predictor. A natural question then concerns the distributional properties of the resulting risk estimators, i.e. their performance as estimators of generalization error, their performance in terms of identifying a good predictor (model selection), and also the impact of the particular cross-validation procedure (e.g. the choice of V in V-fold cross-validation, the use of V-fold vs. Monte Carlo cross-validation). Aside from empirical assessment of different estimation procedures, most of the previous theoretical work has focused primarily on the distributional properties of leave-one-out cross-validation (see Stone (1974) and Stone (1977)).

There is a rich literature on leave-one-out cross-validation in nonparametric univariate regression. For example, Silverman (1984) proposes a fast approximation of the leave-one out cross-validation method in spline regression. We refer to Härdle (1993) for an overview on the leave-one-out cross-validation method in kernel regression. In particular, Härdle and Marron (1985a) and Härdle and Marron (1985b) establish an asymptotic optimality result for leave-one-out cross-validation for choosing the smoothing parameter in nonparametric kernel regression (see p. 158, Härdle (1993)).

Györfi, Kohler, Krzyżak and Walk (2002) established a finite sample result for the single-split cross-validation selector for the squared error loss function. Their theorem was generalized in Dudoit and van der Laan (2003) to general cross-validation schemes and a general class of loss functions. Dudoit and van der Laan (2003) examine the distributional properties of cross-validated risk estimators in the context of both predictor selection and predictor performance assessment for a general class of loss functions.

Finite sample inequalities and asymptotic optimality results for likelihood-based V-fold cross-validation, or equivalently, cross-validation for the purpose of selection among density estimators of the density of the observed data, are established in van der Laan, Dudoit, Keles (2004), in which we also provide an overview of the literature on likelihood based cross-validation (see also Silverman (1984)), which is omitted here for the sake of space.

These cross-validation methods are focused on regression or density estimation. In van der Laan and Dudoit (2003) we generalized cross-validation to a selector among candidate estimators of any parameter which is represented as the minimizer over the parameter space of an expectation of a loss function of the experimental unit and a candidate parameter value, where the loss function is possibly indexed by an unknown nuisance parameter of the true data generating distribution. It is illustrated that this unified loss-based cross-validation approach solves a wide range of estimator selection problems, including estimator selection based on censored data, and, in particular, estimator selection in causal inference problems (in causal inference the observed data structure is modeled as a missing data structure on potential counterfactual random variables). Given a loss function for the full data structure, it is shown that the loss function for censored data structures can be defined as the Inverse Probability of Censoring Weighted (IPCW) or double robust IPCW full data loss function, where these mappings from full data functions to functions of the censored data are defined in van der Laan and Robins (2002). Various applications of this unified cross-validation methodology for estimator selection are selection among regression estimators (Dudoit and van der Laan (2003)), estimator selection with right censored data (Keles, van der Laan, and Dudoit (2003)), likelihood-based cross-validation (van der Laan, Hubbard, and Jewell (2004)), and tree-based estimation and model selection with censored data (Molinaro, Dudoit, and van der Laan (2004)).

For many problems it is easy to generate estimating functions identifying the parameter of interest, while it requires an additional possibly involved step to compute a corresponding loss function. Therefore, the new estimating function based cross-validation methodology presented in this article provides an important new general method for estimator selection.

Organization: The organization of this article is as follows. Firstly, in Section 2 we present our proposed estimating function based cross-validation selector. In Section 3 we present some examples in order to illustrate the methodology. The first example considers the case that the parameter of interest is a mean $Er(X)$ for some given function r, such as the cumulative distribution function at a point, and the candidate estimators are substitution estimators based on kernel density estimators. Under some conditions on r, we show that our method data adaptively undersmoothes the bandwidth so that the cross-validation selected substitution estimators are still asymptotically efficient. In our accompanying technical report we provide a general theorem proving this result for general pathwise differentiable parameters, and it is illustrated with additional examples. In Section 4 we derive a general finite sample inequality for its performance relative to the oracle selector, where we note that this result is most relevant for parameters which cannot be estimated at the root-n rate. We also present the wished corollaries of this general finite sample inequality and the corresponding asymptotic implications for our selector relative to the oracle selector. The fundamental lemmas are presented and proved in the appendix.

5.2. Estimating Function Based Cross-Validation

Suppose we observe n i.i.d. random variables $X_1, \ldots, X_n$ with common distribution P_0. Let $\mathcal{M}$ be the statistical model: that is, it is known that $P_0 \in \mathcal{M}$. Suppose that $\Psi : \mathcal{M} \to D(\mathcal{S})$ is the parameter of interest, where $D(\mathcal{S})$ denotes a space of real valued functions on a set $\mathcal{S}$. For example, if $\mathcal{S} = \{1, \ldots, d\}$, then $D(\mathcal{S}) = \mathbb{R}^d$ is simply the Euclidean space, but if $\mathcal{S}$ is a Euclidean set in $\mathbb{R}^d$, then $D(\mathcal{S})$ denotes the class of real valued d-variate functions. Let $\boldsymbol{\Psi} \equiv \{\Psi(P) : P \in \mathcal{M}\} \subset D(\mathcal{S})$ denote the parameter space.

Let $(X, \psi, \gamma) \to D_b(X \mid \psi, \gamma)$ be an estimating function indexed by a b ranging over a countable set $\mathcal{B}$. We note that an estimating function is simply a well defined real valued function on the tensor product of a support of the observed data structure X, the parameter space $\boldsymbol{\Psi}$, and a nuisance parameter space. Suppose that this set of estimating functions are unbiased in the sense that

$$E_0 D_b(X \mid \Psi(P_0), \Gamma(P_0)) = 0 \text{ for all } b \in \mathcal{B},$$

where $\Gamma : \mathcal{M} \to \{\Gamma(P) : P \in \mathcal{M}\}$ is the nuisance parameter, and $\gamma_0 = \Gamma(P_0)$ denotes its true value. Let $D(X \mid \psi, \gamma) \equiv (D_b(X \mid \psi, \gamma) : b \in \mathcal{B})$ denote the vector-valued (possibly infinite dimensional) estimating function. The heuristic of our method requires that the estimating functions D_b are appropriately standardized so that for each $b \in \mathcal{B}$

$$P_0 D_b(X \mid \psi, \Gamma(P_0)) = -(\psi_b - \psi_{b0}) + o(\mid \psi_b - \psi_{b0} \mid), \qquad (5.1)$$

for real valued parameters ψ_b of ψ, where ψ_{b0} denotes the true parameter value. That is, formally, $\psi_b = \Phi_b(\psi, P_0)$, and $\psi_{b0} = \Phi_b(\psi_0, P_0)$ for some real valued mapping Φ_b. Here we used the notation $P_0 f = \int f(x) dP_0(x)$, and if $f = (f_1, \ldots, f_d)$ is a vector function, then $P_0 f = (P_0 f_1, \ldots, P_0 f_d)$. In the next subsection, we present a general method for construction of such a vector-valued estimating function $D(X \mid \psi, \gamma)$ in which the estimating function D_b is directly derived from the b-specific efficient influence curve for a pathwise differentiable parameter $\Psi_b : \mathcal{M} \to \mathbb{R}$, $b \in \mathcal{B}$. In particular, we will point out that (5.1) can be arranged to hold exactly (no remainder) in a large class of problems in which (e.g.) Ψ_b is a linear parameter on a convex model $\mathcal{M}$, thereby also covering censored data models in which the full data model is convex.

Let $\| \cdot \|$ denote a particular norm on vectors $(x(b) : b \in \mathcal{B})$ with real valued components $x(b)$ of the dimension $\mid \mathcal{B} \mid$. If $\mathcal{B}$ is infinite (but countable), then $\| \cdot \|$ denotes a norm on the infinite dimensional Euclidean space $\mathbb{R}^\infty$. For example, we could use the weighted Euclidean norm

$$\|x\| = \sqrt{\sum_b w(b) x(b)^2}$$

for a known weight function $b \to w(b) \geq 0$ such that $\sum_b w(b) < \infty$. We define a risk function at P_0 as a norm of the expectation (under P_0) of the vector estimating function $D(X \mid \psi, \Gamma(P_0)) = (D_b(X \mid \psi, \Gamma(P_0)) : b)$:

$$\Theta(\psi \mid P_0) \equiv \|P_0 D(X \mid \psi, \Gamma(P_0))\|.$$

For example, we can define a risk function at P_0 as the weighted Euclidean norm of the expectation (under P_0) of the estimating function at candidate $\psi \in \boldsymbol{\Psi}$:

$$\Theta(\psi \mid P_0) \equiv \sqrt{\sum_{b \in \mathcal{B}} w(b) P_0^2 D_b(X \mid \psi, \Gamma(P_0))},$$

where $b \to w(b) \geq 0$ is a known weight function. We note that, if $\mathcal{B}$ is infinite, then $\Theta(\psi \mid P_0)$ needs to be defined as an infinite sum, and thereby as a limit.

Note that $\Theta(\psi_0 \mid P_0) = 0$ so that the corresponding measure of dissimilarity $d(\psi, \psi_0) \equiv \Theta(\psi \mid P_0) - \Theta(\psi_0 \mid P_0)$ between ψ and ψ_0 is indeed always non-negative and minimized at ψ_0:

$$d(\psi, \psi_0) \geq 0, \text{ and } d(\psi, \psi_0) = 0 \text{ if } \psi = \psi_0.$$

In addition, because of Property (5.1), we have for $\psi = \Psi(P)$ close to ψ_0,

$$\Theta(\psi \mid P_0) = \|(\psi_b : b) - (\psi_{b0} : b) + (o(\mid \psi_b - \psi_{b0} \mid) : b)\|, \tag{5.2}$$

where $\psi_b = \Phi_b(\psi, P_0)$ and $\psi_{b0} = \Phi_b(\psi_0, P_0)$. For example, if we use the Euclidean norm, then locally one expects to have

$$\Theta(\psi \mid P_0) \approx \sqrt{\sum_{b \in \mathcal{B}} w(b)(\psi_b - \psi_{b0})^2}.$$

In particular, if (5.1) holds exactly, then we have equality:

$$\Theta(\psi \mid P_0) = \|(\psi_b : b) - (\psi_{b0} : b)\|.$$

This first order expansion of this risk function $\psi \to \Theta(\psi \mid P_0)$ at ψ_0 suggests that $\Theta(\psi \mid P_0)$ is a sensible risk function for the purpose of estimation of $\psi_0 = \Psi(P_0)$, and, in particular, for selecting among candidate estimators of ψ_0.

Specifically, let $B_n \in \{0, 1\}^n$ be the random variable defining the cross-validation scheme, where $B_n(i) = 1$ indicates that observation i is a member of the validation sample, and $B_n(i) = 0$ indicates that observation i is a member of the training sample. Let $p \equiv P(B_n(i) = 1)$ denote the proportion of the learning sample which constitutes the validation sample. It is assumed that B_n is independent of the learning sample $(X_1, \ldots, X_n)$. Given a realization b_n of B_n, let P^0_{n,b_n} and P^1_{n,b_n} denote the empirical distribution of the training sample and validation sample, respectively.

Given candidate estimators $P_n \to \hat{\Psi}_k(P_n)$, $k = 1, \ldots, K(n)$, the cross-validated risk function is now defined as

$$\hat{\Theta}_{n(1-p)}(k) \equiv E_{B_n} \|P^1_{n,B_n} D(\cdot \mid \hat{\Psi}_k(P^0_{n,B_n}), \hat{\Gamma}(P^0_{n,B_n}))\|.$$

For example, in case we use the Euclidean norm, then we have

$$\hat{\Theta}_{n(1-p)}(k) \equiv E_{B_n} \sqrt{\sum_{b \in \mathcal{B}} \left(\sum_{i:B_n(i)=1} D_b(X_i \mid \hat{\Psi}_k(P^0_{n,B_n}), \hat{\Gamma}(P^0_{n,B_n}))/np \right)^2 w(b)}.$$

This cross-validated risk function defines our proposed cross-validation selector

$$\hat{k} = K(P_n) \equiv \arg\min_k \hat{\Theta}_{n(1-p)}(k).$$

This finishes the description of our proposed cross-validation selector among candidate estimators of a parameter $\psi_0 = \Psi(P_0)$.

Benchmark selector.

A natural way to benchmark the selector $\hat{k}$ is to define the following true conditional risk function

$$\tilde{\Theta}_{n(1-p)}(k) = E_{B_n} \| P_0 D(\cdot \mid \hat{\Psi}_k(P^0_{n,B_n}), \gamma_0) \|. \tag{5.3}$$

If we use the Euclidean norm, then this equals

$$\tilde{\Theta}_{n(1-p)}(k) = E_{B_n} \sqrt{\sum_{b \in \mathcal{B}} \left(P_0 D_b(\cdot \mid \hat{\Psi}_k(P^0_{n,B_n}), \gamma_0) \right)^2 w(b)}. \tag{5.4}$$

We can now define the corresponding oracle selector

$$\tilde{k} = \tilde{k}_{n(1-p)} = \tilde{K}_{n(1-p)}(P_n) \equiv \arg\min_k \tilde{\Theta}_{n(1-p)}(k)$$

for selecting among estimators based on a sample of size $n(1-p)$. In particular, if (5.1) holds exactly, then

$$\tilde{k} = \arg\min_k E_{B_n} \| (\Phi_b(\hat{\Psi}_k(P^0_{n,B_n}), P_0) : b) - (\Phi_b(\psi_0, P_0) : b) \|.$$

Finally, we also define the wished oracle selector

$$\tilde{k}_n = \tilde{K}(P_n) \equiv \arg\min_k \tilde{\Theta}_n(k),$$

where

$$\tilde{\Theta}_n(k) \equiv \| P_0 D_b(\cdot \mid \hat{\Psi}_k(P_n), \gamma_0) \|,$$

which compares estimators based on the whole learning sample P_n. In the case of the Euclidean norm this equals

$$\tilde{\Theta}_n(k) \equiv \sqrt{\sum_{b \in \mathcal{B}} \left(P_0 D_b(\cdot \mid \hat{\Psi}_k(P_n), \gamma_0) \right)^2 w(b)}.$$

5.2.1. *Method for construction of vector-estimating function.*

In this subsection we present a general method for constructing such a vector estimating function $D(X \mid \psi, \gamma)$. Firstly, one specifies a collection of real valued pathwise differentiable parameters $\Psi_b : \mathcal{M} \to \mathbb{R}$ indexed by $b \in \mathcal{B}$, so that for all $P \in \mathcal{M}$, $(\Psi_b(P) : b \in \mathcal{B})$ identifies $\Psi(P)$ uniquely. For example, if $\Psi(P) = (\Psi_1(P), \dots, \Psi_d(P))$ is itself already a Euclidean pathwise differentiable parameter, then we would simply define $\Psi_b(P)$ as the b-th component of $\Psi(P)$, $b = 1, \ldots, d$. On the other hand, if $\Psi(P)$ is an infinite dimensional function, then one could, for example, define $\Psi_b(P)$ as the inner product of $\Psi(P)$ with a b-specific basis function. It is assumed that for each $b \in \mathcal{B}$, $\Psi_b(P) = \Phi_b(\Psi(P), P)$ for all $P \in \mathcal{M}$ for some mapping Φ_b. The definition of pathwise differentiability states that Ψ_b is pathwise differentiable at P, relative to a specified set of one-dimensional differentiable submodels $\{P_{\epsilon,s} : \epsilon \in (-\delta, \delta)\} \subset \mathcal{M}$, satisfying $P_{0,s} = P$, with score at $\epsilon = 0$ equal to s, $s \in \mathcal{S} \subset L_0^2(P)$, if for all these submodels

$$\left. \frac{d}{d\epsilon} \Psi_b(P_{\epsilon,s}) \right|_{\epsilon=0} = \langle \ell_b, s \rangle_P \equiv E_P \ell_b(X) s(X)$$

for some $\ell_b \in L_0^2(P)$. We recall that $L_0^2(P)$ is the Hilbert space of real valued functions of X with mean zero endowed with inner product $\langle h_1, h_2 \rangle_P \equiv E_P h_1(X) h_2(X)$ being the covariance operator. Here ℓ_b is called a gradient of the pathwise derivative, whose projection onto the tangent space, that is, the closure of the linear span of $\mathcal{S}$ within the Hilbert space $(L_0^2(P), \langle \cdot, \cdot \rangle_P)$, is unique, and this projection is called the canonical gradient. The canonical gradient is also called the efficient influence curve since a regular asymptotically linear estimator with influence curve equal to canonical gradient is asymptotically efficient, by efficiency theory (Bickel, Ritov Klaassen, and Wellner (1997)). We refer to Bickel, Ritov, Klaassen, and Wellner (1997) for a comprehensive treatment of efficiency theory and illustration with many semi-parametric models, and, in the context of censored data, we refer to van der Laan and Robins (2002) who provide general representations of the class of all gradients, and, of the canonical gradient/efficient influence curve. Let $\Gamma_b(X \mid P_0)$ be a particular gradient, such as this unique canonical gradient. This gives us now a class of functions $\Gamma_b(X \mid P_0)$, $b \in \mathcal{B}$, and it is known that $E_{P_0} \Gamma_b(X \mid P_0) = 0$ for all $b \in \mathcal{B}$, and all $P_0 \in \mathcal{M}$.

Given a gradient representation $\Gamma_b(X, P)$ for all $P \in \mathcal{M}$, it is often not hard to define an actual estimating function $(X, \psi_b, \rho_b) \to D_b(X \mid \psi_b, \rho_b)$ for the parameter ψ_b, possibly depending on a nuisance parameter ρ_b such that $D_b(X \mid \Psi_b(P_0), \rho_{b0}) = \Gamma_b(X \mid P_0)$ for all $P_0 \in \mathcal{M}$, where ρ_{b0} denotes the true value of the nuisance parameter. Since a gradient is, by definition, orthogonal to all nuisance scores, that is, scores of one-dimensional sub-models for which $\frac{d}{d\epsilon} \Psi_b(P_{\epsilon,s})|_{\epsilon=0} = 0$, this estimating function $D_b(X \mid \psi_b, \rho_b)$ for Ψ_b is *minimally dependent on nuisance parameters*: we refer to Chapter 1, Sections 1.4, Lemma 1.2 and 1.3 in van der Laan and Robins (2002) for formal results establishing that the directional derivatives w.r.t. ρ_b are zero, given that ψ_b and ρ_b are variation independent parameters.

In addition, as a consequence of the fact that $\Gamma_b(X \mid P_0)$ is a gradient of the pathwise derivative of the parameter $\Psi_b : \mathcal{M} \to \mathbb{R}$, the estimating function will have the property

$$E_0 D_b(X \mid \psi_b, \rho_{b0}) = -(\psi_b - \psi_{b0}) + o(|\psi_b - \psi_{b0}|). \qquad (5.5)$$

This is a general property for gradients (also called influence curves) of a pathwise derivative, and, in particular, for the canonical gradient (see Lemmas 1.2 and 1.3 in van der Laan and Robins (2002)). In particular, for linear parameters Ψ_b on convex models, one obtains an exact equality (see Klaassen (1987), Chapter 2 van der Laan (1996), van der Laan (1995), and van der Laan (1998)):

$$E_0 D_b(X \mid \psi_b, \rho_{b0}) = -(\psi_b - \psi_{b0}). \qquad (5.6)$$

The fact that the derivative of $E_0 D_b(X \mid \psi_b, \rho_{b0})$ at $\psi_b = \psi_{b0}$ equals minus the identity provides us with the motivation for our proposed risk function.

Finally, let $(X, \psi, \gamma) \to D_b(X \mid \psi, \gamma)$ be an estimating function satisfying

$$D_b(X \mid \Psi(P_0), \Gamma(P_0)) = D_b(X \mid \psi_{b0}, \rho_{b0}) = \Gamma_b(X \mid P_0),$$

for all $b \in \mathcal{B}$, $P_0 \in \mathcal{M}$. At this step, one can use that $\Psi_b(P) = \Phi_b(\Psi(P), P)$ in order to represent an estimating function in ψ_b (as implied by $D_b(X \mid \psi_b, \rho_b)$) in terms of an estimating function in terms of ψ.

By now, we have succeeded in deriving a class of unbiased estimating functions $(X, \psi, \gamma) \to D_b(X \mid \psi, \gamma)$, with nuisance parameter γ, indexed by $b \in \mathcal{B}$, satisfying the desired Property (5.2).

5.3. Some Examples

5.3.1. *Example: substitution estimators based on kernel density estimators*

Let $X \sim f_0$, the model is nonparametric, the parameter of interest is $\psi_0 = E_0 r(X)$ for some function r, and suppose we observe n i.i.d. observations $X_1, \ldots, X_n$ of X. The estimating function for ψ_0, derived from the efficient influence curve $D(X \mid P_0) = r(X) - \psi_0$ at P_0, is given by $D(X \mid \psi) = r(X) - \psi$. Let $\hat{\Psi}_h(P_n)$ be the mean of $r(X)$ w.r.t. a kernel density estimator $1/nh \sum_i K((X_i - \cdot)/h)$, with density kernel K, and let $\hat{\Psi}_0(P_n) = \int r(x) dP_n(x)$ be the empirical mean of $r(X)$. Suppose that the kernel K does not have mean zero so that the bias of the estimators $\hat{\Psi}_h(P_n)$ is linear in h: if we work with orthogonal kernels, we would define h as a power of the bandwidth so that h still represents the bias of the kernel density estimator. In this case, the estimators $\hat{\Psi}_h(P_n)$ have an asymptotic bias $O(h)$, which thus only disappears at $\sqrt{n}$-rate if $h = o(1/\sqrt{n})$. Our cross-validation criterion is defined as

$$\hat{\Theta}_{n(1-p)}(h)^2 = E_{B_n}\left(\hat{\Psi}_0(P^1_{n,B_n}) - \hat{\Psi}_h(P^0_{n,B_n})\right)^2, \qquad (5.7)$$

and the cross-validation selector h_n of h is its minimizer. For example, if B_n represents a leave-one out cross-validation scheme, then this criteria would resemble standard leave-one out cross-validation, except where the outcome is replaced by r:

$$\hat{\Theta}_{n(1-p)}(h)^2 = \frac{1}{n}\sum_{i=1}^n\left(r(X_i) - \hat{\Psi}_h(P_{n,-i})\right)^2.$$

Note that (5.7) represents a slight modification of our general proposal in the sense that we put the E_{B_n} inside the square root (but still outside the squares). It simplifies the algebraic manipulations needed to establish the desired result for general pathwise differentiable parameters as presented in our accompanying technical report. We have $\hat{\Theta}_{n(1-p)}(h)^2 = E_{B_n}\left(\psi_0 - \hat{\Psi}_h(P^0_{n,B_n})\right)^2$. That is, the oracle selector $\tilde{k}_{n(1-p)}$ corresponds with selecting the estimator whose training sample realizations are closest to the true value ψ_0.

In this example $\hat{\Psi}_0$ is an estimator which does not use any smoothing. The purpose of smoothing in this example (that is, selecting a $h > 0$) is to obtain a finite sample improvement relative to $\hat{\Psi}_0$. For example, if it is known that the true cumulative distribution function is very smooth, then it makes sense to use a smooth estimator, even though the discrete empirical cumulative distribution function is already asymptotically efficient in the nonparametric model. However, in order to remain efficient the bandwidth will have to be chosen so that the asymptotic bias is of smaller order than $1/\sqrt{n}$: that is, we wish to show that $h_n = o_P(1/\sqrt{n})$.

We note

$$\hat{\Psi}_h(P_n) = \frac{1}{nh}\sum_{i=1}^n\int r(x)K((X_i - x)/h)dx = \frac{1}{n}\sum_{i=1}^n r_h(X_i),$$

where $r_h(X) \equiv \int r(X_i + yh)K(y)dy$. We also note that the derivative of

$\hat{\Theta}_{n(1-p)}(h)^2$ w.r.t. h is given by

$$U(h, P_n) = 2E_{B_n} \left(\hat{\Psi}_0(P^1_{n,B_n}) - \hat{\Psi}_h(P^0_{n,B_n}) \right) P^0_{n,B_n} \dot{r}_h,$$

where $\dot{r}_h \equiv \frac{d}{dh} \int r(\cdot + yh)K(y)dy$. We will follow the proof of Theorem 1 in our technical report in order to illustrate it in this example. We have $U(h_n, P_n) = 0$, and we note that $U(h_0 = 0, P_0) = 0$, where $U(h, P_0)$ is defined by replacing P^1_{n,B_n} and P^0_{n,B_n} by P_0. The equations $U(0, P_0) = 0$ and $U(h_n, P_n) = 0$ provides us with a basis for establishing that $h_n = o_P(1/\sqrt{n})$, and thereby that $\hat{\Psi}_{h_n}$ is still asymptotically efficient.

Firstly, we note that

$$U(h_n, P_0) - U(0, P_0) = -\{U(h_n, P_n) - U(h_n, P_0)\}.$$

Since

$$U(h, P) = 2 \left(\hat{\Psi}_0(P) - \hat{\Psi}_h(P) \right) P\dot{r}_h,$$

and $\lim_{h\to 0} P_0\dot{r}_h = P_0\dot{r}_0$ is assumed to exist and to be positive, it follows that

$$\frac{d}{dh}U(h, P_0)\bigg|_{h=0} = -2 \left(\lim_{h\to 0} P_0\dot{r}_h \right)^2 < 0,$$

which verifies that the derivative of $h \to U(h, P_0)$ at $h = 0$ is bounded away from zero. Below, we will show that h_n converges to zero in probability. Then, it follows

$$h_n = \left(-\frac{d}{dh}U(h, P_0)\bigg|_{h=0} \right)^{-1} \{U(h_n, P_n) - U(h_n, P_0)\} + o(h_n).$$

Below, we will also show that

$$\{U(h_n, P_n) - U(h_n, P_0)\} - \{U(0, P_n) - U(0, P_0)\} = o_P(1/\sqrt{n}) + o_P(h_n). \tag{5.8}$$

Then, it follows that

$$h_n = \left(-\frac{d}{dh}U(h, P_0)\bigg|_{h=0} \right)^{-1} \{U(0, P_n) - U(0, P_0)\} + o_P(h_n) + o_P(1/\sqrt{n}).$$

Now, we note that

$$U(0, P_n) = E_{B_n} \left(P^1_{n,B_n}r - P^0_{n,B_n}r \right) P^0_{n,B_n} \dot{r}_0.$$

Substitute in the latter expression $P^0_{n,B_n}\dot{r}_0 = (P^0_{n,B_n} - P_0)\dot{r}_0 + P_0\dot{r}_0$, and note that the term resulting from $P_0\dot{r}_0$ equals exactly zero. Thus,

$$U(0, P_n) = E_{B_n} \left((P^1_{n,B_n} - P_0)r - (P^0_{n,B_n} - P_0)r \right) (P^0_{n,B_n} - P_0)\dot{r}_0,$$

which is indeed $o_P(1/\sqrt{n})$ if $P_0 \dot{r}_0 < \infty$ and $P_0 r^2 < \infty$. To conclude, this shows that $h_n = o_P(h_n) + o_P(1/\sqrt{n})$, and hence the result $h_n = o_P(1/\sqrt{n})$.

Convergence of h_n to 0: We need to verify that $h_n = o_P(1)$. If r has compact support, it follows that $h_n \leq M$ for some $M < \infty$. As a consequence, by compactness of $[0, M]$, for each subsequence of h_n, there exists a subsequence (say) h_k which converges to a h_∞ for k converging to infinity. Since $U(h_k, P_k) = 0$ and $U(h_k, P_k) - U(h_k, P_0)$ converges to zero as k converges to infinity under the already needed Donsker class condition specified below, it follows that $U(h_k, P_0)$ converges to zero. In addition, it also follows that $U(h_k, P_0)$ converges to $U(h_\infty, P_0)$, which shows that $U(h_\infty, P_0) = 0$. This shows that $h_\infty = 0$. This proves that h_n converges to zero a.s., and thus, in particular, in probability.

Verification of (5.8): This second order difference can be written as a sum with the following four terms:

$$E_{B_n}(P^0_{n,B_n} - P_0)(r_{h_n} - r_0)P^0_{n,B_n}\dot{r}_{h_n}$$
$$E_{B_n}\left((P^1_{n,B_n} - P_0)r_0 - (P^0_{n,B_n} - P_0)r_{h_n}\right)P^0_{n,B_n}(\dot{r}_{h_n} - \dot{r}_0)$$
$$E_{B_n}P^0_{n,B_n}(r_{h_n} - r_0)(P^0_{n,B_n} - P_0)\dot{r}_{h_n}$$
$$E_{B_n}\left(P^1_{n,B_n}r_0 - P^0_{n,B_n}r_{h_n}\right)(P^0_{n,B_n} - P_0)(\dot{r}_{h_n} - \dot{r}_0).$$

Of course, $r_0 = r$. If $\int(r_{h_n} - r_0)^2(x)dP_0(x)$ converges to zero in probability (as follows from $h_n = o_P(1)$ and continuity of r), and if $\{r_h - r_0 : h \in [0,1]\}$ is a P_0-Donsker class, then it follows (van der Vaart and Wellner (1996)) that this $(P^0_{n,B_n} - P_0)(r_{h_n} - r_0) = o_P(1/\sqrt{n})$. Examples of P_0-Donsker classes are provided in van der Vaart and Wellner (1996): e.g. if r_h has variation smaller than a universal $M < \infty$, then $\{r_h - r_0 : h\}$ is a P_0-Donsker class.

We also assume that $\{\dot{r}_h : h > 0\}$ is a Glivenko-Cantelli class so that $\sup_h(P^0_{n,B_n} - P_0)\dot{r}_h = o_P(1)$. This proves that the first three terms are $o_P(1/\sqrt{n})$: for technical convenience, we assume a V-fold cross-validation scheme so that E_{B_n} only yields a sum of V terms, which each can be analyzed separately. Regarding the fourth term, first write

$$P^1_{n,B_n}r_0 - P^0_{n,B_n}r_{h_n} = (P^1_{n,B_n} - P_0)r_0 - (P^0_{n,B_n} - P_0)r_{h_n} + P_0(r_0 - r_{h_n}).$$

The terms resulting from the first term is $o_P(1/\sqrt{n})$ since $(P^1_{n,B_n} - P_0)r_0 = O_P(1/\sqrt{n})$ and $(P^0_{n,B_n} - P_0)(\dot{r}_{h_n} - \dot{r}_0) = o_P(1)$. Similarly, this follows for the term resulting from the second term. The term resulting from $P_0(r_0 - r_{h_n})$ is given by:

$$P_0(r_0 - r_{h_n})(P_n - P_0)(\dot{r}_{h_n} - \dot{r}_0).$$

It follows that $P_0(r_{h_n} - r_0) = O_P(h_n)$ so that the last term is $o_P(h_n)$.

Thus under these empirical process conditions on $\dot{r}_h$, we have proved that $h_n = o_P(1/\sqrt{n})$, and, consequently, that the substitution estimators based on an integrated kernel density estimator using bandwidth h_n will be asymptotically efficient. We will state this as a result below.

Inspection of the proof shows that we can replace the condition that $\{\dot{r}_h : h \in (0, M]\}$ is a P_0-Glivenko-Cantelli class by $(P_n - P_0)\dot{r}_{h_n} = o_P(1)$, which allows sharper results. Specifically, in our accompanying technical report we prove the desired result for the case that $r(x) = I_{[0,x_0]}(x)$, $\psi_0 = F_0(x_0)$ so that $\dot{r}_h = K((\cdot - x_0)/h)/h$ (which does not generate a Glivenko-Cantelli class), which proves that the data adaptively smoothed empirical distribution function is still asymptotically efficient.

Theorem 5.1: *Let X be a real valued random variable with density f_0 with compact support contained in $[0, M]$, and, given a function r which is continuous F_0-a.e., let $\psi_0 = E_0 r(X)$ be its parameter of interest. Suppose we observe n i.i.d. observations $X_1, \ldots, X_n$ of X. Let $\hat{\Psi}_b(P_n)$ be the mean of $r(X)$ w.r.t. a kernel density estimator $1/nb \sum_i K((X_i - \cdot)/b)$, with density kernel K, and let $\hat{\Psi}_0(P_n) = \int r(x) dP_n(x)$ be the empirical mean of $r(X)$. Let $h \to b(h)$ be a 1-1 parametrization with inverse $b \to h(b)$ satisfying*

$$\frac{d}{dh} P_0 \int r(\cdot + yb(h)) K(y) dy \bigg|_{h=0} \neq 0.$$

Let $r_h(X) \equiv \int r(X + yb(h)) K(y) dy$, and $\dot{r}_h \equiv \frac{d}{dh} r_h$, $h > 0$. Let

$$\hat{\Theta}_{n(1-p)}(h)^2 = E_{B_n} \left(\hat{\Psi}_0(P^1_{n,B_n}) - \hat{\Psi}_h(P^0_{n,B_n}) \right)^2,$$

and h_n be its minimizer over the interval $[0, M]$. Let B_n correspond with V-fold cross-validation for a fixed V. Assume that $\{\dot{r}_h : h \in (0, M]\}$ is a P_0-Glivenko-Cantelli class and $\{r_h : h \in [0, M]\}$ is a P_0-Donsker class. Then $h_n = o_P(1/\sqrt{n})$.

We remark here that in the above theorem, $h(b)$ represents the order of the bias of the integrated kernel density estimator with bandwidth b. For example, if the kernel K does not have mean zero, then one can choose $h = b$. Also, if r_h has variation over $[0, M]$ smaller than a universal $C < \infty$, then $\{r_h : h \in [0, M]\}$ is a P_0-Donsker class.

Our general results in the following section can be applied to the next two examples.

5.3.2. *Example: density estimation*

Let $X \sim P_0$ be a univariate random variable with density f_0. Suppose that the model is nonparametric, the parameter of interest is the density itself: $\Psi(P) = f$, $\psi_0 = \Psi(P_0) = f_0$. Let ϕ_b, $b = 1, \ldots$ be a countable orthonormal basis in the Hilbert space $L^2(dx)$ of square integrable functions w.r.t Lebesgue measure, where dx denotes the Lebesgue measure. Let $\Psi_b(P) = P\phi_b = \int \phi_b(x)f(x)dx$. Now, the efficient influence curve of $\Psi_b(P)$ is given by $\Gamma_b(X \mid P_0) = \phi_b(X) - E_0\phi_b(X)$. A corresponding estimating function is given by $D_b(X \mid \psi) = \phi_b(X) - \int \phi_b(x)\psi(x)dx$. Thus $\hat{\Theta}_{n(1-p)}(k) = E_{B_n}\|(P^1_{n,B_n}\phi_b - \int \phi_b(x)\hat{\Psi}_k(P^0_{n,B_n})(x)dx : b)\|$. Since $P_0 D_b(\cdot \mid \psi) = \int \phi_b(x)(\psi_0 - \psi)(x)dx = \psi_{b0} - \psi_b$, we have

$$\tilde{\Theta}_{n(1-p)}(k) = E_{B_n}\left\| \left(\int \phi_b(x)(\hat{\Psi}_k(P^0_{n,B_n}) - \psi_0)(x)dx : b \right) \right\|.$$

If we use the Euclidean norm, then

$$\tilde{\Theta}_{n(1-p)}(k) = \sqrt{\int \left\{ \hat{\Psi}_k(P^0_{n,B_n})(x) - \psi_0(x) \right\}^2 dx}$$

is simply the $L^2(dx)$-norm between the candidate density estimator $\hat{\Psi}_k(P^0_{n,B_n})$ and the true density ψ_0.

5.3.3. *Example: hazard estimation*

Let $X \sim P_0$ be a univariate random variable with density f_0, survival function S_0 and hazard $\lambda_0 = f_0/S_0$. Suppose that the model is nonparametric, and the parameter of interest is the hazard: $\Psi(P) = \lambda = f/S$, $\psi_0 = \Psi(P_0) = f_0/S_0$. Let ϕ_b, $b = 1, \ldots$ be a countable orthonormal basis in $L^2(dx)$. Let $\Psi_b(P) = \int \phi_b(x)\lambda(x)dx$, where λ denotes the hazard corresponding with probability distribution P. Now, the efficient influence curve of the real valued parameter $\Psi_b(P)$ at P_0 is given by

$$\Gamma_b(X \mid P_0) = \frac{\phi_b(X)}{S_0(X)} - \int_0^X \frac{\phi_b(x)}{S_0(x)}\lambda_0(x)dx.$$

A corresponding estimating function for a candidate hazard ψ is thus defined as

$$D_b(X \mid \psi, S_0) = \frac{\phi_b(X)}{S_0(X)} - \int_0^X \frac{\phi_b(x)}{S_0(x)}\psi(x)dx,$$

which is thus indexed by a root-n estimable nuisance parameter S_0. Thus, the cross-validation criterion $\hat{\Theta}_{n(1-p)}(k)$ is defined as

$$
E_{B_n} \left\| \left(P^1_{n,B_n} \phi_b / S^0_{n,B_n} - \int P^1_{n,B_n} I(x < \cdot) \frac{\phi_b(x)}{S^0_{n,B_n}(x)} \hat{\Psi}_k(P^0_{n,B_n})(x)dx : b \right) \right\|
$$

where S^0_{n,B_n} denotes an estimator of the survival function based on the training sample P^0_{n,B_n}. Since $P_0 D_b(\cdot \mid \psi, S_0) = \int \phi_b(x)(\psi_0 - \psi)(x)dx = \psi_{b0} - \psi_b$, we have that the target criterion equals

$$
\tilde{\Theta}_{n(1-p)}(k) = E_{B_n} \left\| \left(\int \phi_b(x)(\hat{\Psi}_k(P^0_{n,B_n}) - \psi_0)(x)dx : b \right) \right\|.
$$

If we use the Euclidean norm, then

$$
\tilde{\Theta}_{n(1-p)}(k) = \sqrt{\int \left\{ \hat{\Psi}_k(P^0_{n,B_n})(x) - \psi_0(x) \right\}^2 dx}
$$

is simply the $L^2(dx)$-norm between the candidate hazard estimator $\hat{\Psi}_k(P^0_{n,B_n})$ and the true hazard ψ_0. Finally, we note that our cross-validation selector for selection among hazard estimators would in first order not be affected by the nuisance parameter since the nuisance parameter S_0 can be estimated at a parametric rate, while the minimax rates for hazard estimation is worse than the root-n rate for all smoothness classes.

5.4. General Finite Sample Result

Recall

$$
\tilde{\Theta}_{n(1-p)}(k) \equiv E_{B_n} \| P_0 D(\cdot \mid \hat{\Psi}_k(P^0_{n,B_n}), \gamma_0) \|,
$$

$$
\hat{\Theta}_{n(1-p)}(k) \equiv E_{B_n} \| P^1_{n,B_n} D(\cdot \mid \hat{\Psi}_k(P^0_{n,B_n}), \hat{\Gamma}(P^0_{n,B_n})) \|,
$$

$$
\tilde{k} = \mathrm{argmin}_{k=1,\ldots,K(n)} \tilde{\Theta}_{n(1-p)}(k), \quad \hat{k} = \arg \min_{k=1,\ldots,K(n)} \hat{\Theta}_{n(1-p)}(k).
$$

For notational convenience, we define $D^0_{n,B_n}(X \mid \psi) \equiv D(X \mid \psi, \hat{\Gamma}(P^0_{n,B_n}))$ and $D(X \mid \psi) \equiv D(X \mid \psi, \gamma_0)$. We also define $G^1_{n,B_n} = \sqrt{np}(P^1_{n,B_n} - P_0)$ as the centered empirical process based on the validation sample identified by the sample split B_n.

The following theorem provides us with our most general finite sample result. It compares our cross-validation selector with the oracle selector in terms of the criterion $\tilde{\Theta}_{n(1-p)}$ which measures in first order the norm of the estimator minus the true parameter. Our subsequent results are derived by establishing bounds on the remainder terms in this theorem.

Theorem 5.2: We have

$$\tilde{\Theta}_{n(1-p)}(\hat{k}) \leq \tilde{\Theta}_{n(1-p)}(\tilde{k}) \tag{5.9}$$

$$+\frac{1}{\sqrt{np}} E_{B_n} \left\{ \|G^1_{n,B_n} D^0_{n,B_n}(\cdot \mid \hat{\Psi}_{\tilde{k}}(P^0_{n,B_n}))\| + \|G^1_{n,B_n} D(\cdot \mid \hat{\Psi}_{\hat{k}}(P^0_{n,B_n}))\| \right\}$$

$$+\frac{1}{\sqrt{np}} E_{B_n} \left\{ \|G^1_{n,B_n} (D^0_{n,B_n} - D)(\cdot \mid \hat{\Psi}_{\hat{k}}(P^0_{n,B_n}))\| \right\}$$

$$+2 \max_{k \in \{1,\dots,K(n)\}} E_{B_n} \|P_0(D^0_{n,B_n} - D)(\cdot \mid \hat{\Psi}_k(P^0_{n,B_n}))\|.$$

Proof: Firstly, we note that by the triangle inequality property of a norm, we have

$$E_{B_n} \|P_0 D(\cdot \mid \hat{\Psi}_{\hat{k}}(P^0_{n,B_n}))\| \leq E_{B_n} \|P^1_{n,B_n} D(\cdot \mid \hat{\Psi}_{\hat{k}}(P^0_{n,B_n}))\|$$

$$+\frac{1}{\sqrt{np}} E_{B_n} \|G^1_{n,B_n} D(\cdot \mid \hat{\Psi}_{\hat{k}}(P^0_{n,B_n}))\|.$$

The left-hand side equals $\tilde{\Theta}_{n(1-p)}(\hat{k})$, and the last term on the right-hand side represents one of the empirical process terms on the right-hand side of the Inequality (5.9) to be proved. We will now study the other term, and bound it by a sum of five terms, which results in the inequality (5.9). Repeated application (first and third inequality below) of the triangle inequality property of a norm, and the fact that by definition of $\hat{k}$ $\hat{\Theta}_{n(1-p)}(\hat{k}) \leq \hat{\Theta}_{n(1-p)}(\tilde{k})$ (second inequality below), provides us with the following series of inequalities:

$$E_{B_n} \|P^1_{n,B_n} D(\cdot \mid \hat{\Psi}_{\hat{k}}(P^0_{n,B_n}))\|$$

$$\leq E_{B_n} \|P^1_{n,B_n} D^0_{n,B_n}(\cdot \mid \hat{\Psi}_{\hat{k}}(P^0_{n,B_n}))\|$$

$$+E_{B_n} \|P^1_{n,B_n} (D^0_{n,B_n} - D)(\cdot \mid \hat{\Psi}_{\hat{k}}(P^0_{n,B_n}))\|$$

$$\leq E_{B_n} \|P^1_{n,B_n} D^0_{n,B_n}(\cdot \mid \hat{\Psi}_{\tilde{k}}(P^0_{n,B_n}))\|$$

$$+E_{B_n} \|P^1_{n,B_n} (D^0_{n,B_n} - D)(\cdot \mid \hat{\Psi}_{\hat{k}}(P^0_{n,B_n}))\|$$

$$\leq E_{B_n} \|(P^1_{n,B_n} - P_0) D^0_{n,B_n}(\cdot \mid \hat{\Psi}_{\tilde{k}}(P^0_{n,B_n}))\|$$

$$+E_{B_n} \|P_0 D^0_{n,B_n}(\cdot \mid \hat{\Psi}_{\tilde{k}}(P^0_{n,B_n}))\|$$

$$+E_{B_n} \|(P^1_{n,B_n} - P_0)(D^0_{n,B_n} - D)(\cdot \mid \hat{\Psi}_{\hat{k}}(P^0_{n,B_n}))\|$$

$$+E_{B_n} \|P_0(D^0_{n,B_n} - D)(\cdot \mid \hat{\Psi}_{\hat{k}}(P^0_{n,B_n}))\|.$$

Finally, again by the triangle inequality property, we have that the second term of this sum of 4 terms can be bounded as follows:

$$E_{B_n}\|P_0 D^0_{n,B_n}(\cdot \mid \hat{\Psi}_{\tilde{k}}(P^0_{n,B_n}))\| \leq E_{B_n}\|P_0(D^0_{n,B_n} - D)(\cdot \mid \hat{\Psi}_{\tilde{k}}(P^0_{n,B_n}))\|$$
$$+ E_{B_n}\|P_0 D(\cdot \mid \hat{\Psi}_{\tilde{k}}(P^0_{n,B_n}))\|,$$

where the latter equals $\tilde{\Theta}_{n(1-p)}(\tilde{k})$. Collection of all 5 terms yields the proof of the theorem. □

5.4.1. *Corollaries of Theorem 5.2*

We will present corollaries of Theorem 5.2 for the following three norms.

Definition 5.1: For a countable sequence of real numbers a_i and weights $w_b \geq 0$ with $\sum_{b=1}^{\infty} w_b = 1$, define the following norms:
$\|(a_1, a_2, ...)\|_q = (\sum_{b=1}^{\infty} w_b |a_b|^q)^{1/q}$, $q = 1, 2$, and $\|(a_1, a_2, ...)\|_\infty = \sup_{b \geq 1} |a_b|$.

The following corollary of Theorem 5.2 establishes that our cross-validation selector performs as well as the oracle selector $\tilde{k}$ up till a term of order $\log K(n)/np$ and a term r_n due to the estimation of the nuisance parameter.

We refer to van der Vaart and Wellner (1996) (Sections 2.2 and 2.5) for the definition of covering numbers $N(\cdot, \cdot, \cdot)$ and uniform entropies of function classes.

Corollary 5.1: *Let*

$$r_n \equiv 2 \max_{k \in \{1, ..., K(n)\}} E_{B_n}\|P_0(D^0_{n,B_n} - D)(\cdot \mid \hat{\Psi}_k(P^0_{n,B_n}))\|.$$

Assume that $\sup_{b \in B} |D_b(\cdot|\cdot)| \leq M < \infty$ *a.s. Define* $h_{n,K(n)}$ *as the maximum of the entropies* $\int_0^\infty \sup_Q \sqrt{\log N(\epsilon M^2, \mathcal{F}_n, L_2(Q))} d\epsilon$ *corresponding with the following choices of function classes:* $\mathcal{F}_n = \{D_b(\cdot|\hat{\Psi}_k(P^0_{n,B_n})) : b, k\}$, $\mathcal{F}_n = \{(D^0_{n,B_n})_b(\cdot|\hat{\Psi}_k(P^0_{n,B_n})) : b, k\}$, *and* $\mathcal{F}_n = \{(D^0_{n,B_n} - D)_b(\cdot|\hat{\Psi}_k(P^0_{n,B_n})) : b, k\}$, *where* $1 \leq b < \infty$, $1 \leq k \leq K(n)$, *and almost every* P^0_{n,B_n}.

Then for $\|\cdot\| = \|\cdot\|_q$, $q = 1, 2$, *we have for a universal constant* c *(only depending on* M)

$$E[\tilde{\Theta}_{n(1-p)}(\hat{k})] \leq E[\tilde{\Theta}_{n(1-p)}(\tilde{k})] + c\sqrt{\log K(n)}/\sqrt{np} + E[r_n].$$

For $\|\cdot\| = \|\cdot\|_\infty$, *we have*

$$E[\tilde{\Theta}_{n(1-p)}(\hat{k})] \leq E[\tilde{\Theta}_{n(1-p)}(\tilde{k})] + M^2 h_{n,K(n)}/\sqrt{np} + E[r_n].$$

Proof: Firstly, we take expectations on both sides of the inequality in Theorem 5.2. For the second, third, and fourth terms on the right side of the inequality (call them $E_{n,B_n}[U_{n,i}]$, i=1,2,3), we note that by Fubini's theorem,

$$E[E_{n,B_n}U_{n,i}] = E[U_{n,i}] = E_{B_n,P^0_{n,B_n}}E_{P^1_{n,B_n}}U_{n,i}.$$

Finally, we apply the inequalities from Lemma 5.8 below to the inner expectation for these three terms to obtain the desired result.   □

5.4.2. *Asymptotic implications*

The next corollary shows that the cross-validation selector $\hat{k}$ is asymptotically equivalent to the oracle selector $\tilde{k}_{n(1-p)}$ in the case that 1) the rate of convergence achieved by the oracle selector is worse than the almost parametric rate $\log K(n)/n$, and 2) the number $K(n)$ of candidate estimators is polynomial in n.

Our observation is based on the following trivial lemma.

Lemma 5.1: *Suppose that $a_n, b_n, c_n \in R$ are such that $0 < a_n \leq b_n \leq a_n + c_n$, and that $\limsup \frac{c_n}{a_n} = 0$. Then $\lim_{n \to \infty} a_n/b_n = 1$.*

Corollary 5.2: *Let r_n and $h_{n,K(n)}$ be defined as in Corollary 5.1, and assume that, in addition to the assumptions of Corollary 5.1, we have*

$$\frac{\max(r_n, \sqrt{\log K(n)/np})}{E[\tilde{\Theta}_{n(1-p)}(\tilde{k})]} \to 0 \ for \ n \to \infty. \tag{5.10}$$

Then

$$\lim \frac{E[\tilde{\Theta}_{n(1-p)}(\tilde{k})]}{E[\tilde{\Theta}_{n(1-p)}(\hat{k})]} = 1.$$

Proof: This is immediate from Corollary 5.1 and Lemma 5.1.   □

We note that the proportion p in Corollary 5.2 can be selected to converge to zero with sample size at a rate $p(n)$ so that $\log K(n)/np(n)$ remains of smaller order than $E[\tilde{\Theta}_{n(1-p)}(\tilde{k})]$. In this case, Corollary 5.2 provides also conditions under which the cross-validation selector is asymptotically equivalent with the oracle selector $\tilde{k}_n$.

5.4.3. *The Oracle selector in convex linear models*

Finally, we state here that our target criterion $\tilde{\Theta}_{n(1-p)}(\psi)$ actually reduces to the norm of the difference $\psi_0 - \psi$ in the case that the parameter Ψ is linear, and the parameter space is convex. This was made explicit in our examples in Section 3.

Corollary 5.3: *Suppose that*

$$P_0 D_b(\cdot \mid \psi, \gamma_0) = \psi_{b0} - \psi_b \text{ for all } b \in \mathcal{B}.$$

This holds, in particular, if $D_b(X \mid \psi_0, \gamma_0)$ is a gradient of a real valued linear parameter $\Psi_b : \mathcal{M} \to \mathbb{R}$ at P_0 (for all $P_0 \in \mathcal{M}$), $\mathcal{M}$ is convex, and the regularity conditions of Theorem 2.2 in van der Laan (1996) hold. Let ψ be representing the vector $(\psi_b : b \in \mathcal{B})$, so that $\|\psi - \psi_0\| \equiv \|(\psi_b : b) - (\psi_{b0} : b)\|$ is defined as the norm $\|\cdot\|$ of the corresponding vector.

Then in Corollaries 5.1 and 5.2, $E[\tilde{\Theta}_{n(1-p)}(\tilde{k})]$ and $E[\tilde{\Theta}_{n(1-p)}(\hat{k})]$ may be replaced by $E\|\hat{\Psi}_{\tilde{k}}(P^0_{n,B_n}) - \Psi(P_0)\|$ and $E\|\hat{\Psi}_{\hat{k}}(P^0_{n,B_n}) - \Psi(P_0)\|$, so that the oracle selector $\tilde{k}$ is the minimizer over $k \in \{1, 2, ..., K(n)\}$ of the random quantity $E_{B_n}\|\hat{\Psi}_k(P^0_{n,B_n}) - \Psi(P_0)\|$.

Proof: This is trivial. □

5.5. Appendix

In this section we shall give some useful results needed in the proofs of the previous results.

5.5.1. *Some useful lemmas*

Our proof of finite sample results is based on Bernstein's inequality, which we state here as a lemma for ease of reference. A proof is given in Lemma A.2, p. 564 in Györfi, Kohler, Krzyżak and Walk (2002).

Lemma 5.2: *Bernstein's inequality. Let Z_i, $i = 1, \ldots, n$, be independent real valued random variables such that $Z_i \in [a, b]$ with probability one. Let $0 < \sum_{i=1}^{n} VAR(Z_i)/n \le \sigma^2$. Then, for all $\epsilon > 0$,*

$$Pr\left(\frac{1}{n}\sum_{i=1}^{n}(Z_i - EZ_i) > \epsilon\right) \le \exp\left(-\frac{1}{2}\frac{n\epsilon^2}{\sigma^2 + \epsilon(b-a)/3}\right).$$

This implies

$$Pr\left(\frac{1}{n}\mid\sum_{i=1}^{n}(Z_i - EZ_i)\mid > \epsilon\right) \le 2\exp\left(-\frac{1}{2}\frac{n\epsilon^2}{\sigma^2 + \epsilon(b-a)/3}\right).$$

We have the following immediate corollary of Bernstein's inequality, which allows us to obtain the wished tail probabilities for products Z_{1n}^2, $Z_{1n}Z_{2n}$.

Lemma 5.3: Bernstein's inequality. *Given arbitrary random variables* (Z_{1n}, Z_{2n}), *we have*

$$P(\mid Z_{1n}Z_{2n} \mid \geq s) \leq P(\mid Z_{1n} \mid \geq \sqrt{s}) + P(\mid Z_{2n} \mid \geq \sqrt{s}).$$

This allows us to obtain explicit tail probabilities from tail probabilities derived for Z_{1n} *and* Z_{2n} *separately. In particular, if* (Z_{1i}, Z_{2i}), $i = 1, \cdots, n$, *are independent bivariate random variables such that* $Z_{ji} \in [a, b]$, $j = 1, 2$, *with probability one,* $0 < \sum_{i=1}^{n} VAR(Z_{ji})/n \leq \sigma^2$, $j = 1, 2$, *then, for all* $\epsilon > 0$,

$$Pr\left(\mid \frac{1}{n}\sum_{i=1}^{n}(Z_{1i} - EZ_{1i})\frac{1}{n}\sum_{i=1}^{n}(Z_{2i} - EZ_{2i}) \mid > \epsilon\right)$$
$$\leq 4\exp\left(-\frac{1}{2}\frac{n\epsilon}{\sigma^2 + \sqrt{\epsilon}(b-a)/3}\right).$$

These bounds can be directly translated into bounds on the corresponding expectations. In particular, we can apply the following simple lemma.

Lemma 5.4: *Let* Z_n *be a random variable satisfying that* $P(Z_n \geq s) \leq C(n)\exp(-ns/c)$ *for all* $s \geq 0$. *Then* $EZ_n \leq \frac{c(\log C(n)+1)}{n}$.

5.5.2. Lemmas for CThe board on mathematical sciences has evolvedorollary 5.1

Lemma 5.5: *Let* $f_1, ..., f_{K(n)}$ *be functions with the same domain and range dimension as* $X \rightarrow D(X|\psi, \gamma)$, *where* f_k^b *represents the* b^{th} *component of the* k^{th} *function. Assume that* $|f_k^b| \leq M < \infty$ *for* $1 \leq b, k < \infty$. *Then* $E[\max_{f_k^b \in \{f_1^b,...,f_{K(n)}^b\}} |G_{n,B_n}^1 f_k^b|] \leq cM\sqrt{\log K(n)}$, *for a universal contant c.*

Proof: This is trivially implied by Formula (2.5.5) in van der Vaart and Wellner (1996). □

Lemma 5.6: *Under the assumptions and notation of Lemma 5.5, we have* $E[\max_{f_k^b \in \{f_1^b,...,f_{K(n)}^b\}} |(P_{n,B_n}^1 - P)f_k^b|^2] \leq c\log K(n)/n$.

Proof: First, we use the Bonferroni inequality to bound the tail probability (probability of exceeding s) of the quantity inside the expectation by $K(n)\max_k Pr[|(P_{n,B_n}^1 - P)f_k^b|^2 \geq s]$. The latter equals

$K(n) \max_k Pr[|(P^1_{n,B_n} - P)f^b_k| \geq \sqrt{s}]$, which can be bounded by $K(n) \exp(-sn/c)$ for some constant c, by Bernstein's inequality. Finally, by Lemma 5.4, this bound implies the desired result. □

Lemma 5.7: *Consider the assumptions and notation of Lemma 5.5. Let $\mathcal{F}_n \equiv \{f^b_k : 1 \leq k \leq K(n), 1 \leq b \leq \infty\}$, and $h_{n,K(n)} \equiv \int_0^\infty \sqrt{\log \sup_Q N(\epsilon M^2, \mathcal{F}_n, L_2(Q))} d\epsilon$. Then*

$$E[\sup_{f \in \mathcal{F}_n} |G^1_{n,B_n} f|] \leq cM^2 h_{n,K(n)} \text{ for some universal constant } c.$$

Proof. This is a direct application of Lemma 2.14.1 of Chapter 2 in van der Vaart and Wellner (1996). □

Lemma 5.8: *Under the assumptions of Lemmas 5.5, 5.6, and 5.7, we have for $q = 1, 2$*

$$E_{P^1_{n,B_n}} [\max_{f_k \in \{f_1, \ldots, f_{K(n)}\}} \|G^1_{n,B_n} f_k\|_q] \leq c\sqrt{\log K(n)},$$

and

$$E_{P^1_{n,B_n}} [\max_{f_k \in \{f_1, \ldots, f_{K(n)}\}} \|G^1_{n,B_n} f_k\|_\infty] \leq M^2 h_{n,K(n)}.$$

Proof: We have

$$\begin{aligned}
&E_{P^1_{n,B_n}} [\max_{f_k \in \{f_1, \ldots, f_{K(n)}\}} \|G^1_{n,B_n} f_k\|_q] \\
&\leq E_{P^1_{n,B_n}} \| \max_{f_k \in \{f_1, \ldots, f_{K(n)}\}} | G^1_{n,B_n} f^b_k | \|_q \\
&\leq \|E_{P^1_{n,B_n}} \max_{f_k \in \{f_1, \ldots, f_{K(n)}\}} |G^1_{n,B_n} f^b_k|\|_q,
\end{aligned}$$

where the latter inequality follows by Jensen's inequality. The desired result for $q = 1, 2$ follows now from Lemmas 5.5 and 5.6. We also have

$$E_{P^1_{n,B_n}} [\max_{f_k \in \{f_1, \ldots, f_{K(n)}\}} \|G^1_{n,B_n} f_k\|_\infty] = E_{P^1_{n,B_n}} \sup_{f \in \mathcal{F}_n} |G^1_{n,B_n} f|$$

where $\mathcal{F}_n \equiv \{f^b_k : 1 \leq k \leq K(n), 1 \leq b \leq \infty\}$. The desired result for the supremum norm now follows from Lemma 5.7. □

Acknowledgements

This work was supported by NIH grants NIH R01 GM67233 and NIH R01 GM071397. We thank Rob Strawderman for helpful email discussions. We also wish to thank the referee for his/her helpful comments.

References

Andersen, P. K., Borgan, O., Gill, R. D., Keiding, N. (1992). Statistical models based on counting processes, Springer-Verlag, New York.

Bein, E., Hubbard, A. E. and van der Laan, M. J. (2005). Estimating function based cross-validation to estimate the causal effect in randomized trials with non-compliance, Technical report, Division of Biostatistics, University of California, Berkeley.

Bickel, P. J., C.A.J. Klaassen, C. A. J., Ritov, Y., and Wellner, J. A. (1997). *Efficient and Adaptive Estimation for Semiparametric Models*, Springer Verlag.

Breiman, L. (1992). The little bootstrap and other methods for dimensionality selection in regression: x-fixed prediction error, *Journal of the American Statistical Assocation* **87**(419), 738–754.

Breiman, L. (1996a). Heuristics of instability and stabilization in model selection, *Annals of Statistics* **24** (6), 2350–2383.

Breiman, L. (1996b). Out of bag estimation, Technical report, Department of Statistics, University of California, Berkeley.

Breiman, L., Friedman, J. H., Olshen, R. A., and Stone, C. J. (1984). *Classification and regression trees*, The Wadsworth Statistics/Probability series, Wadsworth International Group.

Breiman, L. and Spector, P. (1992). Submodel selection and evaluation in regression: the x-random case, *International Statistical Review* **60**, 291–319.

Cosslett, S. R. (2004). Efficient semiparametric estimation of censored and truncated regressions via a smoothed self-consistency equation, *Econometrica*, **72**(4), 1277–1284.

Dudoit, S. and van der Laan, M. J. (2003). Asymptotics of cross-validated risk estimation in model selection and performance assessment, Technical Report **126**, Division of Biostatistics, University of California, Berkeley, Feb. 2003, URL: www.bepress.com/ucbbiostat/paper126, to appear in The Indian Journal of Statistical Methodology.

Efron, B. and Tibshirani, R. J. (1993). *An Introduction to the Bootstrap*, Chapman & Hall/CRC.

Györfi, L., Kohler, M., Krzyżak, A., and Walk, H. (2002). *A Distribution-Free Theory of Nonparametric Regression*, Springer-Verlag, New York.

Györfi, L., Schäfer, D., and Walk, H. (2002). Relative Stability of global errors of nonparametric function estimators, *IEEE Transactions of Information Theory*, **48**(8), 2230: 2242.

Härdle, W. (1993). *Applied Nonparametric Regression*, Cambridge University Press.

Härdle, W. and Marron, J. S. (1985a). Asymptotic Equivalence of some bandwidth selectors in nonparametric regression, *Biometrika*, **72**, 481–484.

Härdle, W. and Marron, J. S. (1985b). Optimal bandwidth selectin in nonparametric regression function estimation, *Annals of Statistics*, **13**, 1465–1481.

Hastie, T., Tibshirani, R. J., and Friedman, J. H. (2001). *The Elements of Statis-*

tical Learning: Data Mining, Inference, and Prediction, Springer-Verlag.

Keles, S., van der Laan, M. J. and Dudoit, S. (2003). Asymptotically optimal model selection method with right-censored outcomes, Technical Report 124, Division of Biostatistics, University of California, Berkeley, Sept. 2003, URL: www.bepress.com/ucbbiostat/paper124, to appear in *Bernoulli*.

Klaassen, C. A. J. (1987). Consistent estimation of the influence function of locally asymptotically linear estimators, *Annals of Statistics*, **15**, 1548–1562.

Molinaro, A. M., Dudoit, S. and van der Laan, M. J. (2004). Tree-based multivariate regression and density estimation with right-censored data, In S. Dudoit, R. C. Gentleman, and M. J. van der Laan (eds), Special Issue on Multivariate Methods in Genomic Data Analysis, *Journal of Multivariate Analysis*, Vol. 90, No. 1, p. 154-177.

Ripley, B. D. (1996). *Pattern recognition and neural networks*, Cambridge University Press, Cambridge, New York.

Silverman, B. W. (1984). A fast and efficient cross-validation method for smoothing parameter choice in spline regression, *Journal of the American Statistical Association*, **79** (387), 584–589.

Stone, M. (1974). Cross-validatory choice and assessment of statistics predictions, *Journal of the Royal Statistical Society, Series B*, **36** (2), 111–147.

Stone, M. (1977). Asymptotics for and against cross-validation, *Biometrika*, **64** (1), 29–35.

van der Laan, M. J. (1996). Efficient and Inefficient Estimation in Semiparametric Models. CWI-tract **114**, Centre for Mathematics and Computer Science, Amsterdam, the Netherlands, 1996.

van der Laan, M. J. (1995). An Identity for the Nonparametric Maximum Likelihood Estimator in Missing Data and Biased Sampling Models. *Bernoulli* **1(4)**, pp. 335–341.

van der Laan, M. J. (1998). Identity for NPMLE in Censored Data Models, *Lifetime Data Models* **4**, 83–102 (1998).

van der Laan, M. J., and Dudoit, S. (2003) *Unified Cross-Validation Methodology For Selection among Estimators, and a General Cross-validated Adaptive epsilon-Net Estimator: Finite Sample Oracle Inequalities and Examples*, Working Paper #130, Division of Biostatistics, UC Berkeley.

van der Laan, M. J., Dudoit, S., Keles, S.(2004). Asymptotic Optimality of Likelihood-Based Cross-Validation, *Statistical Applications in Genetics and Molecular Biology* Vol. 3: No. 1, Article 4. http://www.bepress.com/sagmb/vol3/iss1/art4, 2004.

van der Laan, M. J., Hubbard, A., Jewell, N. P. (2004). Estimation of Treatment Effects in Randomized Trials with Noncompliance and a Dichotomous Outcome, technical report , Division of Biostatistics, http://www.bepress.edu/ucbbiostat, submitted for publication in the Journal of the Royal Statistical Society B.

van der Laan, M. J., Robins, J. M. (2002). *Unified Methods for Censored Longitudinal Data and Causality*, Springer Verlag, New York.

van der Vaart, A. W., and Wellner, J. A. (1996). *Weak Convergence and Empirical Processes: with Applications to Statistics*, Springer Verlag, New York.

Part II

Nonparametric Methods

CHAPTER 6

Powerful Choices: Tuning Parameter Selection Based on Power

Kjell A. Doksum and Chad M. Schafer

Department of Statistics
University of Wisconsin, Madison
Madison, WI 53706
doksum@stat.wisc.edu

Department of Statistics
Carnegie Mellon University
Pittsburgh, PA 15213
cschafer@stat.cmu.edu

We consider procedures which select the bandwidth in local linear regression by maximizing the limiting power for local Pitman alternatives to the hypothesis that $\mu(x) \equiv \mathrm{E}(Y \mid X = x)$ is constant. The focus is on achieving high power near a covariate value x_0 and we consider tests based on data with X restricted to an interval containing x_0 with bandwidth h. The power optimal bandwidth is shown to tend to zero as sample size goes to infinity if and only if the sequence of Pitman alternatives is such that the length of the interval centered at x_0 on which $\mu(x) = \mu_n(x)$ is nonconstant converges to zero as $n \to \infty$. We show that tests which are based on local linear fits over asymmetric intervals of the form $[x_0 - (1 - \lambda)h, \ x_0 + (1 + \lambda)h]$, where $-1 \leq \lambda \leq 1$, rather than the symmetric intervals $[x_0 - h, \ x_0 + h]$ will give better asymptotic power. A simple procedure for selecting h and λ consists of using order statistics intervals containing x_0. Examples illustrate that the effect of these choices are not trivial: power optimal bandwidth can give much higher power than bandwidth chosen to minimize mean squared error. Because we focus on power, rather than plotting estimates of $\mu(x)$ we plot a correlation curve $\widehat{\rho}(x)$ which indicates the strength of the dependence between Y and X near each $X = x$. Extensions to more general hypotheses are discussed.

6.1. Introduction: Local Testing and Asymptotic Power

We consider (X_1, Y_1), (X_2, Y_2), $\ldots$, (X_n, Y_n) i.i.d. as $(X, Y) \sim P$ and write $Y = \mu(X) + \epsilon$ where $\mu(X) \equiv \mathrm{E}(Y \mid X)$ and $\beta(x) \equiv d\mu(x)/dx$ are assumed to exist, and $\epsilon \equiv Y - \mu(X)$. We are interested in a particular covariate value x_0 and ask if there is a relationship between the response Y and the covariate X for X in some neighborhood of x_0. More formally, we test the hypothesis H that $\beta(x) = 0$ for all x against the alternative K that $\beta(x) \neq 0$ for x in some neighborhood of x_0. Our focus is on achieving high power for a covariate value x_0 for a unit (patient, component, DNA sequence, etc.) of interest. We want to know if a perturbation of x will affect the mean response for units with covariate values near x_0.

Our test statistic is the t-statistic $t_h(x_0)$ based on the locally linear estimate $\widehat{\beta}_h(x_0)$ of $\beta(x_0)$ obtained as the slope of the least squares estimate of $\beta(x_0)$ computed for data in the local data set

$$D_h(x_0) \equiv \{(x_i, y_i) \colon x_i \in N_h(x_0)\}, \quad \text{with} \ \ N_h(x_0) \equiv [x_0 - h, \ x_0 + h].$$

The bandwidth h serves as a lense with different lenses providing insights into the relationship between Y and X (Chaudhuri and Marron (1999) and Chaudhuri and Marron (2000) did "estimation" lenses). The h that maximizes the power over $N_h(x_0)$ provides a customized lense for a unit with covariate value x_0. We discuss global alternatives where $\sup_x |\beta(x)| > 0$ in Remark 6.7. The many available tests for this alternative will have increased power if they are based on variable bandwidths.

We will next show that the power of the test is very sensitive to the choice of h and that this choice at first appears to be inversely related to the usual choice of h based on mean squared error (MSE). In fact, for any alternative with $\beta(x)$ nonzero on an interval centered at x_0 that does not shrink to a point as n goes to infinity, the asymptotic power is maximized by selecting an $h = h_n$ that does not converge to zero as $n \to \infty$.

Because $\widehat{\beta}_h(x_0)$ is asymptotically normal, the asymptotic power of the test based on it for local Pitman alternatives will be determined by its efficacy (EFF) for one-sided alternatives $\beta(x) > 0$ and by its absolute efficacy for two-sided alternatives $\beta(x) \neq 0$ (Kendall and Stuart (1961), Section 25.5, Lehmann (1999)):

$$\mathrm{EFF}_h \equiv \mathrm{EFF}\left(\widehat{\beta}_h(x_0)\right) = \mathrm{E}_K\left[\widehat{\beta}_h(x_0)\right] \bigg/ \left(\mathrm{Var}_H\left[\widehat{\beta}_h(x_0)\right]\right)^{1/2}.$$

Consider the alternative $\mu(x) = \alpha + \gamma\, r(x)$ for a differentiable function $r(\cdot)$. We know (e.g. Fan and Gijbels (1996), p. 62) that for $h \to 0$, conditionally on $\mathbf{X} = (X_1, X_2, \ldots, X_n)$,

$$\mathrm{E}_K\left[\widehat{\beta}_h(x_0) \mid \mathbf{X}\right] = \gamma r'(x_0) + c_1 h^2 + o_{\mathrm{P}}\left(h^2\right) \quad \text{and}$$

$$\mathrm{Var}_H\left[\widehat{\beta}_h(x_0) \mid \mathbf{X}\right] = c_2 n^{-1} h^{-3} + o_{\mathrm{P}}\left(n^{-1} h^{-3}\right)$$

for appropriate constants c_1 and c_2. These expressions yield an approximate efficacy of order $n^{1/2} h^{3/2}$ which is maximized by h not tending to zero (where the expressions are not valid). This shows that h tending to zero does not yield optimal power and that maximizing power is typically very different from minimizing some variant of mean squared error: for example,

$$\mathrm{MSE}(h) \equiv \mathrm{E}\left[(\widehat{\mu}_h(X) - \mu(X))^2 \; \mathbf{1}\{X \in (x_0 - h, \, x_0 + h)\}\right],$$

where $\widehat{\mu}_h(\cdot)$ is the local linear estimate of $\mu(\cdot)$. This subject has been discussed for global alternatives by many authors, sometimes with the opposite conclusion. For a discussion, see e.g. Hart (1997), Zhang (2003a), Zhang (2003b), and Stute and Zhang (2005). In the next section we look more closely at the optimal h and find that for certain least favorable distributions, the optimal h does indeed tend to zero.

Remark 6.1: Why the t-test?

(a) The t-test is asymptotically most powerful in the class of tests using data from $D_h(x_0)$ for a class of perturbed normal linear models defined for $x_i \in N_h(x_0)$ by $Y_i = \gamma(x_i - x_0) + \gamma^2 r(x_i) + \epsilon$, where $\sum(x_i - x_0) = 0$. Here, ϵ has the distribution $\Phi(t/\sigma_0)$, where $\gamma = \gamma_n = O(n^{-1/2})$ is a Pitman alternative, h is fixed, and $r(\cdot)$ is a general function satisfying regularity conditions. To establish the asymptotic power optimality, suppose all parameters are known except γ and consider the Rao score test for testing $H : \gamma = 0$ versus $K : \gamma > 0$.

It is easy to see that this statistic is equivalent to $\sum(x_i - x_0)Y_i$ whose asymptotic power is the same as that of the t-test. It is known (e.g. Bickel and Doksum (1966), Section 5.4.4) that the score test is asymptotically most powerful level α. In the presence of nuisance parameters, a modified Rao score test retains many of its power properties, see Neyman (1959) and Bickel, Ritov, and Stoker (2006). The idea of asymptotic optimality for perturbed models is from Bell and Doksum (1966). The asymptotic power optimality of the t-test is in agreement with the result that when bias does not matter, the uniform kernel is optimal (e.g. Gasser, Müller and Mammitizsch (1985) and Fan and Gijbels (1996)). But see Blyth (1993), who shows that for certain models, a kernel asymptotically equivalent to the derivative of the Epanechnikov kernel maximizes the power asymptotically.

While for alternatives of this type the use of a test based on the t-test is asymptotically most powerful, there undoubtedly exist other situations in which a test based on an appropriate nonparametric estimate of $\mu(\cdot)$ gives better power. We will explore this issue in future work.

(b) The t-test is the natural model selector for choosing between the models $\mu(x) = \beta_0 + \epsilon$ and $\mu(x) = \beta_0 + \beta_1 x + \epsilon'$ in the sense that to minimize mean squared prediction error, we select the former model if the absolute value of the usual t-statistic is less than $\sqrt{2}$ (see e.g. Linhart and Zucchini (1986)). This is easily seen to be equivalent to minimizing the mean squared estimation error. It also is the rule of choice if we apply the methods of Claeskens and Hjorth (2003) who choose the model that minimizes the asymptotic mean squared estimation error for Pitman sequences of alternatives.

Remark 6.2: For a given test statistic T, the numerator of the efficacy is usually (e.g. Lehmann (1999), pg. 172, and Kendall and Stuart (1961)) defined in terms of the derivative of $E_\theta(T)$ evaluated at the hypothesized parameter value θ_0. We consider the "pre-derivative" because we want to investigate how h influences efficacy.

Remark 6.3: Hajek (1962), Behnen and Hušková (1984), Behnen and Neuhaus (1989), Sen (1996), among others, proposed and analyzed procedures for selecting the asymptotically optimal linear rank test by maximizing the estimated efficacy.

Remark 6.4: The approach of selecting bandwidth to maximize efficacy can also be used to test a parametric model of the form $E(Y|X = x) = g(x; \boldsymbol{\beta})$ against a nonparametric model. Simply begin by subtracting off from Y the appropriate model fit under the (null hypothesis) parametric model. See Remark 6.13.

6.2. Maximizing Asymptotic Power

We start with fixed $h > 0$. Then the limiting power for the two-sided alternative is a one-to-one function of the absolute value of

$$\tau_h(x_0) \equiv \lim_{n \to \infty} n^{-1/2} \operatorname{EFF}\left(\widehat{\beta}_h(x_0)\right).$$

Thus the maximizer of the asymptotic power is $h_0 \equiv \arg\max_h |\tau_h(x_0)|$. Our estimate of h_0 is the maximizer of the absolute t-statistic, $\widehat{h} \equiv$

$\arg\max_h |t_h(x_0)|$ or its equivalents given in Remark 6.5. To investigate the limit of the t-statistic, $t_h(x_0)$, for the data $D_h(x_0)$ we write it as signal/noise where

$$\text{signal} \equiv \widehat{\beta}_h(x_0) = \widehat{\text{Cov}}_h(X, Y) \Big/ \widehat{\text{Var}}_h(X) \quad \text{and}$$

$$\text{noise}^2 \equiv \widehat{\text{Var}}\left(\widehat{\beta}_h(x_0)\right) = \widehat{\text{E}}_h\left(\epsilon_h^2\right) \Big/ n_h \widehat{\text{Var}}_h(X).$$

Here $\widehat{\text{Var}}_h(X)$ and $\widehat{\text{Cov}}_h(X, Y)$ denote the sample variance and covariance for the data $D_h(x_0)$,

$$n_h \equiv \sum_{i=1}^{n} \mathbf{1}\{x_i \in N_h(x_0)\} \quad \text{and} \quad \widehat{\text{E}}_h\left(\epsilon_h^2\right) = \text{RSS}_h \Big/ (n_h - 2),$$

where RSS_h is the residual sum of squares $\sum(y_i - \widehat{\mu}_{L,h}(x_i))^2$ for the linear fit $\widehat{\mu}_{L,h}(x_i)$ to y_i based on $D_h(x_0)$.

For fixed h, as $n \to \infty$,

$$\widehat{\beta}_h(x_0) \xrightarrow{P} \beta_h(x_0) \equiv \text{Cov}_h(X, Y) / \text{Var}_h(X),$$

$$(n_h/n) \xrightarrow{P} P_h(x_0) \equiv P(X \in N_h(x_0)), \quad \text{and}$$

$$\widehat{\text{E}}_h\left(\epsilon_h^2\right) \xrightarrow{P} \text{E}_h\left(\epsilon_{L,h}^2\right) \equiv \text{E}_h[Y - \mu_{L,h}(X)]^2,$$

where E_h, Var_h, and Cov_h denote expected value, variance, and covariance conditional on $X \in N_h(x_0)$, and where $\mu_{L,h}(x) = \alpha_h + \beta_h X$ with α_h and β_h the minimizers of $\text{E}_h[Y - (\alpha + \beta X)]^2$. It follows that as $n \to \infty$,

$$n^{-1/2} t_h(x_0) \xrightarrow{P} \tau_h^*(x_0) \equiv \beta_h(x_0) \{P_h(x_0) \text{Var}_h(X)\}^{1/2} \Big/ \left\{\text{E}_h\left(\epsilon_{L,h}^2\right)\right\}^{1/2}$$

$$= \left[\text{SD}_H(Y) \Big/ \left(\text{E}_h\left(\epsilon_{L,h}^2\right)\right)^{1/2}\right] \tau_h(x_0). \tag{6.1}$$

For sequences of Pitman alternatives where $\mu(x) = \mu_n(x)$ converges to a constant for all x as $n \to \infty$, we assume that as $n \to \infty$

$$\text{E}_h\left(\epsilon_{L,h}^2\right) \to \sigma^2 \equiv \text{Var}_H(Y).$$

In this case, $\tau_h^*(x_0) = \tau(x_0)$. It is clear that $\tau_h^*(x_0) \to 0$ as $h \to 0$ because both $P_h(x_0)$ and $\text{Var}_h(X)$ tend to zero while $\text{E}_h(\epsilon_{L,h}^2)$ does not tend to zero except in the trivial case $Y = \mu_{L,h}(x)$ for $x \in N_h(x_0)$. On the other hand, if the density of X has support $[a, b]$ and $x_0 \in [a, b]$, then for $h' \equiv \max\{x_0 - a, b - x_0\}$,

$$\lim_{h \to h'} \tau_h^*(x_0) = \beta_L \left\{\text{Var}(X) \Big/ \text{E}(\epsilon_L^2)\right\}^{1/2},$$

where $\epsilon_L = Y - (\alpha_L + \beta_L X)$ with α_L and β_L the minimizers of $\mathrm{E}[Y - (\alpha + \beta X)]^2$, that is, $\beta_L = \mathrm{Cov}(X, Y)/\mathrm{Var}(X)$, $\alpha_L = \mathrm{E}(Y) - \beta_L \mathrm{E}(X)$. Thus, when X has finite support $[a, b]$ and $x_0 \in [a, b]$, the maximum of $\tau_h^*(x_0)$ over $h \in [0, h']$ exists and it is greater than zero. A similar argument shows that $h = 0$ does not maximize $\tau_h^*(x_0)$ when X has infinite support.

Remark 6.5: Instead of the t-statistic $t_h(x_0)$, we could use the efficacy estimate

$$\widehat{\mathrm{EFF}}\left(\widehat{\beta}_h(x_0)\right) = r_h(x_0),$$

where $r_h(x_0)$ is the sample correlation for data from $D_h(x_0)$. The formula

$$t_h^2(x_0) = (n-2)\, r_h^2(x_0) \Big/ \left(1 - r_h^2(x_0)\right),$$

where the right hand side is increasing in $|r_h(x_0)|$, establishes the finite sample equivalence of $|t_h(x_0)|$ and $|r_h(x_0)|$. They are also similar to $|\rho(x_0)|$ where

$$\widehat{\rho}_h(x_0) = \frac{\widehat{\beta}_h(x_0)\,\widehat{\mathrm{SD}}(X)}{\sqrt{\widehat{\beta}_h^2(x_0)\,\widehat{\mathrm{Var}}(X) + \widehat{\mathrm{Var}}_h(Y)}}$$

is the estimate of the correlation curve $\rho(\cdot)$ discussed by Bjerve and Doksum (1993), Doksum, Blyth, Bradlow, Meng, and Zhao (1994), and Doksum and Froda (2000). Here $\widehat{\rho}_h(x)$ is a standardized version of $r_h(x_0)$ which is calibrated to converge to the correlation coefficient ρ in bivariate normal models.

6.2.1. *Optimal bandwidth for vertical Pitman sequences*

If we consider Pitman alternatives of the form $Y = \alpha + \gamma r(X) + \epsilon$ where $\gamma = \gamma_n = c/\sqrt{n}$, $c \neq 0$, and $|r'(x)| > 0$ in a fixed interval containing x_0, then

$$n^{-1/2}\mathrm{EFF}\left[\widehat{\beta}_h(x_0)\right] \rightarrow \tau_h(x_0),$$

where $\tau_h(x_0)$ is as in Equation (6.1), and the optimal h will be bounded away from zero as before. Thus, the power optimal bandwidth does not tend to zero for sequences of alternatives where, as $n \to \infty$, $\|\beta\|_\infty \to 0$ and $\|\beta\|^{-\infty} \not\to 0$ with $\|\beta\|_\infty \equiv \sup_x |\beta(x)|$ and

$$\|\beta\|^{-\infty} \equiv x^+(\beta) - x^-(\beta) \equiv \sup_x \{x\colon |\beta(x)| > 0\} - \inf_x \{x\colon |\beta(x)| > 0\}.$$

We refer to $\|\beta\|_\infty$ as the "vertical distance" between H and the alternative and $\|\beta\|^{-\infty}$ as the "horizontal discrepancy."

Next we consider "horizontal" sequences of alternatives where $\|\beta\|^{-\infty} \to 0$ and $\|\beta\|_\infty$ may or may not tend to zero as $n \to \infty$ and find that now the power optimal bandwidth tends to zero.

6.2.2. *Optimal bandwidth for horizontal Pitman sequences*

We now consider Pitman alternatives of the form

$$K_n : Y = \alpha + \gamma \, W\left(\frac{X - x_0}{\theta}\right) + \epsilon, \qquad (6.2)$$

where X and ϵ are uncorrelated, ϵ has mean zero and variance σ^2, and $W(\cdot)$ has support $[-1, 1]$. We assume that X is continuous, in which case $\mu(\cdot)$ is the constant α with probability one when $\theta = 0$. Thus the hypothesis holds with probability one when $\gamma\theta = 0$. We consider models where $\theta = \theta_n \to 0$ as $n \to \infty$, and γ may or may not depend on θ and n. For these alternatives the neighborhood where $|\beta(x)| > 0$ shrinks to achieve a Pitman balanced model where the power converges to a limit between the level α and 1 as $n \to \infty$. Note, however, that the alternative does not depend on h. We are in a situation where "nature" picks the neighborhood size θ, and the statistician picks the bandwidth h. This is in contrast to Blyth (1993) and Aït-Sahalia, Bickel and Stoker (2001) who let the alternative depend on h.

Note that for $h > 0$ fixed, γ bounded above, and $\theta \to 0$,

$$\text{Cov}_h(X, Y) = \gamma \text{Cov}_h\left(X, W\left(\frac{X - x_0}{\theta}\right)\right) \to 0$$

because $W((X - x_0)/\theta)$ tends to zero in probability as $\theta \to 0$ and any random variable is uncorrelated with a constant. This heuristic can be verified by the change of variable $s = (x - x_0)/\theta$, $x = x_0 + \theta s$. More precisely,

Proposition 6.1: *Assume that the density $f(\cdot)$ of X has a bounded derivative at x_0. If $h > 0$ is fixed, then as $\gamma\theta \to 0$ in the Model (6.2),*
(a) $\text{Cov}_h(X, Y) = O(\gamma\theta)$; (b) $n^{-1/2}\text{EFF}(\hat{\beta}_h(x_0)) \to 0$; (c) $n^{-1/2}t_h(x_0) \xrightarrow{P} 0$.

Proof: Since ϵ is assumed uncorrelated with X, $\text{Cov}_h(X, Y) =$

$\text{Cov}_h(X, \mu(X))$ and hence

$$\text{Cov}_h(X, Y) = \gamma \text{Cov}_h\left(X, W\left(\frac{X - x_0}{\theta}\right)\right)$$

$$= \gamma \int_{-h}^{h} (x - \text{E}_h(X)) W\left(\frac{x - x_0}{\theta}\right) \left[f(x)\Big/ P_h(x_0)\right] dx$$

$$= \frac{\gamma\theta}{P_h(x_0)} \int_{-1}^{1} (x_0 + s\theta - \text{E}_h(X)) W(s) f(x_0 + s\theta) ds \quad (6.3)$$

$$= \frac{\gamma\theta}{P_h(x_0)} \left[(x_0 - \text{E}_h(X)) f(x_0) \int_{-1}^{1} W(s) ds + O(\theta)\right].$$

Then Result (b) follows because

$$\text{E}_K\left(\widehat{\beta}_h(x_0)\right) \to \text{Cov}_h(X, Y)\Big/ \text{Var}_h(X)$$

and the other factors in EFF_h are fixed as $\theta \to 0$. Similarly (c) follows from Equation (6.1). □

The above proposition shows that for Model (6.2), fixed h leads to small $|\text{EFF}_h|$. Thus we turn to the $h \to 0$ case. If $h > \theta$ then observations (X, Y) with X outside $[x_0 - \theta, x_0 + \theta]$ do not contribute to the estimation of $\beta_h(x_0)$. Thus choosing a smaller h may be better even though a smaller h leads to a larger variance for $\widehat{\beta}_h(x_0)$. This heuristic is made precise in the next result which provides conditions where $h = \theta$ is the optimal choice among h satisfying $h \geq \theta$.

Define

$$m_j(W) \equiv \int_{-1}^{1} s^j W(s) ds \quad j = 0, 1, 2.$$

Theorem 6.1: Assume that the density $f(\cdot)$ of X has a bounded, continuous second derivative at x_0 and that $f(x_0) > 0$. Then, in Model (6.2), as $\theta \to 0$ and $h \to 0$ with $h \geq \theta$, the following holds.

(a) $\text{Cov}_h(X, Y) = \dfrac{\gamma\theta}{2hf(x_0)} \Big\{ -\dfrac{1}{3} m_0(W) f'(x_0) h^2$

$\qquad + f(x_0) m_1(W) \theta + f'(x_0) m_2(W) \theta^2 + o(h^2) + o(\theta^2) \Big\}.$

(b) If $m_1(W) \neq 0$ and $m_2(W) \neq 0$,

$$\tau_{h,\theta}(x_0) \equiv \lim_{n \to \infty} n^{-1/2} \, \text{EFF}\left(\widehat{\beta}_h(x_0)\right)$$

$$= \sigma^{-1}\gamma\theta h^{-3/2} \, [f(x_0)]^{-1/2} \left\{ -\frac{1}{3}m_0(W) \, f'(x_0) \, h^2 \right.$$

$$\left. + f(x_0) \, m_1(W) \, \theta + f'(x_0) \, m_2(W) \, \theta^2 + o(h^2) + o(\theta^2) \right\} \quad (6.4)$$

$$= O\left(\sigma^{-1}\gamma\theta^2 h^{-3/2}\right)$$

and $|\tau_{h,\theta}(x_0)|$ is maximized subject to $h \geq \theta$ by $h = \theta$.
(c) If $m_1(W) = 0$,

$$\lim_{n \to \infty} n^{-1/2} \, \text{EFF}\left(\widehat{\beta}_h(x_0)\right) = O\left(\sigma^{-1}\gamma\theta^3 h^{-3/2}\right)$$

and $|\tau_{h,\theta}(x_0)|$ is maximized subject to $h \geq \theta$ by $h = \theta$.

Proof: Part (a): This is Lemma 6.3, part (e).
Part (b): We have

$$\lim_{n \to \infty} n^{-1/2} \, \text{EFF}_h = \sigma^{-1}\text{Cov}_h(X, Y) \left[(P_h(x_0))\Big/\text{Var}_h(X)\right]^{1/2}.$$

By Lemma 6.3, part (d), we get

$$P_h(x_0)/\text{Var}_h(X) = 6f(x_0) \, h^{-1} + O(1).$$

Equation (6.4) follows from this and part (a). Thus we can write

$$\lim_{n \to \infty} n^{-1/2} \, \text{EFF}_h \asymp \sigma^{-1} \left[c_1\gamma\theta h^{1/2} + c_2\gamma\theta^2 h^{-3/2} + c_3\gamma\theta^3 h^{-3/2}\right] \quad (6.5)$$

for appropriate constants c_1, c_2, and c_3 given in part (b). The square of the expression on the right of Equation (6.5) is dominated by $[\sigma^{-1}\theta^2 h^{-3/2}c_2]^2$ and maximized for $h \geq \theta$ by $h = \theta$.

Part (c): Apply the argument in Part (b). $\qquad\square$

It remains to investigate the case where $h \leq \theta$. We next show that the optimal h cannot be such that $h/\theta \to 0$.

Proposition 6.2: *Assume the conditions of Theorem 6.1 and that $W''(\cdot)$ exists and is bounded in a neighborhood of zero. Then, as $h/\theta \to 0$,*

$$\lim_{n \to \infty} n^{-1/2} \, \text{EFF}_h = \begin{cases} o\left(\sigma^{-1}\gamma\theta^2 h^{-3/2}\right), & \text{if } W'(0) \neq 0 \\ o\left(\sigma^{-1}\gamma\theta^3 h^{-3/2}\right), & \text{if } W'(0) = 0 \end{cases}.$$

Proof: Set $a \equiv h/\theta$ and define

$$m_{j,a}(W) \equiv \int_{-a}^{a} s^j W(s)\, ds, \quad j = 0, 1, 2.$$

Equation (6.4) in Theorem 6.1, part (b), with $m_j(W)$ replaced by $m_{j,a}(W)$ is valid. A Taylor expansion gives

$$m_{1,a}(W) = [2W'(0) + aW''(0)]\, a^3 + o(a^3).$$

The result follows for the $W'(0) \neq 0$ case. The other case is similar.   □

The case where for the smallest optimal h, $h/\theta \to a_0$ as $\theta \to 0$ with $a_0 \in (0, 1)$ remains. It is possible that the optimal h satisfies this condition, for instance if $W(\cdot)$ is steep on $[-1/2, 1/2]$ and nearly constant otherwise. In this case Theorem 6.1 holds with $m_j(W)$ replaced by $m_{j,a}(W)$.

The concept of efficacy is useful when $\lim_{n\to\infty} \mathrm{EFF}_h$ is finite because then the limiting power is between the level α and one. From (6.5) we see that when $h = \theta$ and $m_1(W) \neq 0$ this is the case when $\gamma\theta^{1/2}\sigma^{-1} = cn^{-1/2}$ for some $c > 0$. We obtain the following corollary which follows by extending Theorem 6.1 to sequences of $\gamma\theta^{1/2}/\sigma$. For extensions to uniform convergence for curve estimates, see Einmahl and Mason (2005).

Corollary 6.1: *Assume Model (6.2) and the conditions of Theorem 6.1 and Proposition 6.2.*
(a) Suppose that $m_{1,b}(W) \neq 0$ for some b in $(0, 1]$. Then there exists $a_0 \in (0, 1]$ such that the smallest h that maximizes

$$\lim_{n\to\infty} n^{-1/2}\, \mathrm{EFF}\left(\widehat{\beta}_h(x_0)\right)$$

satisfies $h/\theta \to a_0$ for $\theta \to 0$. Moreover, $m_{1,a_0}(W) \neq 0$, and if $\gamma\theta^{1/2}/\sigma = cn^{-1/2}$, then

$$\sup_h \left[\lim_{n\to\infty} \mathrm{EFF}_h\right] = \sqrt{(3/2)\, f(x_0)}\, m_{1,a_0}(W)\, a_0^{-3/2} c + o\left(\gamma\theta^{1/2}/\sigma\right).$$

(b) The smallest optimal h in part (a) equals θ if and only if $a^{-3/2}m_{1,a}(W)$ is maximized subject to $a \leq 1$ by $a = 1$.
(c) If $m_{1,b}(W) = 0$ for all $b \in (0, 1]$, that is $W(\cdot)$ is symmetric about zero, and if $m_{2,b}(W) \neq 0$ for some $b \in (0, 1]$, then there exists $a_0 \in (0, 1]$ such that the smallest h that maximizes

$$\lim_{n\to\infty} n^{-1/2}\, \mathrm{EFF}\left(\widehat{\beta}_h(x_0)\right)$$

satisfies $h/\theta \to a_0$ for $\theta \to 0$. Moreover $m_{2,a_0}(W) \neq 0$, and if $(\gamma \theta^{3/2}/\sigma) = cn^{-1/2}$, then

$$\sup_h \left[\lim_{n \to \infty} \mathrm{EFF}_h \right] = \sqrt{3/2} \, [f(x_0)]^{-1/2} \, f'(x_0) \, \{ m_{2,a_0}(W)$$
$$- \, m_{0,a_0}(W) \, /3 \} \, a_0^{1/2} c + o\left(\gamma \theta^{3/2}/\sigma \right).$$

Remark 6.6: Doksum (1966) considered minimax linear rank tests for models where $Y \overset{\mathcal{L}}{=} X + V(X)$ with $V(X) \geq 0$ and $\|F_Y - F_X\|_\infty \geq \theta$ and found that the least favorable distribution for testing $H : V(\cdot) = 0$ has $Y = X + V_0(X)$ with X uniform$(0,1)$ and $V_0(x) = [a + \theta - x]_+ \mathbf{1}\{x \geq \xi\}$, with $\xi \in [0, 1 - \theta]$. He considered horizontal Pitman alternatives with $\|V_0\|^{-\infty} = \theta \to 0$ as $n \to \infty$. Fan (1992), and Fan and Gijbels (1996) considered minimax kernel estimates for models where $Y = \mu(X) + \epsilon$ and found that the least favorable distribution for estimation of $\mu(x_0)$ using asymptotic mean squared error has $Y = \mu_0(X) + \epsilon$ with

$$\mu_0(x) = \frac{1}{2} b_n^2 \left[1 - c \left(\frac{x - x_0}{b_n} \right)^2 \right]_+$$

where $b_n = c_0 n^{-1/5}$, for some positive constants c and c_0. Again, $\|\mu_0\|^{-\infty} \to 0$ as $n \to \infty$. Similar results were obtained for the white noise model by Donoho and Liu (1991a) and Donoho and Liu (1991b). Lepski and Spokoiny (1999), building on Ingster (1982), considered minimax testing using kernel estimates of $\mu(x)$ and a model with normal errors ϵ and $\mathrm{Var}(\epsilon) \to 0$ as $n \to \infty$. Their least favorable distributions (page 345) are random linear combinations of functions of the form

$$\mu_j(x) = \left(h^{1/2} \int W^2(t) \, dt \right)^{-1} W\left(\frac{x - t_j}{h} \right),$$

which seems to indicate a model that depends on h. Here each $\mu_j(x)$ has $\|\mu_j\|^{-\infty} \to 0$ as $n \to \infty$.

Remark 6.7: Global Alternatives. The idea of selecting h by maximizing the estimated efficacy of the t-statistic extends to global alternatives where the alternative is $\|\mu(\cdot) - \mu_Y\|_\infty > 0$ or $\|\beta\|_\infty > 0$. Efficacies can be built from test statistics in the literature such as those based on ratios of residual sums of squares for the two models under consideration or on

$$\sum_{i=1}^{n} \left[\widehat{\mu}_{2,h}(X_i) - \widehat{\mu}_{1,h}(X_i) \right]^2 V(X_i),$$

where $\widehat{\mu}_{2,h}(\cdot)$ and $\widehat{\mu}_{1,h}(\cdot)$ and estimates of $\mu(\cdot)$ for two models and $V(\cdot)$ is a weight function (see Azzalini, Bowman, and Härdle (1989), Raz (1990), Doksum and Samarov (1995), Hart (1997), Aït-Sahalia, Bickel and Stoker (2001), Fan, Zhang and Zhang (2001), Zhang (2003b)).

Note that varying bandwidths can also be used for these types of situations. In the case of

$$\sum_{i=1}^{n} [\widehat{\mu}_{2,h}(X_i) - \mu_{1,h}(X_i)]^2 V(X_i),$$

we could think of

$$T_h(x_0) = [\widehat{\mu}_{2,h}(x_0) - \widehat{\mu}_{1,h}(x_0)] V^{1/2}(x_0)$$

as a local test statistic and select h to maximize its squared efficacy. For instance, for testing $\beta(x) = 0$ versus $\|\beta\|_\infty > 0$ using $\widehat{\mu}_{1,h}(X_i) = \bar{Y}$ and $\widehat{\mu}_{2,h}(X)$ a locally linear kernel estimate, the efficacy of $T_h(x_0)$ with $V(\cdot) = 1$ will be approximately the same as the efficacy we obtained from $\widehat{\beta}_h(x_0)$. Let $h(x_0)$ denote the maximizer of the squared efficacy of $T_h(x_0)$ and set $h_i = h(X_i)$. Then the final test statistic would be

$$\sum_{i=1}^{n} \left[\widehat{\mu}_{2,h_i}(X_i) - \bar{Y} \right]^2 V(X_i).$$

Fan, Zhang and Zhang (2001) and Zhang (2003b) proposed a "multiscale" approach to the varying coefficient model where they select h by maximizing a standardized version of the generalized likelihood ratio (GLR) statistic which is defined as the logarithm of the ratio of the residual sums of squares for the two models under consideration. The standardization consists of subtracting the null hypothesis asymptotic mean and dividing by the null hypothesis asymptotic standard deviation. That is, for situations where the degrees of freedom of the asymptotic chi-square distribution of the GLR statistic tends to infinity they are maximizing the efficacy of the GLR statistic.

Other extensions would be to select h to maximize

$$T_{h,\lambda}\left(\mathbf{x}_0\right) = \sum_{j=1}^{g} t_{h,\lambda}^2\left(x_{0j}\right) \bigg/ g \quad \text{or}$$

$$T_{h,\lambda}^*\left(\mathbf{x}_0\right) = \sum_{i=1}^{g} \widehat{\beta}_{h,\lambda}^2\left(x_{0i}\right) \bigg/ \text{SE}\left(\sum_{j=1}^{g} \widehat{\beta}_{h,\lambda}^2(x_{0j}) \right)$$

where $\mathbf{x}_0 \equiv (x_{01}, x_{02}, \ldots, x_{0g})^T$ is a vector of grid points. Finally, we could select h by, for some weight function $v(\cdot)$, maximizing the integrated efficacy $\text{IEFF} \equiv \int \text{EFF}(x) \, v(x) \, dx$.

Remark 6.8: Hall and Heckman (2000) used the maximum of local t-statistics to test the global hypothesis that $\mu(\cdot)$ is monotone. Their estimated local regression slope is the least squares estimate based on k nearest neighbors and the maximum is over all intervals with k at least 2 and at most m. They established unbiasedness and consistency of their test rule under certain conditions.

Remark 6.9: Hall and Hart (1990) found that for a nonparametric two sample problem with null hypothesis of the form $H_0 : \mu_1(x) = \mu_2(x)$ for all x, a test statistic which is a scaled version of $n^{-1} \sum [\widehat{\mu}_{2h}(X_i) - \widehat{\mu}_{1h}(X_i)]^2$ has good power properties when the bandwidths in $\widehat{\mu}_{jh}(x), j = 1, 2$ satisfy $h/n \to p > 0$ as $n \to \infty$ provided

$$s(\mu_1, \mu_2) \equiv \int_0^1 \left\{ \int_t^{t+p} [\mu_2(x) - \mu_1(x)] \, f(x) \, dx \right\}^2 dt > 0$$

when $f(x)$ has support $(0, 1)$. Note that the asymptotic behavior of $n^{1/2} s^{1/2}(\mu_1, \mu_2)$ for sequences of Pitman alternatives depends completely on whether the length of the set on which $[\mu_2(x) - \mu_1(x)]$ differs from zero tends to zero as $n \to \infty$, just as in our case, where the limiting behavior of the scaled efficacy depends on whether the length of $\{x : |\beta(x)| > 0\}$ tends to zero.

Remark 6.10: Godtliebsen, Marron, and Chaudhuri (2004) consider two-dimensional $\mathbf{X}$ and use the sum of two squared local directional t-statistics. We study optimal bandwidths for this case in a subsequent paper.

6.2.3. *Asymmetric windows*

In the previous section we observed that the limiting power depends crucially on $m_1(W)$ and on $m_{1,a}(W)$. If these are zero, the limiting efficacy is of smaller order than if they are not. A simple method for increasing the limiting power is to use locally linear methods over asymmetric windows of the form

$$N_{h,\lambda}(x_0) \equiv [x_0 - (1 - \lambda) \, h, \, x_0 + (1 + \lambda) \, h], \quad h > 0, \quad -1 \le \lambda \le 1.$$

Now the test statistic is the slope $\widehat{\beta}_{h,\lambda}(x_0)$ of the least squares line for the data $D_{h,\lambda}(x_0) \equiv \{(x_i, y_i) : x_i \in N_{h,\lambda}(x_0)\}$. The efficacy $\text{EFF}_{h,\lambda}$ will be computed conditionally on $X \in N_{h,\lambda}(x_0)$ and in Model (6.2) with $h/\theta = a \in (0, 2]$ the expansion of $\mathrm{Cov}_{h,\lambda}(X, Y)$ will be in terms of the integrals

$$m_{j,a,\lambda}(x_0) \equiv \int_{-(1-\lambda)a}^{(1+\lambda)a} s^j W(s)\, ds, \quad \text{if } (1+\lambda)\, a \leq 1, \ (1-\lambda)\, a \leq 1, \quad (6.6)$$

$$\equiv \int_{-(1-\lambda)a}^{1} s^j W(s)\, ds, \quad \text{if } (1+\lambda)\, a > 1, \ (1-\lambda)\, a \leq 1, \quad (6.7)$$

$$\equiv \int_{-1}^{(1+\lambda)a} s^j W(s)\, ds, \quad \text{if } (1+\lambda)\, a \leq 1, \ (1-\lambda)\, a > 1. \quad (6.8)$$

If W is symmetric, the term of order $(\gamma\theta^2/\sigma)h^{-3/2}$ in (6.5) will not vanish and we will maintain high test efficiency. Even if W is not symmetric, absolute efficacy is always increased because

$$\sup_{h,\lambda} |\mathrm{EFF}_{h,\lambda}| \geq \sup_{h} |\mathrm{EFF}_{h,0}|.$$

Note that with asymmetric windows we need $h \geq 2\theta$ to make sure the intervals $N_{h,\lambda}(x_0)$ cover the support of $W(\cdot)$.

Theorem 6.2: Assume that the density f of X has a bounded, continuous second derivative at x_0 and that $f(x_0) > 0$. Also assume that we can find $\lambda \in [0, 1]$ such that $m_{j,0,\lambda}(W) \neq 0$ for either $j = 0$ or $j = 1$. Then, in Model (6.2), as $\theta \to 0$ and $h \to 0$ with $h \geq 2\theta$, the following are true.

(a) $\mathrm{Cov}_{h,\lambda}(X, Y) = \dfrac{\gamma\theta}{2h} \{ m_{1,a,\lambda}(W)\, \theta - m_{0,a,\lambda}(W)\, \lambda h + o(h) + o(\theta) \}.$

(b) $\tau_{h,\theta,\lambda}(x_0) \equiv \lim\limits_{n\to\infty} n^{-1/2}\, \mathrm{EFF}\left(\widehat{\beta}_{h,\lambda}(x_0) \right) = \sqrt{3/2}\, \sigma^{-1} \gamma\theta h^{-3/2}\, [f(x_0)]^{1/2}$
$$\times \{ m_{1,a,\lambda}(W)\, \theta - m_{0,a,\lambda}(W)\, \lambda h + o(\theta) + o(h) \}$$

and $|\tau_{h,\theta}(x_0)|$ is maximized subject to $h \geq 2\theta$ by $h = 2\theta$.

Proof: Part (a): In Lemma 6.5, part (b), we show that

$$P_{h,\lambda}(x_0) = 2f(x_0)\, h + 2f'(x_0)\, \lambda h^2 + o(h^2) \quad \text{and}$$

$$E_{h,\lambda}(X) = x_0 + \lambda h + [f'(x_0)/f(x_0)]\, h^2/3 + o(h^2).$$

Other terms are from equation (6.3) with ranges of integration in equations (6.6)-(6.8) rather than from -1 to 1; and the proof of Lemma 6.3, part (e). **Part (b):** We have

$$\lim_{n\to\infty} n^{-1/2}\, \mathrm{EFF}_h = \sigma^{-1} \mathrm{Cov}_{h,\lambda}(X, Y) \left[(P_{h,\lambda}(x_0)) \Big/ \mathrm{Var}_{h,\lambda}(X) \right]^{1/2}.$$

By Lemma 6.5, part (d), we get

$$P_{h,\lambda}(x_0)/\text{Var}_{h,\lambda}(X) = 6f(x_0)h^{-1} + O(1).$$

Equation (6.4) follows from this and part (a). □

Corollary 6.2: *Under the conditions of Theorem 6.2, for the Model (6.2), and $\theta \to 0$, then, if $\gamma\theta^{1/2}/\sigma = cn^{-1/2}$,*

$$\sup_{h\geq 2\theta} \left[\lim_{n\to\infty} \text{EFF}_h \right] = \sqrt{3/2}\,[f(x_0)]^{1/2}\,[m_{1,a,\lambda}(W) - 2m_{0,a,\lambda}(W)\,\lambda]\,c + o(\theta).$$

The optimal h cannot be such that $h/\theta \to 0$.

Proposition 6.3: *Assume the conditions of Theorem 6.2 with $\lambda \in (0,1)$ and that $W''(\cdot)$ exists and is bounded in a neighborhood of zero. Then, as $h/\theta \to 0$,*

$$\lim_{n\to\infty} n^{-1/2}\text{EFF}\left(\widehat{\beta}_h(x_0)\right) = o\left(\sigma^{-1}\gamma\theta h^{-3/2}\{m_{1,a,\lambda}(W)\,\theta - m_{0,a,\lambda}(W)\,\lambda h\}\right).$$

Proof: The proof is similar to the proof of Proposition 6.2. □

Corollary 6.3: *Assume Model (6.2) and the conditions of Theorem 6.2 and Proposition 6.3 hold. Then there exists $a_0 \in (0,2]$ such that the smallest h that maximizes*

$$\lim_{n\to\infty} n^{-1/2}\,\text{EFF}_{h,\lambda}\left(\widehat{\beta}_{h,\lambda}(x_0)\right)$$

satisfies $h/\theta \to a_0$ for $\theta \to 0$. Moreover, if $\gamma\theta^{1/2}/\sigma = cn^{-1/2}$, then

$$\sup_{h} \left[\lim_{n\to\infty} \text{EFF}_{h,\lambda} \right] = \sqrt{(3/2)\,f(x_0)}\,[m_{1,a_0,\lambda}(W) - m_{0,a_0,\lambda}(W)\,a_0\lambda]$$

$$\times a_0^{-3/2}c + o\left(\gamma\theta^{1/2}/\sigma\right).$$

Remark 6.11: Computations. We select h and λ to maximize the absolute t-statistic $|t_{h,\lambda}(x_0)|$ computed for the data from $D_{h,\lambda}(x_0)$. This is not a challenge computationally since this is a search for the largest absolute t-statistic over **all** possible intervals that include x_0. We approximate this search by using all order statistic intervals of the form $(x_{(i)}, x_{(j)})$ that contain x_0 and at least k points from $\{x_1, x_2, \ldots, x_n\}$ where $k \approx 0.05n$ is a good choice.

Remark 6.12: Critical Values. Under the hypothesis that X and Y are independent, we get a distribution free critical value by using the permutation distribution of the maximum absolute t-statistic obtained by maximizing t-statistics over local neighborhoods. By permuting $(Y_1, Y_2, \ldots, Y_n)$, leaving $(X_1, X_2, \ldots, X_n)$ fixed and computing the maximum t-statistic for these permuted data $(X_i, Y_i), 1 \leq i \leq n$, we get $n!$ equally likely (under H) values of the maximum absolute t-statistic. A subset of the $n!$ permutations are chosen at random to reduce computational complexity. The dashed histogram in Figure 6.1 shows the approximated distribution using neighborhoods centered at x_0. We use 1000 random permutations. For each permutation of the data set, a set of ten bandwidths are used, ranging from the smallest which will ensure at least 20 data points in the neighborhood up to a bandwidth which selects all available data. The solid histogram in Figure 6.1 shows the approximated distribution using asymmetric neighborhoods which include x_0. The set of candidate endpoints of the neighborhoods are the quantiles $\widehat{x}_\alpha = \widehat{F}^{-1}(\alpha)$ of the observed X variable, with α uniformly spaced from zero up to one. These lead to 45 total neighborhoods, not all of which will include x_0. Since it searches over a larger class of neighborhoods, it is not surprising that the simulated distribution using asymmetric neighborhoods has a somewhat larger right tail than that for symmetric intervals.

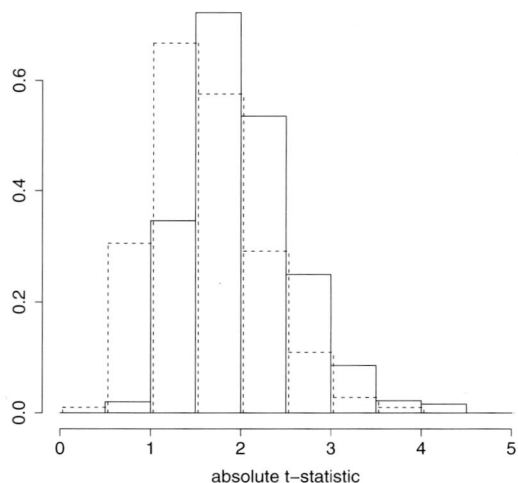

Figure 6.1.   Approximate distribution of maximum absolute t-statistic under the null hypothesis, when using symmetric neighborhoods (dashed) and asymmetric neighborhoods (solid).

Remark 6.13: Testing a Parametric Hypothesis. Suppose that we want to test a linear model against the alternative that near x_0 a nonlinear model leads to a larger signal to noise ratio. Thus, near x_0, a local t-statistic may be significant when the global t-statistic is not, or visa versa. If we introduce $Y_i' \equiv Y_i - (\widehat{\alpha} + \widehat{\beta} X_i)$ where $\widehat{\alpha} + \widehat{\beta} x$ is the global least squares fit, and if we have in mind horizontal alternatives where the optimal h tends to zero, the previous results apply because $(\widehat{\alpha} - \alpha)$ and $(\widehat{\beta} - \beta)$ will converge to zero at a faster rate than $(\widehat{\beta}_h(x_0) - \beta_h(x_0))$ and thus testing a linear model for Y_i is asymptotically equivalent to testing that $E(Y'|X = x)$ is constant. More generally, we may want to test a parametric model of the form $E(Y|X = x) = g(x; \boldsymbol{\beta})$ against the alternative that we get higher power near x_0 by using a nonparametric model. If we set $Y' = Y_i - g(x; \widehat{\boldsymbol{\beta}})$ for suitable $\sqrt{n}$ consistent $\widehat{\boldsymbol{\beta}}$ and smooth $g(\cdot)$, the remark about the linear hypothesis still applies.

6.3. Examples

This section illustrates the behavior of the bandwidth selection procedure based on maximizing absolute efficacy, and compares it with methods based on MSE, using simulated and real (currency exchange) data sets. We utilize the model

$$\mu_\nu(x) = \nu x + \exp\left(-40\left(x - 0.5\right)^2\right), \quad 0 \le x \le 1,$$

and focus on the "flat bump" ($\nu = 0$) and "sloped bump" ($\nu = 1.25$) models shown in Figure 6.2. Assume $X_1, X_2, \ldots, X_n$ are uniform on $(0, 1)$, and that $Y_i - \mu(X_i)$ is normal with mean zero and variance σ^2. Given a sample of data, MSE will be approximated using leave-one-out cross validation.

6.3.1. *Symmetric windows*

The left plot in Figure 6.3 shows both theoretical MSE(h) and limiting efficacy ($|\tau_h(x_0)|$) for a range of h in the sloped bump model, with $x_0 = 0.6$, $n = 1000$, and $\sigma^2 = 0.05$. The mean squared error is minimized by choosing $h = 0.07$, while $|\tau_h(x_0)|$ is maximized when $h = 0.6$. This discrepancy is to be expected: bias of the local linear model estimator for $\mu(x)$ is a critical component of the mean squared error, and the linear fit clearly becomes quickly inappropriate for x outside of the range $[0.53, 0.67]$. Alternatively, imagine one were to ask: "What interval centered at 0.6 would give the best chance of rejecting the hypothesis that $\mu(x)$ is constant?" For sufficiently large σ^2, the procedure tells us to use the entire range of data for the most

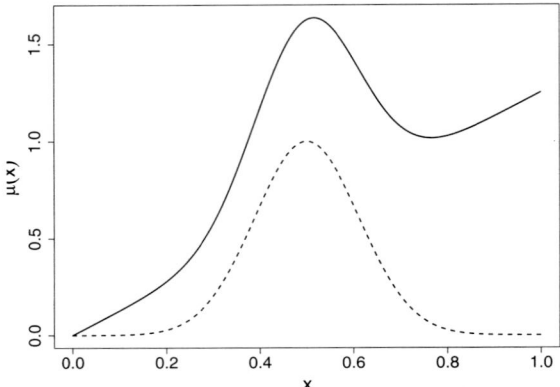

Figure 6.2.   The "sloped bump" (solid) and "flat bump" (dashed) models.

power to reject the hypothesis that $\mu(\cdot)$ is constant. Here, bias is not at issue, we are instead searching for prevalent linear features in the bivariate relationship. There is a large drop in power if we use the MSE optimal h rather than the power optimal bandwidth.

There is, however, a local maximum in the left plot near $h = 0.18$. Switching to the flat bump model, we see that the chosen bandwidth is indeed $h \approx 0.18$; see the right plot of Figure 6.3. The point is the following: When the overall linear trend is present and $\sigma^2 = 0.05$, the local downslope of the bump was not a sufficiently significant feature relative to the overall slope. Once $\sigma^2 < 0.008$, the power optimal bandwidth is chosen small. This illustrates the different behavior for "vertical" and "horizontal" alternatives.

The optimal bandwidth as chosen to maximize $|\tau_h(x_0)|$ does not depend on n, but it does depend on σ^2. This is appropriate given σ^2 is a feature of the bivariate distribution of X and Y, while n is not. Using the flat bump model, Figure 6.4 shows how the theoretically optimal bandwidth chosen by both minimizing MSE and maximizing $|\tau_h(x_0)|$ varies with sample size, and Figure 6.5 shows the same as σ^2 varies. Note that as σ^2 increases, the power optimal bandwidth also grows. Figures 6.4 and 6.5 also depict the bandwidth selection procedures on simulated data sets. A pair of dots connected by a vertical dotted line give the bandwidth selected by minimizing the leave-one-out cross validation (filled circle) and by maximizing the absolute t-statistic ($|t_h(x_0)|$) (open circle), using the same simulated data set.

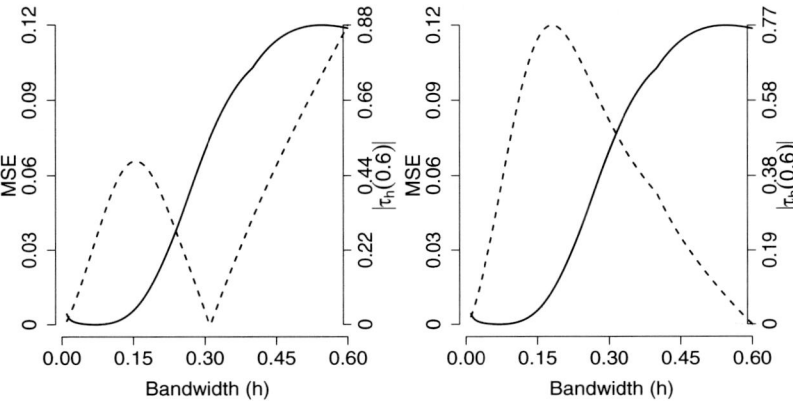

Figure 6.3. Plot of MSE (solid) and $|\tau_h(x_0)|$ (dashed) for $x_0 = 0.6$, $n = 1000$, and $\sigma^2 = 0.05$ when using the sloped bump model (left plot) and the flat bump model (right plot).

6.3.2. *Asymmetric windows*

We now construct the optimal asymmetric windows, as first mentioned in Remark 6.11. Adjacent points along the x axis will often "share" the same optimal neighborhood. Figure 6.6 shows the result of maximizing $|t_{h,\lambda}(x_0)|$ for each value of x_0. For this case, we use the sloped bump model with $n = 1000$ and $\sigma^2 = 0.05$. For any chosen x_0, one can read up to find the neighborhood for that x_0 by finding black horizontal line segment which intersects that vertical slice. Once the black segment is found, however, note that the corresponding neighborhood is in fact the **entire** length of the line segment, both the black and gray portions. For example, if $x_0 = 0.75$, the power optimal neighborhood extends from zero up to approximately 0.88. Figure 6.7 depicts the neighborhoods chosen in the same way, except now based on maximizing the absolute t-statistic when using a simulated data set of size $n = 1000$.

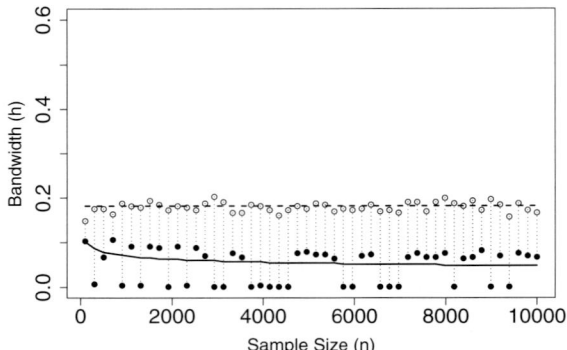

Figure 6.4.   Plot of bandwidth found by minimizing theoretical MSE (solid) and max-
imizing theoretical $|\tau_h(x_0)|$ (dashed) for $x_0 = 0.6$ and $\sigma^2 = 0.05$ when using the flat
bump model. The dots represent the chosen bandwidth based on maximizing the absolute t-statistic (open) and minimizing the leave-one-out cross validation MSE (filled)
using a randomly generated data set of the given size.

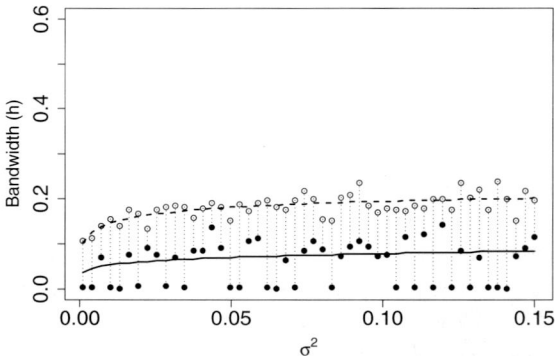

Figure 6.5.   Plot of bandwidth found by minimizing theoretical MSE (solid) and maxi-
mizing theoretical $|\tau_h(x_0)|$ (dashed) for $x_0 = 0.6$ and $n = 1000$ when using the flat bump
model. The dots represent the chosen bandwidth based on maximizing the absolute t-
statistic (open) and minimizing the leave-one-out cross validation MSE (filled) using a
randomly generated data set of the given error variance (σ^2).

6.3.3. *Application to currency exchange data*

Figure 6.8 shows the results of this approach applied to daily Japanese Yen
to Dollar currency exchange rate data for the period January 1, 1992 to

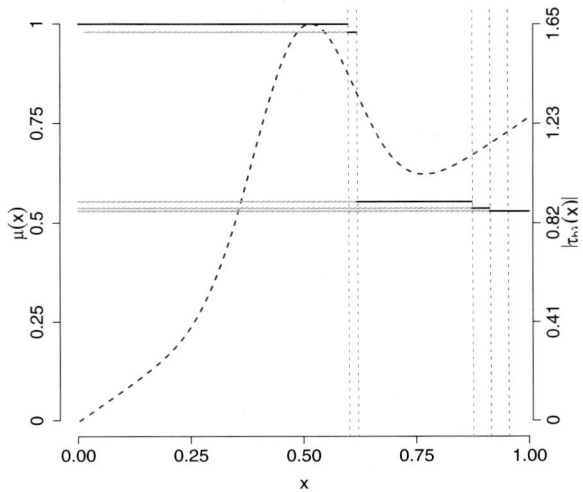

Figure 6.6.   Plot of bandwidth chosen by maximizing $|\tau_{h,\lambda}(x_0)|$ for all x_0, for the case $n = 1000$, and $\sigma^2 = 0.05$ when using the sloped bump model.

April 16, 1995. The variable along the X axis is the standardized return, i.e. the logarithm the ratio of today's exchange rate to the yesterday's rate, standardized to have mean zero and variance one. The response variable is the logarithm of the volume of currency exchange for that day. Figure 6.9 displays, as the solid line, the values of $\widehat{\mu}_{h,\lambda}(x_0)$ for all values of x_0, when the bandwidth is chosen by maximizing $|t_{h,\lambda}(x_0)|$. The plot also shows, as a dashed line, an estimate of formed using local linear regression with tri-cube weighting function (the `loess()` function implemented in R). The smoothing parameter, called span, is chosen by minimizing the generalized cross validation; the chosen value of 0.96 means that the local neighborhood is chosen sufficiently large to include 96% of the data.

Figure 6.10 shows the estimated correlation curve, $\widehat{\rho}_{h,\lambda}(\cdot)$, as described in Remark 6.5. Again, the windows required for the local slope and variance estimates are chosen to maximize the absolute t-statistics.

This problem is well-suited to our approach. We seek those features of the bivariate relationship which these data have the most power to reveal. In Figures 6.8 and 6.10, we observe that there are two major features, one representing when there is a decrease in the exchange rate relative to

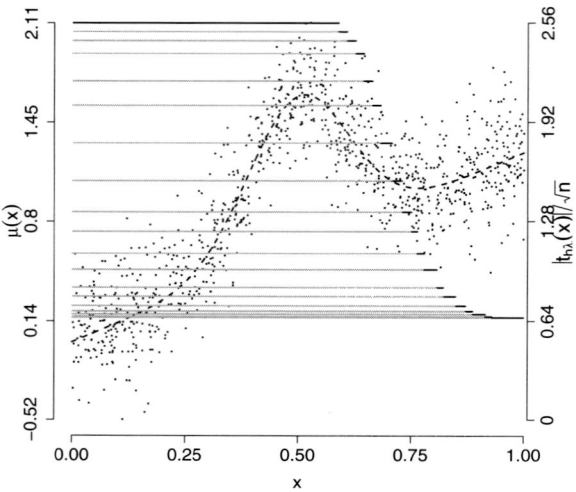

Figure 6.7.   Plot of bandwidth chosen by maximizing the absolute t-statistic for all x_0, using one simulated data set, for the case $n = 1000$, and $\sigma^2 = 0.05$ when using the sloped bump model.

the previous day (return less than one), and another feature representing when there is an increase in the exchange rate relative to the previous day. This finding is consistent with the discussion in Karpoff (1987), who cites several studies which show a positive correlation between absolute return and volume.

6.4. Appendix

Lemma 6.1: *Let $A(x) \equiv \int_{-\infty}^{x} g(t)\,dt$ for some integrable function $g(\cdot)$. If $g''(\cdot)$ is bounded and continuous in a neighborhood of x, then*

$$A(x + h) - A(x - h) = 2A'(x)\,h + A'''(x)\,h^3/3 + o(h^3) \qquad (6.9)$$
$$= 2g(x)\,h + g''(x)\,h^3/3 + o(h^3)\,.$$

Proof:

$$A(x_0 \pm h) = A(x_0) + A'(x_0)\,(\pm h) + A''(x_0)\,h^2/2 + A'''(x_0)\,(\pm h^3)/6 + o(h^3)\,.\square$$

Lemma 6.2: *For constants c_1, c_2, c_3, and c_4 with $c_1 \neq 0$,*

(a) $\qquad (c_1 + c_2 h^2)^{-1} = c_1^{-1} - c_2 c_1^{-2} h^2 + o(h^2)\quad$ *and*

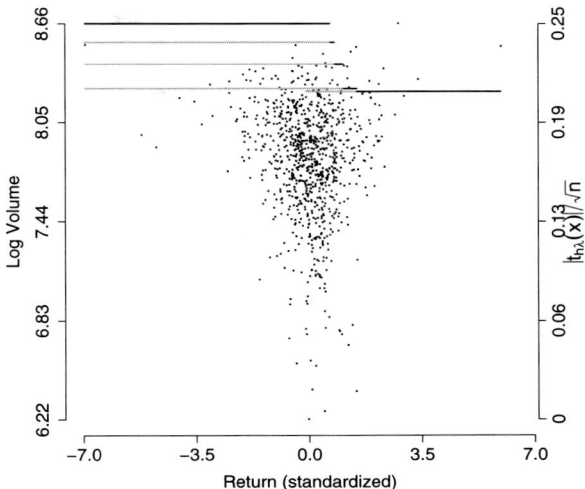

Figure 6.8. Results of analysis of Japanese Yen to Dollar exchange rates. Plot shows bandwidths chosen by maximizing the absolute t-statistics for all x_0.

(b) $\qquad \left(c_1 + c_2 h^2\right)^{-1} \left(c_3 + c_4 h^2\right) = c_1^{-1} \left(c_3 + c_4 h^2\right) - c_2 c_3 c_1^{-2} h^2 + o\left(h^2\right).$

Proof: For part (a), Taylor expand $g(t) = (c_1 + c_2 t)^{-1}$ around $t = 0$. Part (b) follows directly from (a). $\qquad \square$

Lemma 6.3: *If $f''(\cdot)$ is continuous and bounded in a neighborhood of x_0, then*

(a) $\qquad P_h(x_0) = 2f(x_0)\, h + f''(x_0)\, h^3/3 + o\left(h^3\right),$

(b) $\qquad E_h(X) = x_0 + [f'(x_0)/f(x_0)]\, h^2/3 + o\left(h^2\right),$

(c) $\qquad \mathrm{Var}_h(X) = h^2/3 + o\left(h^2\right),$

(d) $\qquad [P_h(x_0)/\mathrm{Var}_h(X)] = 6f(x_0)\, h^{-1} + O(1),\ and$

(e) $\qquad \mathrm{Cov}_h(X,Y) = \dfrac{\gamma\theta}{2hf(x_0)} \Big\{ -\dfrac{1}{3}m_0(W)\, f'(x_0)\, h^2 + O\left(\theta h^2\right)$
$$+ f(x_0)\, m_1(W)\, \theta + f'(x_0)\, m_2(W)\, \theta^2 + o\left(h^2\right) + o\left(\theta^2\right) \Big\}.$$

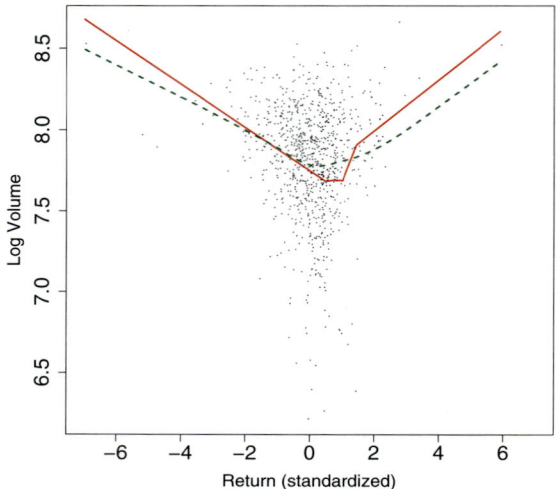

Figure 6.9.   Results of analysis of Japanese Yen to Dollar exchange rates. Plot shows $\hat{\mu}_{h,\lambda}(x)$ for various values of x for both bandwidth selected by maximizing the absolute t-statistics (solid), compared with fit using the R function `loess()` with smoothing parameter `span = 0.96` (dashed).

Proof: Part (a): Use Equation (6.9) with $g(\cdot) = f(\cdot)$.
Part (b): Set $Z \equiv X - x_0$, then $\mathrm{E}_h(X) = x_0 + \mathrm{E}_h(Z)$ with $f_Z(z) = f(z+x_0)$. Write $f_0(z) = f_Z(z)$, then

$$P_h(x_0)\,\mathrm{E}_h(Z) = \int_{-h}^{h} z f_0(z)\,dz = A(h) - A(-h)\,.$$

Now Equation (6.9) with $g(z) = z f_0(z)$, $g(0) = 0$, $g''(0) = 2f'(x_0)$ implies that

$$\mathrm{E}_h(Z) = \left[2f'(x_0)\,h^3/3 + o\!\left(h^3\right)\right]\Big/ P_h(x_0) = (f'(x_0)\,/f(x_0))\,h^2/3 + o\!\left(h^2\right)$$

by Lemma 6.2. The result follows.
Part (c): Set $Z \equiv X - x_0$, then $\mathrm{Var}(X|N_h(x_0)) = \mathrm{Var}(Z|N_h(0))$ and $f_Z(0) = f(x_0)$. Write $f_0(z)$ for $f_Z(z)$. We have

$$\mathrm{E}_h\!\left(Z^2\right) = \mathrm{E}\!\left(Z^2|N_h(x_0)\right) = \frac{1}{P_h(x_0)} \int_{-h}^{h} z^2 f_0(z)\,dz = \frac{[A(h) - A(-h)]}{P_h(x_0)}\,.$$

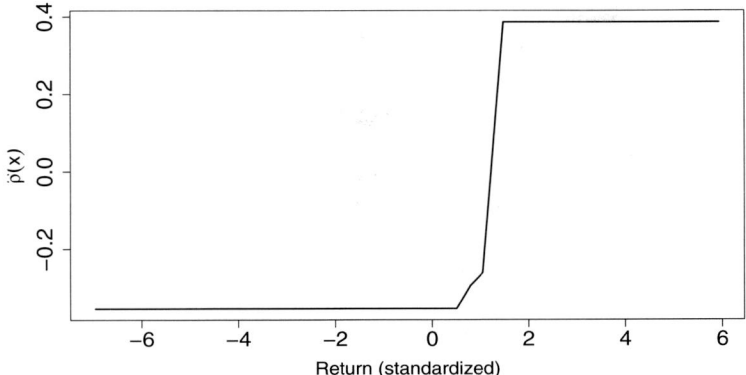

Figure 6.10. Results of analysis of Japanese Yen to Dollar exchange rates. Plot shows estimated correlation curve, $\hat{\rho}_{h,\lambda}(\cdot)$ with window selected by maximizing the absolute t-statistics, as described in Remark 6.5.

Now use Lemma 4.1 with $g(z) = z^2 f_0(z)$. We have $g(0) = g'(0) = 0$, $g''(0) = 2f_0(0) = 2f(x_0)$,

$$\mathrm{E}_h\left(Z^2\right) = \frac{2f(x_0)\,h^2/3}{2f(x_0) + f''(x_0)\,h^2/3} = h^2/3 + o\!\left(h^2\right),$$

$$\mathrm{Var}_h(Z) = \mathrm{E}_h\left(Z^2\right) - \left[\mathrm{E}_h(Z)\right]^2 = h^2/3 + o\!\left(h^2\right).$$

Part (d):

$$[P_h(x_0)\,/\mathrm{Var}_h(X)] = \frac{2f(x_0)\,h + f''(x_0)\,h^3/3 + o\!\left(h^3\right)}{h^2/3 + o\!\left(h^2\right)} = 6f(x_0)\,h^{-1} + O(1).$$

Part (e): Begin with the expression for $\mathrm{Cov}_h(X,Y)$ given in Equation (6.3) and substitute in $P_h(x_0)$ and $\mathrm{E}_h(X)$ given in parts (a) and (b) of this lemma. This gives the following:

$$\mathrm{Cov}_h(X, Y) = \gamma\theta\left(2f(x_0)\,h + f''(x_0)\,h^3/3 + o\!\left(h^3\right)\right)^{-1}$$

$$\times\left[\int_{-1}^{1} s\theta W(s)\,f(x_0 + s\theta)\,ds\right.$$

$$\left. - \int_{-1}^{1} \frac{f'(x_0)}{f(x_0)}\frac{h^2}{3} W(s)\,f(x_0 + s\theta)\,ds + o\!\left(h^2\right)\right].$$

Then using the Taylor expansion of $f(\cdot)$ around x_0 we see that

$$\int_{-1}^{1} s\theta W(s)\, f(x_0 + s\theta)\, ds = f(x_0)\, \theta m_1(W) + f'(x_0)\, \theta^2 m_2(W) + o(\theta^2) \quad \text{and}$$

$$\int_{-1}^{1} \frac{f'(x_0)}{f(x_0)} \frac{h^2}{3} W(s)\, f(x_0 + s)\, ds = m_0(W)\, f'(x_0)\, h^2/3 + o(h^2) + o(\theta^2)$$

where the terms of this second equation which contain $h^2\theta$ or $h^2\theta^2$ are $o(h^2)$ since we are considering behavior under both θ and h going to zero. Combining, we see that

$$\begin{aligned}
\mathrm{Cov}_h(X, Y) = &\left(2f(x_0)\, h + o(h^2)\right)^{-1} \\
&\times \left(f(x_0)\, \theta m_1(W) + f'(x_0)\, m_2(W)\, \theta^2\right. \\
&\left. - m_0(W)\, f'(x_0)\, h^2/3 + o(h^2) + o(\theta^2)\right),
\end{aligned}$$

and the result follows immediately.   □

Lemma 6.4: Let $A(x) \equiv \int_{-\infty}^{x} g(t)\, dt$ for some integrable function $g(\cdot)$. Then

$$\begin{aligned}
D \equiv\ & A(x + (1 + \lambda)\, h) - A(x - (1 - \lambda)\, h) \\
=\ & 2A'(x)\, h + 2A''(x)\, \lambda h^2 + A'''(x)\, (1 + 3\lambda^2)\, h^3/3 + o(h^3) \\
=\ & 2g(x)\, h + 2g'(x)\, \lambda h^2 + g''(x)\, (1 + 3\lambda^2)\, h^3/3 + o(h^3)
\end{aligned}$$

provided $g''(\cdot)$ is bounded and continuous in a neighborhood of x.

Proof: W ehave

$$\begin{aligned}
A(x + (1 + \lambda)\, h) =\ & A(x) + A'(x)\, (1 + \lambda)\, h + A''(x)\, (1 + \lambda)^2\, h^2/2 \\
& + A'''(x)\, (1 + \lambda)^3\, h^3/6 + o(h^3). \\
A(x - (1 - \lambda)\, h) =\ & A(x) - A'(x)\, (1 - \lambda)\, h + A''(x)\, (1 - \lambda)^2\, h^2/2 \\
& - A'''(x)\, (1 - \lambda)^3\, h^3/6 + o(h^3). \quad \square
\end{aligned}$$

Lemma 6.5: If $f''(\cdot)$ is continuous and bounded in a neighborhood of x_0, then

(a)
$$P_{h,\lambda}(x_0) = 2f(x_0)\, h + 2f'(x_0)\, \lambda h^2 + o(h^2),$$

(b)
$$\mathrm{E}_{h,\lambda}(X) = x_0 + \lambda h + [f'(x_0)/f(x_0)]\, h^2/3 + o(h^2),$$

(c)
$$\mathrm{Var}_{h,\lambda}(X) = h^2/3 + o(h^2), \text{ and}$$

(d)
$$[P_{h,\lambda}(x_0)/\mathrm{Var}_{h,\lambda}(X)] = 6f(x_0)\, h^{-1} + O(1).$$

Proof: Part (a): Use Lemma 6.4 with $g(\cdot) = f(\cdot)$.

Part (b): Again set $Z \equiv X - x_0$, then

$$\mathrm{E}_{h,\lambda}(Z) = \left(2f(x_0)\,\lambda h^2 + 2f'(x_0)/3\left(1 + 3\lambda^2\right)h^3 + o\!\left(h^3\right)\right)\Big/\left(P_{h,\lambda}(x_0)\right)$$

$$= \frac{2f(x_0)\,\lambda h + 2f'(x_0)\left(1 + 3\lambda^2\right)h^2/3 + o\!\left(h^2\right)}{2f(x_0) + 2f'(x_0)\,\lambda h + o(h)}$$

$$= \lambda h + \left[f'(x_0)/f(x_0)\right]\left(1 + 3\lambda^2\right)h^2/3 - \left[f'(x_0)/f(x_0)\right]\lambda^2 h^2$$

$$= \lambda h + \left[f'(x_0)/f(x_0)\right]h^2/3 + o\!\left(h^2\right).$$

Part (c): $\mathrm{E}_{h,\lambda}\!\left(Z^2\right) = \dfrac{1}{P_{h,\lambda}(x_0)}\left[2f(x_0)\left(1 + 3\lambda^2\right)h^3/3\right]$

$$= \left(1 + 3\lambda^2\right)h^2/3 = h^2/3 + \lambda^2 h^2 + o\!\left(h^2\right).$$

$$\mathrm{Var}_{h,\lambda}(Z) = h^2/3 + \lambda^2 h^2 - \lambda^2 h^2 + o\!\left(h^2\right) = h^2/3 + o\!\left(h^2\right).$$

Part (d):

$$\frac{P_{h,\lambda}(x_0)}{\mathrm{Var}_{h,\lambda}(X)} = \frac{2f(x_0)\,h + 2f'(x_0)\,\lambda h^2 + o\!\left(h^2\right)}{h^2/3 + o(h^2)} = 6f(x_0)\,h^{-1} + O(1).\ \square$$

Acknowledgements

Doksum's research work is partially supported by NSF grants DMS-9971309 and DMS-0505651. Schafer's research work is partially supported by NSF grant DMS-9971301. We are grateful to Alex Samarov for many helpful comments.

References

Aït-Sahalia, Y., Bickel, P. J., and Stoker, T. M. (2001) Goodness-of-fit Tests for Kernel Regression with an Application to Option Implied Volatilities. *J. of Econ.*, **105** 363–412.

Azzalini, A., Bowman, A. W. and Hardle, W. (1989). On the Use of Nonparametric Regression for Model Checking *Biometrika*, **76** 1–11.

Behnen, K. and Hušková, M. (1984). A simple algorithm for the adaptation of scores and power behavior of the corresponding rank test. *Comm. in Stat. - Theory and Methods*, **13** 305–325.

Behnen, K. and Neuhaus, G. (1989). Rank Tests with Estimated Scores and their Applications. Teubner, Stuttgart.

Bell, C. B. and Doksum, K. A. (1966). "Optimal" One-Sample Distribution-Free Tests and Their Two-Sample Extensions. *Ann. Math. Stat.*, **37** 120–132.

Bickel, P. J. and Doksum, K. A. (2001). Mathematical Statistics. Basic Ideas and Selected Topics, Volume I. Prentice Hall, New Jersey.

Bickel, P. J., Ritov, Y. and Stoker, T. M. (2006). Tailor-made Tests for Goodness-of-Fit to Semiparametric Hypotheses. *Ann. Statist.*, **34**. To Appear.

Bjerve, S. and Doksum, K. (1993). Correlation Curves: Measures of Association as Functions of Covariate Values. *Ann. Statist.*, **21** 890–902.

Blyth, S. (1993). Optimal Kernel Weights Under a Power Criterion. *J. Amer. Statist. Assoc.*, **88** 1284–1286.

Chaudhuri, P. and Marron, J. S. (1999). SiZer for Exploration of Structures in Curves. *J. Amer. Statist. Assoc.*, **94** 807–823.

Chaudhuri, P. and Marron, J. S. (2000). Scale Space View of Curve Estimation. *Ann. Statist.*, **28** 408–428.

Claeskens, G. and Hjorth, N. L. (2003). The Focused Information Criterion. *J. Amer. Statist. Assoc.*, **98** 900–916.

Doksum, K. (1966). Asymptotically Minimax Distribution-free Procedures. *Ann. Math. Stat.*, **37** 619–628.

Doksum, K., Blyth, S., Bradlow, E., Meng, X.-l. and Zhao, H. (1994). Correlation Curves as Local Measures of Variance Explained by Regression. *J. Amer. Statist. Assoc.*, **89** 571–582.

Doksum, K. and Samarov, A. M. (1995). Nonparametric Estimation of Global Functionals and a Measure of the Explanatory Power of Covariates in Regression. *Ann. Statist.*, **23** 1443–1473.

DOKSUM, K. A. and FRODA, S. (2000). Neighborhood Correlation. *J. of Statist. Planning and Inference*, **91** 267–294.

Donoho, D. L. and Liu, R. C. (1991a). Geometrizing Rates of Convergence, II. *Ann. Statist.*, **19** 633–667.

Donoho, D. L. and Liu, R. C. (1991b). Geometrizing Rates of Convergence, III. *Ann. Statist.*, **19** 668–701.

Einmahl, V. and Mason, D. (2005). Uniform in Bandwidth Consistency of Kernel-type Function Estimators. *Ann. Statist.*, **33** 1380–1403.

Fan, J. (1992). Design-adaptive Nonparametric Regression. *J. Amer. Statist. Assoc.*, **87** 998–1004.

Fan, J. (1993). Local Linear Regression Smoothers and Their Minimax Efficiencies. *Ann. Statist.*, **21** 196–216.

Fan, J., Zhang, C. and Zhang, J. (2001). Generalized Likelihood Ratio Statistics and Wilks Phenomenon. *Ann. Statist.*, **29** 153–193.

Fan, J. and Gijbels, I. (1996). Local Polynomial Modelling and Its Applications. Chapman & Hall, London.

Gasser, T., Müller, H.-G. and Mammitzsch, V. (1985). Kernels for Nonparametric Curve Estimation. *J. Roy. Stat. Soc., Ser. B*, **47** 238–252.

Godtliebsen, F., Marron, J. and Chaudhuri, P. (2004). Statistical Significance of Features in Digital Images. *Image and Vision Computing*, **22** 1093–1104.

Hajek, J. (1962). Asymptotically Most Powerful Rank-Order Tests. *Ann. Math. Stat.*, **33** 1124–1147.

Hall, P. and Hart, J. D. (1990). Bootstrap Test for Difference Between Means in Nonparametric Regression. *J. Amer. Statist. Assoc.*, **85** 1039–1049.

Hall, P. and Heckman, N. E. (2000). Testing for Monotonicity of a Regression Mean by Calibrating for Linear Functions. *Ann. Statist.*, **28** 20–39.

Hart, J. (1997). Nonparametric Smoothing and Lack-of-Fit Tests. Springer-Verlag, New York.

Ingster, Y. I. (1982). Minimax nonparametric detection of signals in white Gaussian noise. *Problems in Information Transmission*, **18** 130–140.

Karpoff, J. M. (1987). The Relation Between Price Changes and Trading Volume: A Survey. *J. of Fin. and Quant. Anal.*, **22** 109–126.

Kendall, M. and Stuart, A. (1961). The Advanced Theory of Statistics, Volume II. Charles Griffin & Company, London.

Lehmann, E. (1999). Elements of Large Sample Theory. Springer, New York.

Lepski, O. V. and Spokoiny, V. (1999). Minimax Nonparametric Hypothesis Testing: The Case of an Inhomogeneous Alternative. *Bernoulli*, **5** 333–358.

Linhart, H. and Zucchini, W. (1986). Model Selection. John Wiley & Sons, New York.

Neyman, J. (1959). Optimal Asymptotic Tests of Composite Hypotheses. In *Probability and Statistics: The Harold Cramer Volume* (U. Greenlander, ed.). John Wiley & Sons, New York, 213–234.

Raz, J. (1990). Testing for No Effect When Estimating a Smooth Function by Nonparametric Regression: A Randomization Approach. *J. Amer. Statist. Assoc.*, **85** 132–138.

Sen, P. K. (1996). Regression Rank Scores Estimation in ANOCOVA. *Ann. Statist.*, **24** 1586–1601.

Stute, W. and Zhu, L. X. (2005). Nonparametric Checks For Single-Index Models. *Ann. Statist.*, **33** 1048–1083.

Zhang, C. M. (2003a). Adaptive Tests of Regression Functions via Multi-scale Generalized Likelihood Ratios. *Canadian J. Statist.*, **31** 151–171.

Zhang, C. M. (2003b). Calibrating the Degrees of Freedom for Automatic Data Smoothing and Effective Curve Checking. *J. Amer. Statist. Assoc.*, **98** 609–628.

CHAPTER 7

Nonparametric Assessment of Atypicality

Peter Hall and Jim W. Kay

Mathematical Sciences Institute
Australian National University
Canberra, ACT 0200, Australia
Peter.Hall@maths.anu.edu.au

Department of Statistics
University of Glasgow
Glasgow G12 8QQ, Scotland, UK
jim@stats.gla.ac.uk

Atypicality indices have been used recently in a number of applications to assess how unusual are new observations relative to a given population. We introduce a new nonparametric estimator of an index of atypicality in which kernel density estimation is employed in the original definition, due to Aitchison, which becomes a functional of the unknown density. This estimator overcomes difficulties due to bias that arise if the kernel density estimator is simply plugged in to this functional. Detailed arguments are used to determine the pointwise bias and mean-squared error properties of the estimator, which are shown to depend on the location of the new observation, and the optimal rate of convergence of the bandwidth is determined. Rather surprisingly, it emerges that this optimal rate is the same as that which optimises performance of the kernel estimator of the unknown density and so we recommend that a standard bandwidth selector is used for estimating atypicality using the new estimator. The numerical behavior of the new estimator is studied using simulation, illustrating that its behavior depends on the location of the new observation, and applied in the assessment of atypicality using real data.

7.1. Introduction

We consider a nonparametric approach to the problem of assessing how atypical is a new observation of a univariate continuous random variable X relative to the previously observed values of X. The particular index of atypicality considered was introduced by Aitchison and Dunsmore (1975), with motivation provided by problems in clinical medicine, and formulae were given for estimative and predictive indices on the assumption of multivariate normality; see also Aitchison, Habbema and Kay (1977) and Moran and Murphy (1979). These atypicality indices have been used in interesting applications more recently; see, for example, Albert (1981) and Albert and Harris (1987), who used the predictive version, Christlieb, Wisotzki, and Graßhoff (2002), who used the estimative version, and Jones and Williams-Thorpe (2001), who used one minus the predictive atypicality index which was arrived at from a frequentist argument suggested by Moran and Murphy. See also McLachlan (1992). The use of atypicality indices is useful in practice as it provides an assessment of whether a particular object, be it a patient or a stone-axe or a star, belongs to some reference population based on observations such as total cholesterol level, chemical composition or a spectrum, respectively. An observation with a high atypicality index of 0.95 or greater might be considered unusual and require further investigation, and a value of 0.99 or more would receive special scrutiny.

Given two values, x_1 and x_2, of X, x_1 is said to be more typical of the distribution of X than x_2 if and only if $f(x_1) > f(x_2)$, where f denotes the probability density of X. Given a point x_0, Aitchison defined the atypicality index for x_0 to be

$$\alpha(x_0) = \Pr\{f(X) > f(x_0)\} = \int_{\{x:f(x)>f(x_0)\}} f(x)dx, \qquad (7.1)$$

where $\alpha(x_0)$ takes values between 0 and 1 for each real number x_0 and it measures how "atypical" x_0 is of the full range of values in the support of the distribution of X. Thus, x_0 would be considered highly typical if $\alpha(x_0) = 0$, and highly atypical if $\alpha(x_0) = 1$. In particular, $\alpha(x_0) = 0$ if and only if x_0 is at a "largest mode" of f; such a data value can be said to be more typical than any other value. At the other extreme, $\alpha(x_0) = 1$ if and only if x_0 lies outside the support of the distribution of f. We consider a nonparametric version of the atypicality index in which f is estimated by the kernel density method. An alternative nonparametric measure of atypicality was defined by Hermans et al. (1982) who used the estimated height of the density estimate at x_0 and a more interpretable version of this approach has been used by Collins and Krzanowski (2002).

In Section 2, we discuss a number of measures of atypicality. Simply plugging the kernel density estimate into (7.1) leads to performance problems due to bias, as does the leave-one-out version of this approach. A new estimator is proposed which overcomes these difficulties. The theoretical properties of the proposed estimator are presented in Section 3, with different results emerging depending on whether x_0 is at the level of a turning point of f. We consider the case when x_0 is not such a point and find, somewhat surprisingly, that the optimal convergence rate of bandwidth for the minimisation of pointwise mean squared error is the same as that which optimises the performance of the kernel density estimator of f. In Section 4, we discuss the numerical properties of the proposed estimator by means of simulations and demonstrate that the behaviour of the estimator can depend on the location of x_0. Our methods are applied in the assessment of the atypicality of the observations from a real data set. A proof of the first theorem in Section 3 is outlined in Section 5. Detailed arguments for this result, and for the second theorem, are available in a longer version of the paper (Hall and Kay (2005)).

7.2. Estimating Atypicality

Suppose we have a random sample $\mathcal{X} = \{X_1, \ldots, X_n\}$, drawn from the population with density f. A naive estimator of the atypicality index, defined at (7.1), can be obtained by replacing f there by a nonparametric density estimator constructed from $\mathcal{X}$, for example the kernel estimator

$$\hat{f}(x) = \frac{1}{nh} \sum_{i=1}^{n} K\left(\frac{x - X_j}{h}\right).$$

Here, K is a kernel function, usually a symmetric, compactly supported probability density, and $h > 0$ is a bandwidth. The resulting atypicality-index estimator is

$$\tilde{\alpha}_1(x_0) = \int_{\{x \,:\, \hat{f}(x) > \hat{f}(x_0)\}} \hat{f}(x)\, dx = \int I\{\hat{f}(x) > \hat{f}(x_0)\}\, \hat{f}(x)\, dx, \quad (7.2)$$

where $I(\mathcal{E})$ denotes the indicator of an event $\mathcal{E}$.

Although $\tilde{\alpha}_1$ is appealing on the grounds of its close resemblance to the true α, it can be expected to suffer performance difficulties which result from using the same density estimator in both the indicator function and the multiplier, $\hat{f}(x)$, in the second integral at (7.2). For example, the correlation between these two quantities can be shown to add extra terms to the formula for bias of $\tilde{\alpha}_1$.

This difficulty can be removed by replacing $\tilde{\alpha}_1$ by its leave-one-out form, $\hat{\alpha}_2$ say:

$$\tilde{\alpha}_2(x_0) = \frac{1}{nh} \sum_{i=1}^{n} \int I\{\hat{f}_{-i}(x) > \hat{f}_{-i}(x_0)\} K\left(\frac{x - X_i}{h}\right) dx, \qquad (7.3)$$

where

$$\hat{f}_{-i}(x) = \frac{1}{(n-1)h} \sum_{j\,:\,j\neq i} K\left(\frac{x - X_j}{h}\right)$$

is the version of $\hat{f}$ computed from the $(n-1)$-sample $\mathcal{X}_i = \mathcal{X} \setminus \{X_i\}$.

However, $\tilde{\alpha}_2$ is still unsatisfactory, since it contains bias terms which result from both the indicator function and the factor $K\{(x - X_i)/h\}$ at (7.3). Each of these is of size at least h^2; see section 3 for discussion of the bias arising from the indicator function alone. The bias contribution from the factor $K\{(x - X_i)/h\}$ can be reduced to zero by removing the kernel and substituting X_i into the argument of the indicator function, resulting in the estimator $\tilde{\alpha}_3$:

$$\tilde{\alpha}_3(x_0) = \frac{1}{n} \sum_{i=1}^{n} I\{\hat{f}_{-i}(X_i) > \hat{f}_{-i}(x_0)\}. \qquad (7.4)$$

While $\tilde{\alpha}_3(x_0)$ has attractive bias properties, it has the aesthetically unappealing feature of not being a smooth function of x_0. That difficulty can be overcome by employing a compromise between the estimators at (7.3) and (7.4):

$$\hat{\alpha}(x_0) = \hat{\alpha}(x_0 \mid h, H)$$

$$= \frac{1}{nH} > \sum_{i=1}^{n} \int I\{\hat{f}_{-i}(x) > \hat{f}_{-i}(x_0)\} K\left(\frac{x - X_i}{H}\right) dx, \qquad (7.5)$$

where $H \geq 0$ is a new bandwidth, different from the bandwidth h used to construct $\hat{f}_{-i}$. Letting $H \downarrow 0$ we find that $\hat{\alpha}(x_0 \mid h, H) \to \tilde{\alpha}_3(x_0)$, which we take as the definition of $\hat{\alpha}(x_0 \mid h, 0)$. Of course, $\hat{\alpha}(x_0 \mid h, h) = \tilde{\alpha}_2(x_0)$. The estimators $\tilde{\alpha}_3$ and $\hat{\alpha}$ could also be used in the case where X is multivariate. However, details in the multivariate problem are very different from those in the present case, since two multivariate density estimators intersect along a curve, not at a point.

As we noted in connection with $\tilde{\alpha}_2$, the choice $H = h$ is unsuitable because it results in extra bias terms. To overcome this problem we should choose H to be an order of magnitude smaller than h. While this might seem to require the selection of two bandwidths, that is not really the case. Indeed, having chosen h for the estimator $\hat{\alpha}(x_0 \mid h, 0)$, we simply increase H slightly, by visual inspection, to smooth out the small bumps of $\hat{\alpha}(x_0 \mid h, 0)$ as a function of x_0. This approach, which produces a bandwidth, H, of order n^{-1}, is employed in nonparametric kernel distribution estimation as an alternative to other approaches to bandwidth choice. In both that case and here, the concise size of the subsidiary bandwidth H does not affect first-order properties of mean squared error, provided only that H is of smaller order than h. Indeed, we shall show that for almost all choices of x_0 the optimal size of h is $n^{-1/5}$, much larger than the size of H that would be selected for bump-elimination, and the same as the size chosen by traditional bandwidth selectors.

The latter property suggests the following elementary approach to the choice of h and H for the estimator $\hat{\alpha}(x_0)$. Using a conventional bandwidth selector, for example cross-validation or a plug-in rule, choose h to optimise performance of $\hat{f}$ as an estimator of f, and use the same h to construct the function $\hat{\alpha}$. Then select H to smooth out bumps in $\hat{\alpha}$. It can be seen from formulae for bias and variance in Section 3 that performance of the estimator $\hat{\alpha}(x_0)$ will deteriorate as x_0 gets closer to a point x for which $f(x)$ is at the same level as a turning point, either a local minimum or a local maximum, of f. However, the attractiveness of using a standard bandwidth selector to choose h is very strong.

It seems far from obvious that the order of bandwidth that optimises performance of $\hat{f}$ as an estimator of f, should also give the optimal rate of convergence of $\hat{\alpha}$ to α. Indeed, the problem of estimating α has many of the features of a semiparametric, rather than a nonparametric, problem, and so one might even expect $\hat{\alpha}$ to be root-n consistent for α if h were chosen appropriately. For example, $\hat{\alpha}$ is constructed by integrating a functional of $\hat{f}$, and the operation of integration provides additional smoothing. This leads one to expect that, by choosing h to be an order of magnitude smaller than the bandwidth required for estimating f by $\hat{f}$, the rate of convergence could be improved. It is surprising that this is not the case.

Khas'minskii and Ibragimov (1986a), Khas'minskii and Ibragimov (1986b) and Ibragimov and Khas'minskii (1991) have given a general theory of efficient estimation of nonlinear functionals. However, the results in the present paper do not seem to be derivable from that work.

7.3. Theoretical Properties

7.3.1. *Properties of bias*

Unlike standard problems in kernel density estimation, where bias is essentially a monotone increasing function of bandwidth, bias here can be a convex function of bandwidth. In particular, using too small a value of bandwidth can produce an estimator with unduly large values of both bias and variance. As we shall show, the optimal bandwidth, h, for minimising the bias of $\hat{a}(x_0)$ is usually of size $n^{-1/3}$, although in cases where local extrema of the density come into play, it can be proved to be as large as $n^{-1/5}$. On the other hand, in some instances the bias function is not (approximately) convex, and there the optimal bandwidth is much smaller. The assumptions we shall make of f are standard:

f'' has two continuous derivatives on the real line; the support   (7.6)

of f is a set $\mathcal{S}$, say, the measure of which may be finite or infinite;

and in the interior of $\mathcal{S}$, f' vanishes at at most a finite number of points.

It can be shown that values x_0 for which there exists x_1 such that $f(x_1) = f(x_0)$ and $f'(x_1) = 0$ can be associated with relatively slow convergence rates of $\hat{a}(x_0)$ to $\alpha(x_0)$. Such points are isolated, however, and so for the sake of brevity we shall focus on circumstances where there are no turning points at the same level as x_0.

Of K and h we shall assume the following:

K is a symmetric, compactly supported probability density;

K'' exists, is continuous and of bounded variation on $\mathbb{R}$;   (7.7)

$h = h(n) = o(n^{-1/7})$, $n \to \infty$; $n^{\epsilon-1} = O(h)$, for some $\epsilon > 0$.   (7.8)

A little notation is needed to describe properties of bias, which are more complex here than in standard settings. If $f(x_0) > 0$ then, provided f satisfies (7.6), there exists only a finite set, $\mathcal{C} = \mathcal{C}(x_0)$ say, of values x_1, including $x_1 = x_0$, for which $f(x_1) = f(x_0)$. Assume x_0 is chosen so that for none of these values, including x_0 itself, it is true that $f'(x_0) = 0$.

Put $\kappa = \int K^2$ and $\kappa_2 = \int u^2 K(u)\,du$, and define

$$S_1(x_0) = \sum_{x_1 \in \mathcal{C}} \frac{\kappa_2\, f(x_0)\,\{f''(x_1) - f''(x_0)\}}{2\,|f'(x_1)|},$$

$$S_2(x_0) = \sum_{x_1 \in \mathcal{C}:x_1 \neq x_0} \mathrm{sgn}\,\{f'(x_1)\}\, \frac{\kappa\, f(x_0)}{f'(x_0)}\left\{\frac{f(x_0)\, f''(x_1)}{f'(x_1)^2} - \frac{3}{2}\right\}.$$

Define ϕ and Φ to be the standard normal density and distribution functions, respectively. Let $\lambda_j(v) = v^{-2} \int \{K(u+v) - K(u)\}^2 u^j \, du$, $\rho = \lambda_1/\lambda_0$, $\xi = (nh^3)^{1/2}$ and

$$S_3(x_0) = \xi \int_0^\infty \frac{f'(x_0)}{\{f(x_0)\,\lambda_0(v)\}^{1/2}} \left\{ \frac{f''(x_0)}{f'(x_0)} v + \frac{f'(x_0)}{f(x_0)} \rho(v) \right\}$$
$$\times \phi\left[\frac{\xi f'(x_0)}{\{f(x_0)\,\lambda_0\}^{1/2}} \right] dv$$
$$- 2\,|f'(x_0)| \int_0^\infty v\,\Phi\left[-\frac{\xi\,|f'(x_0)|}{\{f(x_0)\,\lambda_0(v)\}^{1/2}} \right] dv \,.$$

Clearly, $S_1(x_0)$ and $S_2(x_0)$ are functions of x_0 alone; they depend on neither n nor h. On the other hand, $S_3(x_0)$ depends on n and h. It can be shown to satisfy:

$$S_3(x_0) \begin{cases} \sim \text{const.} \, (nh^3)^{-1} & \text{if} \quad nh^3 \to 0 \\ \to \text{const.} & \text{if} \quad nh^3 \to \text{const.} \neq 0 \\ \to 0 & \text{if} \quad nh^3 \to \infty \,. \end{cases} \tag{7.9}$$

To obtain the first limit relation in (7.9), change variable in the integrals over v that define $S_3(x_0)$, and in particular write $v = \xi^{-1} u$. Then, noting that $\lambda_j(v)$ is asymptotic to $(-v)^{j-2} \int K^2$ as $v \to \infty$, the first limit relation in (7.9) can be proved. The second relation in (7.9) is straightforward to derive, and the third follows from the fact that $\phi[\xi\,f'(x_0)\,\{f(x_0)\,\lambda(x_0)\}^{-1/2}]$ converges rapidly to zero as $\xi \to \infty$.

Here and below, "const." denotes a generic constant, differing at different appearances and depending on neither n nor h, and nonzero whenever "const." appears in an asymptotic relation, such as $S_{13}(x_0) \sim$ const. $(nh^3)^{-1}$. The convergences in (7.9) can occur along arbitrary subsequences; they need not be along the full sequence of values of n.

Let $\hat{\alpha}(x_0) = \hat{\alpha}(x_0 \,|\, h, H)$ denote the estimator given by (7.5).

Theorem 7.1: *Assume conditions* (7.7)–(7.9), *that* $f(x_0) > 0$, *that none of the points* $x_1 \in \mathcal{C}(x_0)$ *satisfies* $f'(x_1) = 0$, *and that the second bandwidth,* $H = H(n)$, *satisfies* $0 \leq H = o(h)$ *as* $n \to \infty$. *Then,*

$$E\{\hat{\alpha}(x_0)\} = \alpha(x_0) + h^2\,S_1(x_0) + (nh)^{-1}\,S_2(x_0) \tag{7.10}$$
$$+ h^2\,S_3(x_0) + o\{h^2 + (nh)^{-1}\} \,.$$

Neglecting, for a moment, the relatively complex term involving $S_3(x_0)$, and assuming that neither $S_1(x_0)$ nor $S_2(x_0)$ vanishes, we see from (10) that when $\mathcal{C}_2(x_0)$ is empty, $|E\{\hat{\alpha}(x_0)\} - \alpha(x_0)|$ is essentially a convex function of h. That is, absolute bias becomes large, at the rate h^2, as h increases; and it also increases, this time at the rate $(nh)^{-1}$, as h becomes small. The optimal bandwidth for minimizing bias is therefore generally of size $n^{-1/3}$. Consequently, minimum bias is typically of size $n^{-2/3}$.

These properties continue to hold if we include the term $S_3(x_0)$ in our analysis. Indeed, (7.9) implies that (i) $S_3(x_0)$ is asymptotically constant when h is of the optimal size mentioned at the end of the previous paragraph; (ii) for smaller order of h, $h^2 S_3(x_0)$, makes a contribution that is of the same size as the term $(nh)^{-1} S_2(x_0)$ appearing elsewhere in (10); and (iii) for larger orders of h, $h^2 S_3(x_0)$ makes a negligible contribution to (10).

7.3.2. Mean squared error

For simplicity we treat the setting where x_0 is not at the level of a turning point of f and on the present occasion we confine attention to the case where h is of optimal size for minimising mean squared error. That is, $h \asymp n^{-1/5}$, meaning that the ratio of the left- and right-hand sides is bounded away from zero and infinity as $n \to \infty$.

Let (N_1, N_2) be bivariate normal with zero means, unit variances and coefficient of correlation equal to $\frac{1}{2}$, and put $\kappa(j) = \int (K^{(j)})^2$,

$$a_1 = \int_{-\infty}^{\infty} \int_{-\infty}^{\infty} \left\{ P(N_1 \leq u_1, N_2 \leq u_2) - \Phi(u_1)\,\Phi(u_2) \right\} du_1\, du_2 \,,$$

$$a_2 = \int_{-\infty}^{\infty} \int_{-\infty}^{\infty} \left[\Phi\{\min(u_1, u_2)\} - \Phi(u_1)\,\Phi(u_2) \right] du_1\, du_2 \,,$$

$$S_4(x_0) = \left[2\,\kappa\, a_1 \left\{ \sum_{x_1 \in C : x_1 \neq x_0} f'(x_1)^{-1} \right\}^2 \right.$$
$$\left. + 2\,\kappa\,(a_2 - a_1) \sum_{x_1 \in C : x_1 \neq x_0} f'(x_1)^{-2} + \frac{6\,\kappa(1)^2\, a_2}{\kappa(2)\, f'(x_0)^2} \right] f(x_0)^3 \,.$$

Theorem 7.2: *Assume the conditions of Theorem 7.1, except that (7.8) is replaced by the constraint that $h \asymp n^{-1/5}$. Then, $\mathrm{var}\{\hat{\alpha}(x_0)\} \sim (nh)^{-1} S_4(x_0)$, as $n \to \infty$.*

Combining the results of Theorems 7.1 and 7.2, and Equation (7.9), we deduce that when $h \asymp n^{-1/5}$,

$$E\{\hat{\alpha}(x_0) - \alpha(x_0)\}^2 = h^4 S_1(x_0)^2 + (nh)^{-1} S_4(x_0) \qquad (7.11)$$
$$+ o\{h^4 + (nh)^{-1}\} \,.$$

Theorems 7.1 and 7.2, and hence also Result (7.11), can be generalized so that they continue to apply when $n^{1/5}h \to 0$ or $n^{1/5}h$ diverges to infinity. In that setting it can be seen that, provided $S_1(x_0) \neq 0$, the left-hand side of (11) must be of strictly larger order than $n^{-4/5}$ if h is of strictly larger order than $n^{-1/5}$, and again of strictly larger order than $n^{-4/5}$ if h is of strictly smaller order than $n^{-1/5}$.

Therefore, provided $S_1(x_0)$ does not vanish, the optimal choice of bandwidth is of size $n^{-1/5}$. This motivates the suggestion, in Section 2, that a conventional bandwidth selector be employed, based on optimizing performance of $\hat{f}$ as an estimator of f. This will not give the best possible constant in the rate of convergence. However, except in cases where $f(x_0)$ is at the same height as the value of f at a turning point, it will give the best rate of convergence of $\hat{\alpha}(x_0)$ to $\alpha(x_0)$.

Under the conditions imposed on f and x_0 in Theorems 7.1 and 7.2, $S_4(x_0)$ will not vanish, and $S_1(x_0)$ will vanish if and only if $f''(x_1) = f''(x_0)$ at each point x_1 for which $f(x_1) = f(x_0)$. This can happen if, for example, f is unimodal, if x_0 and x_1 are on opposite sides of the mode, and if f is perfectly quadratic between x_0 and x_1. In such cases, using a bandwidth h that is of strictly larger order than $n^{-1/5}$ leads to improved performance.

7.4. Numerical Properties

The performance of the estimator $\tilde{\alpha}_3(x_0)$ of atypicality at a point x_0 was investigated as follows. The true density function was taken to be a $(0.5, 0.5)$ mixture of the Normal distributions $N(-0.5, 0.4^2)$ and $N(0.5, 0.2^2)$. The true atypicalities were computed using numerical methods for the thirteen values of x_0 which are displayed in Figure 1. Note that the true density at each of the points P_2, P_8, P_{10}, P_{12} is at the level of a turning point of f, and so we would expect the performance of the atypicality estimator to be different for these points.

7.4.1. *Estimation performance in relation to bandwidth*

Gaussian kernels were employed in the kernel estimator. Five-hundred simple random samples of sizes 25, 50 and 100, respectively, were drawn from the true density. The atypicality of the thirteen values of x_0 were estimated using the estimator $\tilde{\alpha}_3(x_0)$ for 41 values of the bandwidth parameter. These values were equally spaced between -2.5 and 1.5 (or 2 in some cases) on a $\log_{10}$ scale.

For each choice of x_0 and bandwidth, estimates of the pointwise mean

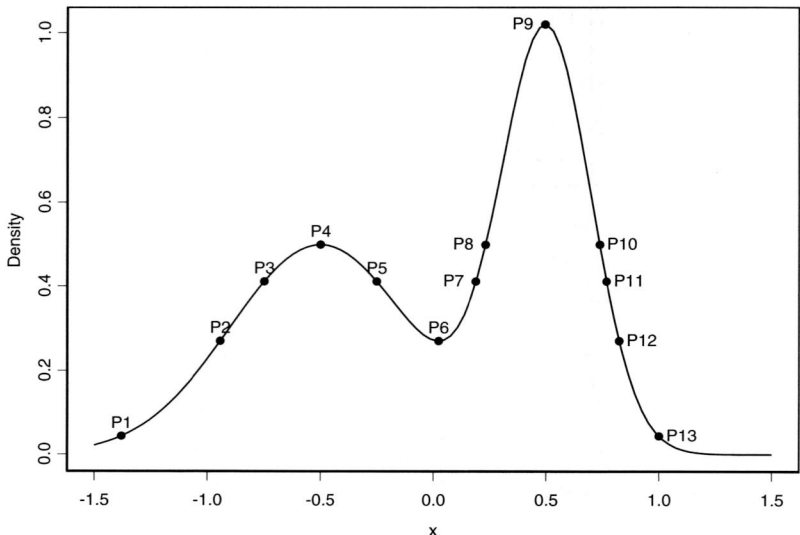

Figure 7.1.   True density function with thirteen points indicated.

squared error (MSE) of the estimator $\tilde{\alpha}_3$ were computed. The results for eight of the points are plotted in Figures 7.2 and 7.3.

It is clear from the figures that the mean squared error near the minima generally decreases as the sample size increases. This observation is supported by the fact that, at point P_2, for sample sizes of 500, 1000 and 2000 the mean squared error was estimated as $3.3 \times 10^{-4}, 2.1 \times 10^{-4}$ and 1.1×10^{-4}, respectively. The minima of MSE for points P_2, P_4 and P_{12} are not well-determined and a wide range of values of the bandwidth give essentially the same performance. The minima for points P_4 and P_{12} are much larger than for the other six points; this is to be expected for P_{12}, while at P_4 there are estimation difficulties at this mode due to the shape of f.

The MSE for the points not considered in Figures 7.2 and 7.3 are given in Table 1. The behavior of MSE at P_1 is similar to that of P_2. At sample sizes 25 and 50, the behavior at P_3 is similar to that of P_4, but a more

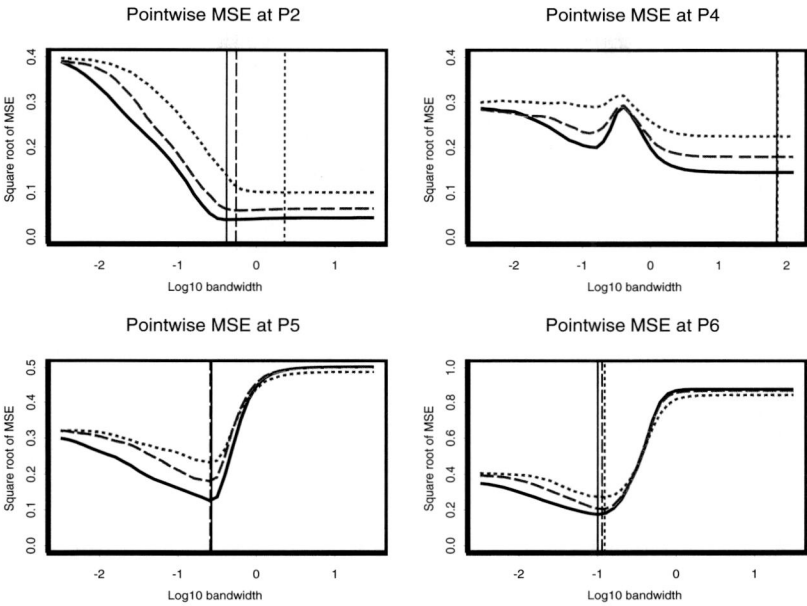

Figure 7.2. Square root of pointwise mean squared error is plotted against $\log_{10}$ bandwidth for points P_2, P_4, P_5, P_6 and the optimal values indicated by vertical lines for three sample sizes: 25 (dotted), 50 (dashed) and 100 (full).

reasonable minimizer results at sample size 100. The optimal bandwidths and shape of the MSE function for P_7 is similar to that of P_6, while P_9 is similar to P_{10}. The minimizing bandwidths for P_{13} are close together and similar to those of P_7. Due to the fact that we recommend the use of a standard bandwidth selector for the estimation of atypicality it is of interest to compare the optimal values of bandwidth with the distributions of bandwidths obtained for the samples using the direct-plug-in method of Sheather and Jones (1991)(SJ). When the sample size is 100, the 500 SJ $\log_{10}$bandwidth range from -0.93 to -0.63, with median at -0.81. The estimated optimal values for P_3, P_8, P_9 and P_{10} lie inside the SJ range; for points P_5, P_6, P_7 and P_{13} the values are outside the range but close to the endpoints; for points P_1, P_2, P_4, P_{11} and P_{12} the estimated optimal bandwidths are larger than the maximum SJ value, dramatically so for P_4 and P_{12}.

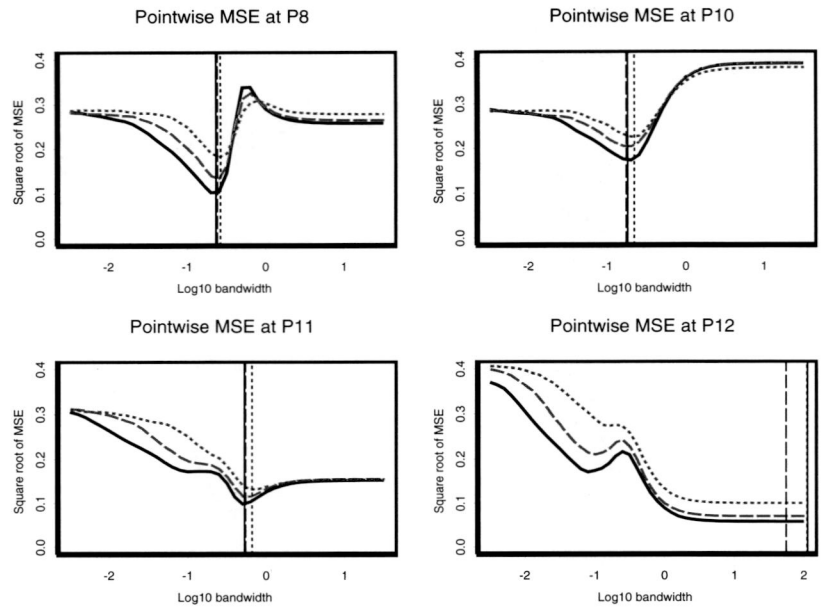

Figure 7.3.   Square root of pointwise mean squared error is plotted against $\log_{10}$ bandwidth for points $P_8, P_{10}, P_{11}, P_{12}$ and the optimal values indicated by vertical lines for three sample sizes: 25 (dotted), 50 (dashed) and 100 (full).

7.4.2. *Estimating performance in relation to* x_0

In order to investigate the influence of x_0 on the performance of the estimator $\tilde{\alpha}_3$, the atypicality indices were computed for each of the 500 simple random samples from f of sample size 100 on a grid of 151 equally-spaced values of x_0 in the interval (-1.5, 1.5). The Sheather-Jones direct-plug-in estimates of bandwidth were computed for each of the 500 samples and used in estimating the atypicalities. The true values of the atypicalities for points on this grid were also computed, along with the estimated pointwise mean-squared error, bias and standard error of $\tilde{\alpha}_3$.

In Figure 7.4 we notice that there is quite a lot of variability, especially in the vicinity of the left maximum, and this suggests that the performance of the $\tilde{\alpha}_3$ estimator might be poor in the range of x_0 values from P_2 to P_8. It is clear from the figure (top-right) that there are large biases, especially at P_4 and P_6, but to a lesser extent near P_{10}, P_{11} and P_{12}. It seems likely

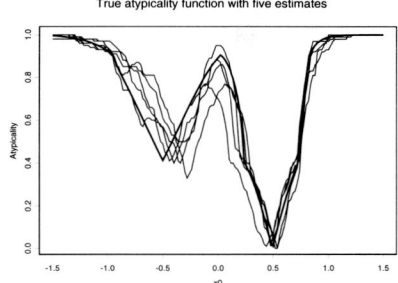
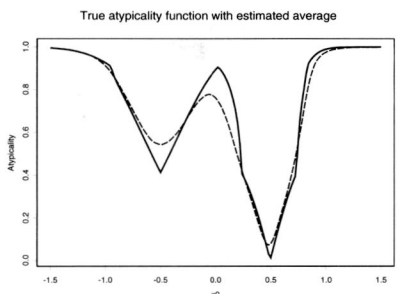

Figure 7.4. *Left.* The true atypicality function $\alpha(x_0)$(bold) is shown for a range of values of x_0 together with the estimated atypicality functions based on five of the simple random samples from f. *Right.* The true atypicality function $\alpha(x_0)$(bold) is shown for a range of values of x_0 together with the average of the 500 estimated atypicality functions based on the simple random samples from f. The estimator $\hat{\alpha}_3$ was used.

that part of the poor performance is due to the difficulty in estimating the left-hand mode.

The performance of the estimator is better near the more pronounced right-hand mode, although there are some problems there too. In Figure 7.5 we see that, of the points $P_1 - P_{13}$ of special interest, the mean-squared error is largest near points P_4, P_6, P_7 and $P_{10} - P_{12}$. Points P_2 and P_9 have a similar MSE, while points P_1 and P_{13} have very low MSE. Therefore the behaviour at points P_{12} and P_{10} supports the expectation of difficulties, while this is less clear for points P_2 and P_8. The estimated bias is largest in magnitude for points P_4, P_6, P_7 and P_{12}, and of them point P_{12} is expected to be linked with performance problems. At points P_{11} and P_{10} there is also a fairly large bias, while the bias is low at points P_1 - P_3, P_5, P_8 and P_{13}. Most of points $P_1 - P_{13}$ have a large estimated standard error, including P_{12}, P_{10}.

7.4.3. *Illustration of $\hat{\alpha}$: an application*

We now consider an application and consider the effect of using the second bandwidth H employed in the estimator $\hat{\alpha}$. The data, taken from Aitchison and Dunsmore (1975), consist of 21 distinct values of logged cortisone in patients who have been diagnosed as having bilateral hyperplasia, a form of the hypertensive condition Cushing's syndrome. At each of 45 equally-

Table 7.1. Square root of pointwise mean squared error is plotted against $\log_{10}$ bandwidth for points $P_8, P_{10}, P_{11}, P_{12}$ and the optimal values indicated by vertical lines for three sample sizes: 25 (dotted), 50 (dashed) and 100 (full).

x_0	n	$\log_{10}$ Bandwidth					
		-2.0	-1.0	-0.5	0.0	0.5	1.5
	25	0.201	0.084	0.021	0.019	0.018	0.019
P_1	50	0.190	0.036	0.013	0.013	0.013	0.013
	100	0.133	0.018	0.009	0.009	0.009	0.009
	25	0.302	0.240	0.230	0.189	0.183	0.182
P_3	50	0.282	0.199	0.187	0.163	0.154	0.152
	100	0.270	0.155	0.165	0.155	0.140	0.138
	25	0.323	0.259	0.337	0.576	0.562	0.559
P_7	50	0.288	0.205	0.342	0.584	0.559	0.557
	100	0.278	0.184	0.341	0.586	0.557	0.554
	25	0.400	0.239	0.264	0.463	0.492	0.495
P_9	50	0.395	0.174	0.222	0.466	0.500	0.504
	100	0.298	0.119	0.173	0.462	0.497	0.501
	25	0.222	0.088	0.142	0.118	0.091	0.089
P_{13}	50	0.181	0.040	0.080	0.085	0.065	0.064
	100	0.129	0.021	0.059	0.078	0.059	0.057

spaced values of logged cortisone (covering the range of the sample data) the values of $\tilde{\alpha}_3$ and $\hat{\alpha}$ were computed and the results are shown in Figure 7.6. The direct-plug-in method of Sheather and Jones (1991) was used to select the bandwidth h. The values of $\hat{\alpha}$ were computed using Monte Carlo approximation, based on 500 simulations. Figure 7.6 shows the $\hat{\alpha}$ and $\tilde{\alpha}_3$ estimates of atypicality. When the second bandwidth H is taken to be one-fifth of the SJ bandwidth, the values obtained are very close to those resulting from the use of $\tilde{\alpha}_3$, while producing a smoother, more aesthetically pleasing, estimate. When the second bandwidth was taken to be equal to the SJ value, which gives estimator $\tilde{\alpha}_2$, the estimates are clearly over-smoothed, illustrating the bias predicted in Section 2 of this chapter for this estimator. Two other values of H were considered; when H was taken to be one-hundredth of the SJ bandwidth, the two estimators gave virtually identical results, but when H was taken to be one-tenth of the SJ bandwidth the resulting estimates were very close to those shown in the left-hand plot of Figure 7.6. The values of $\hat{\alpha}$ for the 21 patients in the sample range from 0.029 to 0.962, with the second largest value being 0.888. The atypicality

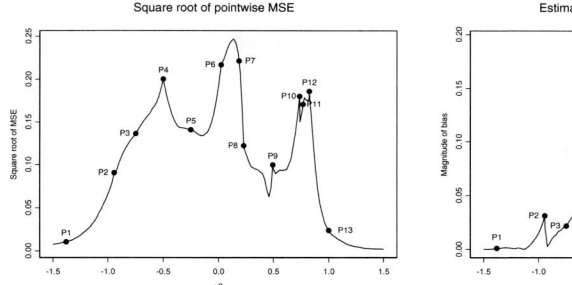

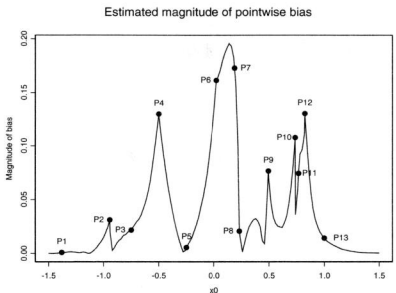

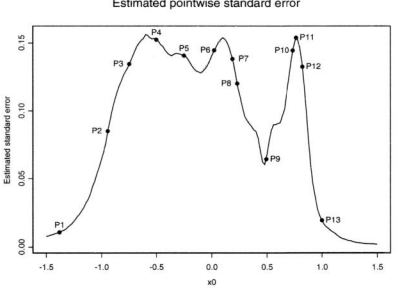

Figure 7.5. For a range of values of x_0, this shows the square root of pointwise MSE (top-left), the estimated magnitude of the pointwise bias (top-right) and the estimated standard error (bottom) of $\tilde{\alpha}_3$.

value of 0.962 is obtained for the smallest observation, with $x_0 = -2.53$, and this observation should be subjected to further scrutiny.

7.5. Outline of Proof of Theorem 7.1

The proof will be divided into three steps, which will respectively establish:
(I) an expansion of $I_1(\mathbb{R})$, where, for any subset $\mathcal{R}$ of the real line,

$$I_1(\mathcal{R}) \equiv \int_{\mathcal{R}} I\{E\hat{f}(x) > E\hat{f}(x_0)\}\, f(x)\, dx - \int_{\mathcal{R}} I\{f(x) > f(x_0)\}\, f(x)\, dx\, ;$$

(II) an expansion of $I_2(\mathbb{R})$, where

$$I_2(\mathcal{R}) \equiv \int_{\mathcal{R}} P\{\hat{f}(x) > \hat{f}(x_0)\}\, f(x)\, dx - \int_{\mathcal{R}} I\{E\hat{f}(x) > E\hat{f}(x_0)\}\, f(x)\, dx\, ;$$

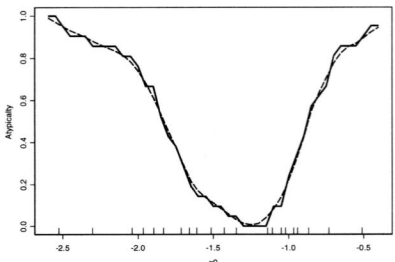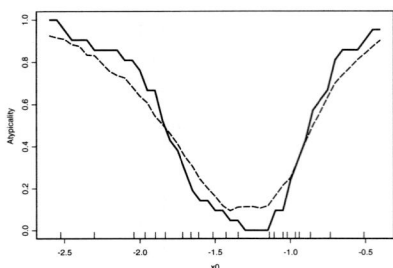

Figure 7.6. Values of $\hat{\alpha}$ (dotted) and $\tilde{\alpha}_3$ for a range of values of x_0, with the second bandwidth taken as the SJ bandwidth divided by 5 (left) and equal to the SJ bandwidth (right). The observed data values are marked on the x_0 axis.

and (III) the bias expansion (7.10), by combining the results of Steps (I) and (II).

7.5.1. Step (I): Expansion of $I_1(\mathbb{R})$

Let $\delta = \delta(n) > 0$ decrease slowly to zero as h decreases. It is sufficient that

$$\delta \log n \to \infty \quad \text{and} \quad (nh)^{-1/2} \log n = o(\delta^2),$$

as $h \to 0$. Furthermore, let

$$\mathcal{N} = \mathcal{N}(x_1) = \{x : |x - x_1| \leq \delta\} \tag{7.12}$$

denote a shrinking neighbourhood of a point x_1 for which $f(x_1) = f(x_0)$. We shall state asymptotic properties of $I_1(\mathcal{N})$. Two cases need to be considered: (a) $x_1 \neq x_0$ and $f'(x_1) \neq 0$, and (b) $x_1 = x_0$ and $f'(x_0) \neq 0$. Define $d_1 = \frac{1}{2} \kappa_2 \{f''(x_1) - f''(x_0)\}$. In case (a),

$$I_1(\mathcal{N}) = \pm \frac{d_1 f(x_0)}{f'(x_1)} h^2 + o(h^2). \tag{7.13}$$

In case (b),

$$I_1(\mathcal{N}) = o(h^2). \tag{7.14}$$

Let x_0 be any point for which $f(x_0) > 0$, and let $\mathcal{C}$ denote the set of points x_1 such that $f(x_1) = f(x_0)$. Clearly, $\mathcal{C}$ contains x_0. Recall the definition of $\mathcal{N}(x_1)$ at (7.12). It may be proved that

$$I_1(\mathcal{R}) = \sum_{x_1 \in \mathcal{C}} I_1\{\mathcal{N}(x_1)\} \quad \text{for all sufficiently large } n, \qquad (7.15)$$

and (7.13) and (7.14) may be used to develop a formula for the series on the right-hand side. Together, they give, respectively, the contributions to $S_1(x_0)$ from $x_1 \neq x_0$, and the vanishing contribution to $S_1(x_0)$ from $x_1 = x_0$.

7.5.2. Step (II): Expansion of $I_2(\mathbb{R})$

Let $x_1 \in \mathcal{C}$ and let $\mathcal{N} = \mathcal{N}(x_1)$ be as at (7.12). We treat separately the two cases considered in Step (I).

(a) $x_1 \neq x_0$ and $f'(x_1) \neq 0$. Define $c_1 = \{2\,\kappa\,f(x_0)\}^{-1/2}\,f'(x_1)$ and

$$c_2 = \frac{\{\kappa\,f(x_0)\}^{1/2}}{2^{1/2}\,f'(x_1)} \left\{ \frac{f''(x_1)}{f'(x_1)} - \frac{f'(x_1)}{2\,f(x_0)} \right\}.$$

Then

$$I_2(\mathcal{N}) = (nh)^{-1}\,|c_1|^{-1}\left\{ c_2\,f(x_0) - \tfrac{1}{2}\,c_1^{-1}\,f'(x_1) \right\} + o\{h^2 + (nh)^{-1}\}. \quad (7.16)$$

(b) $x_1 = x_0$ and $f'(x_0) \neq 0$. Define $\xi = (nh^3)^{1/2}$,

$$t_1(v) = \frac{s(v)\,f'(x_0)}{\{f(x_0)\,\lambda_0(v)\}^{1/2}}, \quad t_2(v) = \tfrac{1}{2}\left\{ \frac{f''(x_0)}{f'(x_0)}\,v + \frac{f'(x_0)}{f(x_0)}\,\rho(v) \right\}.$$

Then it may be proved that, with $\mathcal{N} = \mathcal{N}(x_1)$,

$$I_2(\mathcal{N}) = 2\,\xi\,h^2 \int_0^\infty t_1(v)\,t_2(v)\,\phi\{\xi\,t_1(v)\}\,dv \qquad (7.17)$$

$$-2\,h^2\,|f'(x_0)| \int_0^\infty v\,\Phi\!\left[-\frac{\xi\,|f'(x_0)|}{\{f(x_0)\,\lambda_0(v)\}^{1/2}} \right] dv$$

$$+o\{h^2 + (nh)^{-1}\}.$$

Furthermore, for each $B > 0$,

$$I_2(\mathbb{R}) = \sum_{x_1 \in \mathcal{C}} I_2\{\mathcal{N}(x_1)\} + O(n^{-B}). \qquad (7.18)$$

7.5.3. Step (III): Combining steps (I) and (II)

In a slight abuse of notation, let $I_1(\mathcal{R})$ and $I_2(\mathcal{R})$ denote the versions of those quantities in the case of a sample of size $n - 1$, rather than n. All our asymptotic results, such as the expansions of $I_j(\mathbb{R})$ and of $I_j\{\mathcal{N}(x_1)\}$

for $j = 1, 2$, obtained in steps (I) and (II), hold without change. In this notation, and for all $B > 0$,

$$E\{\hat{\alpha}(x_0 \mid h, 0)\} = I_1(\mathbb{R}) + I_2(\mathbb{R})$$
$$= \sum_{x_1 \in \mathcal{C}} I_1\{\mathcal{N}(x_1)\} + \sum_{x_1 \in \mathcal{C}} I_2\{\mathcal{N}(x_1)\} + O(n^{-B}),$$

the second identity following from (7.15) and (7.18). This result, and (7.13), (7.14), (7.16) and (7.17), give (7.10) in the case $H = 0$.

To treat the case $0 < H = o(h)$, observe that by standard calculations,

$$E\{\hat{\alpha}(x_0 \mid h, H)\} = \int P\{\hat{f}_{-1}(x) > \hat{f}_{-1}(x_0)\} \, dx \int K(u) \, f(x - Hu) \, du$$

$$= \int P\{\hat{f}_{-1}(x) > \hat{f}_{-1}(x_0)\} \, f(x) \, dx + O(H^2)$$

$$= E\{\hat{\alpha}(x_0 \mid h, 0)\} + O(H^2).$$

Using this result, (7.10) for $0 < H = o(h)$ follows from its counterpart in the case $H = 0$.

References

Aitchison, J. and Dunsmore, I. R. (1975). *Statistical Prediction Analysis.* Cambridge University Press.

Aitchison, J., Habbema, J.D.F. and Kay, J.W. (1977). A critical comparison of two methods of discrimination. *Applied Statistics* **26** 15-25.

Albert, A. (1981). Atypicality indices as reference values for laboratory data. *Amer. J. Clin. Pathol.* **76** 421-425.

Albert, A. and Harris, E. K. (1987). *Multivariate Interpretation of Clinical Laboratory Data.* Marcel Dekker, New York.

Christlieb, N., Wisotzki, L. and Graßhoff, G. (2002). Statistical methods of automatic spectral classification and their application to the Hamburg/ESO Survey. *Astronomy and Astrophysics* **391** 397-406.

Collins, G. S. and Krzanowski, W. J. (2002) Nonparametric discriminant analysis of phytoplankton species using data from analytical flow cytometry. *Cytometry* **48** 26-33.

Hall, P. and Kay, J. W. (2005) *Nonparametric assessment of atypicality.* Technical Report No. 05-02. University of Glasgow.
www.stats.gla.ac.uk/Research/TechRep2005/hallkay.pdf.

Ibragimov, I. A. and Khas'minskii, R. Z. (1991). Asymptotically normal families of distributions and efficient estimation. *Ann. Statist.* **19** 1681-1724.

Hermans, J., Habbema, J.D.F., Kasanmoentalib, T.K.D. and Raatgever, J.W. (1982). *Manual For the ALLOC80 Discriminant Analysis Program.* Department of Medical Statistics. University of Leiden.

Jones, M.C. and Williams-Thorpe, O. (2001). An illustration of the use of an atypicality index in provenancing British stone axes. *Archaeometry* **43** 1-18.

Khas'minskii, R.Z. and Ibragimov, I.A. (1986a). Some new results in the nonparametric estimation of functionals. *Transactions of the Tenth Prague Conference on Information Theory, Statistical Decision Functions, Random Processes* **A** 31-40, Reidel, Dordrecht.

Khas'minskii, R.Z. and Ibragimov, I.A. (1986b). Asymptotically efficient nonparametric estimation of functionals of a spectral density function. *Probab. Theory Related Fields* **73** 447-461.

McLachlan, G.J. (1992). *Discriminant Analysis and Statistical Pattern Recognition*. Wiley, New York.

Moran, M.A. and Murphy, B.J. (1979). A closer look at two alternative methods of statistical discrimination. *Applied Statistics* **28** 223-232.

Sheather, S. J. and Jones, M. C. (1991). A reliable data-based bandwidth selection methodfor kernel density estimation. *J. Roy. Statist. Soc B* **53** 683-690.

CHAPTER 8

Selective Review on Wavelets in Statistics

Yazhen Wang

Department of Statistics
University of Connecticut
Storrs, CT 06269
yzwang@stat.uccnn.edu

This article provides a brief review on main wavelet statistical methodologies developed in past fifteen years. The wavelet methods are used for estimating regression functions, detecting and estimating change-points, solving statistical inverse problems, and studying stochastic processes and self-similarity.

8.1. Introduction

Wavelets are viewed as a synthesis of ideas originated in engineering, physics, and pure mathematics. In past fifteen years, the field has grown at an explosive rate, and wavelet articles have appeared in a remarkable variety of publications. Wavelets are applied to a diverse set of problems and have made significant technology advancement.

Wavelets enjoy various properties including orthogonality, localization, and fast computational algorithms. Wavelets provide unconditional bases for many useful function spaces and have excellent compression capabilities for functions in the spaces. In other words, these classes of functions have economical descriptions (or sparse representations) in terms of wavelets: we do not need too many terms in the wavelet expansion to approximate a function closely. Successful applications of this property include wavelet shrinkage for signal and image compression and nonparametric regression.

Wavelets enable us to do time-frequency (or time-scale) analysis. Wavelet analysis can "unfold" a signal of interest into a function over the time-frequency plane that tells us "when" which "frequency" occurs. This is similar to music notation, which tells the player which notes (frequency information) to play at any given moment (time information). It allows us to simultaneously locate frequency content and time content of the signal. The time-frequency analysis is a suitable tool for analyzing non-stationary time series and signals (e.g. locally stationary time series and speech signals). Another application of this time-frequency localization is that wavelets can describe the local features (such as jumps) of a function and provide tools for studying change-points in statistics and edge detection in image processing.

Wavelets provide simultaneous quasi-diagonalization of a large class operators. A wavelet-vaguelette decomposition (WVD) can simultaneously quasi-diagonalize dilation-homogeneous operators (Calderon-Zygmund Operators) and a class of functions. The decomposition allows us to develop a new and better method for solving linear inverse problems and modeling non-stationary processes. Because of the localization property, the WVD provides an excellent tool for the study of change-points for indirect data.

Disordered structures and random processes that are self-similar on certain length and time scales are very common in nature. They are found on the largest and the smallest scales: in galaxies and landscapes, in earthquakes and fractures, in aggregates and colloids, in rough surfaces and interfaces, in glasses and polymers, in proteins and other large molecules. Self-similar structures in wavelets make them advantageous in handling self-similar phenomenon.

These properties of wavelets are utilized to develop wavelet based methodologies for solving statistical problems. The rest of the paper is organized as follows. Section 2 introduces wavelets. Sections 3 and 4 review wavelet shrinkage methods for function estimation in the nonparametric regression model and statistical inverse problems, respectively. Section 5 covers change-points for both direct and indirect data. Section 6 features local self-similarity and non-stationary stochastic processes. Section 7 provides a brief description of basis development in high dimensions.

8.2. Wavelets

Wavelet bases are derived from father wavelet and mother wavelet by a process of dyadic dilations and translations. We start by introducing Haar wavelet basis. Let

$$\phi(x) = \begin{cases} 1, 0 \leq x \leq 1 \\ 0, \text{ otherwise,} \end{cases}$$

and

$$\psi(x) = \phi(2\,x) - \phi(2\,x - 1) = \begin{cases} 1, & 0 \le x \le 1/2; \\ -1, & 1/2 < x \le 1; \\ 0, & \text{otherwise.} \end{cases}$$

In wavelet terminology, ϕ and ψ are called father and mother wavelets, respectively.

Define the continuous wavelet transformation of a function $f(x)$ by

$$Tf(s, u) = \frac{1}{\sqrt{s}} \int f(x)\, \psi\left(\frac{u - x}{s}\right) dx,$$

where s and u represent scale (or inverse of frequency) and spatial position (or time), respectively. The two dimensional space defined by the pair of variables (s, u) is called the scale-space (or time-frequency) plane, and the wavelet transformation $Tf(s, u)$ is a bivariate function of scale and position (or time and frequency).

Dyadically dilate and translate $\psi(x)$

$$\psi_{j,k}(x) = 2^{j/2}\, \psi(2^j\, x - k), \quad j = 0, 1, \cdots, \quad k = 0, 1, \cdots, 2^j - 1.$$

It is easy to verify that $\phi(x)$ and $\psi_{j,k}(x)$ form an orthonormal basis in $L^2([0, 1])$. Then any integrable function $f(x)$ on $[0, 1]$ has an expansion

$$f(x) = \beta_{-1,0}\, \phi(x) + \sum_{j=0}^{\infty} \sum_{k=0}^{2^j - 1} \beta_{j,k}\, \psi_{j,k}(x),$$

where the wavelet coefficients

$$\beta_{-1,0} = \int f(x)\, \phi(x)\, dx, \qquad \beta_{j,k} = \int f(x)\, \psi_{j,k}(x)\, dx.$$

Because $\psi_{j,k}$ has compact support of width 2^{-j} located at $k\, 2^{-j}$, $\beta_{j,k}$ corresponds to the continuous wavelet transformation of $f(x)$ at frequency 2^j and spatial position $k\, 2^{-j}$, and describes the information content of $f(x)$ at frequency 2^j and spatial location $k\, 2^{-j}$.

Although the Haar basis was discovered in 1910, it is not very useful in practice. Its basis functions are discontinuous, so it is not effective in approximating smooth functions or localizing frequency components. General wavelet bases are obtained as follows. Replacing Haar mother and father wavelets with a pair of very special functions (ϕ, ψ), we dyadically dilate and translate ψ to obtain a wavelet basis ϕ, $\psi_{j,k}$, $j = 0, 1, \cdots$, $k = 0, 1, \cdots, 2^j - 1$, which form an orthonormal basis in $L^2([0, 1])$ (boundary corrections are needed for some base functions whose supports stretch outside $[0, 1]$). Most useful wavelet bases require the pair (ϕ, ψ) to be smooth and compactly supported. Unlike Haar basis or Fourier basis, such special ϕ and ψ are very complicated to construct and have no explicit forms. However, they can be numerically evaluated by the fast cascade algorithm. See Daubechies (1988) and Daubechies (1992).

In practice we observe a function, a signal or a stochastic process only at finite number of points. The continuous wavelet transformation needs to be discretized as the discrete wavelet transformation which can be written as a linear transformation represented by an orthogonal matrix $\mathcal{W}$. Suppose we have observations $z_1, \cdots, z_n$, $n = 2^J$. Let $\mathbf{z} = (z_1, \cdots, z_n)$. The discrete wavelet transformation of $\mathbf{z}$ is given by $\mathbf{w} = \mathcal{W}\mathbf{z}$. The elements of $\mathbf{w}$ are called discrete wavelet coefficients, with the $n - 1$ elements of $\mathbf{w}$ indexed dyadically by $w_{j,k}$ for $j = 0, 1, \cdots, J - 1$, $k = 0, 1, \cdots, 2^j - 1$, and the remaining element labelled as $w_{-1,0}$. Because $\mathcal{W}$ is orthogonal, we can easily reconstruct $\mathbf{z}$ from its discrete wavelet coefficients $\mathbf{w}$ by the inverse discrete wavelet transformation $\mathbf{z} = \mathcal{W}^T \mathbf{w}$.

The rows of $\mathcal{W}$ correspond to a discretized version of the wavelet basis $(\phi, \psi_{j,k})$, and $w_{j,k}$ relate to a discretized version of the continuous wavelet transformation $\beta_{j,k}$ as follows. If we dyadically index the first $n - 1$ rows of $\mathcal{W}$ by $(2^j + k)$ for $j = 0, 1, \cdots, J - 1$, $k = 0, 1, \cdots, 2^j - 1$, and denote by $W_{j,k}(i)$ the i-th element of the $(2^j + k)$-th row of $\mathcal{W}$, then $n^{1/2} W_{j,k}(i)$ is approximately equal to $\psi_{j,k}(i/n)$. If $z_1, \cdots, z_n$ are observations of $f(x)$ at $x = i/n$, $i = 1, \cdots, n$, then $n^{1/2} w_{j,k} \approx \beta_{j,k}$.

Mallat's pyramidal algorithm requires only $O(n)$ operations for performing discrete wavelet transformation and inverse discrete wavelet transformation. Thus we can fast process wavelet analysis by computing the discrete wavelet coefficients of finite observations and reconstructing the observations from their corresponding discrete wavelet coefficients. See Mallat (1989) and Mallat (1999).

8.3. Nonparametric Regression

8.3.1. *Wavelet shrinkage*

Consider the nonparametric regression model

$$y_i = f(x_i) + \varepsilon_i, \qquad i = 1, \cdots, n = 2^J, \qquad (8.1)$$

where $x_i = i/n$, ε_i are i.i.d. normal random errors, and $f(x)$ is a function. The problem is to estimate $f(x)$ based on data y_i.

Because wavelets provide sparse representations for a wide class of functions, in a series of papers, Donoho and Johnstone proposed wavelet shrinkage to take advantage of wavelets' sparse representations and efficiently estimate regression functions. They established asymptotic minimax for estimating regression functions over a wide class of Besov spaces and showed that wavelet shrinkage estimators can be turned to achieve minimax rates over the entire scale of Besov spaces and significantly outperform linear estimators.

Wavelet shrinkage works as follows. First compute the discrete wavelet coefficients of data $y_1, \cdots, y_n$; second shrink the wavelet coefficients; third construct the estimator of $f(x)$ by using the shrunk wavelet coefficients. Specifically, let $(y_{j,k})$, $(\theta_{j,k})$ and $(\varepsilon_{j,k})$ be the discrete wavelet coefficients of (y_i), $(f(x_i))$ and (ε_i), respectively. Since discrete wavelet transformation is linear, from model (8.1) we yield

$$y_{j,k} = \theta_{j,k} + \varepsilon_{j,k}, \qquad j = 0, 1, \cdots, J - 1, \quad k = 0, 1, \cdots, 2^j - 1, \qquad (8.2)$$

where $\varepsilon_{j,k}$ are i.i.d. normal, due to the orthogonality of discrete wavelet transformation.

Because of wavelet's sparse representations, there are relatively a small number of large $|\theta_{j,k}|$ and a large number of small $|\theta_{j,k}|$. Shrinking rules are used to select only those $y_{j,k}$ whose corresponding $\theta_{j,k}$ are of large magnitude. We use the selected $y_{j,k}$ to recover the large $\theta_{j,k}$ and reconstruct a function from the recovered wavelet coefficients as an estimator of f.

Two shrinking rules are hard threshold rule

$$\hat{\theta}_{j,k} = \delta_h(y_{j,k}, \lambda) = y_{j,k} \mathbf{1}_{\{|y_{j,k}| > \lambda\}},$$

and soft threshold rule

$$\hat{\theta}_{j,k} = \delta_s(y_{j,k}, \lambda) = \text{sign}(y_{j,k}) \, (|y_{j,k}| - \lambda)_+,$$

where λ is threshold. The wavelet estimator $\hat{f}$ is the function constructed using $\hat{\theta}_{j,k}$ as wavelet coefficients.

Various ways are proposed to select threshold λ. Universal threshold is defined by

$$\lambda = \sigma \sqrt{2 \log n},$$

which is used to shrink $y_{j,k}$ at all levels. It is derived from the fact that with probability tending to one, the maximum of $|\varepsilon_{j,k}|$ is bounded by $\sigma \sqrt{2 \log n}$.

SureShrink is to choose a threshold for $y_{j,k}$ at each level. SureShrink threshold λ_j for level j is the value that minimizes $\text{SURE}(y_{j,k}, \lambda_j)$ over all

λ_j, where

$$\text{SURE}(y_{j,k}, \lambda_j) = 2^j - 2 \sum_{k=0}^{2^j-1} 1_{\{|y_{j,k}| \le \lambda_j\}} + \sum_{k=0}^{2^j-1} (|y_{j,k}| \wedge \lambda_j)^2.$$

It is based on the fact that $\text{SURE}(y_{j,k}, \lambda_j)$ is Stein's unbiased estimator of the ℓ_2-risk of the wavelet estimator constructed using $\delta_s(y_{j,k}, \lambda_j)$, shrunk $y_{j,k}$ with soft threshold rule and threshold λ_j.

Cross-validation threshold is to treat threshold λ as a tuning parameter and apply classic cross-validation procedure to select λ. First divide data into two parts, one part with odd indices and one part with even indices; second use one part of the data to form a wavelet estimator with threshold λ and use the other part of the data to validate the prediction error of the wavelet estimator. The cross-validation threshold is defined to be the value that minimizes the prediction error with correction of multiplying a factor $(1 - \log 2/ \log n)^{-1/2}$. See Nason (1996).

FDR threshold (threshold based on false discovery rate) was investigated by Abramovich and Benjamini (1995) and Abramovich and Benjamini, Donoho and Johnstone (2005). It works as follows. Order $y_{j,k}$ in terms of their absolute values from the largest one to the smallest one and denote the ordered values by

$$|y|_{(1)} \ge |y|_{(2)} \ge \cdots \ge |y|_{(i)} \ge \cdots \ge |y|_{(n)},$$

and compute normal quantiles

$$\hat{t}_i = \sigma \, \Phi^{-1} \left(1 - \frac{i \, q}{2 \, n} \right),$$

where q is a small positive number that controls the false discovery rate. Compare $|y|_{(i)}$ to $\hat{t}_i$ and let κ be the largest index i for which $|y|_{(i)} \ge \hat{t}_i$. FDR threshold is taken to be $\hat{t}_\kappa$, and we use it to threshold $y_{j,k}$

$$\hat{\theta}_{j,k} = y_{j,k} 1_{\{|y_{j,k}| \ge \hat{t}_\kappa\}}.$$

Other thresholding methods include block thresholding rules, which shrink wavelet coefficients in groups rather than individually, and make simultaneous decisions to retain or to discard all coefficients within a block. See Cai (1999), Cai (2002) and Hall, Kerkyacharian and Picard (1998). Antoniadis and Fan (2001) offer a penalized least-squares method to these problems, which is also applicable to non-equispace designs. Above wavelet shrinkage methods are for normal random errors. It is important but not straightforward to extend the methods to non-normal data. See Antoniadis and Sapatinas (2001), Donoho and Jin (2005), Koul and Ni (2004) and Kolaczyk and Nowak (2005) for some recent development on wavelet shrinkage for non-normal random errors.

8.3.2. *Bayesian wavelet shrinkage*

Because Bayes solutions for normal type problems have similar desirable property of wavelet shrinkage: heavily shrink small arguments and only slightly shrink the large arguments, Bayesian wavelet shrinkage was proposed as a competitor. By choosing appropriate prior and loss function, the resulting Bayesian wavelet shrinkage can be very close to thresholding, or even match hard- and soft-thresholding rules. Here we mainly review three approaches.

The first one is adaptive Bayesian wavelet shrinkage proposed by Chipman, Kolaczyk, and McCulloch (1997). Assume the variance σ^2 of ε_i is known. From (8.2) the distribution of wavelet coefficients $y_{j,k}$ given $\theta_{j,k}$ is $N(\theta_{j,k}, \sigma^2)$. The prior on $\theta_{j,k}$ is a mixture of two normals

$$\theta_{j,k}|\gamma_j \sim \gamma_j\, N(0, \tau_j^2) + (1 - \gamma_j)\, N(0, \tau_j^2/c_j^2),$$

where γ_j are independent Bernoulli random variables with $P(\gamma_j = 1) = p_j$, and c_j, τ_j, p_j are hyperparameters. As p_j are taken to be very small, and c_j are chosen to be much bigger than 1, the two normals in the prior of $\theta_{j,k}$ are selected to capture the sparse representation of $f(x)$ by wavelets: a small number of large wavelet coefficients modeled by $\gamma_j\, N(0, \tau_j^2)$ and a large number of small coefficients described by $(1 - \gamma_j)\, N(0, \tau_j^2/c_j^2)$.

Under the squared error loss the posterior mean of $\theta_{j,k}$ is

$$\hat{\theta}_{j,k} = y_{j,k} \left(\frac{\tau_j^2}{\sigma^2 + \tau_j^2} P(\gamma_j = 1|y_{j,k}) + \frac{\tau_j^2}{c_j^2\, \sigma^2 + \tau_j^2} P(\gamma_j = 0|y_{j,k}) \right),$$

where

$$P(\gamma_j = 1|y_{j,k}) = \frac{p_j\, \pi(y_{j,k}|\gamma_j = 1)}{p_j\, \pi(y_{j,k}|\gamma_j = 1) + (1 - p_j)\, \pi(y_{j,k}|\gamma_j = 0)},$$

$$\pi(y_{j,k}|\gamma_j = 1) \sim N(0, \sigma^2 + \tau_j^2), \qquad \pi(y_{j,k}|\gamma_j = 0) \sim N(0, \sigma^2 + \tau_j^2/c_j^2).$$

An empirical Bayes method was proposed to tune the hyperparameters. Since the hyperparameters are level dependent, as a shrinker $\hat{\theta}_{j,k}$ is a smooth interpolation between two lines through origin with slopes $\tau_j^2/(\sigma^2 + \tau_j^2)$ and $\tau_j^2/(c_j^2\, \sigma^2 + \tau_j^2)$, and it shrinks $y_{j,k}$ differently across levels. Thus, the Bayesian approach is an adaptive shrinker.

The second method is a full Bayesian approach proposed by Clyde, Paramigiani and Vidakovic (1998). They treat both $\theta_{j,k}$ and σ^2 unknown. Put an inverse chisquare distribution as a prior for σ^2, and select the prior of $\theta_{j,k}$ as a mixture of a normal distribution and a point mass at zero, which capture the strategy of keeping a small number of large wavelet coefficients and excluding a large number of small wavelet coefficients, respectively. That is,

$$\theta_{j,k}|\gamma_j, \sigma^2 \sim \gamma_j\, N(0, \tau_j^2) + (1 - \gamma_j)\, 1_{\{0\}},$$

$$\gamma_j \sim \mathrm{Bin}(1, p_j), \qquad \eta\nu/\sigma^2 \sim \chi_\nu^2,$$

where $c_j, \tau_j, p_j, \eta, \nu$ are hyperparameters. The standard Bayesian mechanism yields the posterior means of $\theta_{j,k}$ given $y_{j,k}$, which lack of explicit form, and Markov Chain Monte Carlo method can be used to numerically evaluate the posterior means.

The third approach is to use weighted absolute error loss studied by Abramovich, Sapatinas and Silverman (1998). The prior of $\theta_{j,k}$ is a mixture of a normal and a point mass at zero,

$$\theta_{j,k} \sim \gamma_j\, N(0, \tau_j^2) + (1 - \gamma_j)\, 1_{\{0\}},$$

where

$$\tau_j^2 = C_1\, 2^{-\alpha j}, \qquad \gamma_j = \min(1, C_2\, 2^{-\beta j}),$$

and C_1, C_2, α, β are hyperparameters. The parameters α and β govern the stochastically decaying speed of $\theta_{j,k}$ and thus control the smoothness of the underlying function $f(x)$. Abramovich, Sapatinas and Silverman (1998) showed that for known σ^2, the Bayesian solution is the posterior median which has the form

$$\mathrm{median}(\theta_{j,k}|y_{j,k}) = \mathrm{sign}(y_{j,k})\max(0, \zeta_{j,k}),$$

where

$$\zeta_{j,k} = |y_{j,k}|\frac{\tau_j^2}{\sigma^2 + \tau_j^2} - \frac{\tau_j\,\sigma}{\sqrt{\sigma^2 + \tau_j^2}}\Phi^{-1}\left(\frac{1 + \min(\omega_{j,k}, 1)}{2}\right),$$

$$\omega_{j,k} = \frac{1 - \gamma_j}{\gamma_j}\frac{\sqrt{\sigma^2 + \tau_j^2}}{\sigma}exp\left\{-\frac{\tau_j^2\, y_{j,k}^2}{2\,\sigma^2\,(\sigma^2 + \tau_j^2)}\right\},$$

and Φ^{-1} is the inverse of the standard normal distribution function.

Empirical Bayesian approach was investigated by Clyde and George (2000) and Johnstone and Silverman (2005).

8.3.3. *Rough function estimation*

In scientific studies objects may be very far from smooth or piecewise smooth. Mathematically, fractal functions are used to model these rough objects, and fractal dimension is defined to quantitatively characterize these functions and measure their roughness. For the problem of estimating a fractal function based on noisy data, it is very critical to preserve its fractal dimension. Smooth or piecewise smooth functions can't well approximate a fractal function, and smoothing noisy data substantially reduces its fractal dimension and degrades its quality, so traditional smoothing methods may be inappropriate for fractal function estimation. Wang (1997a) studied estimation of fractal functions by wavelet shrinkage and showed that as sample size tends to infinity, the fractal dimension of the wavelet estimator over the observed resolution levels converges in probability to the fractal dimension of the underlying function. Therefore, for large samples, wavelet shrinkage can remove most of the noise and produce a wavelet estimate with fractal dimension well preserved. Moreover, the wavelet estimator can reveal that some parts of the function are more or less rougher than others and indicate varying local dimension for the underlying function.

8.3.4. *Dependent data*

So far we consider nonparametric regression with i.i.d. data. However, in many applications, data may be dependent. In fact, correlations between observations that are far apart may decay to zero at a slower rate than we would expect from independent data or short-range dependent data. Slow decay in correlation is often referred to as long-range dependence or long-memory. An alternative definition of long-range dependence for a stationary process is that its spectral density has a pole at zero.

Consider the problem of estimating $f(x)$ based on data from the nonparametric regression model (8.1), where random errors $\varepsilon_1, \cdots, \varepsilon_n$ are zero-mean stationary process with correlation

$$\operatorname{corr}(\varepsilon_i, \varepsilon_j) \asymp |i - j|^{-\alpha}, \qquad |i - j| \to \infty, \qquad \alpha \in (0, 1).$$

As functional Gaussian noise and fractional Brownian motion are often used to model phenomena exhibiting long-range dependence, Wang (1996) proposed the following fractional Gaussian noise model to approximate nonparametric regression with long-range dependent Gaussian errors.

Process $Y(x)$, $x \in [0, 1]$, is observed from the fractional Gaussian noise model

$$Y(dx) = f(x)\,dx + \epsilon^{2-2H}B_H(dx), \tag{8.3}$$

where $f(x)$ is an unknown function, ϵ is the noise level, and $B_H(dx)$ is a fractional Gaussian noise, which is a formal derivative of a standard fractional Brownian motion $B_H(x)$, $H \in (0,1)$.

Wang (1996) developed the decorrelation of a class of Gaussian processes including fractional Gaussian noise and fractional Brownian motion by using the idea of simultaneous diagonalization through WVD and the fact that the Gaussian processes are linked to dilation-homogeneous operators. The WVD simultaneously decorrelates fractional Gaussian noise and fractional Brownian motion and achieves a quasi-decorrelation of Brownian motion and white noise. Such simultaneous decorrelations are used to establish asymptotic minimax risks for the fractional Gaussian noise model and gave explicit forms for their convergence rates. In order to achieve minimax rates, the wavelet shrinkage estimators are required to use level dependent thresholds. Johnstone and Silverman (1997), Johnstone (1999) and von Sachs and Macgibbon (2000) investigated wavelet shrinkage for both short-range and long-range dependent data.

8.4. Inverse Problems

Suppose we wish to recover a function $f(x)$ but observe only a transformation, $(Kf)(x)$, of the underlying function, where K is a linear transformation such as Abel Transform, Convolution transform, or Radon transform. In these problems, K is often non-invertible, in the sense that no inverse of K exists as a bounded linear operator, and we call them ill-posed inverse problems. Inverse problems are very challenging and difficult. The inverse problem in the presence of white noise is defined as follows,

$$y_i = (Kf)(x_i) + \varepsilon_i, \qquad i = 1, \cdots, n, \tag{8.4}$$

where $x_i = i/n$, ε_i are i.i.d. normal errors, K is a transformation, and $f(x)$ is a function. Our goal is to estimate $f(x)$ based on data $y_1, \cdots, y_n$.

The singular value decomposition (SVD) is widely used to solve inverse problems. As a wavelet analogue of SVD, Donoho (1995) created the WVD to solve inverse problems in the presence of white noise. WVD consists of three sets of basis functions, an orthogonal wavelet basis $(\psi_{j,k})$ and two near-orthogonal vaguelette bases $(u_{j,k})$ and $(v_{j,k})$. The three sets of basis functions are linked together as follows. The bases $(u_{j,k})$ and $(v_{j,k})$ are mutually biorthogonal, K transforms $\psi_{j,k}$ into $\kappa_j\,u_{j,k}$, and K* transforms

$(v_{j,k})$ back to $\kappa_j \psi_{j,k}$, where κ_j are quasi-singular values. The WVD has the following reproducing formula

$$f(x) = \sum_{j,k} [Kf, u_{j,k}] \kappa_j^{-1} \psi_{j,k}(x).$$

We can see from above reproducing formula that like SVD, WVD can recover f(x) from observations about $(Kf)(x)$. However, WVD is much better than SVD. It simultaneously represents the operator K in a quasi-diagonal form and effectively represents a large class of functions including functions with spatial variabilities such as jumps. Because of their localization property, the wavelets and vaguelettes in a WVD are much more effective in dealing with local behaviors of f(x) than eigenfunctions in a SVD. The WVD based estimator of $f(x)$ is constructed as follows. First use data y_i to compute empirical vaguelette coefficients $[(y_i), u_{j,k}]$; second shrink the empirical vaguelette coefficients with level dependent threshold λ_j to obtain $\delta([(y_i), u_{j,k}], \lambda_j)$; third reconstruct $f(x)$ by the reproducing formula with $[Kf, u_{j,k}]$ replaced by $\delta([(y_i), u_{j,k}], \lambda_j)$. Donoho (1995) showed that the WVD estimator achieves minimax over a wide range of Besov spaces and outperforms linear estimators.

Kolaczyk (1996) investigated the WVD shrinking method and applied it to tomographic image reconstruction. Wang (1997b) investigated function estimation for inverse problems in the presence of long-memory noise by employing two WVDs — one for the inverse problem which simultaneously quasi-diagonalizes both the operator and the prior information and one for long-range dependence which decorrelates long-memory noise — to convert the linear inverse problem into a new inverse problem with white noise. Their minimax risks converge to zero at rates that depend on both transformation and long-rage dependence, which differ from those for problems with either direct observations or indirect observations with independence or short-range dependence. It is also of interest to point out that by the use of WVD diagonalization, the two problems, a linear inverse problem and estimation with long-range dependence, can be formally converted from one to another. Abramovich and Silverman (1998) considered a vaguelette-wavelet decomposition for solving inverse problems. However, WVD can't be used to solve inverse problems with scale preferred transformation like boxcar deconvolution. Fan and Koo (2002), Johnstone, Kerkyacharian, Picard, and Raimondo (2004) and Johnstone and Raimondo (2004) proposed wavelet methods to solve deconvolution problems.

8.5. Change-points

8.5.1. *Direct data*

Wavelets have a property to "zoom in" on very short lived frequency phenomena, such as transients in signals and singularities in functions, and
hence provide a tool to study localized changes. Wang (1995) and Wang
(1998) proposed wavelet methods to study change-points of functions in one
and two dimensions. The methods are first to compute wavelet transformation of the noisy data and then to compare the wavelet coefficients with
the estimated threshold. It uses the spatial positions at which the wavelet
transformation across fine scale levels exceeds the threshold to detect and
locate change-points such as jumps and sharp cusps.

A function $f(x)$ in one dimension is said to have an α-cusp at x_0 for
$\alpha \in [0, 1)$ if there exists a positive constant C such that, as h tends to zero
from left or right,

$$|f(x_0 + h) - f(x_0)| \geq C|h|^\alpha.$$

$\alpha = 0$ corresponds to that $f(x)$ has a jump at x_0. If f is smooth at u,
$Tf(s, u)$ has the order $s^{3/2}$ as s tends to zero, and if f has a sharp-cusp at
u, the maximum of $|T(s, u)|$ over a neighborhood of u with size proportional
to the scale s converges to zero at a rate no faster than $s^{\alpha+1/2}$ as s tends
to zero.

Suppose we have observations $y_1, \cdots, y_n$ from Model (8.1), and $f(x)$
may have sharp cusps and is smooth otherwise. Our goal is to detect if
$f(x)$ has sharp cusps, and if it has, estimate the locations of the sharp
cusps.

From (8.2) and the property of the wavelet transformation for a function
with sharp cusps, we can see that while at time point $t = k \, 2^{-j}$ where
$f(x)$ is smooth, $\theta_{j,k}$ is of order $2^{-3j/2}$, and nearby a sharp cusp of $f(x)$,
$2^{(\alpha+j)/2} \theta_{j,k}$ has an absolute value bounded below from zero. Thus, at high
resolution levels, $\theta_{j,k}$ dominate $\varepsilon_{j,k}$ nearby where $f(x)$ has sharp cusps, and
is negligible in comparison with $\varepsilon_{j,k}$ where $f(x)$ is smooth. This fact implies
that, at some high resolution levels j_n with $2^{j_n} \sim n^{1/(2\alpha+1)}/\log^2 n$, nearby
sharp cusps of $f(x)$, $y_{j_n,k}$ are dominated by $\theta_{j_n,k}$ whose absolute value is
significantly large, and hence significantly larger than the others. Define

$$T_n = \max\{|y_{j_n,k}| : k = 0, 1, \cdots, 2^{j_n} - 1\}.$$

Then T_n is of much larger order under $f(x)$ with sharp cusps than under
smooth $f(x)$, and thus can be served as a statistic to detect sharp cusps in

$f(x)$. To carry out the detection, we need a threshold value for T_n so that if T_n exceeds the threshold value, we can claim that $f(x)$ has sharp cusps. One choice of threshold is the universal threshold $D_n = \hat{\sigma}\sqrt{2\log n/n}$, where $\hat{\sigma}$ is median of $|y_{j_n,k}|$ divided by 0.6745. It is based on asymptotic theory of the maximum of Gaussian processes and works well when noise ε_i is Gaussian or has light-tailed.

We estimate the sharp cusps of $f(x)$ by the locations of $|y_{j_n,k}|$ that exceed threshold D_n, that is, if $|y_{j_n,k}| > D_n$ for some k, the corresponding jump location is estimated by $\hat{\tau} = k\,2^{-j_n}$.

Wang (1995) and Raimondo (1998) established asymptotic theory for the wavelet method and developed fast algorithms to practically implement the method. They showed that the wavelet estimators of the change-points are optimal or nearly optimal. Wang (1998) proposed the wavelet method for change curves in two dimensions and applied the method to edge detection in image processing. Raimondo and Tajvidi (2003) considered the change-points problem when the random errors ε_i follow a heavy-tail distribution. They designed sharp cusp detection by employing a peak-over-threshold method to model the wavelet coefficients of the heavy-tail noise and using the generalized Pareto distribution to approximate the exceedances of the wavelet coefficients.

8.5.2. *Indirect data*

SVD and traditional change-point methods have great difficulty in dealing with change-points for indirect observations. Tools to detect and locate change-points for indirect observations require two properties. They must extract information about $f(x)$ from indirect observations, and they must characterize the local features of $f(x)$. Eigenfunctions in a SVD, like the Fourier basis, often have trouble in focusing on the local behavior of $f(x)$, while conventional detection techniques based on smoothing cannot recover information about $f(x)$ from indirect observation. In contrast to the huge amount of literature on change-points for direct data, there has been little study of change-points for indirect data.

WVD can not only extract information about $f(x)$ from its indirect observations but also characterize localized features of $f(x)$ near a point. It is very suitable for studying change-points for indirect data. Utilizing this special property of WVD, Wang (1999) proposed a method to detect and estimate change-points of the underlying functions f(x) based on observations about $(Kf)(x)$ from Model (8.4). The method uses a WVD to

extract the information about the wavelet transformation of f(x) from indirect data. Once we have the wavelet transformation of f(x), similar to
wavelet methods for direct data, we can detect and estimate change-points
by the wavelet transformation across fine scale levels. Asymptotic theory
for the detection and estimation was established.

8.6. Local Self-similarity and Non-stationary Stochastic Process

Many natural phenomena exhibit some sort of self-similarity, and scientists
have applied self-similarity models to many areas including image processing (fractal image compression and segmentation), dynamic systems (tur-
bulence), and biology and medicine (physiological time series). Although
self-similar stochastic processes were first introduced in a theoretical context by Kolmogorov in 1941, statisticians were made aware of the practical
applicability of such processes through the work of B. B. Mandelbrot (Mandelbrot and van Ness (1968)).

We say a stochastic process $Y(t)$ is a self-similar process with self-
similarity parameter H if for any positive stretching factor c, the distribution of the rescaled and reindexed process $c^{-H} Y(c\,t)$ is the same as that of
the original process $Y(t)$. That is, for any sequence of time points $t_1, \ldots, t_n$
and any positive constant c, the collections $\{c^{-H} Y(c\,t_1), \ldots, c^{-H} Y(c\,t_n)\}$
and $\{Y(t_1), \ldots, Y(t_n)\}$ have the same probability distribution. As a consequence, the sample path of $Y(t)$ has the qualitative features that are
invariant to magnification or shrinkage, so that stochastically any sections
of the sample path have the same general appearance regardless of the
locations and length of the sections.

The value of the self-similarity parameter or scaling exponent H dictates
the dynamic behavior of a self-similar process $Y(t)$. If $Y(t)$ has finite second
moments, and its associated increment process $X(t) = Y(t) - Y(t-1)$ is
stationary, then H is assumed to be between zero and one, and the value
of H is used to describe the autocorrelation of $X(t)$. For $H > 1/2$, $X(t)$
has slowly decaying autocorrelation which is often referred to as long-range
dependence or long memory, and $X(t)$ is widely used to model phenomena
with power-law correlation and long-memory. For $H < 1/2$, $X(t)$ is characterized by serial correlations that decay rapidly and sum to zero. $H = 1/2$
corresponds to that $X(t)$ is serially uncorrelated. The estimation of H as
a constant has been extensively studied, predominantly in the context of
long memory where it is assumed that $H \in (1/2, 1)$.

Self-similarity models with constant self-similarity parameter H assume that the self-similar features of the underlying phenomenon persist over time. However, many phenomena often display self-similarity only on certain time scale, and/or have self-similarity in different time periods with different self-similarity parameters. That is, self-similar patterns are local and change as the phenomenon itself evolves. To adequately model such phenomena, we need to introduce non-stationary self-similar processes by allowing the scaling exponent to vary as a function of time, and develop a statistical procedure to characterize the exponent's progression. Mallat, Papanicolaou and Zhong (1998) used local cosine bases to model a class of locally stationary processes with approximately convolution covariance operators. With the efficient modeling of covariance operators by local cosine bases, Donoho, Mallat, and von Sachs (1998) investigated estimation of the covariance of locally stationary processes. Neumann and von Sachs (1997) and Nason, von Sachs, and Kroisandt (2000) considered adaptive wavelet estimation of the evolutionary spectra. Nason and Sapatinas (2002) used wavelet packets to model non-stationary time series. Flandrin and Gançlaves (1993) and Flandrin and Gançlaves (1994) proposed locally self-similar processes whose scaling exponents vary with time and discussed their applications. Cavanaugh, Wang and Davis (2003) introduced the following two locally similar processes. The first example is defined by the stochastic integral

$$Y(t) = \int_{-\infty}^{0} \left\{ (t-u)^{H(t)-1/2} - (-u)^{H(t)-1/2} \right\} dB_u + \int_{0}^{t} (t-u)^{H(t)-1/2} dB_u,$$

$$(8.5)$$

where B_u is standard Brownian motion, and $H(t) \in (0,1)$ represents the scaling function. As $Y(t)$ is an extension of fractional Brownian motion and allows its self-similarity parameter to vary over time, we call it *generalized fractional Brownian motion,* gfBm. For gfBM, Cavanaugh, Wang and Davis (2003) provided an explicit forms for its covariance function and showed that gfBm can be locally approximated by fBm, and its sample path has fractal dimension $2 - \min\{H(t) : 0 \le t \le 1\}$ and local dimension $2 - H(t_0)$ at a given time point t_0. The increment process of fBm is stationary, while gfBM has a non-stationary increment (unless $H(t)$ is constant) and if $H(t)$ is smooth, the increment process of gfBm is locally stationary.

The second example is defined through difference equation

$$\Phi(\mathrm{B})(1 - \mathrm{B})^{H(t)-1/2} X(t) = \Theta(\mathrm{B})\epsilon(t), \qquad (8.6)$$

where B is a backshift operator given by $\mathrm{B}X(t) = X(t-1)$, $\Phi(\mathrm{B})$ and $\Theta(\mathrm{B})$ are polynomials in B with roots outside the unit circle, $\epsilon(t)$ is Gaussian white noise, and $H(t) \in (0,1)$ is the scaling function. Similar to gfBM, $X(t)$ is an extension of a fractional autoregressive integrated moving-average process with self-similarity parameter allowing to evolve over time, we thus refer it to as a *generalized* fARIMA or gfARIMA process. A special case of (8.6) is defined by

$$(1 - \mathrm{B})^{H(t)-1/2} X(t) = \epsilon(t).$$

We refer it to as *generalized fractionally integrated noise*, since it is an extension of fractionally integrated noise.

For locally self-similar process, its scaling function often carries important, even decisive, information about the behavior of the process, so it is desirable to develop statistical inference based on sample observations of the process. Wang, Cavanaugh, and Song (2001) and Cavanaugh, Wang and Davis (2003) proposed a procedure based on wavelets for constructing an estimator of the time-varying scaling exponent of a locally self-similar process. They established an approximate local log-linear relationship between the square of the wavelet transformation of $Y(t)$ and the scale for the transformation and used local least-squares regression to estimate $H(t)$. Specifically, let ψ be a mother wavelet and

$$TY(a,t) = a^{-1/2} \int \psi\left(\frac{u-t}{a}\right) Y(u)\, du = a^{1/2} \int \psi(x)\, Y(t+a\,x)\, dx$$

be wavelet transformation of $Y(t)$. Then

$$E\left\{|TY(a,t)|^2\right\} \asymp C_1\, a^{1+2\,H(t)}, \qquad \text{as } a \to 0.$$

Set

$$y_t(a) = \log\left\{|TY(a,t)|^2\right\},\ C_2 = E\left(\log\left[|TY(a,t)|^2 / E\left\{|TY(a,t)|^2\right\}\right]\right),$$
$$\varepsilon_t(a) = \log\left[|TY(a,t)|^2 / E\left\{|TY(a,t)|^2\right\}\right] - C_2.$$

Then we have an approximate regression model

$$y_t(a) \approx c + \{2\,H(t) + 1\}\, \log a + \varepsilon_t(a), \qquad \text{for small scale } a, \qquad (8.7)$$

with $c = \log C_1 + C_2$. The least squares estimator of $H(t)$ is given as follows.

1. Select a sequence of small scales $a_1 > \cdots > a_k$, say $a_j = 2^{-j}$ where $j = 1, \ldots, k$.
2. Define a set of bivariate data (x_j, y_j), $j = 1, \cdots, k$, by setting

$$x_j = \log a_j \qquad \text{and} \qquad y_j = y_t(a_j) \qquad \text{for each } j.$$

3. Evaluate the least-squares estimate of $H(t)$ in (8.7) via

$$\hat{H}(t) = \left\{ \frac{\sum(x_j - \bar{x})\,(y_j - \bar{y})}{\sum(x_j - \bar{x})^2} - 1 \right\} / 2,$$

where $\bar{x} = \sum x_j / k$, $\bar{y} = \sum y_j / k$.

Consistency of $\hat{H}(t)$ was established and extensive simulations were conducted to check the effectiveness of the procedure. The method was applied to several practical examples. However, asymptotic distribution is not available for $\hat{H}(t)$, although simulations indicate that $\hat{H}(t)$ is asymptotically normal. See Craigmile and Percival (2005), Katul, Vidakovic and Alberson (2001) and Vidakovic, Katul and Albertson (2000) for related work.

8.7. Beyond Wavelets

Wavelets are designed to handle point singularities. Functions in one dimension have only point singularities but complex singularity structures may occur in high dimensional functions. For example, a function in three dimensions may be singular along curves or surfaces. Donoho and his collaborators have created high dimensional bases like ridgelets, curvelets and beamlets to better represent complex functions in high dimensions and use them to find some hidden structures from noisy high dimensional data. See Donoho (2000) and other related papers listed at http://www-stat.stanford.edu/~donoho.

Acknowledgments

This work was supported by the NSF grant DMS-0504323.

References

Abramovich, F. and Benjamini, Y. (1995). Thresholding of wavelet coefficients as multiple hypotheses testing procedure. In *Wavelets and Statistics*(eds. Antoniadis and Oppenheim), 5-14. New York: Springer-Verlag.

Abramovich, F., Benjamini, Y., Donoho, D. L. and Johnstone, I. M. (2005). Adapting to Unknown Sparsity by Controlling the False Discovery Rate. To appear in *Ann. Statist.*

Abramovich, F., Sapatinas, T., and Silverman, B. W. (1998). Wavelet thresholding via a Bayesian approach. *J. Roy. Statits. Soc. B* **60**, 725-749.

Abramovich, F. and Silverman, B. W. (1998). Wavelet decomposition approaches to statistical inverse problems. *Biometrika* **85**, 115-129.

Angelini, C. and Vidakovic, B. (2004). Gamma-minimax wavelet shrinkage: a robust incorporation of information about energy of a signal in denoising applications. *Statistica Sinica* **14**, 103-125.

Antoniadis, A. and Fan, J. (2001). Regularization of Wavelet Approximations. *J. Amer. Statist. Assoc.* **96**, 939-967.

Antoniadis, A. and Fryzlewicz, P. (2005). Parametric modeling of threshold across scales in wavelet regression. Manuscript.

Antoniadis, A. and Sapatinas, T.(2001). Wavelet shrinkage for natural exponential families with quadratic variance functions. *Biometrika* **88**, 805-820.

Barber, S. and Nason, G. P. (2004). Real nonparametric regression using complex wavelets. J. R. Statist. Soc. Ser. B **66**, 927-939.

Barber, S., G. P. Nason, G. P. and B. W. Silverman, B. W. (2002). Posterior probability intervals for wavelet thresholding. *J. R. Statist. Soc. Ser. B* **64**, 189-206.

Cai, T. (1999). Adaptive wavelet estimation: a block thresholding and oracle inequality approach. *Ann. Statist.* **27**, 898-924.

Cai, T. (2002). On block thresholding in wavelet regression: Adaptivity, block size, and threshold level. Statistica Sinica 12, 1241-1273.

Cavanaugh, J. E., Wang, Y. and Davis, W. (2003). Local self-similar processes and their wavelet analysis, in *Handbook of Statistics, Volume 21: Stochastic Processes: Modeling and Simulation* (eds. Shanbhag and Rao), Elsevier Science, 93-135.

Chipman, H. A., Kolaczyk, E. D. and McCulloch, R. E. (1997). Adaptive Bayesian wavelet shrinkage. *J. Amer. Statist. Assoc.* **92**, 1413-1421.

Clyde, M. and George, E. (2000). Flexible empirical Bayes estimation for wavelets. *J. Roy. Statist. Soc. B* **62**, 681-698.

Clyde, M.,Parmigiani, G. and Vidakovic, B. (1998). Multiple shrinkage and subset selection in wavelets. *Biometrika* **85**, 391-401.

Craigmile, P. E. and Percival, D. B. (2005). Asymptotic decorrelation of between-scale wavelet coefficients. *IEEE Transactions on Information Theory* **51**, 1039-1048.

Daubechies, I. (1988). Orthonormal bases of compactly supported wavelets. *Comm. Pure Appl. Math.* **41**, 909-996.

Daubechies, I. (1992). Ten Lectures on Wavelets. CBMS-NSF series in applied mathematics, no. 61, SIAM, Philadelphia.

Donoho, D. L. (1995). Nonlinear solutions of linear inverse problems by wavelet-vaguelette decomposition. *Appl. Comput. Harm. Anal.* **2**, 101-126.

Donoho, D. L. (2000). Ten Lectures on Beyond Wavelets. Available from http://www-stat.standord.edu/~donoho

Donoho, D. L. and Jin, J. (2005). Asymptotic minimaxity of FDR for sparse exponential model. To appear in *Ann. Statist.*

Donoho, D. L. and Johnstone, I. M. (1994). Ideal spatial adaptation by wavelets shrinkage. *Biometrika* **81**, 425-455.

Donoho, D. L. and Johnstone, I. M.(1995). Adapting to unknown smoothness via wavelets shrinking. *J. Amer. Statist. Assoc.* **90**, 1200-1224.

Donoho, D. L. and Johnstone, I. M. (1998). Minimax estimation via Wavelets shrinkage. *Ann. Statist.* **26**, 879-921.

Donoho, D. L. and Johnstone, I. M., Kerkyacharian, G. and Picard, D. (1995). Wavelet shrinkage: Asymptopia ? (with discussion). *J. Roy. Statist. Soc.*

B **57** 301-369.

Donoho, D. L., Mallat, S. G. and von Sachs, R. (1998). Estimating covariances of locally stationary processes: convergence of best-basis methods. Technical report, Stanford University.

Fan, J. (1996). Test of significance based on wavelet thresholding and Neyman's truncation. *J. Amer. Statist. Assoc.* **91**, 674-688.

Fan, J. and Koo, J. Y. (2002). Wavelet deconvolution. it IEEE traction on information theory. **48**, 734-747.

Flandrin, P. and Gançlaves, P. (1994). From wavelets to time-scale energy distributions. In *Advances in Wavelet Analysis* (eds. Schumaker and Webb), Academic Press, New York, 309-334.

Fryzlewicz, P. (2005). Bivariate hard threshold in wavelet function estimation. Manuscript.

Gançlaves, P. and Flandrin, P. (1993). Bilinear time-scale analysis applied to local scaling exponents estimation. In *Progress in Wavelet Analysis and Applications* (eds. Meyer and Roques), Frontieres, Paris, 271-276.

Hall, P. and Patil, P. (1996). On the choice of smoothing parameter, threshold and truncation in nonparametric regression by non-linear wavelet methods. *J. Roy. Statist. Soc. B* **58**, 361-378.

Hall, P., Kerkyacharian, G. and Picard, D. (1998). Block threshold rules for curve estimation using kernel and wavelet method. *Ann. Statist.* **26**, 922-942.

Johnstone, I. M. (1999). Wavelet shrinkage for correlated data and inverse problems: Adaptivity results. *Statistica Sinica* **9**, 51-83.

Johnstone, I. M. and Silverman, B. W. (1997). Wavelet threshold estimators for data with correlated noise. *J. Roy. Statist. Soc. B* **59**, 319–351.

Johnstone, I. M. and Silverman, B. W.(2005). Empirical Bayes selection of wavelet thresholds. *Ann. Statist.* **33**, 1700-1752.

Johnstone, I. M., Kerkyacharian, G., Picard, D. and Raimondo, M. (2004). Wavelet deconvolution in a periodic setting (with discussion). *J. Roy. Statist. Soc. B* **66**, 547-573.

Johnstone, I. M. and Raimondo, M. (2004). Periodic boxcar deconvolution and diophantine approximation. *Ann. Statist.* **32**, 1781-1804.

Katul, G., Vidakovic, B. and Albertson, J. (2001). Estimating global and local scaling exponents in turbulent flows using wavelet transformations. *Physics of Fluids* **13**, 241-250.

Kolaczyk, E. D. (1996). A wavelet shrinkage approach to tomographic image reconstruction. J. Amer. Statist. Assoc. 91, 1079-1090.

Kolaczyk, E. D., Ju, J. and Gopal, S. (2005). Multiscale, multigranular statistical image segmentation. To appear in *J. Amer. Statist. Assoc.*

Kolaczyk, E. D. and Nowak, R. D. (2004). Multiscale likelihood analysis and complexity penalized estimation. *Ann. Statist.* **32**, 500-527.

Kolaczyk , E. D. and Nowak, R. D. (2005). Multiscale generalized linear models for nonparametric function estimation. To appear in *Biometrika.*

Mallat, S. G. (1989). A theory for multiresolution signal decomposition: the wavelet representation. *IEEE Trans. Patt. Anal. Mach. Intel.* **11**, 674-693.

Mallat, S. G. (1999). A Wavelet Tour of Signal Processing. Academic Press, San Diego, CA. Second Edition.

Mallat, S. G., Papanicolaou, G. and Zhong, S. (1998). Adaptive covariance estimation of locally stationary processes. *Ann. Statist.* **26**, 1-47.

Mandelbrot, B. B. and van Ness, J. W. (1968). Fractional Brownian motions, fractional noises and applications. *SIAM Rev.* **10**, 422-437.

Nason, G. P. (1996). Wavelet shrinkage using cross-validation. *J. Roy. Statist. Soc. B* **58**, 463-479.

Nason, G. P. and Sapatinas, T. (2002). Wavelet packet transfer function modelling of non-stationary time series. *Statist. Comput.* **12**, 45-56.

Nason, G. P., von Sachs, R. and Kroisandt, G. (2000). Wavelet processes and adaptive estimation of the evolutionary wavelet spectrum. *J. R. Statist. Soc. Ser. B* **62**, 271-292.

Neumann, M. and von Sachs, R.(1997). Wavelet thresholding in anisotropic function classes and application to adaptive estimation of evolutionary spectra. *Ann. Statist.* **25**, 38-76.

Percival, D. P. and Guttorp, P. (1994). Long-memory processes, the Allen variance and wavelets. In *Wavelets in Geophysics*(eds. Foufoula-Georgiou and Kumar), New York, Academic Press, 325-357.

Raimondo, M. (1998). Minimax estimation of sharp change points. *Ann. Statist.* **26**, 1379-1397.

Raimondo, M. and Tajvidi, N. (2003). A peak over threshold model for change-point detection by wavelets. *Statistica Sinica* **14**, 395-412.

Tadesse, M., Ibrahim, J., Vannucci, M. and Gentleman, R. (2005). Wavelet thresholding with Bayesian false discovery rate control. *Biomet.* **61**, 25-35.

Vidakovic, B. (1999). Statistical Modeling by Wavelets. Wiley, New York.

Vidakovic, B., Katul, G. and Albertson, J. (2000). Multiscale denoising of self-similar process. *J. Geophys. Res.-Atmos.* **105**, 27049-27058.

von Sachs, R. and Macgibbon, B. (2000). Non-parametric curve estimation by wavelet thresholding with locally stationary errors. *Scand. J. Statist.* **27**, 475-499.

Wang, Y. (1995). Jump and sharp cusp detection via wavelets. *Biometrika* **82**, 385-397.

Wang, Y. (1996). Function estimation via wavelet shrinkage for long-memory data. *Ann. Statist.* **24**, 466-484.

Wang, Y. (1997a). Fractal function estimation via wavelet shrinkage. *J. Roy. Statist. Soc. B* **59**, 603-613.

Wang, Y. (1997b). Minimax estimation via wavelets for indirect long-memory data. *J. Statist. Plan. Infer.* **64**, 45-55.

Wang, Y. (1998). Change curve estimation via wavelets. *J. Amer. Statist. Assoc.* **93**, 163-172.

Wang, Y. (1999). Change-point analysis via wavelets for indirect data. *Statistica Sinica* **9**, 103-117.

Wang, Y., Cavanaugh, J. E. and and Song C.(2001). Self-similarity index estimation via wavelets for locally self-similar processes. *J. Statist. Plan. Infer.* **99**, 91-110.

CHAPTER 9

Model Diagnostics via Martingale Transforms: A Brief Review

Hira L. Koul

Department of Statistics and Probability
Michigan State University
East Lansing, MI 48824, USA
koul@stt.msu.edu

This chapter gives a brief review of asymptotically distribution free tests of goodness-of-fit hypothesis of an error distribution based on certain residual empirical processes in regression and some generalized conditionally heteroscedastic time series models. We also discuss tests of lack-of-fit hypothesis of a parametric regression or a parametric time series model based on partial sum processes of residuals. The underlying theme is that all of these tests can be based on certain martingale transforms of these processes initiated by Khmaladze.

9.1. Introduction

This chapter discusses some recent developments in the goodness-of-fit and lack-of-fit testing hypotheses based on residual empirical processes and partial sum processes. The former problem pertains to testing for a particular distribution of the errors in some regression and time series models while the latter pertains to fitting a particular parametric model to the regression and/or autoregressive function without the knowledge of the error distribution. Needless to say there is a vast literature available on these problems. For that reason this review is *a priori* incomplete and apologies to those authors who are not cited in here. The emphasis here is to illustrate the usefulness of some residual empirical processes and the Khmaladze transformation in constructing asymptotically distribution free tests for these problems.

9.1.1. *Goodness-of-fit hypotheses*

A classical problem in statistics is to test if a random sample comes from a given distribution or from a given distribution up to unknown location-scale parameters. These models are generic to many other statistical models including the celebrated regression and autoregressive and generalize autoregressive conditionally heteroscedastic (ARCH-GARCH) models where one is testing that innovations are from a given distribution. It will be argued here that omnibus tests for the goodness-of-fit hypothesis in the one sample location-scale problem based on the residual empirical process have perfect analogs in the above models in terms of certain weighted residual empirical processes. And more importantly the Khamaladze's martingale transformation of the residual empirical process that yields asymptotically distribution free tests for the one sample location-scale model does the same thing for the regression and ARCH-GARCH models when applied to these residual empirical processes.

9.1.2. *Location-scale model*

Consider the one sample location-scale model where one observes a random sample $Y_1, \cdots, Y_n$ from a univariate distribution function (d.f.) F on the real line $\mathbb{R}$, and let F_0 be a known continuous d.f. on $\mathbb{R}$. The problems of interest are to test $H_0 : F = F_0$, or more generally,

$$H : F(y) = F_0(\frac{y - \mu}{\sigma}), \quad \text{for all } y \in \mathbb{R}, \text{ some } \mu \in \mathbb{R}, \text{ and some } \sigma > 0.$$

Let μ_n and σ_n denote $n^{1/2}$-consistent estimators of μ, σ, respectively, under H. Let F_n ($\hat{F}_n$) denote the empirical d.f. based on Y_i, $1 \leq i \leq n$ (($Y_i - \mu_n)/\sigma_n$, $1 \leq i \leq n$) and $D(y) = n^{1/2}(F_n(y) - F_0(y))$, $\hat{D}(y) = n^{1/2}(\hat{F}_n(y) - F_0(y))$, $y \in \mathbb{R}$.

It is well-known that there exist tests based on F_n for testing H_0 that are distribution free for all sample sizes, i.e. their null distribution is known and does not depend F_0, for all sample sizes. See, e.g. Durbin (1973a) and Durbin (1973b), the book edited by D'Agostino and Stephens (1986), and references therein. Examples of such tests are the classical Kolmogorov and Cramér-von Mises tests. Consequently these tests can be easily implemented. In comparison, similar tests based on $\hat{F}_n$ for testing H are not even asymptotically distribution free. Their asymptotic null distributions depend not only on F_0 but also on the estimators μ_n and σ_n, see, e.g. Durbin (1973a, b). One proposal made by Rao (1972) is to use maximum likelihood estimators μ_n, σ_n under H_0 based only on one half of the sample in constructing $\hat{F}_n$. These tests are then asymptotically distribution free (ADF) for all those F_0 that have finite and positive definite Fisher information matrix for the location-scale parameters. But as is obvious, different half samples used in constructing these estimators will possibly lead to different inference. Thus it is desirable to have an alternative way of obtaining ADF tests.

Khmaladze (1979) and Khmaladze (1981) proposed a martingale transform of $\hat{F}_n$ whose asymptotic distribution is the same as that of $B \circ F_0$, where B is the standard Brownian motion on $[0, \infty)$, so that the above types of tests are ADF for testing H. For some reason this methodology has not been adopted in the statistical practice. It has not even appeared in any graduate text books so far. Part of the purpose of this review is to show how this methodology can be implemented even in more complicated problems in which the location-scale structure is generic. Such models include the heteroscedastic regression models and ARCH-GARCH models where one is interested in a goodness-of-fit hypothesis pertaining to the innovation d.f. For an Euclidean vector a, let a' denote its transpose.

To describe a generic version of the Khamaladze transformation, let $\mathcal{E}$ be a subset of an Euclidean space, $Z \in \mathcal{E}$ be a r.v., U be a uniformly distributed r.v. on $[0, 1]$, independent of Z, and $g_z(u) : \mathcal{E} \times [0, 1] \mapsto \mathbb{R}^q$, $E\|g_Z(U)\|^2 < \infty$. Let $\ell = (1, g)'$, $A_z(t) = E_0 \ell_z(U) \ell_z(U)' I(U \geq t)$, $z \in \mathcal{E}$, $0 \leq t \leq 1$. Assume $A_z(t)$ is positive definite for all $z \in \mathcal{E}, 0 \leq t < 1$. Write $A_{z,t}^{-1}$ for the inverse matrix $(A_z(t))^{-1}$. Define the functional process for a $\phi \in L_2 = L_2([0, 1])$, $\mathcal{J}_\phi(z, u) = E\phi(U)A_{z,U}^{-1}\ell_z(U)I(U \leq u)$, and

$$\mathcal{T}_\phi(z, u) = \phi(u) - \mathcal{J}_\phi(z, u)'\ell_z(u), \quad 0 \leq u < 1. \tag{9.1}$$

Let $\mathcal{H} = \{\phi \in L_2, \int \phi(v)\ell(v)dv = 0\}$. The following lemma is similar to the Proposition 4.1 in Khmaladze and Koul (2004).

Lemma 9.1: *The functional $\mathcal{T}_\phi$ is norm preserving from L_2 to $\mathcal{H}$, i.e. for any ϕ, ϕ_1, $\phi_2 \in L_2$,*

$$\text{(a) } E\mathcal{T}_{\phi_1}(Z, U)\ell_Z(U)' = 0, \tag{9.2}$$
$$\text{(b) } Cov(\mathcal{T}_{\phi_1}(Z, U), \mathcal{T}_{\phi_2}(Z, U)) = E\phi_1(U)\phi_2(U).$$

Sketch of a proof. Since Z and U are independent, it suffices to prove these claims conditionally on $Z = z$, for every $z \in \mathcal{E}$. Fix a $z \in \mathcal{E}$. Thus the

Claim (9.2)(a) follows because

$$EJ_\phi(z,U)'\ell_z(U)\ell_z(U)' = \int_0^1 \int_{s\le u} \phi(s)\ell_z(s)'A_{z,s}^{-1}ds\ell_z(u)\ell_z(u)'du$$

$$= \int_0^1 \phi(s)\ell_z(s)'A_{z,s}^{-1} \int_{u\ge s} \ell_z(u)\ell_z(u)'duds = \int_0^1 \phi(s)\ell_z(s)'ds.$$

The fact (9.2)(a) in particular implies that $ET_\phi(x,U) = 0$, for all $\phi \in L_2$. Hence, the L.H.S. of (9.2)(b) equals $E\phi_1(U)\phi_2(U) - B_1 - B_2 + C$, where

$$B_1 = E\phi_1(U) \int \phi_2(u)\ell_z(u)'A_{z,u}^{-1}\ell_z(U)I(U \ge u)du,$$

$$= \int \int_{u\le v} \phi_1(v)\phi_2(u)\ell_z(u)'A_{z,u}^{-1}\ell(v)du\,dv,$$

$$B_2 = E\phi_2(U) \int \phi_1(v)\ell(v)'A_{z,v}^{-1}\ell_z(U)I(U \ge v)dv,$$

$$= \int \int_{u\ge v} \phi_1(v)\phi_2(u)\ell(v)'A_{z,v}^{-1}\ell_z(u)dv\,du,$$

and

$$C = E \int \int \phi_1(v)\phi_2(u)\ell(v)'A_{z,v}^{-1}\ell_z(U)\ell_z(U)'I(U \ge v \vee u)A_{z,u}^{-1}\ell_z(u)dv\,du$$

$$= \int \int \phi_1(v)\phi_2(u)\ell(v)'A_{z,v}^{-1}A_{z,v\vee u}A_{z,u}^{-1}\ell_z(u)dv\,du.$$

Using the continuity of the uniform distribution on $[0,1]$, write the double integral in C as a sum of the two double integrals. In the first, $v \le u$ and in the second $u \le v$. The first term equals B_2 while the second term equals B_1, so that $C = B_1 + B_2$, as was to be proved.

See Khmaladze (1993) for the justification of interchanging the integrals in the above argument, and Tsigroshvili (1998) for a discussion about the above transformation when $A_{z,t}^{-1}$ may not exist. Sometimes the inverse A_{x,t_0}^{-1} for some $t_0 < 1$. In this case $A_{z,t}^{-1}$ exists for all $t \le t_0$ and the above transform is well-defined for all $t \le t_0$.

Write $T_\phi(u)$ when there is no Z present. An immediate consequence of the above lemma is that if $U_1, \cdots, U_n$ are i.i.d. copies of U, then the finite dimensional behavior of the process $\sum_{i=1}^n T_\phi(U_i)/\sqrt{n}$, $\phi \in L_2$, is the same as that of the Brownian motion B_ϕ on $[0,1]$, indexed by a $\phi \in L_2$. In particular, applying this transformation to the class of indicator functions $\phi_t(u) = I(u \leq t)$, we obtain that the process $\sum_{i=1}^{n} \mathcal{I}_{I(U_i \leq t)}/\sqrt{n}$ is a Brownian motion in $0 \leq t \leq 1$.

Now consider the testing problem H. Assume the following.

F_0 has a uniformly continuous density f_0, $f_0 > 0$ a.e. and $\qquad$ (9.3)
$$\sup_{y \in \mathbb{R}} |y| f_0(y) < \infty.$$

$$n^{1/2}(\mu_n - \mu)/\sigma = O_p(1), \qquad n^{1/2}(\sigma_n - \sigma)/\sigma = O_p(1). \qquad (9.4)$$

In the sequel, for any sequences of stochastic processes $\mathcal{Y}_n$ and $\mathcal{Z}_n$ defined over a subset $\mathcal{S}$ of an Euclidean space, $\mathcal{Y}_n = \mathcal{Z}_n + u_p(1)$ means that $\sup_{y \in \mathcal{S}} |\mathcal{Y}_n(y) - \mathcal{Z}_n(y)| = o_p(1)$.

From Durbin (1973a) it follows that under (9.3), (9.4), and H,

$$\hat{D}(y) = D(y) + \frac{n^{1/2}(\mu_n - \mu)}{\sigma} f_0(y) + \frac{n^{1/2}(\sigma_n - \sigma)}{\sigma} y f_0(y) + u_p(1). \qquad (9.5)$$

The above limiting process simplifies further considerably if we use asymptotically efficient estimators of μ and σ at F_0. To describe these assume further that

F_0 has absolutely continuous Lebesgue density f_0, with a.e. $\qquad$ (9.6)
derivative $\dot{f}_0$ satisfying $\int \{ [\frac{\dot{f}_0(y)}{f_0(y)}]^2 + [1 + y\frac{\dot{f}_0(y)}{f_0(y)}]^2 \} dF_0(y) < \infty.$

We shall discuss the transformation of $\hat{D}(F_0^{-1})$. Accordingly, let

$$p_0(t) = f_0(F_0^{-1}(t)), \qquad \dot{p}_0(t) = \frac{\dot{f}_0(F_0^{-1}(t))}{f_0(F_0^{-1}(t))},$$
$$q_0(t) = F_0^{-1}(t) p_0(t), \qquad \dot{q}_0(t) = 1 + F_0^{-1}(t) p_0(t),$$
$$g = (\dot{p}_0, \dot{q}_0)', \qquad \ell = (1, g)', \qquad \mathcal{I}(t) = E g(U) g(U)' I(U \geq t), \quad 0 \leq t \leq 1,$$
$$\varepsilon = \frac{Y - \mu}{\sigma}, \qquad \varepsilon_i = \frac{Y_i - \mu}{\sigma}, \quad 1 \leq i \leq n.$$

Note that the Fisher information matrix for the one sample location-scale model at F_0 is $\mathcal{I}(0)/\sigma^2$. Assume that

$$\mathcal{I}(0) = E \begin{pmatrix} \dot{p}_0^2(U) & \dot{p}_0(U)\dot{q}_0(U) \\ \dot{p}_0(U)\dot{q}_0(U) & \dot{q}_0^2(U) \end{pmatrix} \quad \text{is positive definite.}$$

Let $\theta_n = \sigma^{-1} n^{1/2}(\mu_n - \mu, \sigma_n - \sigma)$ and suppose, μ_n, σ_n are taken to be so that under H,

$$\theta_n = -\mathcal{I}^{-1}(0) n^{-1/2} \sum_{i=1}^{n} \begin{pmatrix} \dot{p}_0(F_0(\varepsilon_i)) \\ \dot{q}_0(F_0(\varepsilon_i)) \end{pmatrix} + o_p(1). \qquad (9.7)$$

Then, from the above results it follows that under H, $\hat{D}(F_0^{-1})$ converges weakly to a continuous Gaussian process on $[0,1]$, with mean zero and the covariance function

$$\rho(s,t) = s \wedge t - st - \Big(p_0(s), q_0(s)\Big)\mathcal{I}^{-1}(0)\binom{p_0(t)}{q_0(t)}, \quad 0 \le s, t \le 1.$$

Some simplification of this covariance occurs if F_0 is symmetric around zero. In this case the Fisher Information matrix is a diagonal matrix and the above covariance simplifies to

$$\rho_1(s,t) = s \wedge t - st - a^{-1}p_0(s)p_0(t) - b^{-1}q_0(s)q_0(t), \quad 0 \le s, t \le 1, \quad (9.8)$$

where $a = \int_0^1 \dot{p}_0^2(u)du$, $b = \int_0^1 \dot{q}_0^2(u)du$.

The distribution of the Gaussian process with the covariance function ρ or ρ_1 depends on F_0 in a complicated way. We shall now apply the above transformation that converts such a Gaussian process to a time transformed Brownian motion. Since there in no Z variable present in this problem, in (9.1), take $\ell_z(u) \equiv \ell_{1\sigma}(u) = (1, \sigma^{-1}\dot{p}_0(u), \sigma^{-1}\dot{q}_0(u))'$. Denote the corresponding A_z matrix by $A_{1\sigma}$. Then

$$A_{1\sigma}(t) = \begin{pmatrix} 1-t & -p_0(t)/\sigma & -q_0(t)/\sigma \\ -p_0(t)/\sigma & & \\ -q_0(t)/\sigma & & \mathcal{I}(t)/\sigma^2 \end{pmatrix}.$$

Note that under (9.6), the south east 2×2 matrix of $A_{1\sigma}(0)$ equals the Fisher information matrix. Clearly $A_{1\sigma}(t)$ is non-negative definite for all t. We need to assume that

$$A_{1\sigma}^{-1}(t) \quad \text{exists for all } 0 \le t < 1. \quad (9.9)$$

Then Khmaladze transform of the indicator function $\phi_t(u) = I(u \le t)$ is

$$\mathcal{T}_{\phi_t}(u) = I(u \le t) - \int_0^{t \wedge u} \ell_{1\sigma}(s)' A_{1\sigma}^{-1}(s)\, ds\, \ell_{1\sigma}(u), \quad u \in [0,1].$$

The corresponding transformation of $D(F_0^{-1})$ based on ε_i is

$$\mathcal{U}_{1n}(t, (\mu, \sigma)) = n^{-1/2} \sum_{i=1}^n \mathcal{T}_{\phi_t}(F_0(\varepsilon_i)), \quad 0 \le t \le 1.$$

Even though this converges weakly to the Brownian motion, it is not useful from the inference point of view. In order to implement tests based on this we need to replace the parameters μ and σ by their estimates. Let $\hat{U}_i = F_0((Y_i - \mu_n)/\sigma_n)$. Also, let ℓ_{1n} and $A_{1n}(t)$ stand for $\ell_{1\sigma}$ and $A_{1\sigma}(t)$, respectively, after σ is replaced by σ_n in these entities. Then the transform of $\hat{D}(F_0^{-1})$ and the process of interest is

$$\hat{\mathcal{U}}_{1n}(t) = \mathcal{U}_{1n}(t, (\mu_n, \sigma_n))$$
$$= n^{-1/2} \sum_{i=1}^{n} [I(\hat{U}_i \leq t) - \int_0^{t \wedge \hat{U}_i} \ell_{1n}(v)' A_{1n}^{-1}(v) dv \, \ell_{1n}(\hat{U}_i)].$$

From Khmaladze (1981) it follows that $\hat{\mathcal{U}}_{1n} \Longrightarrow B$, in uniform metric. We remark here that (9.7) is used only for motivating the above transformation. In fact one only need the estimators μ_n, σ_n to satisfy (9.4) in order for this weak convergence result to hold. Consequently, any test of H based on a continuous function of $\hat{\mathcal{U}}_{1n}$ will be ADF for testing H.

For the later use we shall next describe analogs of the above transformation when fitting a distribution up to an unknown location parameter only or up to an unknown scale parameter only, i.e. for testing $\widetilde{H}_0 : F(y) = F_0(y - \mu)$, for all $y \in \mathbb{R}$ and some $\mu \in \mathbb{R}$, or testing $\widetilde{H} : F(y) = F_0(y/\sigma)$, for all $y \in \mathbb{R}$ and for some $\sigma > 0$. In the former the scale parameter is known and taken to be unity while in the latter the location parameter is known and taken to be zero.

Consider $\widetilde{H}_0$ first. Let $\hat{\mu}$ be a $n^{1/2}$-consistent estimator of μ under $\widetilde{H}_0$, $\widetilde{U}_i = F_0(Y_i - \hat{\mu})$, and let $\widetilde{D}(t) = n^{-1/2} \sum_{i=1}^{n} \{I(\widetilde{U}_i \leq t) - t\}$. Tests of $\widetilde{H}_0$ are to be based on $\widetilde{D}$. As above, if $n^{1/2}(\hat{\mu} - \mu) = -n^{-1/2} a^{-1} \sum_{i=1}^{n} \dot{p}_0(Y_i - \mu) + o_p(1)$, then $\widetilde{D}$ converges weakly to a continuous Gaussian process with the covariance function

$$\rho_{loc}(s, t) = s \wedge t - st - a^{-1} p_0(s) p_0(t), \quad 0 \leq s, t \leq 1.$$

Now, in (9.1), take $\ell_z(t) \equiv \ell_2(t) = (1, \dot{p}_0(t))'$, $A_z(t) \equiv A_2(t) = E\ell_2(U)\ell_2(U)I(U \geq t)$, $0 \leq t \leq 1$. Assume $A_2^{-1}(t)$ exists for all $0 \leq t < 1$. Note that here there are no unknown parameters in ℓ_2 and hence in A_2. Then, the process that yields the ADF tests for $\widetilde{H}_0$ here is

$$\widetilde{\mathcal{U}}_{2n}(t) = \frac{1}{\sqrt{n}} \sum_{i=1}^{n} \left\{ I(\widetilde{U}_i \leq t) - \int_0^{t \wedge \widetilde{U}_i} \ell_2(s)' A_2^{-1}(s) \, ds \, \ell_2(\widetilde{U}_i) \right\}. \qquad (9.10)$$

Again, the estimator $\hat{\mu}$ needs only to satisfy (9.4) with $\sigma \equiv 1$, under $\widetilde{H}_0$, in order for tests based on this process to be ADF.

Next, consider $\widetilde{H}$. Let $\hat{\sigma}$ be a $n^{1/2}$-consistent estimator of σ under $\widetilde{H}$ and now let $\bar{U}_i = F_0(Y_i/\hat{\sigma})$ and $\mathcal{D}_n(t) = n^{-1/2} \sum_{i=1}^{n} \{I(\bar{U}_i \leq t) - t\}$. As above, if $n^{1/2}(\hat{\sigma} - \sigma)/\sigma = n^{-1/2} b^{-1} \sum_{i=1}^{n} \dot{q}_0(Y_i/\sigma) + o_p(1)$, then $\mathcal{D}_n$ converges weakly to a continuous Gaussian process with the covariance function

$$\rho_{scale}(s,t) = s \wedge t - st - b^{-1}q_0(s)q_0(t), \quad 0 \le s, t \le 1.$$

Now, in (9.1), take $\ell_z(t) \equiv \ell_{2\sigma}(t) = (1, \dot{q}_0(t)/\sigma)'$, $A_z(t) = A_{2\sigma}(t) = E\ell_{2\sigma}(U)\ell_{2\sigma}(U)I(U \ge t)$, $0 \le t \le 1$. Assume $A_{2\sigma}^{-1}(t)$ exists for all $0 \le t < 1$. Let $\ell_{2n} = \ell_{2\hat{\sigma}}$ and $A_{2n}(t) = A_{2\hat{\sigma}}(t)$. Then, the process that is used here to construct ADF tests for $\widetilde{H}$ is

$$\bar{\mathcal{U}}_{2n}(t) = \frac{1}{\sqrt{n}} \sum_{i=1}^{n} \left\{ I(\bar{U}_i \le t) - \int_0^{t \wedge \bar{U}_i} \ell_{2n}(s)' A_{2n}^{-1}(s) ds\, \ell_{2n}(\bar{U}_i) \right\}.$$

9.1.3. Regression models

Now consider the linear regression model. Let ξ be a p-dimensional random vector of design variables, $X = (1, \xi')'$ and suppose the response variable Y has finite expectation and $E(Y|\xi) = \beta'X$, for some $\beta \in \mathbb{R}^{p+1}$. In other words for some $\beta \in \mathbb{R}^p$ and $\sigma > 0$, we have the Model

$$Y = \beta'X + \sigma\varepsilon, \quad \varepsilon \text{ independent of } X. \tag{9.11}$$

Let F denote the distribution of ε. Here the problem of interest is to test $H_0 : F = F_0$ based on the observations (X_i, Y_i), $1 \le i \le n$, from this model. Let β_n, σ_n be $\sqrt{n}$-consistent estimators of β, σ of (9.11) under H_0. Assume that $E\|\xi\|^2 < \infty$ and EXX' is positive definite. Then for all sufficiently large n, $S_n = n^{-1}\sum_{i=1}^n X_i X_i'$ will be almost surely positive definite. Let

$$W_1(y) = n^{-1/2} \sum_{i=1}^{n} [I(Y_i - \beta_n'X_i \le y\sigma_n) - F_0(y)],$$

$$W_X(y) = n^{-1/2} S_n^{-1/2} \sum_{i=1}^{n} X_i [I(\varepsilon_i \le y) - F_0(y)],$$

$$\hat{W}_X(y) = n^{-1/2} S_n^{-1/2} \sum_{i=1}^{n} X_i [I(I(Y_i - \beta_n'X_i \le y\sigma_n)) - F_0(y)], \quad y \in \mathbb{R}.$$

The process W_1 provides an analog of the one sample empirical process $\hat{D}$ suitable in linear regression. Using the methods of proof of Koul (2002), one can verify that under (9.3), H_0, and (9.11),

$$W_1(y) = D(y) + \sigma^{-1} n^{1/2} [\bar{X}'(\beta_n - \beta) f_0(y) + (\sigma_n - \sigma) y f_0(y)] + u_p(1),$$

$$\hat{W}_X(y) = W_X(y) + \sigma^{-1} n^{1/2} [(\beta_n - \beta) f_0(y) + S_n^{-1/2} \bar{X}_n (\sigma_n - \sigma) y f_0(y)]$$
$$+ u_p(1).$$

Loynes (1980) gives an analog of the first result above in nonlinear parametric regression models set up.

From these approximations one observes that if ξ has mean zero then $\bar{X}_n = (1, \bar{\xi}_n)' \to_p (1, 0, \cdots, 0)'$ and only the intercept parameter among the regression parameters appears in the above approximation of W_1. On the other hand, if the scale parameter σ is known, then upon taking $\sigma_n = \sigma = 1$ one sees that the asymptotic null behavior of W_X depends on the design only through β_n and is analogous to that of $p+1$ independent $\widetilde{D}$ processes in the one sample location model.

The Khmaladze transform of these processes are

$$\mathcal{K}_1(t) = n^{-1/2} \sum_{i=1}^{n} [I(r_i \leq t) - \int_0^{t \wedge r_i} \ell_{1n}(s) A_{1n}^{-1}(s)\, ds \ell_{1n}(r_i)]$$

$$\mathcal{K}_X(t) = n^{-1/2} S_n^{-1/2} \sum_{i=1}^{n} X_i [I(r_i \leq t) - \int_0^{t \wedge r_i} \ell_{1n}(s) A_{1n}^{-1}(s)\, ds \ell_{1n}(r_i)],$$

where $r_i = F_0((Y_i - \beta_n' X_i)/\sigma_n)$. One can verify that under (9.6), $\mathcal{K}_1 \Longrightarrow B_0$, in $D([0,1])$, $\mathcal{K}_n \Longrightarrow \boldsymbol{B}$ in $D^{p+1}[0,1]$ and in uniform metric, where $\boldsymbol{B} = (B_0, \cdots, B_p)'$ is a vector of $(p+1)$ independent Browninan motions on $[0, \infty)$. Thus, for example, $\sup_{0 \leq t \leq 1} \|\mathcal{K}_X(t)\| \Longrightarrow \sup_{0 \leq t \leq 1} \|\boldsymbol{B}(t)\|$. The distribution of this latter r.v. has been tabulate by Kiefer (1959). See also Chapter 6 in Koul (2002) for an augmented version of this table.

More generally, let $\{m_\beta(x), \tau_\beta(x) > 0, \beta \in \Omega \subset \mathbb{R}^q, x \in \mathbb{R}^p\}$ be given families of parametric functions and consider the nonlinear parametric heteroscedastic regression model

$$Y = m_\beta(X) + \tau_\beta(X)\eta, \tag{9.12}$$

where the error r.v. η is independent of X, and $E\eta = 0$, $E\eta^2 = 1$. Let F now stand for the d.f. of η. Based on a random sample (X_i, Y_i), $1 \leq i \leq n$, from this model the problem is to test $H_0 : F = F_0$. Assume that $m_\beta(x)$ is differentiable w.r.t. β for all x with vector of derivatives $\dot{m}_\beta(x)$ satisfying

$$\eta^{-2} E \sup_{\|b-\beta\| \leq \epsilon} [m_b(X) - m_\beta(X) - \dot{m}_\beta(X)'(b - \beta)]^2 \to 0, \quad \text{as } \epsilon \to 0,$$

$$E\|\dot{m}_\beta(X)'\dot{m}_\beta(X)\|^2 < \infty, \quad \Sigma_\beta = E\dot{m}_\beta(X)\dot{m}_\beta(X)' \text{ is positive definite.}$$

Also assume that $\sqrt{n}$-consistent estimators β_n of β exists under (9.12) such that $\Sigma_{\beta_n} = \Sigma_\beta + o_p(1)$, $\Sigma_{\beta_n}^{-1/2} = \Sigma_\beta^{-1/2} + o_p(1)$, and the following holds:

$$\max_{1 \leq i \leq n} n^{1/2} |(\tau_{\beta_n}(X_i)/\tau_\beta(X_i)) - 1| = O_p(1). \tag{9.13}$$

We now need to take $Z = X$, $\ell_z \equiv \ell_{1\tau_\beta(x)}$, $A_z \equiv A_{1\tau_\beta(x)}$ in (9.1) to obtain an ADF goodness-of-fit test. Note that this is the first place we have an explicit dependence of these entities on an x. Denote the corresponding transform with $\phi = \phi_t$, $U = F_0((Y - m_\beta(X))/\tau_\beta(X))$ by $T_t((X, U), \beta)$. Then, ADF tests of H_0 are to be based on the processes $n^{-1/2} \sum_{i=1}^{n} T_t((X_i, U_i), \beta_n)$ or $(n\Sigma_{\beta_n})^{-1/2} \sum_{i=1}^{n} \dot{m}_{\beta_n}(X_i) T_t((X_i, U_i), \beta_n)$. If we let $\tau_{ni} = \tau_{\beta_n}(X_i)$, $\hat{V}_i = F_0((Y_i - m_{\beta_n}(X_i))/\tau_{ni})$, $\ell_{3ni} = \ell_{1\tau_{ni}}$, and $A_{3ni} = A_{1\tau_{ni}}$, then these processes are equivalent to

$$\mathcal{Z}_1(t) = n^{-1/2} \sum_{i=1}^{n} \left[I(\hat{V}_i \leq t) - \int_0^{t \wedge \hat{V}_i} \ell_{3ni}(s)' A_{3ni}^{-1}(s) \, ds \, \ell_{3ni}(\hat{V}_i) \right],$$

$$\mathcal{Z}_X(t) = (n\Sigma_{\beta_n})^{-1/2} \sum_{i=1}^{n} \dot{m}_{\beta_n}(X_i) \left[I(\hat{V}_i \leq t) \right.$$
$$\left. - \int_0^{t \wedge \hat{V}_i} \ell_{3ni}(s)' A_{3ni}^{-1}(s) \, ds \, \ell_{3ni}(\hat{V}_i) \right].$$

Again the asymptotic null behavior of these process is the same as that of $\mathcal{K}_1$ and $\mathcal{K}_X$. As discussed in Khmaladze and Koul (2004) there is no loss of the asymptotic power against a large class of nonparametric contiguous alternatives in using these transformed tests in comparison with the tests based on non-transformed residual empirical processes.

It is of interest to investigate the analogs of the above transforms when either the regression function or the heteroscedasticity function is not smooth. In this case it is unclear as to what plays the role of $\dot{m}_\beta$ which is needed to define the analog of the above weighted empirical process W_X and $\mathcal{Z}_X$. However one does not need this to analyze the analog of the residual empirical process W_1 given by

$$\tilde{W}_1(y) = n^{-1/2} \sum_{i=1}^{n} [I(Y_i - m_{\beta_n}(X_i) \leq y\tau_{ni}) - F_0(y)].$$

In the special case of a homoscedastic segmented regression model where $\tau \equiv 1 \equiv \tau_{ni}$ is known, $p = 1$, $q = 5$, and $m_\beta(x) = \beta_1' h_c(x)$, with $\beta_1 = (a_0, b_0, a_1, b_1)'$, $h_c(x) = (I(x \leq c, xI(x \leq c, I(x > c), xI(x > c)'$, for some $\beta = (\beta_1', c)' \in \mathbb{R}^4 \times [-\infty, \infty]$ with the jump $a_1 - a_0 + (b_1 - b_0)c \neq 0$, it was noted in Koul (2000) that the estimation of c has no effect on the asymptotic behavior of the residual empirical process $\tilde{W}_1$. In fact, the following holds. Assume $EX^2 < \infty$, the d.f. G of X has a positive and continuous density at c and that F_0 satisfies (9.3). Let $\mu_j(c) = EX^j I(X \leq c)$, $\bar{\mu}_j(c) = EX^j I(X >$

c), $j = 1, 2$, and

$$\Sigma_c = E h_c(X) h_c(X)' = \begin{pmatrix} G(c) & \mu_1(c) & 0 & 0 \\ \mu_1(c) & \mu_2(c) & 0 & 0 \\ 0 & 0 & \bar{G}(c) & \bar{\mu}_1(c) \\ 0 & 0 & \bar{\mu}_1(c) & \bar{\mu}_2(c) \end{pmatrix}, \quad (9.14)$$

$$\Gamma_c = (G(c), \mu_1(c), \bar{G}(c), \bar{\mu}_1(c))'.$$

Also, suppose that there exists an estimator θ_{1n} of θ_1 such that for some constant $k \neq 0$ and for some real valued function ψ with $E\psi(\varepsilon) = 0$ and $E\psi^2(\varepsilon) < \infty$,

$$n^{1/2}(\theta_{1n} - \theta_1) = k \Sigma_c^{-1} n^{-1/2} \sum_{i=1}^{n} h_c(X_i)\psi(\varepsilon_i) + o_p(1).$$

Then, under H_0,

$$\tilde{W}_1(y) = D(y) + k \Gamma_c' \Sigma_c^{-1} n^{-1/2} \sum_{i=1}^{n} h_c(X_i)\psi(\varepsilon_i) f_0(y) + u_p(1).$$

Because $\Gamma_c' \Sigma_c^{-1} = (1, 0, 1, 0)$, the coefficient of $f_0(y)$ in the above approximation is $c\, n^{-1/2} \sum_{i=1}^{n} \psi(\varepsilon_i)$. In other words we obtain that

$$\tilde{W}_1(y) = D(y) + k\, n^{-1/2} \sum_{i=1}^{n} \psi(\varepsilon_i) f_0(y) + u_p(1).$$

This is similar to (9.5) with $\sigma_n = \sigma = 1$ and μ_n as an M-estimator in there. Thus the Khmaladze transform given at (9.10) with $\tilde{U}_i$ replaced by $F_0(Y_i - \theta_{1n}' h_{c_n}(X_i))$ provides ADF tests here, where c_n is an n-consistent estimator of c. Examples of estimators c_n, θ_{1n} satisfying these conditions include a class of M-estimators, cf. Koul, Qian, and Surgailis (2003).

To develop an analog of the above transformation for the case of multi-segmented homoscedastic regression models is straightforward. But the problems of obtaining analogs of the above tests in the case of segmented heteroscedastic regression models where τ_β may be smooth or non-smooth or when m_β and τ_β are non-smooth in an arbitrary fashion are open at the time of this writing.

A more general problem is that of testing a goodness-of-fit hypothesis of an error d.f. in nonparametric regression models where $Y = \mu(X) + \sigma(X)\eta$, with μ and $\sigma > 0$ as purely nonparametric functions. Clearly, analogues of $\tilde{W}_1$ based on nonparametric residuals can be used to construct tests of such a hypothesis, but as is apparent from the work of Akritas and van Keilegom (2001), the asymptotic distributions of this process depends on the null error d.f. in a very complicated way. Verifying the weak convergence of an analog of $\mathcal{Z}_1$ based on full residuals is an open problem at this time.

Recently, Cheng (2005) made the following suggestion useful in homoscedastic nonparametric regression models. Use a part of the sample to estimate the regression function and the other part to form either the residual empirical or an error density estimate based on these residuals. Among other things, she proved a suitably standardized so constructed residual empirical process converges to the Brownian bridge and that the asymptotic distribution of the maximum of a suitably normalized deviation of the density estimator from the null density is the same as in the case of the one sample i.i.d. set up established by Bickel and Rosenblatt (1973).

9.1.4. *ARCH-GARCH models*

Let $\{Y_i : i \in \mathbb{Z} = 0, \pm 1, \cdots \}$ be a strictly stationary and ergodic real time series, r be a known positive integer and $\Theta_1 (\Theta_2)$ be a subset of $\mathbb{R}^q (\mathbb{R}^r)$ and let $\Theta = \Theta_1 \times \Theta_2$. In the models of interest one observes the process Y_i such that for some sequences of past measurable functions μ_i from Θ_1 to $\mathbb{R}$ and h_i from Θ to $\mathbb{R}^+ = (0, \infty)$,

$$\eta_i = (Y_i - \mu_i(\theta_1))/\sqrt{h_i(\theta)}, \qquad i \in \mathbb{Z}, \tag{9.15}$$

are i.i.d. standardized r.v.'s having d.f. F, independent of the past. Here, the "past measurable" means that for every $s = (s_1, s_2) \in \Theta$, $s_1 \in \Theta_1$, the functions $\mu_i(s_1)$ and $h_i(s)$ are $\mathcal{F}_{i-1}$ measurable, where $\mathcal{F}_i$ is the σ-field generated by $\{\eta_i, \eta_{i-1}, \cdots, Y_0, Y_{-1}, \cdots \}$, $i \in \mathbb{Z}$. The Model (9.15) includes autoregressive (AR), generalized autoregressive conditionally heteroscedastic (GARCH) and ARMA-GARCH models. For example, The ARMA (1,1)-GARCH(1,1) model is given by the relation

$$Y_i = aY_{i-1} + b\varepsilon_{i-1} + \varepsilon_i, \qquad \varepsilon_i = \eta_i \sqrt{h_i}, \tag{9.16}$$
$$h_i = \alpha_0 + \alpha\varepsilon_{i-1}^2 + \beta h_{i-1}.$$

This model is an example of the model (9.15) with $q = 2$, $r = 3$, $\theta_1 = (a, b)'$, $\theta_2 = (\alpha_0, \alpha, \beta)'$, $\theta = (\theta_1, \theta_2)'$, $\varepsilon_i(\theta_1) = Y_i - aY_{i-1} - b\varepsilon_{i-1}(\theta_1)$, $\mu_i(\theta_1) \equiv Y_i - \varepsilon_i(\theta_1)$, and $h_i(\theta) = \alpha_0 + \alpha\varepsilon_{i-1}^2(\theta_1) + \beta h_{i-1}(\theta)$. The parameter space here is a compact subset of $\mathbb{R}^5$ whose members satisfy certain conditions given in Ling and Li (1997) under which this process is stationary and ergodic.

A GARCH (p_1, p_2) model is defined by the equations

$$Y_i = \eta_i \sqrt{h_i} \text{ and } h_i = \alpha_0 + \sum_{j=1}^{p_1} \alpha_j Y_{j-i}^2 + \sum_{j=1}^{p_2} \beta_j h_{j-i} \qquad (9.17)$$

where p_1, p_2 are known positive integers. Clearly this model is also an example of the model (9.15) with $\mu_i \equiv 0 = q$, h_i as given above, and $\theta = \theta_2 = (\alpha_0, \alpha_1, \cdots, \alpha_{p_1}, \beta_1, \cdots, \beta_{p_2})'$ and $r = p_1 + p_2 + 1$. The parameter space is now a compact subset of $\mathbb{R}^r$ satisfying certain conditions that ensures the stationarity and ergodicity of this process, see, e.g. Ling and Li (1997) and Ling and McAleer (2002).

Boldin (1982) and Koul (1991) observed that for the zero mean linear AR models the tests of $H_0 : F = F_0$ based on the residual empirical process are ADF provided F_0 has zero mean, finite variance and satisfies (9.3). The condition of having zero mean is crucial for the validity of this ADF property. Boldin (1990) and Koul (1992) observed that similar facts hold true for the moving-average (MA) models. This is not true of many nonlinear time series models like threshold AR models as noted in Koul (1996), non-stationary AR models in Ling (1998), ARCH/GARCH models in Boldin (1998) and Boldin (2002), Horváth, Kokoszka and Teyssiere (2001), and Berkes and Horváth (2003). For ARCH/GARCH models, Horváth, Kokoszka and Teyssiere (2004) proposed a parametric bootstrap method for testing Gaussianity of the errors, but only studied its validity via Monte Carlo experiments. Lee and Na (2002) and Bachmann and Dette (2005) establish the Bickel-Rosenblatt result for L_2 distances between the kernel density estimator based on residuals and the error density in linear AR(1) models. Bai (2003) gives an interesting review of the Khmaladze methodology in the time series context using ordinary residual empirical process.

We shall now describe certain weighted empirical processes useful for testing H_0 in these models, especially when the effect of the autoregressive functions μ_i is less prevalent asymptotically than that of the moving average heteroscedasticity functions h_i as happens to be the case in the above two examples (9.16) and (9.17) under H_0 when F_0 is symmetric around zero. Roughly speaking, the weights in these processes are asymptotically orthogonal to the space generated by the slopes of the functions μ_i. Moreover, under H_0 and under some conditions, this vector of processes converges weakly to a vector of independent copies of a Gaussian process, each having the covariance function ρ_{scale} above.

More precisely, let Y_0 denote either the vector $(Y_0, Y_{-1}, \cdots, Y_{1-p})'$ or the vector $(Y_0, Y_{-1}, \cdots ,)'$. In either case Y_0 is chosen to be independent of η_i, $i \geq 1$. Let $\{Y_1, \cdots , Y_n\}$ be observations obeying the Model (9.15). Assume that the functions μ_i, h_i are twice continuously differentiable, for all i. Let $\varepsilon_i(s_1) = Y_i - \mu_i(s_1)$, $\eta_i(s) = \varepsilon_i(s_1)/\sqrt{h_i(s)}$. Let $\dot{\mu}_i(s_1) = \partial \mu_i(s_1)/\partial s_1$, $s_1 \in \Theta_1$. Moreover, for an $s \in \Theta$, write

$$\dot{\mu}_i(s) = \begin{pmatrix} \dot{\mu}_i(s_1) \\ 0 \end{pmatrix}, \quad \dot{h}_i(s) = \begin{pmatrix} \dot{h}_{1i}(s) \\ \dot{h}_{2i}(s) \end{pmatrix}, \quad \dot{h}_{ki}(s) = \frac{\partial h_i(s)}{\partial s_k}, \quad k = 1, 2.$$

Let θ be the true parameter and $\eta_i = \eta_i(\theta)$. Assume (9.6) holds and that

$$E \dot{p}_0(U) \dot{q}_0(U) = 0.$$

This condition is satisfied, for example, when (9.6) holds and when F_0 is symmetric around zero. With a, b as in (9.8), recall that $\mathcal{I}(0)$ is now the diagonal matrix with a, b on the diagonal. Let

$$W_i(s) = \begin{pmatrix} \dfrac{\dot{\mu}_i(s_1)}{h_i^{1/2}(s)} & \dfrac{\dot{h}_{1i}(s)}{h_i(s)} \\ 0 & \dfrac{\dot{h}_{2i}(s)}{h_i(s)} \end{pmatrix} = \begin{pmatrix} W_{11,i}(s) & W_{12,i}(s) \\ 0 & W_{22,i}(s) \end{pmatrix}, \quad \text{say,}$$

$$\mathbf{I}(\theta) = E_\theta W_1(\theta) \mathcal{I}(0) W_1(\theta)'.$$

Under H_0, $\mathcal{I}(0)$ is known and $\mathbf{I}(\theta)$ is the Fisher information under the Model (9.15). We shall assume that under H_0, $\mathbf{I}(\theta)$ is positive definite. Let $\hat{\theta}_n$ be an estimator of θ under (9.15) satisfying

$$n^{1/2}(\hat{\theta}_n - \theta) = -\mathbf{I}(\theta)^{-1} n^{-1/2} \sum_{i=1}^n W_i(\theta) \begin{pmatrix} \dot{p}_0(F_0(\eta_i)) \\ \dot{q}_0(F_0(\eta_i)) \end{pmatrix} + o_p(1).$$

The vector of weighted empirical process suitable here is $K(t, \theta_n)$, where

$$K_n(t, s) = \frac{1}{\sqrt{n}} \sum_{i=1}^n W_{22,i}(s)[I(F_0(\eta_i(s)) \leq t) - t], \quad t \in [0, 1], \ s \in \Theta.$$

The tests of H_0 are to be based on $K_n(t, \hat{\theta}_n)$. For example, the Kolmogorov-Smirnov test statistic is

$$\widetilde{K}_n = \sup_{0 \leq t \leq 1} \|\hat{\mathcal{I}}_n^{-1/2} K_n(t, \hat{\theta}_n)\|, \quad \hat{\mathcal{I}}_n = \frac{1}{n} \sum_{i=1}^n W_{22,i}(\hat{\theta}_n) W_{22,i}(\hat{\theta}_n)'.$$

Suppose

$$E_\theta \Big\{ [W_{11,1}(\theta), \ W_{12,1}(\theta)] W_{22,1}(\theta)' \Big\} = 0.$$

This condition reduces the effect of the location related parameters, v.i.z., of θ_1 and $\mu_1(\theta_1)$, on the asymptotic behavior of the above process. Under some additional integrability conditions involving μ_i, h_i, and under H_0 and (9.6), the following holds:

$$K(t, \hat{\theta}_n) = \frac{1}{\sqrt{n}} \sum_{i=1}^{n} W_{22,i}(\theta) \left[I(F_0(\eta_i) \leq t) - t - \frac{q_0(t)}{b} \dot{q}_0(F_0(\eta_i)) \right] + u_p(1).$$

Let $Z_n(t)$ denote the vector of the leading process on the right hand side above. Let $Z(t)$ be a vector of r independent mean zero Gaussian processes on $[0, 1]$ with $\text{Cov}(Z(s), Z(t)) = \rho_{scale}(s, t) I_{r \times r}$. Let $\Gamma_\theta = E_\theta W_{22,1}(\theta) W_{22,1}(\theta)'$. Using a conditioning argument and a weak convergence result, one readily obtains that $\text{Cov}(Z_n(s), Z_n(t)) = \rho_{scale}(s, t) \Gamma_\theta$, and that $\Gamma_\theta^{-1/2} K(t, \hat{\theta}_n) \Longrightarrow Z(t)$. Since, $\|\hat{I}_n - \Gamma_\theta\| = o_p(1)$, it follows that the asymptotic behavior of the vector of processes $\hat{I}_n^{-1/2} K(t, \hat{\theta}_n)$ is exactly the same as that of r independent processes in the one sample scale model. Hence, $\widetilde{K}_n \to_d \sup_{0 \leq t \leq 1} \|Z(t)\|$. These results are proved in Koul and Ling (2006) where all of the needed assumptions are verified for the two examples (9.16) and (9.17).

We shall now describe the Khamaldze transform of the above process here. Let $\ell_{2\sigma}$ and $A_{2\sigma}$ be as in the definition of $\bar{\mathcal{U}}_{2n}$. Let $\sigma_{ni} = \sqrt{h_i(\theta_n)}$, $\widetilde{V}_i = F_0((Y_i - \mu_i(\theta_{1n}))/\sigma_{ni})$, $\ell_{2ni} = \ell_{2\sigma_{ni}}$, $A_{2ni} = A_{2\sigma_{ni}}$ and for $0 \leq t \leq 1$, let

$$\mathcal{L}_n(t) = \frac{1}{\sqrt{n}} \sum_{i=1}^{n} W_{22,i}(\theta_n) \left\{ I(\widetilde{V}_i \leq t) - \int_0^{t \wedge \widetilde{V}_i} \ell_{2ni}(s)' A_{2ni}^{-1}(s) ds \, \ell_{2ni}(\widetilde{V}_i) \right\}.$$

It is conjectured that $\Gamma_\theta^{-1/2} \mathcal{L}_n \Longrightarrow B$, under H_0. Consequently, the tests based on $\sup_{0 \leq t \leq 1} \|\hat{I}_n^{-1/2} \mathcal{L}_n(t)\| \to_d \sup_{0 \leq t \leq 1} \|B(t)\|$, where now B is a vector of r independent Brownian motions on $[0, \infty)$.

At present the asymptotic behavior of the residual empirical processes in the general Model (9.15) when either μ_i or h_i are non-smooth parametric functions or purely nonparametric functions is not known. However, in some self exciting threshold AR models of Tong (1990), results similar to those in segmented linear regression models mentioned in the previous section hold, cf. Koul, Stute and Li (2005).

9.2. Lack-of-fit Tests

Another classical problem in statistics is to assess the effect of a p-dimensional vector X on the response variable Y. Assuming Y has finite expectation, this is often done in terms of the function $\mu(x) = E(Y|X = x)$. If X is a vector of independent variables then this function is known as the regression function and if X is a vector of lagged Y variables then it is called the autoregressive function. In practice one often has a parametric model $\mathcal{M} = \{m_\beta(x); \beta \in \Omega \subset \mathbb{R}^q, x \in \mathbb{R}^p\}$ available and one is interested in testing $H_1 : \mu \in \mathcal{M}$, against the alternative that H_1 is not true. The monographs of Hart (1997) and Li (2004) and review paper of MacKinnon (1992) discuss numerous lack-of-fit tests in regression and time series contexts. Fan, Zhang and Zhang (2001) discuss tests of lack-of-fit of a regression model based on generalized likelihood ratios. This principle is also applied to time series and other model diagnostics in the monograph by Fan and Yao (2003). Koul and Ni (2004) discuss lack-of-fit tests based on certain minimum L_2 distances between a kernel type nonparametric regression function estimate and the model $\mathcal{M}$. Here our focus is to review some lack-of-fit tests based on certain partial sum processes and their Khmaladze type transforms in regression and autoregressive set up.

9.2.1. *Regression model checks*

Here one observes i.i.d. vectors (X_i, Y_i), $1 \leq i \leq n$ from the distribution of (X, Y) and the problem of interest is to test $H_1 : \mu \in \mathcal{M}$, i.e. $H_1 : \mu(x) = m_\beta(x)$, for some $\beta \in \Omega \subset \mathbb{R}^q$, and for all x. One of the processes useful for this problem is the partial sum process

$$V(x) = n^{-1/2} \sum_{i=1}^n (Y_i - \mu(X_i)) I(\gamma(X_i) \leq x),$$

$$V_\beta(x) = n^{-1/2} \sum_{i=1}^n (Y_i - m_\beta(X_i)) I(\gamma(X_i) \leq x), \quad \beta \in \Omega, \, x \in \mathbb{R}^r,$$

where γ is a measurable function on R^p to R^r, r a positive integer. These processes are also known as marked residual empirical processes with the marks $Y_i - \mu(X_i)$ and $Y_i - m_\beta(X_i)$, respectively. Tests of the simple hypothesis $\mu = \mu_0$, where μ_0 is a known regression function, are to be based on V^0, the V process with $\mu = \mu_0$, while that of H_1 are to be based on $\hat{V}(x) = V_{\beta_n}(x)$, where β_n is a $n^{1/2}$-consistent estimator of β, under H_1.

Several well known lack-of-fit tests are based on the analogues of $\hat{V}$ process. Von Neuman (1941) used its analog to test for the constancy of the regression function on $[0, 1]$ when $p = 1 = r$, $\gamma(x) = x$, and $X_i = i/n$. In this case the residuals become $r_i = Y_i - \bar{Y}$. For the same problem, Buckley (1991) showed that the test base on $2(n-1)B_n / \sum_{i=1}^n (Y_i - Y_{i-1})^2$ is locally most powerful in a certain Bayesian model, where $B_n = \int \hat{V}^2(x) dG_n(x) =$

$n^{-2} \sum_{j=1}^{n} (\sum_{k=1}^{j} r_k)^2$, with G_n denoting the empirical of X_i's in general while here $G_n(x) = [nx]/n$, $0 \leq x \leq 1$.

For random X_i's omnibus tests of H_1 based on the $\hat{V}$ process have been investigated by several authors. Su and Wei (1991) use this process with $\gamma(x) \equiv x$ to fit a generalized linear regression model. Stute (1997) discusses the asymptotic distribution theory of this process in detail when $p = 1$ and $\gamma(x) \equiv x$ while bootstrap method is used to implement various tests based on this process in Stute, González Manteiga, and Presedo Quindimil (1998) when $p \geq 1$, $q \geq 1$, $m_\beta(x) = \beta' g(x)$, $x \in \mathbb{R}^p$, with g a vector of q functions, and $\gamma(x) = x$. Stute, Thies, and Zhu (1998) discusses the analog of the Khmaladze transform of $\hat{V}$ when $p = 1$, $\gamma(x) = x$, while its analog for fitting the generalized linear model $m_\beta(x) = \nu_\theta(\beta' x)$, and $\gamma(x) = \beta' x$, $x \in \mathbb{R}^p, p \geq 1$, is investigated in Stute and Zhu (2002), where ν_θ is a known parametric link function. Other authors that have proposed tests based on this process include An and Cheng (1991) and Diebolt and Zuber (1999). Khmaladze and Koul (2004) discuss the transformed statistics in the case $p \geq 1$, $\gamma(x) = x$, using innovation martingale ideas of Khmaladze (1993).

Hart (1997) discusses numerous tests based on the nonparametric regression function estimators. An advantage of a test based on $\hat{V}$ is that it avoids the problem of choosing a window width and it can distinguish the alternatives within $\sqrt{n}$ neighborhoods of the null model. Neither methodologies over come the curse of high dimensionality although there is some evidence via a simulation study in Khmaladze and Koul (2004) that tests proposed there preserve the large sample level for moderate samples rather well. Koenker and Xiao (2002) studied tests based on the Khmaladze transformation of a regression quantile process to test the hypothesis that the effect of the covariate vector X on the location and/or on the location-scale of the conditional quantiles of Y, given X, is linear in X.

We shall now describe the transformation of the process $\mathcal{V}_\beta = V_\beta \circ G^{-1}$, where G is the d.f. of $\gamma(X)$. In this transformation the role of Z and ℓ_Z of the generic transformation (9.1) is played by the error $Z = Y - \mu_\beta(X)$ and the vector of derivatives $\dot{m}_\beta$ of m_β w.r.t. β, assumed to exist, i.e., $\ell_Z(U) \equiv Z h_\beta(U) = Z \dot{m}_\beta(G^{-1}(U))$ in the definition of $\mathcal{T}_\phi$. Hence, with

$\phi(u) \equiv \phi_t(u) = I(u \leq t),$

$$\mathcal{T}_{\phi_t}(Z, U) \equiv Z\{I(U \leq t) - \int_{\infty}^{U \wedge t} h_\beta(s)' A_\beta^{-1}(s) I(U \geq s) h_\beta(U) ds\},$$

$$\mathcal{S}(t, \beta, G) = n^{-1/2} \sum_{i=1}^{n} (Y_i - m_\beta(X_i)) \mathcal{T}_{\phi_t}(U_i)$$

$$= \mathcal{V}_\beta(t) - \int_0^{G^{-1}(t)} h_\beta(s)' A_\beta^{-1}(s) \int_{z \geq s} h_\beta(z) d\mathcal{V}_\beta(z) ds$$

where now $A_\beta(t) = E h_\beta(U) h_\beta(U) I(U \geq t)$ is assumed to have inverse for every $0 \leq t < 1$. If instead $A_\beta^{-1}(t_0)$ exists for some $t_0 < 1$ then the above transformation is defined only on $[0, t_0]$. Stute, Thies, and Zhu (1998) were the first to use this transformed process to implement ADF tests based on $\mathcal{S}(t, \beta_n, G_n)$ in the case $p = 1$, $\gamma(x) \equiv x$, where G_n is the empirical of $X_i, 1 \leq i \leq n$. These authors also discuss an analog of this test to fit a heteroscedastic parametric regression model.

9.2.2. *Fitting an autoregressive model*

As is well known in an autoregressive model of order p, Y_i is regressed upon the previous lag p vector $X_i = (Y_{i-1}, \cdots, Y_{i-p})'$. In the context of an autoregressive time series, tests of a closely related problem of fitting a homoscedastic parametric autoregressive function of order 1 based on an analog of the above process have been investigated by An and Cheng (1991), Kim and Lee (2002), Koul and Stute (1999), and Stute, González Manteiga, Presedo Quindimil and Koul (2005).

In all of these papers the null model m_β is assumed to be smooth. Thus at first it is not apparent as to how to modify this transformation to fit a non-smooth model like a SETAR model of Tong (1990) where $m_\beta(Y_{i-1}) = \beta_1' h_c(Y_{i-1})$, with h_c as in (9.14), is neither smooth in β nor in x. Koul, Stute and Li (2005) observe that the asymptotic behavior of $\hat{\mathcal{V}}_n$ is not affected by the threshold parameter estimation when the jump size between different linear AR segments is fixed, just as in segmented regression model. They then investigate an analogue of the above transformation. But the problem of how to implement a lack-of-fit test of a general non-smooth autoregressive function using the above methodology is open.

An and Cheng (1991) and Kim and Lee (2002) use some heavy mixing conditions to establish their asymptotic results. Koul and Stute (1999) establish the weak convergence of a general marked empirical process under some weak regularity assumptions. Using the fact that in an AR model, error is orthogonal to the autoregressive function, martingale arguments become appropriate tools to design test statistics and study their associated distribution properties.

Then there are test procedures which require nonparametric estimation of μ. See, e.g. Hjellvik and Tjøstheim (1995), Poggi and Portier (1997) and McKeague and Zhang (1994), among others. Typically, tests based on smoothing methodology heavily rely on the chosen bandwidth, say, so that the attained level and power of the test may drastically change from bandwidth to bandwidth. However, the tests based on generalized likelihood principle overcome some of this deficiency, cf. Fan and Yao (2003). Fan and Zhang (2004) discuss generalized likelihood ratio tests for fitting a parametric spectral density to a stationary time series.

9.3. Censoring

The Khamaldze transformation method has also been adapted to fitting a survival distribution under random censoring on the right, cf. Nikabadze and Stute (1997), where one does not only have a finite dimensional nuisance parameter but also an infinite dimensional nuisance parameter via the censoring distribution. Here we describe another application of the above methodology to the interval censoring case 1 where a goodness-of-fit testing problem automatically becomes a lack-of-fit testing problem.

Consider a situation where there is a need to find if a tumor is present in a mice or not. Thus at the time T of inspection one knows whether tumor is present or not. If it is present then one concludes that the time of the onset of the tumor X satisfies $X \leq T$, and if it is absent then $X > T$. This is an example of the interval censored case 1 data, where the event onset time X is not observed, cf. Keiding (1991) and Groeneboom and Wellner (1992). Instead one observes T and $\delta = I(X \leq T)$. Let F denote the d.f. of X on $[0, \infty)$. The problem of interest is to test $H_2 : F = F_\theta$ for some $\theta \in \Theta$, where $\{F_\theta, \theta \in \Theta\}$ is a parametric family of d.f.'s on $[0, \infty)$, based on n i.i.d. observations (δ_i, T_i) from the distribution of (δ, T). Assume X is independent of T.

In principle tests of H_2 can be based on any of the classical distance measures between $\{F_\theta, \theta \in \Theta\}$ and the nonparametric maximum likelihood estimate $\hat{F}_n$ of F. But the asymptotic null distributions of such statistics appear to be intractable because of the complicated and non-standard nature of the weak limits of the finite dimensional distributions of suitably standardized $\hat{F}_n$, cf. Groeneboom and Wellner (1992).

Now note that $E(\delta|T) = F(T)$ and $\sigma^2(T) = Var(\delta|T) = F(T)(1 - F(T))$. Thus, the regression function of δ, given T, is $F(T)$, and this regression is heteroscedastic. Hence this goodness-of-fit testing problem is converted to that of a lack-of-fit testing problem and one can readily adopt the above methodology here. Let $\sigma_\theta^2(T) = F_\theta(T)(1 - F_\theta(T))$. Assume $\dot{F}_\theta$ exists and the matrix

$$A_\theta(t) = E \frac{\dot{F}_\theta(T)\dot{F}_\theta(T)'}{\sigma_\theta^2(T)} I(T \geq t),$$

is positive definite for all $0 \leq t < \infty$. Then the suitable transformed process here is

$$\mathcal{U}_n(t, \theta, G) = V_n(t, \theta) - \int_{s \leq t} \frac{\dot{F}_\theta(s)'}{\sigma_\theta(s)} A_\theta^{-1}(s) \left[\int_{u \geq s} \frac{\dot{F}_\theta(u)}{\sigma_\theta(u)} V_n(du, \theta) \right] dG(s),$$

where now G denotes the d.f. of T and

$$V_n(t, \vartheta) = n^{-1/2} \sum_{i=1}^{n} \frac{(\delta_i - F_\vartheta(T_i))}{\sigma_\vartheta(T_i)} I(T_i \leq t), \qquad 0 \leq t \leq \infty, \vartheta \in \Theta.$$

Note that this process is well defined at $t = \infty$, because $\sigma_\theta(T) = 0$ if and only if $(\delta - F_\theta(T)) = 0$, a.s. Koul and Yi (2005) give some assumptions when tests based on $\widehat{\mathcal{U}}_n(t) = \mathcal{U}_n(t, \hat{\theta}, G_n)$ are ADF for H_2, where $\hat{\theta}$ is a $n^{1/2}$-consistent estimator of θ under H_2 and G_n is the empirical of T_i, $1 \leq i \leq n$. Some finite sample simulations are also included in there.

When fitting the d.f. F_0 up to a scale parameter, i.e. when $F_\theta(x) = F_0(\theta x)$, $\theta > 0$, $x \geq 0$, the statistic $\widehat{\mathcal{U}}_n$ has the following form that may be used to compute it.

$$\widehat{\mathcal{U}}_n(t) = \sqrt{n} \sum_{i=1}^{n} \left[I(T_i \leq t) - n^{-1} \sum_{j=1}^{n} \frac{\frac{T_j T_i f_0(\hat{\theta}T_j) f_0(\hat{\theta}T_i)}{\sigma_0(\hat{\theta}T_j)\sigma_0(\hat{\theta}T_i)} I(T_j \leq t \wedge T_i)}{n^{-1} \sum_{k=1}^{n} \frac{T_k^2 f_0^2(\hat{\theta}T_k)}{\sigma_0^2(\hat{\theta}T_k)} I(T_k \geq T_j)} \right] \hat{e}_i,$$

where $\sigma_0^2 = F_0(1 - F_0)$, f_0 is a derivative of F_0, a known d.f. and $\hat{e}_i = (\delta_i - F_0(\hat{\theta}T_i))/\sigma_0(\hat{\theta}T_i)$.

9.4. Khamaladze Transform or Bootstrap

Given the availability of fast computers and bootstrap methodology it is natural to ask why not use either Monte Carlo or bootstrap methodology to implement the classical tests. Koul and Sakhanenko (2005) compare Monte-Carlo, naive bootstrap, and the smooth bootstrap methods of implementing the Kolmogorov-Smirnov test with the Khmaladze transformed test of an error d.f. in a nonlinear regression model. It is observed there that the transformed test has far better level performance than several of the bootstrap methods considered, and that naive bootstrap does worse in the preservation of the level while smooth bootstrap is slightly better but not as good as the transformed test.

References

Akritas, M. G., van Keilegom, I. (2001). Non-parametric estimation of the residual distribution. *Scand. J. Statist.* **28**, no. 3, 549–567.

An, H. Z. and Cheng, B. (1991). A Kolmogorov-Smirnov type statistic with application to test for nonlinearity in time series. *Int. Statist. Rev.*, **59**, 287-307.

Bachmann, D. and Dette, H. (2005). A note on the Bickel-Rosenblatt test in autoregressive time. *Statist. & Probab. Lett.* **73**, no. 3, 221-234.

Bai, J. (2003). Testing parametric conditional distributions of dynamic models. *Review of Economics and Statistics*, **85**, 531-549.

Berkes, I., Horváth, L. (2003). Limit results for the empirical process of squared residuals in GARCH models. *Stochastic Process. Appl.* **105**, no. 2, 271–298.

Bickel, P. J. and Rosenblatt, M. (1973). On some global measures of the deviations of density function estimates. *Ann. Statist.* **1**, 1071–1095.

Boldin, M. V. (1982). Estimation of the distribution of noise in autoregression scheme. *Theor. Probab. Appl.*, **27**, 866-871.

Boldin, M. V. (1990). On hypothesis testing in a moving average scheme by the Kolmogorov-Smirnov and the ω^2 tests. *Theory Probab. Appl.*, **34(4)**, 699–704.

Boldin, M. V. (1998). On residual empirical distributions in ARCH models with applications to testing and estimation. *Mitteilungen aus dem Mathem. Seminar*, Giessen. **235**.

Boldin, M. V. (2002). On sequential residual empirical processes in heteroscedastic time series. *Math. Methods Statist.* **11**, no. 4, 453–464.

Buckley, M. J. (1991). Detecting a smooth signal: optimality of cusum based procedures. *Biometrika* **78**, no. 2, 253–262.

Cheng, F. (2005). Asymptotic distributions of error density and distribution function estimators in nonparametric regression. *J. Statist. Plann. Inference*, **128**, no. **2**, 327–349.

Diebolt, J. and Zuber, J. (1999). Goodness-of-fit tests for nonlinear heteroscedastic regression models. *Statist. Probab. Lett.*, **42**, 53–60.

Durbin, J. (1973a). Weak convergence of the sample distribution function when parameters are estimated. *Ann. Statist.* **1**, 279–290.

Durbin, J. (1973b). *Distribution theory for test based on the sample d.f.* SIAM, Philadelphia. .

D'Agostino, R. B. and Stephens, M. A. (1986) *Goodness-of-fit techniques.* Statistics: Textbooks and Monographs, **68**. Marcel Dekker, Inc., New York.

Fan, J., Zhang, C. M., and Zhang, J. (2001). Generalized likelihood ratio statistics and Wilks phenomenon. *The Annals of Statistics*, **29**, 153-193.

Fan, J. and Zhang, W. (2004). Generalized likelihood ratio tests for spectral density. *Biometrika*, **91**, 195-209.

Fan, J.; Yao, Q. (2003). *Nonlinear Time Series: Nonparametric and Parametric Methods.* Springer Series in Statisics, Springer, New York.

Groeneboom, P., Wellner, J. A. (1992). *Information bounds and nonparametric maximum likelihood estimation.* DMV Seminar, 19. Birkhuser Verlag, Basel,

Hart, J. D. (1997). *Nonparametric smoothing and lack-of-fit tests.* Springer Series in Statistics. Springer-Verlag, New York.

Horváth, L., Kokoszka, P., and Teyssiere, G. (2001). Empirical process of the squared residuals of an ARCH sequence. *Ann. Statist.* **29**, 445-469.

Horváth, L,, Kokoszka, P., and Teyssiere, G. (2004). Bootstrap misspecification tests for ARCH based on the empirical process of squared residuals. *J. Stat. Comput. Simul.* **74**, no. 7, 469–485.

Hjellvik, V.; Tjøstheim, D. (1995). Nonparametric test of linearity for time series. *Biometrika* **82**, 351-368.

Hjellvik, V. and Tjøstheim, D. (1996). Nonparametric statistics for testing of linearity and serial independence. *J. Nonparametric Statist.* **6**, 223-251.

Keiding, N. (1991). Age-specific incidence and prevalence: A statistical perspective (with discussion) *J. Roy. Statist. Soc. Ser. A*, **154**, 371-412.

Khmaladze, E. V. (1979). The use of ω^2 tests for testing parametric hypotheses. *Theory of Probab. & Appl.*, **24**(2), 283 - 301.

Khmaladze, E. V. (1981). Martingale approach in the theory of goodness-of-fit tests. *Theory Probab. Appl.*, **26**, 240-257.

Khmaladze, E. V. (1993). Goodness of fit problem and scanning innovation martingales. *Ann. Statist.*, **21**, 798-829.

Khmaladze, E. V. and Koul, H. L. (2004). Martingale transforms goodness-of-fit tests in regression models. *Ann. Statist.* **32**, no. 3, 995–1034.

Kiefer, J. (1959). K-sample analogues of the Kolmogorov-Smirnov and Cramr-V. Mises tests. *Ann. Math. Statist.* **30**, 420–447.

Kim, Y. and Lee, S. (2002). On the Kolmogorov-Smirnov type test for testing nonlinearity in time series. *Comm. Statist. Theory Methods*, **31**, no. 2, 299–309.

Koenker, R. and Xiao, J. (2002). Inference on the quantile regression process. *Econometrica* **70**, 1583-1612.

Koul, H. L. (1991). A weak convergence result useful in robust autoregression. *J. Statist. Plann. Inference*, **29**, 291–308.

Koul, H. L. (1992). *Weighted Empiricals and Linear Models.* IMS Lecture Notes, **21**. Hayward, California.

Koul, H. L. (1996). Asymptotics of some estimators and sequential residual empiricals in nonlinear time series. *Ann. Statist.*, **24**, 380-404.

Koul, H. L. (2000). Fitting a two phase linear regression model. Special issue dedicated to Professor Vasant P. Bhapkar. *J. Indian Statist. Assoc.* **38**, no. 2, 331–353.

Koul, H. L. (2002). *Weighted empirical processes in dynamic nonlinear models.* Second edition of *Weighted empiricals and linear models* [Inst. Math. Statist., Hayward, CA, 1992; Lecture Notes in Statistics, **166**. Springer-Verlag, New York.

Koul, H. L. and Ling, S. (2006). Fitting an error distribution in some heteroscedastic time series models. To appear in *Ann. Statist.*, April.

Koul, H. L. and Ni, P. (2004). Minimum distance regression model checking. *J. Statist. Plann. Inference*, **119**, no. 1, 109–141.

Koul, H. L., Qian, L., and Surgailis, D. (2003). Asymptotics of M-estimators in two-phase linear regression models. *Stochastic Process. Appl.* **103**, no. 1, 123–154.

Koul, H. L. and Sakhanenko, L. (2005). Goodness-of-fit testing in regression: A finite sample comparison of bootstrap methodology and Khmaladze transformation. *Statist. Prob. Lett.* **74**, no. 3, 290-302.

Koul, H. L. and Stute, W. (1999). Nonparametric model checks in time series. *Ann. Statist.* **27**, 204-237.

Koul, H. L., Stute, W. and Li, F. (2005). Model diagnosis for SETAR time series. *Statistca Sinica.* **15**, no. 3, 795-817.

Koul, H. L. and Yi, T. (2005). Goodness-of-fit testing in interval censoreing case 1. To appear in *Statist. Prob. Lett.*

Lee, S. and Na, S. (2002). On the Bickel-Rosenblatt test for first-order autoregressive models. *Statist. Probab. Lett.* **56**, no. 1, 23–35.

Li, W. K. (2004). *Diagnostic Checks in Time Series.* Chapman and Hall.

Ling, S. (1998). Weak convergence of the sequential empirical processes of residuals in nonstationary autoregressive models. *Ann. Statist.* **26**, 741–754.

Ling, S. and Li, W. K. (1997). On fractionally integrated autoregressive moving-average time series models with conditional heteroscedasticity. *J. Amer. Statist. Assoc.*, **92**, 1184-1192.

Ling, S. and McAleer, M. (2002). Necessary and sufficient moment conditions for the GARCH(r, s) and asymmetric power GARCH(r, s) models. *Econometric Theory*, **18**, 722-729.

Loynes, R. M. (1980). The empirical distribution function of residuals from generalised regression. *Ann. Statist.*, **8**, 285–298.

MacKinnon, J. G. (1992). Model specification tests and artificial regression. *J. Econometric Literature*, **30** (March), 102-146.

McKeague, I. W. and Zhang, M. J. (1994). Identification of nonlinear time series from the first order cumulative characteristics. *Ann. Statist.*, **22**, 495-514.

Nikabadze, A. and Stute, W. (1997). Model checks under random censorship. *Statist. Probab. Lett.* **32**, no. 3, 249–259.

Poggi, J. M. and Portier, B. (1997). A test of linearity for functional autoregressive models. *J. Time Ser. Anal.* **18**, no. 6, 615–639.

Rao, K. C. (1972). The Kolmogoroff, Crámer-von Mises Chi squares statistics for goodness -of-fit tests in the parametric case. (abstract). *Bull. Inst. Math. Statist.*, **1**, p 87.

Stute, W. (1997). Nonparametric model checks for regression. *Ann. Statist.*, **25**,

613–641.

Stute, W., González Manteiga, W., and Presedo Quindimil, M. (1998). Bootstrap approximations in model checks for regression. *J. Amer. Statist. Assoc.* **93**, no. 441, 141–149.

Stute, W., Presedo Quindimil, M., and González Manteiga, W.; Koul, H. L. (2005). Model checks of higher order time series. Preprint.

Stute, W., Thies, S. and Zhu, Li-Xing. (1998). Model checks for regression: an innovation process approach. *Ann. Statist.*, **26**, 1916–1934.

Stute, W. and Zhu, L. (2002). Model checks for generalized linear models. *Scand. J. Statist.* **29**, no. 3, 535–545.

Su, J. Q. and Wei, L. J. (1991). A lack-of-fit test for the mean function in a generalized linear model. *J. Amer. Statist. Assoc.* **86**, 420-426.

Tong, H. (1990). *Non-Linear Time Series: A Dynamical System Approach.* Oxford University Press. Oxford, UK.

Tsigroshvili, Z. (1998). Some notes on goodness-of-fit tests and innovation martingales. (English. English, Georgian summary) *Proc. A. Razmadze Math. Inst.* **117**, 89–102.

von Neumann, J. (1941). Distribution of the ratio of the mean square successive difference to the variance. *Ann. Math. Statist.* **12**, 367–395.

Statistical Learning and Bootstrap

CHAPTER 10

Boosting Algorithms: with an Application to Bootstrapping Multivariate Time Series

Peter Bühlmann and Roman W. Lutz

Seminar für Statistik, ETH Zürich
CH-8092 Zürich, Switzerland
buhlmann@stat.math.ethz.ch

We describe boosting as an estimation method within specified parametric, potentially very high-dimensional models. Besides some short review and historical remarks, we focus on high-dimensional, multivariate linear models (overcomplete dictionaries) for independent and time series data. In particular, we propose a new bootstrap method for high-multivariate, linear time series. We demonstrate its usefulness and we describe relations to some of Bickel's contributions in time series and the bootstrap.

10.1. Introduction

Since its inception in a practical form in Freund and Schapire (1996), boosting has attracted a lot of attention both in the machine learning and statistics literature. This is in part due to its excellent reputation as a prediction method. The gradient descent view of boosting as articulated in Breiman (1998) and Breiman (1999) and Friedman, Hastie and Tibshirani (2000) provides a basis for the understanding and new variants of boosting. As an implication, boosting is not only a black-box prediction tool but also an estimation method in specified classes of models, allowing for interpretation of specific model-terms.

We focus here on boosting with the squared error loss, mainly for the multivariate case. Based on it, we propose here a new time series bootstrap method, and we will make a link to some of Bickel's contributions and ideas in time series and the bootstrap. The boosting approach for multivariate linear time series addresses one of the problems which the first author discussed with Peter Bickel a decade ago.

Throughout the paper, we assume that the data are realizations of random variables

$$(X_1, Y_1), \ldots, (X_n, Y_n)$$

from a stationary process with p-dimensional predictor variables X_i and q-dimensional response variables Y_i; the jth component of a p-dimensional x will be denoted by $x^{(j)}$. Most often, we will consider $X_i \in \mathbb{R}^p$ and $Y_i \in \mathbb{R}^q$.

10.1.1. *AdaBoost*

AdaBoost (Freund and Schapire (1996)) is an ensemble algorithm for binary classification with $Y_i \in \{0, 1\}$. It is (still) the most popular boosting algorithm and it exhibits a remarkable performance in numerous empirical studies. It works by specifying a base classifier ("weak learner") which is repeatedly applied to iteratively reweighted data, yielding an ensemble of classifiers $\hat{g}^{[1]}(\cdot), \ldots, \hat{g}^{[m]}(\cdot)$, where each $\hat{g}^{[k]}(\cdot) : \mathbb{R}^p \to \{0, 1\}$. That is:

$$
\begin{array}{lll}
\text{reweighted data 1} & \xrightarrow{\text{base procedure}} & \hat{g}^{[1]}(\cdot) \\
\text{reweighted data 2} & \xrightarrow{\text{base procedure}} & \hat{g}^{[2]}(\cdot) \\
\cdots & & \cdots \\
\text{reweighted data m} & \xrightarrow{\text{base procedure}} & \hat{g}^{[m]}(\cdot).
\end{array}
$$

Finally, the AdaBoost classifier is

$$\hat{\mathcal{C}}^{[m]}_{AdaBoost}(\cdot) = \text{sign}(\sum_{j=1}^{m} c_j \hat{g}^{[m]}(\cdot)), \tag{10.1}$$

where c_j are linear combination weights, depending on the in-sample performance of the classifier $\hat{g}^{[j]}(\cdot)$. Thus, the AdaBoost classifier is a weighted majority vote among the ensemble of individual classifiers. A key issue is how to reweigh the original data; once we have reweighted data, one simply applies the base procedure to it as if it would be the original dataset. A statistically motivated description can be found in Friedman, Hastie and Tibshirani (2000).

From the description above, AdaBoost involves three specifications: (i) the base procedure ("weak learner"), (ii) the construction of reweighted data, (iii) the size of the ensemble m. Regarding (i), most popular are classification trees; issue (ii) is defined by the AdaBoost description (cf. Friedman, Hastie and Tibshirani (2000)); and the value m in (iii) is a simple one-dimensional tuning parameter.

10.2. Boosting and Functional Gradient Descent

Breiman (1998) and Breiman (1999) showed that the somewhat mysterious AdaBoost algorithm can be represented as a steepest descent algorithm in function space which we call functional gradient descent (FGD). This great result opened the door to use boosting in other settings than classification. For simplicity, we focus first on the univariate case with 1-dimensional response variables Y_i ($q = 1$).

In the sequel, boosting and functional gradient descent (FGD) are used as a terminology for the same method or algorithm. The goal is to estimate a function

$$f_0(\cdot) = \operatorname{argmin}_{f(\cdot)} \mathbb{E}[\rho(Y, f(X))], \tag{10.2}$$

where $\rho(\cdot, \cdot)$ is a real-valued loss function which is typically convex with respect to the second argument. The function class which we minimize over is not of interest for the moment and hence notationally omitted.

Examples of loss functions and their population minimizers are given in the following table; each case corresponds to a boosting algorithm, as explained in Section 10.2.1.

	L_2Boosting	LogitBoost	AdaBoost
spaces	$y \in \mathbb{R}, f \in \mathbb{R}$	$y \in \{0, 1\}, f \in \mathbb{R}$	$y \in \{0, 1\}, f \in \mathbb{R}$
$\rho(y, f)$	$\|y - f\|^2$	$\log_2(1 + \exp(-2(2y - 1)f))$	$\exp(-(2y - 1)f)$
f_0	$\mathbb{E}[Y\|X = x]$	$\frac{1}{2}\log\left(\frac{p(x)}{1-p(x)}\right)$	$\frac{1}{2}\log\left(\frac{p(x)}{1-p(x)}\right)$

For the last row, $p(x) = \mathbb{P}[Y = 1 | X = x]$.

10.2.1. *The generic boosting algorithm*

Having specified a loss function $\rho(\cdot, \cdot)$ we pursue some sort of empirical minimization: instead of (10.2), we do some constrained minimization of the empirical risk

$$n^{1-} \sum_{i=1}^{n} \rho(Y_i, f(X_i)) \tag{10.3}$$

with respect to $f(\cdot)$. We emphasize here that the constraints will enter non-implicitly in terms of a (boosting) algorithm. This is in contrast to empirical risk minimization over suitable (small enough) function classes or by pursuing penalized empirical risk minimization using e.g. ℓ^2- or ℓ^1-penalties.

The base procedure ("weak learner")

Boosting or FGD pursues constrained minimization of (10.3) by iterative steepest descent in function space. To explain this, we elaborate a bit more on the notion of a base procedure, often called the "weak learner" in the machine learning community. Based on some (pseudo-) response variables $\mathbf{U} = U_1, \ldots, U_n$ and predictor variables $\mathbf{X} = X_1, \ldots, X_n$, the base procedure yields a function estimate

$$\hat{g}(\cdot) = \hat{g}_{(\mathbf{U},\mathbf{X})}(\cdot) : \mathbb{R}^p \to \mathbb{R}.$$

Note that we focus here on function estimates with values in $\mathbb{R}$, rather than classifiers with values in $\{0, 1\}$ as described in Section 10.1.1.

Typically, the function estimate $\hat{g}(x)$ can be thought as an approximation of $\mathbb{E}[U|X = x]$. For example, the base procedure could be a nonparametric kernel estimator (if p is small) or a nonparametric statistical method with some structural restrictions (for $p \geq 2$) such as a regression tree (or class-probability estimates from a classification tree).

Componentwise linear least squares: For cases with $p \gg n$, a useful base procedure is componentwise linear least squares:

$$\hat{g}(x) = \hat{\gamma}_{\hat{\mathcal{S}}} x^{(\hat{\mathcal{S}})},$$

$$\hat{\gamma}_j = \frac{\sum_{i=1}^{n} U_i X_i^{(j)}}{\sum_{i=1}^{n} (X_i^{(j)})^2} \ (j = 1, \ldots, p), \quad \hat{\mathcal{S}} = \mathrm{argmin}_{1 \leq j \leq p} \sum_{i=1}^{n} (U_i - \hat{\gamma}_j X_i^{(j)})^2.$$

This base procedure fits a linear regression with the one predictor variable which reduces residual sum of squares most.

The algorithm

The generic FGD or boosting algorithm is as follows.

Step 1. Initialize $\hat{f}^{[0]} \equiv 0$. Set $m = 0$.

Step 2. Increase m by 1.
Compute the negative gradient and evaluate it at $f = \hat{f}^{[m-1]}(X_i)$:

$$U_i = -\frac{\partial}{\partial f} \rho(Y, f)\big|_{f=\hat{f}^{[m-1]}(X_i)}, \ i = 1, \ldots, n.$$

Step 3. Fit negative gradient vector $U_1, \ldots, U_n$ by using the base procedure, yielding the estimated function

$$\hat{g}^{[m]}(\cdot) = \hat{g}_{\mathbf{U},\mathbf{X}}(\cdot) : \mathbb{R}^p \to \mathbb{R}.$$

The function estimate $\hat{g}^{[m]}(\cdot)$ may be thought of as an approximation of the negative gradient vector $(U_1, \ldots, U_n)$.

Step 4. Do a one-dimensional numerical line-search for the best step-size:

$$\hat{\delta}^{[m]} = \mathrm{argmin}_\delta \sum_{i=1}^{n} \rho(Y_i, \hat{f}^{[m-1]}(X_i) + \delta \hat{g}^{[m]}(X_i)).$$

Step 5. Update $\hat{f}^{[m]} = \hat{f}^{[m-1]}(\cdot) + \nu \cdot \hat{\delta}^{[m]} \hat{g}^{[m]}(\cdot)$ where $0 < \nu \leq 1$ is reducing the step-length for following the approximated negative gradient.

Step 6. Iterate Steps 2-5 until $m = m_{stop}$ is reached for some specified stopping iteration m_{stop}.

The factor ν which reduces the step-length in Step 5 should be chosen "small": our proposal for a default is $\nu = 0.1$. The FGD algorithm does depend on this factor ν, but its choice is not very crucial as long as it is taken to be "small". On the other hand, the stopping iteration m_{stop} is an important tuning parameter of boosting or FGD. Data-driven choices can be done by using cross-validation schemes; computationally much more attractive are internal estimates from information criteria such as AIC, see Section 10.3.3.

By definition, the generic FGD algorithm yields a linear combination of base procedure estimates:

$$\hat{f}^{[m_{stop}]}(\cdot) = \nu \sum_{m=1}^{m_{stop}} \hat{g}^{[m]}(\cdot)$$

which can be interpreted as an estimate from an ensemble scheme, i.e. the final estimator is an average of individual estimates from the base procedure, similar to the formula for AdaBoost in (10.1). Thus, the boosting solution implies the following constraint for minimizing the empirical risk in (10.3): it is a linear combination of fits from the base procedure; in addition, it will be a regularized fit of such linear combinations, see Section 10.2.2.

10.2.2. *Boosting with the squared error loss: L_2-Boosting*

When using the squared error loss $\rho(y, f) = |y - f|^2$, the generic FGD algorithm above takes the simple form of refitting the base procedure to residuals of the previous iteration, cf. Friedman (2001).

Step 1 (initialization and first estimate). Given data $\{(X_i, Y_i); i = 1, \ldots, n\}$, fit the base procedure

$$\hat{f}^{[1]}(\cdot) = \nu \cdot \hat{g}_{(\mathbf{Y}, \mathbf{X})}(\cdot).$$

Set $m = 1$.

Step 2. Increase m by 1.
Compute residuals $U_i = Y_i - \hat{f}^{[m-1]}(X_i)$ $(i = 1, \ldots, n)$ and fit the base procedure to the current residuals. The fit is denoted by $\hat{g}^{[m]}(\cdot) = \hat{g}_{(\mathbf{U}, \mathbf{X})}(\cdot)$. Update

$$\hat{f}^{[m]}(\cdot) = \hat{f}^{[m-1]}(\cdot) + \nu \cdot \hat{g}^{[m]}(\cdot),$$

where $0 < \nu \le 1$ is a pre-specified step-size parameter. (The line-search $\hat{\delta}^{[m]}$ is omitted; in fact, $\hat{\delta}^{[m]} = 1$ if the base procedure is fitted by least squares).

Step 3 (iteration). Repeat Step 2 until some stopping value m_{stop} for the number of iterations is reached.

A glimpse of history

With $m = 2$ (one boosting step), L_2Boosting has already been proposed by Tukey (1977) under the name "twicing".

C.F. Gauss

J.W. Tukey

R.V. Southwell

L_2Boosting with the componentwise least squares base procedure for a fixed collection of basis functions (and using $\nu = 1$) coincides with the matching pursuit algorithm of Mallat and Zhang (1993). Matching pursuit is also known in computational mathematics under the name of "(weak) greedy algorithm" (DeVore and Temlyakov (1996); Temlyakov (2000)). All these methods are also known under the keyword "Gauss-Southwell algorithm" whose origin goes back to Carl Friedrich Gauss, the "Princeps Mathematicorum", and to Sir Richard Southwell. While Gauss is famous, Southwell is less known. Sir Richard Southwell was a faculty member of the Engineering School of Oxford University. In the early 1940s, he developed a powerful iterative procedure, known as the relaxation method, which was successfully applied to a large variety of problems in engineering and physical science. Quoting from a newsletter of the Society of Oxford University Engineers: "Southwell was a first class lecturer and attendance at his lectures was a pleasure".

Gauss realized that a linear system of equations

$$A\beta = b, \ A \in \mathbb{R}^{n \times p}, \ \beta \in \mathbb{R}^p, \ b \in \mathbb{R}^n$$

can be solved for β by iteratively pursuing the solution for one component of β while keeping all others fixed: the iteration goes over the component indices of β: $j = 1, 2, \ldots, p, 1, 2, \ldots, p, 1, 2, \ldots$ This is known as the Gauss-Seidel algorithm; the same idea is also used in backfitting (cf. Burja, Hastie and Tibshirani (1989)), and Bickel, Klaassen, Ritov and Wellner (1993) describe some of its (statistical) properties. Southwell's contribution has been to alter the way the iterations are done. Instead of going systematically through the component-indices of β as described above, he argued for the greedy version: select the component such that a suitable error-norm is minimized. In our context, this translates to select the predictor variable such that residual sum of squares is minimized which is exactly what the componentwise linear least squares base procedure does.

Tukey (1977)'s twicing seems to be the first proposal to formulate the Gauss-Southwell idea in the context of a nonparametric smoothing estimator, beyond the framework of linear models (dictionaries of basis functions).

L_2-Boosting, Lasso and LARS

Efron, Hastie, Johnstone, and Tibshirani (2004) made an intriguing connection between L_2Boosting with componentwise linear least squares and the Lasso (Tibshirani (1996)) which is an ℓ^1-penalized least squares method for linear regression. They consider a version of L_2Boosting, called forward stagewise linear regression (FSLR), and they show that FSLR with infinitesimally small step-sizes produces a set of solutions which is approximately equivalent to the set of Lasso solutions when varying the regularization parameter in Lasso (see also (10.4) below). The approximate equivalence is derived by representing FSLR and Lasso as two different modifications of their computationally clever least angle regression (LARS) algorithm. In special cases where the design matrix satisfies a "positive cone condition", FSLR, Lasso and LARS all coincide (Efron, Hastie, Johnstone, and Tibshirani (2004); p. 425).

As Efron, Hastie, Johnstone, and Tibshirani (2004) Section 8 write, their LARS procedure is not directly applicable to more general base procedures (e.g. regression trees) and to problems which we will present in Section 10.3.

During the iterations of L_2Boosting, we get an interesting set of solutions $\{\hat{f}^{[m]}(\cdot); m = 1, 2, \ldots\}$ and corresponding regression coefficients $\{\hat{\beta}^{[m]} \in \mathbb{R}^p; m = 1, 2, \ldots\}$. Heuristically, due to the results in Efron, Hastie, Johnstone, and Tibshirani (2004), it is "similar" to the set of Lasso solutions $\{\hat{\beta}_\lambda \in \mathbb{R}^p; \lambda \in \mathbb{R}^+\}$ when varying the penalty parameter λ, where

$$\hat{\beta}_\lambda = \operatorname{argmin}_{\beta \in \mathbb{R}^p} \sum_{i=1}^n (Y_i - \sum_{j=1}^p \beta^{(j)} X_i^{(j)})^2 + \lambda \sum_{j=1}^p |\beta^{(j)}|. \qquad (10.4)$$

Computing the set of boosting solutions $\{\hat{f}^{[m]}; m = 1, 2 \ldots\}$ is computationally quite cheap since every boosting step is typically simple: hence, estimating a good stopping iteration m_{stop} via e.g. cross-validation is computationally attractive, and the computational gain is even more impressive when using an internal information criterion such as AIC, see Section 10.3.3. (Of course, for the special case of linear regression, LARS (Efron, Hastie, Johnstone, and Tibshirani (2004)) is computationally even more efficient than boosting). On the other hand, regularized boosting with e.g. an ℓ^1 penalty term (cf. Lugosi and Vayatis (2004)) requires for tuning via cross-validation (selecting the penalty parameter) that the whole algorithm is run repeatedly for many candidate penalty parameters and all training-/test-sets from cross-validation.

10.2.3. *A selective review of theoretical results for boosting*

The difficulty to analyse some boosting method lies in the fact that one has to understand the statistical properties of an algorithm. This is in contrast to theoretical analysis of more explicit estimators such as ℓ^1-penalized versions of boosting or the Lasso. Regarding the latter, in case of linear regression, we have an explicit estimation functional as in (10.4), and the theoretical analysis is *not* considering the numerical algorithm for computing the solution of the convex minimization above.

Consistency results for boosting algorithms with early stopping as described in Section 10.2.1 have been given by Jiang (2004) for AdaBoost, Bickel and Ritov (2004) for general loss functions, Zhang and Yu (2003) for general loss functions, and Bühlmann (2004) for L_2Boosting; Bühlmann and Yu (2003) have shown minimax optimality of L_2Boosting in the toy problem of one-dimensional curve estimation. There are quite a few other theoretical analyses of boosting-type methods which use an ℓ^1-penalty instead of early stopping for regularization. We have outlined in the last paragraph of Section 10.2.2 a computational advantage of early-stopped boosting which distinguishes itself — as a method — from ℓ^1-regularized boosting.

10.3. L_2-Boosting for High-dimensional Multivariate Regression

We are describing here a boosting method for multivariate data, including seemingly unrelated (Zellner (1962) and Zellner (1963)) structures. Consider the multivariate linear regression model with n observations of a q-dimensional response and a p-dimensional predictor. In matrix notation:

$$\mathbf{Y} = \mathbf{X}\mathbf{B} + \mathbf{E}, \tag{10.5}$$

with $\mathbf{Y} \in \mathbb{R}^{n \times q}$, $\mathbf{X} \in \mathbb{R}^{n \times p}$, $\mathbf{B} \in \mathbb{R}^{p \times q}$ and $\mathbf{E} \in \mathbb{R}^{n \times q}$. The multivariate case requires an extension of the notation. We denote by $\mathbf{Y}_i \in \mathbb{R}^q$ the i-th sample point of the response variable (row-vector of $\mathbf{Y}$) and by $\mathbf{Y}^{(k)} \in \mathbb{R}^n$ the k-th component of the response (column-vector of $\mathbf{Y}$); and analogously for $\mathbf{X}$, $\mathbf{B}$ and $\mathbf{E}$. For each $\mathbf{Y}^{(k)}$, $k = 1, \ldots, q$, we have a univariate regression model with the predictor matrix $\mathbf{X}$ and the coefficient vector $\mathbf{B}^{(k)}$. For the row-vectors of the error matrix $\mathbf{E}_i$, $i = 1, \ldots, n$, we assume $\mathbf{E}_i$ i.i.d. $\sim \mathcal{N}(\mathbf{0}, \mathbf{\Sigma})$. Assuming that $\mathbf{X}$ has full rank p (in particular $p \leq n$), the ordinary least squares estimator exists:

$$\hat{\mathbf{B}}_{\mathrm{OLS}} = (\mathbf{X}^T \mathbf{X})^{-1} \mathbf{X}^T \mathbf{Y}$$

which equals the ordinary least squares estimate for each of the q univariate regressions. In particular, it is independent of $\mathbf{\Sigma}$.

10.3.1. *The implementing loss function*

In many examples, p and q are large relative to sample size n and we would like to do a sparse model fit. We construct a boosting method by specifying a loss function and a base procedure. Regarding the former, we use the Gaussian negative log-likelihood:

$$-\ell_{\boldsymbol{\Sigma}}(\mathbf{B}) = -\log((2\pi)^{nq/2}\det(\boldsymbol{\Sigma})^{n/2})$$
$$+ \frac{1}{2}\sum_{i=1}^{n}(\mathbf{Y}_i^T - \mathbf{X}_i^T\mathbf{B})\boldsymbol{\Sigma}^{-1}(\mathbf{Y}_i^T - \mathbf{X}_i^T\mathbf{B})^T.$$

The first term on the right-hand side is a constant (w.r.t. $\mathbf{B}$): we drop it and with $\boldsymbol{\Gamma}^{-1} = \boldsymbol{\Sigma}^{-1}$, the loss function becomes

$$L(\mathbf{B}) = \frac{1}{2}\sum_{i=1}^{n}(\mathbf{Y}_i^T - \mathbf{X}_i^T\mathbf{B})\boldsymbol{\Gamma}^{-1}(\mathbf{Y}_i^T - \mathbf{X}_i^T\mathbf{B})^T. \qquad (10.6)$$

We distinguish here between $\boldsymbol{\Gamma}$ and $\boldsymbol{\Sigma} = \mathrm{Cov}(\mathbf{E}_i)$: $\boldsymbol{\Gamma}$ is the implementing covariance matrix for the loss function. We may use for it an estimate of $\boldsymbol{\Sigma}$ (e.g. from another model-fit such as univariate boosting for each response separately) or we can choose something simpler, e.g. $\boldsymbol{\Gamma} = \mathbf{I}_q$ (in particular if q is large). In case of the latter, the loss function may still be reasonable (if the q components are on the same scale) and the statement in Theorem 10.1 is then with $\boldsymbol{\Gamma} = \mathbf{I}_q$.

The right hand-side can be written as $\sum_{i=1}^{n}\rho(\mathbf{Y}_i^T, \mathbf{B})$ (implicitly involving $\boldsymbol{\Gamma}^{-1}$ and $\mathbf{X}_i^T$), very much like in (10.3).

The maximum likelihood estimator of $\mathbf{B}$ is the same as the OLS solution and is therefore independent of $\boldsymbol{\Sigma}$. The covariance matrix $\boldsymbol{\Sigma}$ becomes only relevant in the seemingly unrelated regressions (SUR; Zellner (1962), and Zellner (1963)) where some covariates influence only a few components of the q-dimensional response.

10.3.2. *The base procedure*

The input data for the base procedure is the design matrix $\mathbf{X}$ and a pseudo-response matrix $\mathbf{U} \in \mathbb{R}^{n \times q}$ (not necessarily equal to $\mathbf{Y}$). The base procedure fits the linear least squares regression with one selected covariate (column of $\mathbf{X}$) and one selected pseudo-response (column of $\mathbf{U}$) so that the loss function in (10.6), with $\mathbf{U}$ instead of $\mathbf{Y}$, is reduced most.

Thus, the base procedure fits one selected matrix element of $\mathbf{B}$:

$$\hat{\mathbf{B}}_{jk} = 0, \ (jk) \neq (\hat{s}\hat{t}), \ \hat{\mathbf{B}}_{\hat{s}\hat{t}} = \hat{b}_{\hat{s}\hat{t}},$$

$$\hat{b}_{jk} = \frac{\sum_{v=1}^{q}(\mathbf{U}^{(v)})^T\mathbf{X}^{(j)}\mathbf{\Gamma}_{vk}^{-1}}{(\mathbf{X}^{(j)})^T\mathbf{X}^{(j)}\mathbf{\Gamma}_{kk}^{-1}},$$

$$(\hat{s}\hat{t}) = \text{argmin}_{1 \leq j \leq p, 1 \leq k \leq q}\{L(\mathbf{B}); \mathbf{B}_{jk} = \hat{b}_{jk}, \mathbf{B}_{rs} = 0 \ (rs \neq jk)\}$$

$$= \text{argmax}_{1 \leq j \leq p, 1 \leq k \leq q}\frac{\left(\sum_{v=1}^{q}(\mathbf{U}^{(v)})^T\mathbf{X}^{(j)}\mathbf{\Gamma}_{kv}^{-1}\right)^2}{(\mathbf{X}^{(j)})^T\mathbf{X}^{(j)}\mathbf{\Gamma}_{kk}^{-1}}. \tag{10.7}$$

Corresponding to the parameter estimate, there is a function estimate $\hat{\mathbf{g}}(\cdot)$: $\mathbb{R}^p \to \mathbb{R}^q$ defined as follows:

$$(\hat{\mathbf{g}})_k(\mathbf{x}) = \begin{cases} \hat{b}_{\hat{s}\hat{t}}x^{(\hat{s})} & \text{if } k = \hat{t} \\ 0 & \text{if } k \neq \hat{t}, \end{cases} \ k = 1, \ldots, q.$$

From (10.7) we see that the coefficient $\hat{b}_{jk}$ is not only influenced by the k-th response but also by other responses, depending on the partial correlations of the errors (via $\mathbf{\Gamma}^{-1}$ if $\mathbf{\Gamma}$ is a reasonable estimate for $\mathbf{\Sigma}$) and the correlations of the other responses with the j-th covariate (i.e. $(\mathbf{U}^{(v)})^T\mathbf{X}^{(j)}$).

10.3.3. *The algorithm*

Our **multivariate L_2Boosting** algorithm is defined as follows.

Step 1 (initialization and first estimate). Fit the base procedure

$$\hat{\mathbf{f}}^{[1]}(\cdot) = \nu \cdot \hat{\mathbf{g}}_{\mathbf{Y},\mathbf{x}}(\cdot),$$

where $\hat{\mathbf{g}}_{\mathbf{Y},\mathbf{x}}(\cdot)$ is the base procedure based on data $\mathbf{Y}, \mathbf{X}$. Set $m = 1$.

Step 2. Increase m by 1. Compute current residuals

$$\mathbf{U}_i^{[m]} = \mathbf{Y}_i - \hat{\mathbf{f}}^{[m-1]}(\mathbf{X}_i) \ (i = 1, \ldots, n)$$

and fit the base procedure from (10.7) to $\mathbf{U}$ and $\mathbf{X}$. The fit is denoted by $\hat{\mathbf{g}}^{[m]}(\cdot)$. Update

$$\hat{\mathbf{f}}^{[m]}(\cdot) = \hat{\mathbf{f}}^{[m-1]}(\cdot) + \nu \cdot \hat{\mathbf{g}}^{[m]}(\cdot), \ 0 < \nu < 1.$$

Step 3 (iteration). Repeat Step 2 until a stopping iteration m_{stop} is met.

We also obtain a sequence of estimates $\hat{\mathbf{B}}^{[m]}$ which correspond to $\hat{\mathbf{f}}^{[m]}(\mathbf{x}) = (\hat{\mathbf{B}}^m)^T\mathbf{x}, \ \mathbf{x} \in \mathbb{R}^p$. Multivariate L_2Boosting with componentwise linear least squares resembles very much the univariate L_2Boosting analogue described in Section 10.2.2. The difference is that we search in addition for the best response-component $k \in \{1, \ldots, q\}$ to improve the loss function $L(\cdot)$. As in the univariate case, the step-size ν should be chosen small, e.g. $\nu = 0.1$.

The number of iterations m_{stop} is a tuning parameter and can be estimated, for example by cross validation or an AIC criterion. The latter is computationally attractive. It relies on the fact that the base procedure in (10.7) involves a linear hat operator and an optimization over the best pair of indices $(\hat{s}\hat{t})$. Then, the boosting fit at iteration m can be represented as a hat operator $\mathcal{B}^{[m]} : \mathbb{R}^{nq} \to \mathbb{R}^{nq}$ which maps the response $\mathbf{Y}$ to the fitted values $\hat{\mathbf{Y}}$. Neglecting the fact that a search over the best pair of indices $(\hat{s}\hat{t})$ has taken place in the repeated use of the base procedure in (10.7), $\mathcal{B}^{[m]}$ becomes a linear operator and its trace can serve as the number of degrees of freedom (d.f.):

$$\text{d.f.} = \text{trace}(\mathcal{B}^{[m]}).$$

With this notion of degrees of freedom, one can then use the corrected AIC (AIC_c), cf. Hurvich *et al.* (1998), to estimate the optimal stopping iteration m_{stop}. Using model selection criteria for stopping the boosting iterations has been successfully demonstrated in Bühlmann (2004) and Bühlmann and Yu (2005); in the context of multivariate boosting, the details are described in Lutz and Bühlmann (2005).

10.3.4. *Properties of multivariate L_2Boosting*

We summarize here some of the results from Lutz and Bühlmann (2005). *An asymptotic result*
First, we describe a consistency result for multivariate L_2Boosting in linear regression where the number of predictors $p = p_n$ and the dimension of the response $q = q_n$ are allowed to grow very fast as sample size n increases. Consider the Model

$$\begin{aligned}
&\mathbf{Y}_i = \mathbf{f}(\mathbf{X}_i) + \mathbf{E}_i, \ i = 1, \ldots, n, \qquad \mathbf{Y}_i, \mathbf{E}_i \in \mathbb{R}^{q_n}, \ \mathbf{X}_i \in \mathbb{R}^{p_n}, \\
&\mathbf{f}(\mathbf{x}) = \mathbf{B}^T\mathbf{x}, \qquad \mathbf{B} \in \mathbb{R}^{p_n \times q_n}, \ \mathbf{x} \in \mathbb{R}^{p_n}, \\
&\mathbf{X}_i \text{ i.i.d. and } \mathbf{E}_i \text{ i.i.d. with } \mathbb{E}[\mathbf{E}_i] = \mathbf{0} \text{ and } \text{Cov}(\mathbf{E}_i) = \mathbf{\Sigma}.
\end{aligned} \tag{10.8}$$

Because p_n and q_n are allowed to grow with n, also the predictors and the responses depend on n, but we often ignore this notationally. To identify the magnitude of $b_{jk} = b_{jk,n} = \mathbf{B}_{n;jk}$, we assume $\mathbb{E}|\mathbf{X}_1^{(j)}|^2 = 1, \ j = 1, \ldots, p_n$. We make the following assumptions:

(A1) The dimension of the predictor and the response in model (10.8) satisfies $p_n = O(\exp(Cn^{1-\xi}))$, $q_n = O(\exp(Cn^{1-\xi}))$ $(n \to \infty)$, for some $0 < \xi < 1, 0 < C < \infty$.

(A2) $\sup_{n \in \mathbb{N}} \sum_{j=1}^{p_n} \sum_{k=1}^{q_n} |b_{jk,n}| < \infty$.

(A3) For the implementing $\mathbf{\Gamma}^{-1}$ in (10.6):
$$\sup_{n \in \mathbb{N}, 1 \leq k \leq q_n} \sum_{\ell=1}^{q_n} |\mathbf{\Gamma}_{k\ell,n}^{-1}| < \infty, \quad \inf_{n \in \mathbb{N}, 1 \leq k \leq q_n} \mathbf{\Gamma}_{kk}^{-1} > 0.$$

(A4) $\sup_{1 \leq j \leq p_n} \|(\mathbf{X}_1^{(j)})\|_\infty < \infty$, where $\|x\|_\infty = \sup_{\omega \in \Omega} |x(\omega)|$ (Ω denotes the underlying probability space).

(A5) $\sup_{1 \leq k \leq q_n} \mathbb{E}|\mathbf{E}_1^{(k)}|^s < \infty$ for some $s > 4/\xi$ with ξ from (A1).

Assumption (A1) allows for very large predictor and response dimensions relative to sample size n. Assumption (A2) is an ℓ^1-norm sparseness condition. Assumption (A4) could be weakened to existence of sufficiently high moments, at the expense of a slower growth in (A1) which could still be as fast as $O(n^\beta)$ for any $0 < \beta < \infty$ (see also Section 10.4).

Theorem 10.1: Consider the Model (10.8) satisfying (A1)-(A5). Then, the multivariate L_2Boosting estimate $\hat{\mathbf{f}}^{[m_n]}$ with the component-wise linear learner from (10.7) satisfies: for some sequence $(m_n)_{n \in \mathcal{N}}$ with $m_n \to \infty$ $(n \to \infty)$ sufficiently slowly,

$$\mathbb{E}_\mathbf{x} \left| \left(\hat{\mathbf{f}}^{[m_n]}(\mathbf{x}) - \mathbf{f}(\mathbf{x}) \right)^T \mathbf{\Gamma}^{-1} \left(\hat{\mathbf{f}}^{[m_n]}(\mathbf{x}) - \mathbf{f}(\mathbf{x}) \right) \right| = o_p(1) \ (n \to \infty),$$

where $\mathbf{x}$ denotes a new observation, independent of and with the same distribution as the training data.

A proof is given in Lutz and Bühlmann (2005). Theorem 10.1 says that multivariate L_2Boosting recovers the true sparse regression function even if the dimensions of the predictor or response grow essentially exponentially with sample size n. For the univariate linear model analogue, such a result has been shown for L_2Boosting in Bühlmann (2004) and for the Lasso in Greenshtein and Ritov (2004).

A summary of some empirical results

In Lutz and Bühlmann (2005), multivariate L_2Boosting has been compared with multivariate forward variable selection and with individual L_2Boosting where each of the response components are fitted separately. A crude summary is as follows.

(i) Multivariate forward variable selection was often worse than boosting. The reason can be attributed to the fact that it often pays off to use a method which does variable selection *and* shrinkage such as boosting or Lasso. This has been observed in various contexts, cf. Tibshirani (1996) for the Lasso, Friedman (2001) and Bühlmann (2004) for boosting. Examples where forward variable selection perform better than boosting are of the following type: only very few effective predictor variables contributing a strong signal and many non-effective predictors with no influence on the responses: e.g. the coefficient matrix $\mathbf{B}$ has only very few rows with large entries and all others are zero.

(ii) Multivariate L_2Boosting was found to be better than individual boosting if the errors are highly correlated, i.e. $\boldsymbol{\Sigma}$ is (strongly) non-diagonal. In contrast to individual boosting and other individual methods like OLS for multivariate regression, multivariate L_2Boosting allows to take an estimate of $\boldsymbol{\Sigma}$ into account via the loss function in (10.6). For three real data sets, where we do not know whether $\boldsymbol{\Sigma}$ is strongly non-diagonal, multivariate boosting and individual boosting were comparable in terms of a crossvalidation score; larger differences may be masked by a substantial noise variance which enters because we consider differences $\hat{\mathbf{Y}} - \mathbf{Y} = (\hat{\mathbf{f}} - \mathbf{f}) - \mathbf{E}$, whereas for simulated data-sets we measure the discrepancy $\hat{\mathbf{f}} - \mathbf{f}$ directly.

10.4. L_2-Boosting for Multivariate Linear Time Series

Obviously, the boosting method from Section 10.3 can be used for vector autoregressive processes

$$\mathbf{X}_t = \sum_{j=1}^{p} \mathbf{A_j}\mathbf{X}_{t-j} + \mathbf{E}_t, \quad t \in \mathbb{Z}, \tag{10.9}$$

where $\mathbf{X}_t \in \mathbb{R}^q$ is the q-dimensional observation at time t, $\mathbf{A_j} \in \mathbb{R}^{q \times q}$ and $\mathbf{E}_t \in \mathbb{R}^q$ i.i.d. with $\mathbb{E}[\mathbf{E}_t] = \mathbf{0}$ and $\mathrm{Cov}(\mathbf{E}_t) = \boldsymbol{\Sigma}$. The model is stationary and causal if all roots of $\det(\mathbf{I} - \sum_{j=1}^{p} \mathbf{A_j}z^j)$ $(z \in \mathbb{C})$ are greater than one in absolute value.

For observations $\mathbf{X}_t$ $(t = 1, \ldots, n)$, the equation in (10.9) can be written as a multivariate regression model as in (10.5) with $\mathbf{Y} = [\mathbf{X}_{p+1}, \ldots, \mathbf{X}_n]^T \in \mathbb{R}^{(n-p) \times q}$, $\mathbf{B} = [\mathbf{A_1}, \ldots, \mathbf{A_p}]^T \in \mathbb{R}^{qp \times q}$ and $\mathbf{X} \in \mathbb{R}^{(n-p) \times qp}$ the corresponding design matrix. The consistency result from Theorem 10.1 carries over to the time series case. We consider the following $q = q_n$-dimensional VAR(∞) Model:

$$\mathbf{X}_t = \sum_{j=1}^{\infty} \mathbf{A_j}\mathbf{X}_{t-j} + \mathbf{E}_t, \quad t \in \mathbb{Z}, \tag{10.10}$$

with $\mathbf{E}_t \in \mathbb{R}^{q_n}$ i.i.d. with $\mathbb{E}[\mathbf{E}_t] = \mathbf{0}$, $\mathrm{Cov}(\mathbf{E}_t) = \boldsymbol{\Sigma}$ and $\mathbf{E}_t$ independent of $\{\mathbf{X}_s;\ s < t\}$. Again, we ignore notationally that the model and its terms depend on n due to the growing dimension q_n. Assume that

(B1) $\{\mathbf{X}_t\}_{t\in\mathbb{Z}}$ in (10.10) is strictly stationary and α-mixing with mixing coefficients $\alpha(\cdot) = \alpha_n(\cdot)$.

(B2) The dimension satisfies: $q = q_n = O(n^\beta)$ for some $0 < \beta < \infty$.

(B3) $\sup_{n\in\mathbb{N}} \sum_{j=1}^{\infty} \sum_{k,r=1}^{q_n} |a_{k,r;j,n}| < \infty$, $\quad a_{k,r;j,n} = (\mathbf{A}_{j;n})_{kr}$.

(B4) The mixing coefficients and moments satisfy: for some $s \in \mathbb{N}$ with $s > 2(1+\beta) - 2$ (β as in (B2)) and $\gamma > 0$

$$\sum_{k=1}^{\infty} (k+1)^{s-1} \alpha_n(k)^{\gamma/(2s+\gamma)} < \infty,$$

$$\sup_{1\le j\le q_n, n\in\mathbb{N}} \mathbb{E}|\mathbf{X}_t^{(j)}|^{4s+2\gamma} < \infty, \qquad \sup_{1\le j\le q_n, n\in\mathbb{N}} \mathbb{E}|\mathbf{E}_t^{(j)}|^{2s+\gamma} < \infty.$$

Theorem 10.2: *Assume the Model (10.10), satisfying the assumptions (B1)-(B4), and require that (A3) holds. Consider multivariate L_2Boosting with componentwise linear least squares (as in Section 10.3) using $p = p_n$ lagged variables (as in Model (10.9)) with $p_n \to \infty$, $p_n = O(n^{1-\kappa})$ ($n \to \infty$), where $2(1+\beta)/(s+2) < \kappa < 1$. Then, the assertion from Theorem 10.1 holds with $f(\mathbf{x}) = \sum_{j=1}^{\infty} \mathbf{A}_j \mathbf{x}_{t-j}$, $\hat{\mathbf{F}}^{(m_n)}(\mathbf{x}) = \sum_{j=1}^{p_n} \hat{\mathbf{A}}_j^{(m_n)} \mathbf{x}_{t-j}$ and $\mathbf{x}$ a new realization from (10.10), independent from the training data.*

A proof is given in Lutz and Bühlmann (2005). Multivariate L_2Boosting is thus a consistent method for vector autoregressive models of infinite order. Assuming invertibility of the autoregressive representation, i.e. $\det(\mathbf{I} - \sum_{j=1}^{\infty} \mathbf{A}_j z^j) \neq 0$ for $|z| \le 1$, the range of consistency of multivariate L_2Boosting covers all causal, linear multivariate processes.

10.4.1. *The modification for stationary model fitting*

It is a desirable feature that a time series fit yields a stationary model. This property holds e.g. for the Yule-Walker estimator in autoregressive models (cf. Brockwell and Davis (1991)) and it allows to simulate stationary processes from the fitted model. The latter will be important for bootstrapping stationary time series data, see Section 10.4.3. Ensuring a stationary model fit is not so easily possible with the Lasso or Ridge regression which are well known regularization methods for regression.

With boosting, it is straightforward to modify the algorithm so that stationarity of the fitted model is ensured. In every iteration, we only allow an update $\hat{f}^{[m]}(\cdot)$ which is stationary. This can be achieved as follows.

Modified multivariate L_2Boosting for stationary model fitting

We modify Step 2 of the algorithm in Section 10.3.3.

Modified Step 2 (a). Check if $\hat{f}^{[m]}(\cdot)$ corresponds to a stationary process. To do so, denote the corresponding estimates by $\hat{\mathbf{A}}_j^{[m]}$ and consider the condition:

$$\det(I - \hat{\mathbf{A}}_1^{[m]}z - \ldots - \hat{\mathbf{A}}_p^{[m]}z^p) \neq 0 \text{ for } |z| \leq 1. \tag{10.11}$$

If (10.11) holds, then accept this $\hat{f}^{[m]}(\cdot)$ and proceed to Step 3 in the multivariate L_2Boosting algorithm from Section 10.3.3.
If (10.11) fails, then go to the modified Step 2 (b) below.

Modified Step 2 (b). Consider the set $V = (\{1,\ldots,p\} \times \{1,\ldots q\}) \setminus (\hat{s},\hat{t})$, where $(\hat{s},\hat{t})$ have been chosen by the base procedure from (10.7) in Step 2, yielding $\hat{\mathbf{g}}^{[m]}$ and a violation of (10.11). Use the best argument $(\hat{s}_{new}, \hat{t}_{new})$ in the base procedure in (10.7) from the restricted set V and update to get a new $\hat{f}^{[m]}(\cdot)$. Go back to Step 2 (a).

10.4.2. *Two problems discussed with Peter Bickel around 1995*

What is a linear process? This question was raised by Peter Bickel when the first author has been staying at Berkeley. We then realized that the class of linear processes is surprisingly large and we pointed out the difficulties for testing whether a process is linear (Bickel and Bühlmann (1996) and Bickel and Bühlmann (1997)). While the latter is more a negative result, the former "supports" the idea of approximations with linear processes in an automatic way. The notion of an automatic approximation, i.e. learned by a machine, has been particularly motivated for bootstrapping linear time series.

We certainly do not rule out the possibility that some time series data should be modeled using nonlinear processes, ideally by incorporating mechanistic understanding about the underlying scientific problem.

How should we model higher-order Markov transition probabilities? When sticking to automatic approximations for a stationary process, it is most common to do this in a Markovian framework of higher order. The transition probabilities or densities for X_t given the previous $X_{t-1}, \ldots$ are then of the form

$$p(x_t|x_{t-1},\ldots) = p(x_t|x_{t-1},\ldots,x_{t-p}). \tag{10.12}$$

Obviously, if p is large we run into the curse of dimensionality.

In the context of categorical processes with values in a finite space, the idea of variable length Markov chains $\{X_t\}_{t\in\mathbb{Z}}$ (Rissanen (1983); Bühlmann and Wyner (1999)) is

$$p(x_t|x_{t-1},\ldots) = p(x_t|x_{t-1},\ldots,x_{t-\ell}),$$
$$\ell = \ell(x_{t-1},\ldots) = \min\{r;\ \mathbb{P}[X_t = x_t|X_{-\infty}^{t-1} = x_{-\infty}^{t-1}]$$
$$= \mathbb{P}[X_t = x_t|X_{t-r}^{t-1} = x_{t-r}^{t-1}]\quad\text{for all } x_t\},$$

where $\ell \equiv 0$ corresponds to independence.

Here, we denote by $x_{t-r}^{t-1} = x_{t-1},\ldots,x_{t-r}$. Thus, the memory-length function $\ell = \ell(\cdot)$ depends on how the past actually looks like. It can be conveniently represented by a tree: Figure 10.1 shows an example for a binary process. The memory-length function can be read top-down from the tree:

$$\ell(x_{t-1},\ldots) = \begin{cases} 2 \text{ if } x_{t-1} = 1, x_{t-2} \in \{0,1\}, x_{-\infty}^{t-3} = \text{ arbitrary,} \\ 2 \text{ if } x_{t-1} = 0, x_{t-2} = 1, x_{-\infty}^{t-3} = \text{ arbitrary,} \\ 3 \text{ if } x_{t-1} = 0, x_{t-2} = 0, x_{t-3} \in \{0,1\}, x_{-\infty}^{t-4} = \text{ arbitrary.} \end{cases}$$

Each terminal node in the tree from Figure 10.1 corresponds to a state in the variable length Markov chain. It becomes clear from the tree repre-

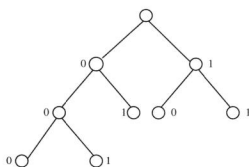

Figure 10.1.   Tree representation of a variable length Markov chain with binary values. The memory-length function can be read top-down from the tree. Each terminal node corresponds to a state in the Markov chain.

sentation that the model for the memory is hierarchical: Peter Bickel kept asking whether one could allow for "holes" in the tree and proceed nonhierarchically. This issue has also shown up in the context of motif finding in computational biology by the need to model long range interactions: for example, Zhao, Huang and Speed (2004) propose permuted variable length Markov models.

From the perspective of boosting or also the Lasso, we can do a computationally efficient approach for modelling non-hierarchical dependence. Take p in (10.12) large and do a sparse model fit: as a result $p(x_t|x_{t-1}, \ldots, x_{t-p}) = p(x_t|\{x_{t-j}; j \in \mathcal{A}\})$ where $\mathcal{A} \subseteq \{1, \ldots, p\}$, i.e. only *some* of the lagged variables will be effective for modelling x_t. What seems quite natural today has not been so obvious 10 years ago. We have described in Section 10.4.1 a boosting approach for non-hierarchical, higher order Markov modeling in the context of multivariate, linear time series; from a methodological point of view, the multivariate-ness causes no additional major complication for our sparse model fitting, while for more classical approaches, the extension to high-multivariate models is often complicated.

10.4.3. L_2-Boost Bootstrap

We propose here a new bootstrap for stationary, multivariate time series, based on vector autoregressive model fitting. It works as follows.

Step 1. Specify the maximal lag p.

Step 2. Use multivariate L_2Boosting for the vector-autoregressive model of order p as in (10.9), with the modification from Section 10.4.1. This yields estimated coefficient matrices $\hat{\mathbf{A}}_1, \ldots, \hat{\mathbf{A}}_p$.
Compute residuals $\hat{\mathbf{E}}_t = \mathbf{X}_t - \sum_{j=1}^p \hat{\mathbf{A}}_j \mathbf{X}_{t-j}$ and consider the centered versions $\tilde{\mathbf{E}}_t = \hat{\mathbf{E}}_t - (n-p)^{-1} \sum_{t=p+1}^n \hat{\mathbf{E}}_t$ $(t = p+1, \ldots, n)$.

Step 3. Do i.i.d. resampling, more than n times, from the empirical c.d.f. of $\tilde{\mathbf{E}}_t$ which yields $\mathbf{E}_1^*, \ldots, \mathbf{E}_{n+k}^*$. Then, generate

$$\mathbf{X}_1^* = \ldots = \mathbf{X}_p^* = 0,$$

$$\mathbf{X}_t^* = \sum_{j=1}^p \hat{\mathbf{A}}_j \mathbf{X}_{t-j}^* + \mathbf{E}_t^*, \quad t = p+1, \ldots, n+k.$$

Use the last stretch $\mathbf{X}_{k+1}^*, \ldots, \mathbf{X}_{n+k}^*$ as a (approximately) stationary realization from the fitted model.

Step 4. Define the bootstrapped estimator by the plug-in rule:

$$\hat{\theta}^* = h_n(\mathbf{X}_{k+1}^*, \ldots, \mathbf{X}_{n+k}^*),$$

where $h_n(\cdot)$ is the function defining the estimator $\hat{\theta} = h_n(\mathbf{X}_1, \ldots, \mathbf{X}_n)$.

The maximal lag p should be chosen "large": due to sparse model fitting, there is often no big decrease in performance when choosing a too large p.

Theorems 10.1 and 10.2 support this proposal. The value k should be chosen "large" to ensure that the initial conditions for the simulation in Step 3 have negligible influence.

Bootstrap inference can now be done in a standard way. Note that in contrast to model-based time series bootstraps, the block bootstrap (Künsch (1989)) is not using the plug-in rule as in Step 4; for a discussion, see also Bühlmann (2002).

Some numerical examples

It is not difficult to present examples where the L_2Boost-Bootstrap is superior than a vector-autoregressive, model based bootstrap using the (non-parsimonious) Yule-Walker estimator with the *AIC*-order selection (AR-YW Bootstrap). The gains of the L_2Boost-Bootstrap over the AR-YW method are most pronounced if q is large relative to sample size and the true model is sparse; but already for moderate q, we see substantial improvements (see below). Sparseness includes dependence structures with "holes": e.g. the true coefficients are $\mathbf{A}_1 \neq \mathbf{0}$, $\mathbf{A}_2 = \mathbf{0}$, $\mathbf{A}_3 \neq \mathbf{0}$, corresponding to a sparse VAR(3).

We present in Figure 10.2 variance estimates with the L_2Boost-Bootstrap and the AR-YW bootstrap for sample partial autocorrelation estimators

$$\widehat{\text{Parcor}}(\mathbf{X}_t^{(r)}, \mathbf{X}_{t-j}^{(s)}) \tag{10.13}$$

for some lags $j \in \{1, 2, 3\}$ and some $r, s \in \{1, \dots, q\}$, $r \neq s$. The sample partial correlation is the sample correlation after having subtracted the linear effects (from OLS regression) of all variables $\mathbf{X}_{t-1}, \dots, \mathbf{X}_{t-j+1}$ in between. The true underlying model is a 5-dimensional VAR(3) ($q = 5$) as in (10.9) with $\mathbf{A}_2 = \mathbf{0}$ and $\mathbf{A}_1$, $\mathbf{A}_3$ sparse having only 7 non-zero entries (out of 25) each. Sample size is chosen as $n = 50$. The number of bootstrap samples is 1000. The reductions in mean squared error of the L_2Boost-Bootstrap variance estimates over the AR-YW bootstrap for the five situations from left to right in Figure 10.2 are substantial:

MSE reductions: 41.9%   62.7%   64.7%   74.4%   61.6%

10.4.4. *Graphical modeling for stochastic processes: an outlook*

Graphical models are very useful for describing conditional dependencies for multivariate observations. For multivariate stationary stochastic processes

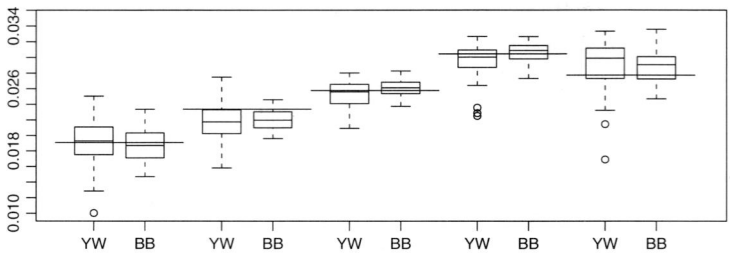

Figure 10.2. Boxplots of bootstrap variance estimates of sample partial autocorrelations as in (10.13) for 5 combinations of various lags j and components r, s: AR-YW bootstrap (YW) and L_2Boost-Bootstrap with AIC_c stopping (BB); true variances indicated by horizontal line. 50 model simulations.

$\{\mathbf{X}_t\}_{t \in \mathbb{Z}}$, an object of interest is the conditional dependence structure among the components $\{\mathbf{X}_t^{(j)}\}_{t \in \mathbb{Z}}$ for $j = 1, \ldots, q$. Various notions of conditional dependencies, associations and causality exist, cf. Brillinger (1996) or Dahlhaus and Eichler (2003).

Without having some mechanistic understanding about the underlying process $\{\mathbf{X}_t\}_{t \in \mathbb{Z}}$, and if q is large relative to sample size n, the Gaussian framework often yields a useful first approximation. It requires knowledge of second order moments of the process $\{\mathbf{X}_t\}_{t \in \mathbb{Z}}$ only, and this may be approximated by (potentially high-order) VAR(p) models as in (10.9). One specific graphical model within this framework is the partial correlation graph: Dahlhaus and Eichler (2003) give a precise description for VAR(p) models in terms of zero-elements in the coefficient matrices $\mathbf{A}_j$. When using our multivariate L_2Boosting from Sections 10.3.3 and 10.4.1, we get a sparse VAR(p) model fit, and we can then immediately read off a boosting estimate for a partial correlation graph. We remark that such a straightforward graph estimate is only possible due to the sparseness of the VAR(p)-model fit with many estimated zeroes.

So far, we have not analyzed some properties of the boosting estimate for (high-dimensional) partial correlation graphs for stochastic processes. But we think that the boosting methods have an interesting potential for graphical modeling, both for multivariate stationary stochastic processes as outlined above as well as for multivariate i.i.d. data.

References

Bickel, P.J. and Bühlmann, P. (1996). What is a linear process? *Proc. Nat. Acad. Sci.* USA **93**, 12128–12131.

Bickel, P. J. and Bühlmann, P. (1997). Closure of linear processes. *J. Theor. Probab.* **10**, 445–479.

Bickel, P. J., Klaassen, C. A. J., Ritov, Y. and Wellner, J. A. (1993). *Efficient and Adaptive Estimation for Semiparametric Models.* Johns Hopkins University Press, Baltimore.

Bickel, P. J. and Ritov, Y. (2004). The golden chain. Discussion on three papers on boosting (auths: Jiang, W.; Lugosi, G. and Vayatis, N.; Zhang, T). *Ann Statist.* **32**, 91–96.

Breiman, L. (1998). Arcing classifiers. Ann. Statist. **26**, 801–849 (with discussion).

Breiman, L. (1999). Prediction games & arcing algorithms. *Neural Computation* **11**, 1493-1517.

Brillinger, D.R. (1996). Remarks concerning graphical models for time series and point processes. Brazil. *Rev. Econometrics* **16**, 1–23.

Brockwell, P. J. and Davis, R. A. (1991). *Time Series: Theory and Methods.* 2nd ed. Springer, New York.

Bühlmann, P. (2002). Bootstraps for time series. Statistical Science **17**, 52–72.

Bühlmann, P. (2004). Boosting for high-dimensional linear models. To appear in Ann. Statist.

Bühlmann, P. and Wyner, A. J. (1999). Variable length Markov chains. *Ann. Statist.* **27**, 480–513.

Bühlmann, P. and Yu, B. (2003). Boosting with the L_2loss: regression and classification. *J. Amer. Statist. Assoc.* **98**, 324–339.

Bühlmann, P. and Yu, B. (2005). Boosting, model selection, lasso and nonnegative garrote. Preprint.

Buja, A., Hastie, T. and Tibshirani, R. J. (1989). Linear smoothers and additive models. *Ann. Statist.* **17**, 453–555 (with discussion).

Dahlhaus, R. and Eichler, M. (2003). Causality and graphical models for time series. In *Highly structured stochastic systems* (P. Green, N. Hjort, and S. Richardson eds.). Oxford University Press, Oxford.

DeVore, R. A. and Temlyakov, V. N. (1996). Some remarks on greedy algorithms. *Adv. Comp. Math.* **5**, 173–187.

Efron, B., Hastie, T., Johnstone, I. and Tibshirani, R. J.(2004). Least angle regression. *Ann. Statist.* **32**, 407–499 (with discussion).

Freund, Y. and Schapire, R. E. (1996). Experiments with a new boosting algorithm. In Machine Learning: Proc. Thirteenth International Conference, pp. 148–156. Morgan Kauffman, San Francisco.

Friedman, J. H. (2001). Greedy function approximation: a gradient boosting machine. *Ann. Statist.* **29**, 1189–1232.

Friedman, J.H., Hastie, T. and Tibshirani, R. J. (2000). Additive logistic regression: a statistical view of boosting. *Ann. Statist.* **28**, 337–407 (with discussion).

Greenshtein, E. and Ritov, Y. (2004). Persistence in high-dimensional linear predictor selection and the virtue of overparametrization. *Bernoulli* **10**, 971–988.

Hurvich, C.M., Simonoff, J.S. and Tsai, C.-L. (1998). Smoothing parameter selection in nonparametric regression using an improved Akaike information criterion. *J. Roy. Statist. Soc., Ser. B*, **60**, 271–293.

Jiang, W. (2004). Process consistency for AdaBoost. *Ann. Statist.* **32**, 13–29 (disc. pp. 85–134).

Künsch, H. R. (1989). The jackknife and the bootstrap for general stationary observations. *Ann. Statist.* **17**, 1217–1241.

Lugosi, G. and Vayatis, N. (2004). On the Bayes-risk consistency of regularized boosting methods. *Ann. Statist.* **32**, 30–55 (disc. pp. 85–134).

Lutz, R.W. and Bühlmann, P. (2005). Boosting for high-multivariate responses in high-dimensional linear regression. To appear in *Statistica Sinica*.

Mallat, S. G. and Zhang, Z. (1993). Matching pursuits with time-frequency dictionaries. *IEEE Trans. Signal Proc.* **41**, 3397–3415.

Rissanen, J. (1983). A universal data compression system. *IEEE Trans. Inform. Theory* **IT-29**, 656–664.

Temlyakov, V. N. (2000). Weak greedy algorithms. *Adv. Comp. Math.* **12**, 213–227.

Tibshirani, R. J.(1996). Regression shrinkage and selection via the lasso. *J. Roy. Statist. Soc., Ser. B*, **58**, 267–288.

Tukey, J. W. (1977). *Exploratory data analysis*. Addison-Wesley, Reading, MA.

Zellner, A. (1962). An efficient method of estimating seemingly unrelated regressions and tests for aggregation bias. *J. Amer. Statist. Assoc.* **57**, 348–368.

Zellner, A. (1963). Estimators for seemingly unrelated regression equations: some exact finite sample results. *J. Amer. Statist. Assoc.* **58**, 977–992 (Corr. **67**, 255).

Zhang, T. and Yu, B. (2003). Boosting with early stopping: convergence and consistency. To appear in *Ann. Statist.*

Zhao, X., Huang, H. and Speed, T. P. (2004). Finding short DNA motifs using permuted Markov models, RECOMB'04.

CHAPTER 11

Bootstrap Methods: A Review

S. N. Lahiri

Department of Statistics
Iowa State University
Ames, IA 50011
snlahiri@iastate.edu

This article gives a review of the literature on bootstrap methods since its introduction by Efron (1979). It describes the bootstrap methodology and its properties in some selected inference problems under independence. It also describes various formulations of the bootstrap for different classes of dependent processes including Markov processes, long range dependent time series and spatial processes. Further, it outlines some important results and open problems in these areas.

11.1. Introduction

Since its inception Efron (1979), the bootstrap method has been adapted and employed in an amazingly large number of inference problems under various data generating mechanisms. The papers in the 2003 special issue of the Statistical Science on the silver anniversary of the bootstrap give a good indication of the wide range of applicability and the huge impact of the bootstrap on different areas of statistics and related subjects. By now, there exist several books and monographs — Hall (1992), Efron and Tibshirani (1993), Shao and Tu (1995), Davison and Hinkley (1997), Lahiri (2003a) — that deal with different aspects of the method, yet fail to exhaust the literature collectively. Thus, a complete review of the literature on bootstrap in a short chapter is impossible. In this chapter, I will highlight some of the key ideas, issues, and results that shaped the development of the bootstrap to this date and indicate some open problems and areas of possible future work.

The basic idea behind the bootstrap method can be described as follows.

Let $X_1, \ldots, X_n$ be a collection of random variables with joint distribution P_n. For estimating a population parameter θ, suppose we have constructed an estimator $\hat{\theta}_n$ based on $X_1, \ldots, X_n$ (e.g. using the maximum likelihood method or the method of moments). A common problem that the statistician must deal with is to assess the accuracy of $\hat{\theta}_n$, for example, by using an estimate of its mean squared error (MSE) or an interval estimate of a given confidence level. However, any such measure of accuracy depends on the sampling distribution of $\hat{\theta}_n$, which is typically unknown in practice and often very complicated. Bootstrap method provides a general recipe for estimating the distribution of $\hat{\theta}_n$ and its functionals without restrictive assumptions on the data generating mechanism.

To describe the basic principle underlying the bootstrap methods, suppose that the data are generated by a collection of random variables $\{X_1, \ldots, X_n\} \equiv \mathbf{X_n}$ with joint distribution P_n. Given $\mathbf{X_n}$, first construct an estimate $\hat{P}_n$ of P_n. Next generate random variables $\{X_1^*, \ldots, X_n^*\} \equiv \mathbf{X_n^*}$ from $\hat{P}_n$. If $\hat{P}_n$ is a reasonably "good" estimator of P_n, then the relation between $\{X_1, \ldots, X_n\}$ and P_n is closely reproduced (in the bootstrap world) by $\{X_1^*, \ldots, X_n^*\}$ and $\hat{P}_n$. Define the bootstrap version θ_n^* of $\hat{\theta}_n$ by replacing $X_1, \ldots, X_n$ with $X_1^*, \ldots, X_n^*$. Then the conditional distribution of θ_n^* (given $\mathbf{X_n}$) gives the bootstrap estimator of the distribution of $\hat{\theta}_n$. To define the bootstrap estimators of functionals of the distribution of $\hat{\theta}_n$, such as the variance or the quantiles of $\hat{\theta}_n$, we may simply use the "plug-in" principle and employ the corresponding functional to the conditional distribution of θ_n^*. Thus, the bootstrap estimator of the variance of $\hat{\theta}_n$ is given by the conditional variance of θ_n^* and the bootstrap estimator of the α-quantile of (the distribution of) $\hat{\theta}_n$ is given by the α-quantile of (the conditional distribution of) $\hat{\theta}_n^*$.

Although the bootstrap principle appears to be a very simple one, it provides a valid answer in a surprisingly large number of situations. The quality of the bootstrap approximation depends on the estimator $\hat{P}_n$ of the joint distribution P_n that we estimate from $X_1, \ldots, X_n$ and on the statistic $\hat{\theta}_n$, whose distribution we wish to estimate. In the simplest case, we have a collection of independent and identically distributed (i.i.d) random variables $X_1, \ldots, X_n$ with common distribution F (say). In this case, an estimator of P_n is chosen as $\hat{F}_n^n$ for some suitable estimator $\hat{F}_n$ of F. In presence of dependence, different estimators of P_n have been constructed depending on the dependence structure of the underlying process, leading to many versions of the bootstrap, such as block bootstrap, sieve bootstrap.

In general, having chosen a particular bootstrap method for a specific application, it is very difficult (and often, impractical) to derive closed form analytical expressions for the bootstrap estimators of various population quantities. This is where the computer plays an indispensable role. Bootstrap estimators of the distribution of $\hat{\theta}_n$ can be computed numerically using Monte-Carlo simulation. First, a large number (usually several hundred) of independent copies of $\hat{\theta}_n^*$ are constructed by repeated resampling. The empirical distribution of these bootstrap replicates gives the desired Monte-Carlo approximation to the true bootstrap distribution of $\hat{\theta}_n^*$. From this point of view, the introduction of the bootstrap has been very timely. Almost none of the interesting applications of the bootstrap would have been possible without the computing power of the present day computers.

The rest of the paper is organized as follows. We conclude this section by collecting some notation to be used in the rest of the paper. In Section 2, we describe some results on bootstrapping i.i.d random variables. In Section 3, we consider bootstrapping non-i.i.d data in presence of a model, such as a regression or an autoregression model. Section 4 is devoted to various types of block bootstrap methods and their properties. Sections 5 and 6 respectively deal with bootstrap methods based on the notions of sieve approximations and data transformations, such as the discrete Fourier transform. Sections 7–9 deal with bootstrap methods for different classes of dependent processes, namely, Markov processes, long memory processes and spatial processes, respectively.

Notation: For random vectors Y, Z, let $\mathcal{L}(Y)$ denote the probability distribution of Y and $\mathcal{L}(Y|Z)$ (or $\mathcal{L}(Y|\Gamma)$) the conditional distribution of Y given Z (or a conditioning σ-field Γ, respectively). Let P_*, E_*, Var_* respectively denote the bootstrap probability, expectation, and variance (conditional on the observations). Let $\mathcal{I} = \sqrt{-1}$ and $\mathbb{Z} = \{0, \pm 1, \pm 2, \ldots\}$. Let Φ denote the standard normal distribution function. For a matrix A, let A' denote its transpose.

11.2. Bootstrap for i.i.d Data

In this section, we consider a selective set of problems that illustrate the advantages of using the bootstrap over traditional methods. These include the cases where the bootstrap outperforms classical approaches by achieving better accuracy and where traditional approaches fail but the bootstrap provides a valid answer. We also point out some cases where a blind application of the bootstrap gives a wrong answer and indicate the known remedies for the problem.

We begin with a brief description of the bootstrap method of Efron (1979) for i.i.d observations. Let $X_1, \ldots, X_n$ be a collection of i.i.d random variables with common distribution function F. Let $\hat{F}_n$ be an estimator of F based on $X_1, \ldots, X_n$. The most common choice of $\hat{F}_n$ is the empirical distribution of $X_1, \ldots, X_n$:

$$F_n(x) = n^{-1} \sum_{i=1}^{n} I(X_i \leq x), \quad x \in \mathbb{R},$$

where $I(\cdot)$ denotes the indicator function. However, other choices of $\hat{F}_n$, such as a smooth estimator of the distribution function may be preferable in some applications. Next, (conditional on $X_1, \ldots, X_n$), let $X_1^*, \ldots, X_m^*$ be i.i.d random variables with common distribution $\hat{F}_n$. Here m denotes the resample size which is typically set equal to the sample size n, but can be different in some applications. Note that for $\hat{F}_n = F_n$, the X_i^*'s are generated by simple random sampling with replacement from $\{X_1, \ldots, X_n\} (\equiv \mathbf{X_n})$. For each $n \geq 1$, let $T_n = t_n(\mathbf{X_n}; \theta)$ be a random variable of interest that depends on the observations $\mathbf{X_n}$ as well as on some population parameter $\theta = \theta(F)$. A common example of T_n is given by $T_n = \sqrt{n}(\bar{X}_n - \mu)/\sigma$, the *normalized* sample mean, where $\mu = EX_1$ and $\sigma^2 = Var(X_1)$ are the population mean and variance and where $\bar{X}_n = n^{-1} \sum_{i=1}^{n} X_i$ denotes the sample mean. The bootstrap version of $T_n = t_n(\mathbf{X_n}; \theta)$ based on a resample of size m is given by

$$T_{m,n}^* = t_m(X_1^*, \ldots, X_m^*; \hat{\theta}_n), \tag{11.1}$$

where $\hat{\theta}_n$ is obtained by replacing F with $\hat{F}_n$ in the definition of θ. The bootstrap approximation to $\mathcal{L}(T_n)$ is given by $\mathcal{L}(T_{m,n}^* | \mathbf{X_n})$ and for a parameter $\varphi_n \equiv \varphi(\mathcal{L}(T_n))$ based on a functional φ, its bootstrap estimator is given by $\hat{\varphi}_n \equiv \varphi(\mathcal{L}(T_{m,n}^* | \mathbf{X_n}))$.

As an example, consider the normalized sample mean $T_n = \sqrt{n}(\bar{X}_n - \mu)/\sigma$. The bootstrap version of this T_n based on a resample of size m and resampling distribution $\hat{F}_n$ is given by $T_{m,n}^* = \sqrt{m}(\bar{X}_m^* - \hat{\mu}_n)/\hat{\sigma}_n$, where $\bar{X}_m^* = m^{-1} \sum_{i=1}^{m} X_i^*$ is the bootstrap sample mean and $\hat{\mu}_n$ and $\hat{\sigma}_n^2$ are respectively the mean and the variance of $\hat{F}_n$. In particular, for $\hat{F}_n = F_n$, $\hat{\mu}_n = \bar{X}_n$, $\hat{\sigma}_n^2 = n^{-1} \sum_{i=1}^{n} (X_i - \bar{X}_n)^2$ ($\equiv s_n^2$, say) and we get the "ordinary" bootstrap version of T_n:

$$T_{m,n}^* = \sqrt{m}(\bar{X}_m^* - \bar{X}_n)/s_n. \tag{11.2}$$

How good is the approximation generated by the bootstrap resampling? The following result answers this for the ordinary bootstrap method applied to the normalized sample mean with $m = n$.

Theorem 11.1: *Let $\{X_n\}_{n\geq 1}$ be a sequence of i.i.d random variables with $EX_1 = \mu$ and $Var(X_1) = \sigma^2 \in (0, \infty)$. Let $T_n = \sqrt{n}(\bar{X}_n - \mu)/\sigma$ and $T^*_{n,n}$ be given by (11.2) with $m = n$. Then,*

(i)

$$\Delta_n \equiv \sup_{x \in \mathbb{R}} \left| P(T_n \leq x) - P_*(T^*_{n,n} \leq x) \right| \to 0 \quad as \quad n \to \infty, \quad a.s.$$

(11.3)

(ii) If, in addition, $E|X_1|^3 < \infty$ and $\mathcal{L}(X_1)$ is non-lattice (i.e. $|E \exp(\iota t)| \neq 1$ for all $t \neq 0$), then $\Delta_n = o(n^{-1/2})$ a.s.

Part (i) of Theorem 11.1 is due to Bickel and Freedman (1981) and Singh (1981) while part (ii) is due to Singh (1981). Theorem 2.1 shows that for the sample mean of i.i.d random variables with a finite variance, the bootstrap approximation is asymptotically consistent. Further, under the conditions of (ii), the bootstrap approximation is superior to the classical normal approximation, which approximates $\mathcal{L}(T_n)$ only at the rate $O(n^{-1/2})$. In the literature, this property of the bootstrap approximation is referred to as *second order correctness*. The bound $o(n^{-1/2})$ in (ii) of the theorem can be made more precise under additional moment conditions. If $EX_1^4 < \infty$ and X_1 is non-lattice, the order of Δ_n is $O(n^{-1})$ in probability and it is $O(n^{-1}(\log \log n)^{1/2})$ in the almost sure sense. Thus, the bootstrap approximation is considerably more accurate than the normal approximation for the normalized sample mean of non-lattice random variables when the fourth moment is finite.

Second order correctness of the bootstrap approximation is also known for the studentized sample mean and for studentized versions of many general classes of estimators, including maximum likelihood estimators (MLEs), M-estimators and L-estimators; see Babu and Singh (1984), Helmers (1991), Lahiri (1994b), and the references therein. Under mild conditions, this implies (cf. Hall (1988)) that one-sided confidence intervals (CIs) constructed using the bootstrap approximation to studentized statistics (known as *bootstrap-t CIs*) incur an error (in coverage probability) of order $O(n^{-1})$ only, but the normal CIs have an error of order $O(n^{-1/2})$. Similar improvements in the coverage error rate can be achieved for non-pivotal statistics by combining bootstrap approximation with analytical corrections (cf. Efron (1987)). For two sided CIs, however, a single application of the bootstrap, even when applied to a studentized statistic, does not typically improve the error rate over normal CIs. This is a consequence of the parity of the second and third order terms in the Edgeworth expansions of such statistics. However, higher order improvements are possible in the two sided case also, by using some modifications such as calibration (Loh (1987)), bootstrap iteration (Lee and Young (1995)), symmetrization (cf. Hall (1992)) and transformation (cf. Hall (1992)). For a discussion of various types of bootstrap CIs and a detailed analysis of their properties, see Hall (1988) and Hall (1992).

Next we shift the focus from smooth functionals of the e.d.f to a non-smooth functional, the sample quantile and consider one of the early success stories of the bootstrap where classical approaches failed. It was well-known for a long time that the (delete-1) jackknife method failed in the problem of estimating the variance of the sample quantile in the sense that the jackknife variance estimator was inconsistent. As a test for the versatility of the bootstrap, it was applied to the variance estimation problem yielding promising numerical results (cf. Efron (1979)). In an important paper, Ghosh, Parr, Singh and Babu (1984) provided a theoretical confirmation of this observation. The main result of Ghosh *et al.* (1984) show that under mild moment and smoothness conditions, the bootstrap estimator of the *variance* of the sample quantile is consistent. On the related problem of estimating the *distribution function* of a sample quantile, Bickel and Freedman (1981) showed that the bootstrap captured the "right" limit distribution not only for a single sample quantile, but for the quantile process as a whole. However, compared to the case of the sample mean, the rate of bootstrap approximation to the distribution of the sample quantile is considerably slow (Singh (1981)). In this problem, better rates can be achieved by resampling from a smooth estimator $\hat{F}_n$ of the distribution function F (Hall, DiCiccio, and Romano (1989), Falk and Reiss (1989)) or by bootstrap iteration (Ho and Lee (2005)).

We conclude this section by pointing out some of cases where the ordinary bootstrap looses its advantage over traditional methods and where it may even fail to give a correct answer. For the first example, we revisit the set up of Theorem 2.1 but now suppose that the random variables are lattice. Theorem 2.1 shows that if $EX_1^2 < \infty$, the bootstrap provides a valid approximation. However, the superiority of the bootstrap approximation over normal approximation may no longer hold under standard metrics, such as the sup-norm metric or the Levy metric. For any two distribution functions F, G on $\mathbb{R}$, the Levy metric is given by $d_L(F, G) = \inf$ $\{\epsilon > 0 : F(x - \epsilon) - \epsilon < G(x) < F(x + \epsilon) + \epsilon$ for all $x \in \mathbb{R}\}$.

Theorem 11.2: *Let T_n, $T^*_{n,n}$ and Δ_n be as in Theorem 11.1, and $E|X_1|^4 <$ ∞. Also, suppose that X_1 is lattice with maximal span $h \in (0, \infty)$ (i.e., $|E \exp(\iota t)| \neq 1$ for all $t \in (0, 2\pi/h)$ and $|E \exp(\iota t)| = 1$ for $t = 2\pi/h$). Then,*

$$\limsup_{n \to \infty} n^{1/2} \Delta_n = h/\sqrt{2\pi\sigma^2} \quad a.s. , \tag{11.4}$$

$$\limsup_{n \to \infty} n^{1/2} d_L(\mathcal{L}(T_n), \mathcal{L}(T^*_{n,n})) = h/[\sigma(1 + \sqrt{2\pi})] \quad a.s. . \tag{11.5}$$

The assertion (11.4) is due to Singh (1981) and (11.5) due to Lahiri (1994a). Theorem 11.2 thus shows that the bootstrap looses its second order correctness for lattice random variables. A practical implication of this result is that for the standard discrete distributions (such as Binomial, Poisson, Geometric, etc. which are lattice with maximal span $h = 1$), inference based on the bootstrap approximation may not be better than that based on the normal approximation.

For the next example, we again consider the sample mean but drop the hypothesis of finite variance. More precisely, let $\{X_n\}_{n \geq 1}$ be a sequence of i.i.d random variables with distribution F that lie in the domain of attraction of a stable law of order $\alpha \in (0, 2)$. In this case, X_1 has a finite mean but infinite variance. Athreya (1987) showed that in this case, the bootstrap approximation to the distribution of the centered and scaled sample mean converged to a random limit when resample size m equaled the sample size n, and thus the bootstrap failed to provide a valid approximation. He also showed that (rather surprisingly) the inconsistency of the bootstrap approximation was overcome by choosing a suitably smaller resample size where $m = o(n)$ as $n \to \infty$. Giné and Zinn (1989) and Arcones and Giné (1989) gave some further results on the effect of the resample size and moments on the validity of the bootstrap for the sample mean. A similar inconsistency phenomenon occurs in the context of bootstrapping the sample maximum (or minimum); a resample size $m = n$ yields a random limit for the bootstrap approximation but choosing "$m = o(n)$ as $n \to \infty$" resolves the problem (Fukuchi (1994)). Indeed, the "m out of n" bootstrap (with $m = o(n)$) is a "quick fix" in many applications (including certain time series models) where the standard choice $m = n$ does not work. A detailed documentation of this phenomenon is given by Bickel, Götze, and van Zwet (1997). See also Politis, Romano and Wolf (1999).

11.3. Model Based Bootstrap

For practical applications, extension of the bootstrap method beyond the i.i.d set up is very important. Indeed, in his seminal paper itself, Efron (1979) considered extensions of the bootstrap method to regression models. Suppose that the data is generated by a multiple linear regression model

$$Y_i = x_i'\beta + \epsilon_i, \quad i = 1, \ldots, n, \tag{11.6}$$

where $x_1, \ldots, x_n$ is a collection of nonrandom $p \times 1$ design vectors, β is the regression parameter, and $\epsilon_i, i = 1, \ldots, n$ are i.i.d random variables with $E\epsilon_1 = 0$ and $E\epsilon_1^2 \in (0, \infty)$. Let $\hat{\beta}_n$ denote the least squares estimator (LSE) of β. Suppose that we wish to approximate the sampling distribution of $T_n = A_n(\hat{\beta}_n - \beta)$, where A_n is a known $p \times p$ scaling matrix. To generate the bootstrap approximation in this problem, first we compute the residuals $e_i = Y_i - x_i'\hat{\beta}_n$ and center them as $\tilde{e}_i = (e_i - \bar{e}_n)$, $i = 1, \ldots, n$, where $\bar{e}_n = n^{-1}\sum_{i=1}^n e_i$. Next generate the bootstrap error variables $\epsilon_1^*, \ldots, \epsilon_n^*$ by random sampling (with replacement) from the collection $\{\tilde{e}_i : i = 1, \ldots, n\}$ and define

$$Y_i^* = x_i'\hat{\beta}_n + \epsilon_i^*, \quad i = 1, \ldots, n.$$

The bootstrap version β_n^* of $\hat{\beta}_n$ is now defined by replacing $Y_1, \ldots, Y_n$ in the definition of $\hat{\beta}_n$ by $Y_1^*, \ldots, Y_n^*$ and the bootstrap version of T_n is given by $T_n^* = A_n(\beta_n^* - \hat{\beta}_n)$. As before, the bootstrap approximation to $\mathcal{L}(T_n)$ is now given by $\mathcal{L}(T_n^*|Y_1, \ldots, Y_n)$.

Freedman (1981) established theoretical validity of the above method and showed that the centering of the residuals was a critical step for the validity of the bootstrap. He also gave a formulation of the bootstrap for the "correlation" model, where the x_i's in (11.6) were random. Shorack (1982) and Lahiri (1992a) considered bootstrapping M-estimators of β in regression Model (3.1) and observed that like the LSE, a similar centering adjustment was also needed for the bootstrapped M-estimator. Shorack (1982) gave a general solution to the problem by suitably redefining the bootstrap estimating equation, while Lahiri (1992a) suggested resampling from an weighted empirical distribution. Second order correctness of both bootstrap methods was established by Lahiri (1992a). Further, validity of the bootstrap for the M-estimated residual-empirical process was proved by Koul and Lahiri (1994). In an important work, Liu and Singh (1992b) investigated robustness of the bootstrap and other resampling plans in regression models. Their work shows that the bootstrap is sensitive to deviations from the homoskedasticity assumption on the error variances, but outperforms competing resampling methods in terms of efficiency in the homoskedastic case. For more on resampling methods in parametric and nonparametric regression problems, see Hall (1992), Davison and Hinkley (1997), Chatterjee and Bose (2002), and the references therein.

The idea of resampling the residuals can be carried over to some common time series models driven by i.i.d innovations. Investigations along this line were initiated in the mid-1980s by Freedman and Peters (1984), Efron and Tibshirani (1986)) and Swanepoel and van Wyk (1986), among others. To describe the adjustments needed in the time series case, consider a stationary autoregressive process of order 1 (AR(1)):

$$X_t = \beta_1 X_{t-1} + \epsilon_t, \quad t = 0, \pm 1, \pm 2, \dots, \tag{11.7}$$

where $\beta_1 \in (-1, 1)$ is the autoregressive parameter and where $\{\epsilon_t\}_{t \in \mathbb{Z}}$ is a sequence of i.i.d random variables with $E\epsilon_1 = 0$ and $E\epsilon_1^2 = 1$. Given the observations $X_1, \dots, X_n$, let $\hat{\beta}_1$ denote the LSE of the autoregressive (AR) parameter β_1. To generate a bootstrap copy of the observations $X_1, \dots, X_n$, we form the residuals $e_t \equiv X_t - \hat{\beta}_1 X_{t-1}$, $t = 2, \dots, n$, and then center them and resample randomly, with replacement from the centered residuals $\{[e_t - (n-1)^{-1} \sum_{i=2}^{n} e_i], t = 2, \dots, n\}$ to get the bootstrap error variables e_t^*, $t = 2, \dots, n$. Finally, we use the structural equation of the Model (11.7) to generate the bootstrap observations $X_1^*, \dots, X_n^*$ as

$$X_t^* = \hat{\beta}_1 X_{t-1}^* + \epsilon_t^*, \quad t = 2, \dots, n, \tag{11.8}$$

where X_1^* is randomly selected from $\{X_1, \dots, X_n\}$. This is referred to as the AR-bootstrap method in the literature.

Note that in our formulation of the AR-bootstrap method above, we have tacitly assumed that the process $\{X_t\}_{t \in \mathbb{Z}}$ has zero mean. In case this information is not available (which is typically the case), we must subtract the sample mean from the observations before carrying out the above steps. Note also that the bootstrap method for the AR(1) model, with or without the centering at the mean, can be easily extended to the more general case of an AR(p) model for any $p \geq 1$. Bose (1988) established second order correctness of this AR-bootstrap method for the centered and scaled LSE of the AR parameter vector for a stationary p-th order AR process. Interestingly, for the unstable AR processes (e.g. $|\beta_1| = 1$ in (11.7)), the AR-bootstrap fails with the usual choice of the resample size $m = n$, but works if $m = o(n)$ as $n \to \infty$ (Datta (1995), Heimann and Kreiss (1996)).

For this and some related problems, Datta and McCormick (1995a) and Sakov and Bickel (1999) provide some guidelines on the choice of m. For descriptions and properties of the bootstrap for nonstationary AR-processes and ARMA processes, see Kreiss and Franke (1992), Datta (1996), Ferretti and Romo (1996), and Chapter 8 of Lahiri (2003a).

11.4. Block Bootstrap

One of the common drawbacks of the model based bootstrap methods is that these are very sensitive to model misspecification. Small deviations from the model assumptions may completely render the bootstrap approximations invalid, e.g. as in the case of the AR bootstrap in the stationary ($|\beta_1| < 1$) and the unstable ($|\beta_1| = 1$) cases. A completely new approach to bootstrapping time series data in absence of a model was put forward by Künsch (1989). Quite early in the bootstrap literature, Singh (1981) showed that resampling single observations, as considered by Efron (1979) for independent data, failed to produce valid approximations in presence of dependence. Building on this, Künsch(1989) advocated the idea of re-sampling blocks of observations at a time. By retaining the neighboring observations together within the blocks, the dependence structure of the random variables at short lag distances is preserved. As a result, resampling blocks allows one to carry over this information to the bootstrap variables. The same resampling plan was also independently suggested by Liu and Singh (1992a), who coined the term "Moving Block Bootstrap" (MBB). To describe it, suppose that $\{X_t\}_{t \in \mathbb{Z}}$ is a stationary weakly dependent time series and that $\{X_1, \ldots, X_n\} \equiv \mathbf{X_n}$ are observed. Let ℓ be an integer satisfying $1 \leq \ell < n$. Define the blocks $\mathcal{B}_1 = (X_1, \ldots, X_\ell)$, $\mathcal{B}_2 = (X_2, \ldots, X_{\ell+1}), \ldots, \mathcal{B}_N = (X_N, \ldots, X_n)$, where $N = n - \ell + 1$. For simplicity, suppose that ℓ divides n. Let $b = n/\ell$. To generate the MBB samples, we select b blocks at random with replacement from the collection $\{\mathcal{B}_1, \ldots, \mathcal{B}_N\}$. Since each resampled block has ℓ elements, concatenating the elements of the b resampled blocks serially yields $b \cdot \ell = n$ bootstrap observations $X_1^*, \ldots, X_n^*$. Note that if we set $\ell = 1$, then the MBB reduces to the ordinary bootstrap method of Efron (1979) for i.i.d data. However, for a valid approximation in the dependent case, it is typically required that

$$\ell^{-1} + n^{-1}\ell = o(1) \quad \text{as} \quad n \to \infty. \tag{11.9}$$

Next suppose that the random variable of interest is of the form $T_n = t_n(\mathbf{X_n}; \theta(\mathbf{P_n}))$ where $P_n = \mathcal{L}(\mathbf{X_n})$. The MBB version of T_n based on blocks of size ℓ is defined as

$$T_n^* = t_n(X_1^*, \ldots, X_n^*; \theta(\hat{P}_n)),$$

where $\hat{P}_n = \mathcal{L}(X_1^*, \ldots, X_n^*|\mathbf{X_n})$ and where we suppress the dependence on ℓ to ease the notation.

To illustrate the construction of T_n^* in a specific example, suppose that T_n is the centered and scaled sample mean $T_n = n^{1/2}(\bar{X}_n - \mu)$. Then the MBB version of T_n is given by $T_n^* = n^{1/2}(\bar{X}_n^* - \tilde{\mu}_n)$, where $\bar{X}_n^*$ is the sample mean of the bootstrap observations and where $\tilde{\mu}_n = E_*(\bar{X}_n^*)$. It is easy to check that

$$\tilde{\mu}_n = N^{-1} \sum_{i=1}^{N} (X_i + \ldots + X_{i+\ell-1})/\ell = N^{-1}\Big[\sum_{i=\ell}^{N} X_i + \sum_{i=1}^{\ell-1} (X_i + X_{n-i+1})\Big],$$

which is different from $\bar{X}_n$ for $\ell > 1$. Lahiri (1991) established second order correctness of the MBB approximation for the normalized sample mean where the bootstrap sample mean is centered at $\tilde{\mu}_n$. The "naive" centering of $\bar{X}_n^*$ at $\bar{X}_n$ is not appropriate as it leads to a loss of accuracy of the MBB approximation (Lahiri (1992b)). Second order correctness of the MBB approximation for studentized statistics has been established independently by Götze and Künsch (1996) for stationary processes and by Lahiri (1996) in multiple linear regression models with dependent errors.

Several variants of the block bootstrap method exist in the literature. One of the early versions of the block bootstrap, implicit in the work of Carlstein (1986), restricts attention to the collection of non-overlapping blocks in the data and resamples from this smaller collection to generate the bootstrap observations. This is known as the *non-overlapping block bootstrap* (NBB). Because the blocks in the NBB construction do not overlap, it is easier to analyze theoretical properties of NBB estimators than those of MBB estimators of a population parameter. Other variants of the block bootstrap include the *circular block bootstrap* (CBB) and the *stationary bootstrap* (SB) of Politis and Romano (1992) and Politis and Romano (1994), the *matched block bootstrap* (MaBB) of Carlstein, Do, Hall, Hesterberg, and Künsch (1998), the *tapered block bootstrap* (TBB) of Paparoditis and Politis (2001a), among others. The CBB and the SB are primarily motivated by the need to remove the uneven weighting of the blocks in the MBB and are based on the idea of periodic extension of the observed segment of the time series. The biases of the variance estimators generated by the MBB, NBB, CBB and the SB are of the order $O(\ell^{-1})$ while the variances are of the order $O(n^{-1}\ell)$, where ℓ denotes the block size and n, the sample

size. It turns out that the MBB and the CBB have asymptotically equivalent performance and are also the most accurate of these four methods.
For relative merits of these four methods, see Lahiri (1999a) and Politis
and White (2003). The MaBB uses a stochastic mechanism to reduce the
edge effects from joining independent blocks in the MBB while the TBB
shrinks the boundary values in a block towards a common value, like the
sample mean, to achieve the same. Although somewhat more complex than
the MBB or the CBB, both the MaBB and the TBB yield more accurate
variance estimators, with biases of the order $O(\ell^{-2})$ and variances of the
order $O(n^{-1}\ell)$.

Performance of the block bootstrap methods crucially depends on the
choice of the block size and on the dependent structure of the process. Explicit formulas for MSE-optimal block sizes for estimating the variances of
smooth functions of sample means are known for the MBB, CBB, NBB
and SB (Hall, Horowitz, and Jing (1995), Lahiri (1999a)). Thus, one can
use these expressions to formulate plug-in estimators of the optimal block
sizes (Politis and White (2003)). However, some major drawbacks of this
approach are that — i) extensive analytical work is needed to derive the
theoretical expressions for the optimal block sizes and that — ii) the derived formulas are applicable only to the estimators considered - there is no
obvious way of extending them to a new set of estimators. For the variance
estimation problem, Bühlmann and Künsch (1999) formulated a method
based on linearization of an estimator using its influence function, which is
somewhat more general than the direct plug-in approach. But perhaps the
most widely used method in this context is given by Hall *et al.* (1995) who
develop a general empirical method for estimating the optimal block sizes
for estimating *both* the variance and the distribution function. The Hall *et
al.* (1995) method uses the subsampling method to construct an estimator
of the MSE as a function of the block size and then minimize it to produce the estimator of the optimal block size. An alternative method based
on the Jackknife-after-bootstrap method (Efron (1992), Lahiri (2002)) has
been recently proposed by Lahiri, Furukawa, and Lee (2003). They call it a
non-parametric plug-in (NPPI) method, as it works like a plug-in method
but at the same time, it does not require the user to find an exact expression
for the optimal block size analytically. The key construction of the NPPI
method combines more than one resampling method suitably and thereby
implicitly estimates the population parameters that appear in the formulas for the optimal block sizes. Further, the NPPI method is applicable to
block bootstrap estimation problems involving the variance, the distribu-

tion function and the quantiles. For further discussion of the block length selection rules for block bootstrap methods, see Chapter 7 of Lahiri (2003a) and the references therein.

11.5. Sieve Bootstrap

Construction of a bootstrap method for weakly dependent processes may be thought of as a two step procedure. Suppose that $\{X_t\}_{t\in\mathbb{Z}}$ is a sequence of stationary random veriables with joint distribution P. To generate a bootstrap approximation, in the first step, one approximates P with another probability distribution that is easier to estimate and then use the data to estimate this approximating distribution. For the MBB, the approximating distribution to P is given by P_ℓ^∞, where P_ℓ is the joint distribution of $X_1, \ldots, X_\ell$. In the second step of the MBB, an estimator of P_ℓ is constructed using the empirical distribution of the blocks. The sieve bootstrap essentially has a similar two-level construction. The basic idea behind the sieve bootstrap is to construct a sequence of probability distributions $\{\tilde{P}_n\}_{n\geq 1}$ that forms a sieve, i.e. the sequence $\{\tilde{P}_n\}_{n\geq 1}$ is such that for each $n \geq 1$, $\tilde{P}_{n+1}$ is a finer approximation to P than $\tilde{P}_n$ and $\tilde{P}_n$ converges to P in a suitable sense.

For linear processes satisfying certain invertability conditions, Bühlmann (1997) developed a sieve bootstrap method based on a sieve of AR models. Let $\{X_t\}_{t\in\mathbb{Z}}$ be a sequence of stationary random variables with an infinite order AR representation:

$$(X_i - \mu) = \sum_{t=1}^{\infty} \gamma_t (X_{i-t} - \mu) + \epsilon_i, \quad i \in \mathbb{Z}, \tag{11.10}$$

where $\mu = EX_1$, $\{\gamma_k\}_{k\geq 1}$ is a sequence of constants satisfying $\sum_{k=1}^{\infty} \gamma_k^2 < \infty$ and $\{\epsilon_i\}_{i\in\mathbb{Z}}$ is a sequence of i.i.d random variables with $E\epsilon_1 = 0, E\epsilon_1^2 = 1$. Under some regularity conditions, a linear process admits such a representation (cf. Anderson, 1971, Chapter 7). Let $\{p_n\}_{n\geq 1}$ be a sequence of integers such that $p_n \to \infty$ but $p_n/n \to 0$ as $n \to \infty$. In the first step, the sieve approximation $\tilde{P}_n$ is generated by using the p_n-th order AR-process:

$$(X_i - \mu) = \sum_{t=1}^{p_n} \gamma_t (X_{i-t} - \mu) + \epsilon_i, \quad i \in \mathbb{Z}. \tag{11.11}$$

In the next step, the AR-bootstrap method is used for the approximating Model (11.11). Thus, the unknown parameters γ_t's are estimated from the data, say using the Yule-Walker estimators, the residuals

$$\hat{\epsilon}_{in} = (X_i - \bar{X}_n) - \sum_{t=1}^{n} \hat{\gamma}_{tn}(X_{i-t} - \bar{X}_n), \quad i = p_n + 1, \ldots, n,$$

are centered and resampled to generate the bootstrap error variables ϵ_i^*, $i = p_n + 1, \ldots, n$, and the bootstrap observations are generated by

$$(X_i^* - \bar{X}_n) = \sum_{t=1}^{p_n} \hat{\gamma}_t(X_{i-t}^* - \bar{X}_n) + \epsilon_i^*, \quad i = p_n + 1, \ldots, n, \qquad (11.12)$$

with the initial vector $(X_1^*, \ldots, X_{p_n}^*)$ randomly selected from the size-p_n blocks of $X_1, \ldots, X_n$. Bühlmann (1997) established consistency of this sieve bootstrap method. Choi and Hall (2000) investigated the issue of rate of convergence and established second order correctness. For another construction of the sieve bootstrap, see Bühlmann (2002).

11.6. Transformation Based Bootstrap

Let $\theta \equiv \theta(P)$ be a parameter of interest that depends on the joint distribution P of the sequence $\{X_t\}_{t \in \mathbb{Z}}$. Suppose that our goal is to approximate the sampling distribution of a normalized or studentized statistic $R_n = r_n(\mathbf{X}_n; \theta)$, where $\mathbf{X}_n = \{X_1, \ldots, X_n\}$. Let $\mathbf{Y}_n = h_n(\mathbf{X}_n)$ be a transformation of $\mathbf{X}_n$ such that the components of $\mathbf{Y}_n$, say, $\{Y_i : i \in \mathcal{I}_n\}$ are "approximately independent". Also suppose that the variable R_n can be expressed (at least to a close approximation) in terms of $\mathbf{Y}_n$ as $R_n = r_{1n}(\mathbf{Y}_n; \theta)$ for some reasonable function r_{1n}. Then, to approximate the distribution of R_n by the *transformation-based bootstrap* (TrBB), we may resample from a *suitable* sub-collection $\{Y_i : i \in \mathcal{J}_n\}$ of $\{Y_i : i \in \mathcal{I}_n\}$ to generate the bootstrap observations $\mathbf{Y}_n^* \equiv \{\mathbf{Y}_i^* : \mathbf{i} \in \mathcal{I}_n\}$ either selecting a single Y-value at a time or selecting a block of Y-values from $\{Y_i : i \in \mathcal{J}_n\}$ as in the MBB, depending on the dependence structure of $\{Y_i : i \in \mathcal{J}_n\}$. The TrBB estimator of the distribution of R_n is then given by the conditional distribution of $R_n^* \equiv r_{1n}(\mathbf{Y}_n^*; \hat{\theta}_n)$ given the data $\mathbf{X}_n$, where $\hat{\theta}_n$ is an estimator of θ.

The most common TrBB method is based on the Fourier transform of the data and is referred to as the *Frequency Domain Bootstrap* (FDB) in the literature. Here we give a short description of the FDB. Given the data $\mathbf{X}_n$, define its Fourier transform

$$Y_n(w) = n^{-1/2} \sum_{t=1}^{n} X_t \exp(-\iota w t), \quad w \in (-\pi, \pi]. \qquad (11.13)$$

It is well known (cf. Brockwell and Davis (1996), Chapter 10), Lahiri (2003b)) that the Fourier transforms $Y_n(\lambda_1), \ldots, Y_n(\lambda_k)$ are *asymptotically independent* for any set of distinct ordinates $-\pi < \lambda_1 < \ldots < \lambda_k \leq \pi$. Further, the original observations $\mathbf{X_n}$ admit a representation in terms of the transformed values $\mathbf{Y_n} = \{\mathbf{Y_n(w_j)} : \mathbf{j} \in \mathcal{I}_\mathbf{n}\}$ as (cf. Brockwell and Davis, 1991),

$$X_t = n^{-1/2} \sum_{j \in \mathcal{I}_n} Y_n(w_j) \exp(\iota t w_j), \quad t = 1, \ldots, n \qquad (11.14)$$

where $w_j = 2\pi j/n$, $\mathcal{I}_n = \{-\lfloor(n-1)\rfloor/2, \ldots, \lfloor(n-1)\rfloor/2\}$, and where for any real number x, $\lfloor x \rfloor$ denotes the largest integer not exceeding x. Thus, one can express a given variable $R_n = r_n(\mathbf{X_n}; \theta)$ also in terms of the transformed values $\mathbf{Y_n}$, and resample from the Y-values to define the FDB version of R_n. For weakly dependent data, variants of the FDB method have been proposed and studied by Hurvich and Zeger (1987), and Franke and Härdle (1992). Under some regularity conditions, Dahlhaus and Janas (1996) established second order correctness of the FDB for a class of estimators called the "ratio statistics". Ratio statistics are defined as the ratio of two "spectral mean" estimators of the form $\int_0^\pi g(w)I_n(w)dw$ where $g : [0, \pi) \to I\!R$ is an integrable function and where $I_n(w) = |Y(w)|^2$ is the periodogram of $\mathbf{X}_n$. It is known (cf. Lahiri (2003a), Section 9.2)) that the FDB for spectral means and ratio estimators is rather sensitive to deviations from the model assumptions. Kreiss and Paparoditis (2003) proposed a modified version of the FDB that provides a valid approximation for a wider class of spectral mean estimators. A completely different version of the TrBB method has been recently proposed by Percival, Sardy, and Davison (2000). They advocate the use of the wavelet transforms (WTs) of the data in place of the Fourier transform. See Percival, Sardy, and Davison (2000) for further details on the WT-based bootstrap.

11.7. Bootstrap for Markov Processes

Extension of the Bootstrap methods from i.i.d random variables to Markov chains was initiated by Kulperger and Prakasa Rao (1989) for the finite state space case. Note that if the chain is stationary, then there are only finitely many unknown parameters that characterize the joint distribution of the chain. Kulperger and Prakasa Rao (1989)'s approach involved estimating these parameters from the data and using the estimated transition matrix and the marginal distribution to generate the bootstrap observations. This approach was extended to the countable case by Athrey and

Fuh (1992). Athrey and Fuh (1992) also suggested a completely new boot-strap method based on the idea of regeneration. By a well-known result (Athrey and Ney (1992)) on Harris recurrent Markov chains, successive returns to a recurrent state gives a decomposition of the chain into i.i.d cycles (of random lengths). The regeneration based bootstrap resamples these i.i.d cycles to generate the bootstrap observations. For the first- and second- order properties of the regeneration based bootstrap, see Athrey and Fuh (1992) and Datta and McCormick (1995b), respectively.

Bootstrap methods for Markov processes based on estimated transition probability functions have been also extended to the case where the state space is Euclidean. In this case, one can use the nonparametric function es-timation methodology to estimate the marginal distribution and the tran-sition probability function. For consistency of the method, see Rajarshi (1990) and for the second order properties of the method, see Horowitz (2003). A "local" version of the method (called the Local Markov Bootstrap or MLB, in short) has been recently put forward by Paparoditis and Politis (2001b). The idea here is to construct the bootstrap chain by sequential drawing — having selected a set of bootstrap observations, the next ob-servation is randomly selected from a "neighborhood of close values" of the observation(s) in the immediate past. Paparoditis and Politis (2001b) showed that the resulting bootstrap chain was stationary and Markov and also that it enjoyed some robustness with regard to the Markovian assump-tion. For more on the properties of the MLB, see Paparoditis and Politis (2001b).

11.8. Bootstrap under Long Range Dependence

Let $\{X_t\}_{t \in \mathbb{Z}}$ be a stationary process with $EX_1^2 \in (0, \infty)$, autocovariance function $r(\cdot)$ and spectral density function $f(\cdot)$. We say that the process $\{X_t\}_{t \in \mathbb{Z}}$ is long range dependent (LRD) if $\sum_{k=1}^{\infty} |r(k)| = \infty$ or if $f(\lambda) \to \infty$ as $\lambda \to 0$; Otherwise, $\{X_t\}_{t \in \mathbb{Z}}$ is said to be short range dependent (SRD). We also use the acronym LRD (SRD) for long (respectively, short) range dependence. Limit behaviors of many common statistics and tests under LRD are different from their behaviors under SRD. For example, the sample mean of n observations from a LRD process may converge to the population mean at a rate *slower* than $O_p(n^{-1/2})$ and similarly, with proper centering and scaling, the sample mean may have a *nonnormal* limit distribution even when the population variance is finite. More specifically, we consider the following result on the sample mean under LRD. Let $\{Z_t\}_{t \in \mathbb{Z}}$ be a zero mean unit variance Gaussian process with an auto-covariance function $r_1(\cdot)$ satisfying

$$r_1(k) \sim Ck^{-\alpha} \quad \text{as} \quad k \to \infty, \tag{11.15}$$

for some $\alpha \in (0, 1)$, where for any two sequences $\{s_n\}_{n \geq 1}$ in $I\!R$ and $\{t_n\}_{n \geq 1}$ in $(0, \infty)$, we write $s_n \sim t_n$ if $s_n/t_n \to 1$ as $n \to \infty$. Note that here $\sum_{k=1}^{\infty} |r_1(k)| = \infty$ and hence the process $\{Z_t\}$ is LRD. Next suppose that the X_t-process derives from the Z_t-process through the transformation

$$X_t = H_q(Z_t), \quad t \in \mathbb{Z}, \tag{11.16}$$

for some integer $q \geq 1$, where $H_q(x)$ is the q-th Hermite polynomial, i.e. for $x \in I\!R$, $H_q(x) = (-1)^q \big(\exp(x^2/2) \big) \frac{d^q}{dx^q} \big(\exp(-x^2/2) \big)$. Results in Taqqu (1975), Taqqu (1979) and Dobrushin and Major (1979) imply the following result on the sample mean, where W_q is a multiple Wiener-Ito integral with respect to the random spectral measure W of the Gaussian white noise process defined as

$$W_q = \frac{1}{A^{q/2}} \int \frac{\exp(\mathcal{I}(x_1 + \ldots + x_q)) - 1}{\mathcal{I}(x_1 + \ldots + x_q)}$$

$$\times \prod_{k=1}^{q} |x_k|^{(\alpha-1)/2} dW(x_1) \ldots dW_q(x_q)$$

with $A = 2\Gamma(\alpha)\cos(\alpha\pi/2)$.

Theorem 11.3: *Suppose that $\{X_t\}_{t \in \mathbb{Z}}$ admits the representation (11.16) for some $q \geq 1$. If $\alpha \in (0, q^{-1})$, then*

$$n^{q\alpha/2}(\bar{X}_n - \mu) \to^d W_q, \tag{11.17}$$

where $\mu = EX_1$.

For $q = 1$, W_q has a normal distribution with mean zero and variance $2/[(1 - \alpha)(2 - \alpha)]$. However, for $q \geq 2$, W_q has a non-normal distribution. Although the bootstrap methods described in the earlier sections are successful in a variety of problems under SRD, they need not provide a valid answer under LRD. The following result gives the behavior of the MBB approximation under LRD:

Theorem 11.4: *Let $\bar{X}_n^*$ denote the MBB sample mean based on blocks of size ℓ and resample size n. Suppose that the conditions of Theorem 8.1 hold and that $n^\delta \ell^{-1} + \ell n^{1-\delta} = o(1)$ as $n \to \infty$ for some $\delta \in (0, 1)$. Then,*

$$\sup_{x \in I\!R} \left| P_* \left(c_n(\bar{X}_n^* - \hat{\mu}) \leq x \right) - P \left(n^{q\alpha/2}(\bar{X}_n - \mu) \leq x \right) \right| = o(1), \quad n \to \infty$$

for some sequence $\{c_n\}_{n \geq 1} \in (0, \infty)$ if and only if $q = 1$.

This theorem is a consequence of the results in Lahiri (2003a). It shows that for any choice of the scaling sequence, the MBB method fails to capture the distribution of the sample mean whenever the limit distribution of $\bar{X}_n$ is non-normal. With minor modifications of the arguments in Lahiri (2003a), it can be shown that the same conclusion also holds for the NBB and the CBB. Intuitively, this may not be very surprising. The heuristic arguments behind the construction of these block bootstrap methods show (cf. Section 5) that all three methods attempt to estimate the initial approximation P_ℓ^∞ to the joint distribution P of $\{X_t\}_{t \in \mathbb{Z}}$, but P_ℓ^∞ itself gives an inadequate approximation to P under LRD. Indeed, for the same reason, the MBB approximation fails even for $q = 1$ with the natural choice of the scaling sequence $c_n = n^{q\alpha/2}$. In this case, the (limit) distribution can be captured by using the MBB only with specially constructed scaling sequences $\{c_n\}_{n \geq 1}$, where $c_n \sim [n/\ell^{1+q\alpha}]^{1/2}$ as $n \to \infty$. However, this is more of a coincidental phenomenon and provides very little confidence with regard to the applicability of the block bootstrap methods to inference problems in general under LRD. Formulation of a suitable bootstrap method that works for both normal and nonnormal cases is still an open problem. For related results on subsampling and empirical likelihood methods under LRD, see Hall, Jing, and Lahiri (1998), Lahiri (2005), Nordman, Lahiri and Sibbertsen (2004), and the references therein.

11.9. Bootstrap for Spatial Data

The use of blocks in the context of bootstrapping spatial coverage patterns was first considered by Hall (1985) and it predated the formal introduction of the block bootstrap methods in time series problems. For spatial data observed on regular grids, Politis and Romano (1993) gave an extension of the MBB for rectangular sampling regions and proved its validity. For irregularly spaced spatial data Politis, Paparoditis, and Romano (1999) and Lahiri (2003c) formulated block bootstrap methods and established its consistency, allowing the sampling regions to have a general shape. Following Lahiri (2003c), we now describe a version of the block bootstrap method that is applicable to regular-gridded data as well as to irregularly spaced spatial data with a possibly nonuniform density. Let $\{Z(s) : s \in \mathbb{R}^d\}$ be a stationary random field that is observed at finitely many points $\{s_1, \ldots, s_n\}$ over a sampling region R_n, say. The points $s_1, \ldots, s_n$ may lie on a regular grid in $\mathbb{R}^d$, such as on the integer grid $\mathbb{Z}^d$, or may be irregularly spaced.

Note that in the former case, it is enough to suppose that the random field $Z(\cdot)$ is defined only on the regular grid, not on the continuum. Let a_n be such that

$$a_n^{-1} + a_n/[\text{vol.}(R_n)]^{1/d} \to 0, \quad \text{as} \quad n \to \infty. \tag{11.18}$$

Here a_n determines the size of the blocks for the spatial block bootstrap method (similar to the block length ℓ in the time series case). Next partition R_n by cubes of volume a_n^d. Since, by (11.18), a_n^d is small compared to the volume of R_n, a number of cubes in the partition are completely contained in R_n; call these the *interior* subregions. The remaining cubes intersect both R_n and its complement; call these the *boundary* cubes and the part of R_n contained in these cubes as *boundary* subregions. Let the total number of interior and boundary subregions be denoted by K_1 and K_2, respectively. For the spatial block bootstrap, we would like to construct a copy of the Z-process on each of these $(K_1 + K_2)$ subregions.

To that end, let G be a regular grid on $I\!\!R^d$, say, the integer grid $Z\!\!\!Z^d$. Consider translates of the scaled unit cube $\mathcal{U}_n \equiv [0, a_n)^d$ by points on the grid G. Let $\{\mathcal{B}_i : i \in \mathcal{I}_n\}$ denote the collection of translates of $\mathcal{U}_n$ that are completely contained in R_n. The $\mathcal{B}_i$'s here correspond to the "observed blocks" in the time series case. Next we select a random sample of size $K_1 + K_2$ with replacement from the collection of $\{\mathcal{B}_i : i \in \mathcal{I}_n\}$. We use the observations from the first K_1 resampled blocks to construct a copy of the Z-process over the interior subregions. For boundary subregions, there are two ways to proceed. Under the first, we may repeat the same procedure as in the case of the interior subregions and use the last K_2 resampled blocks themselves in place of the boundary subregions. This would generate a bootstrap copy of the Z-process over an enlarged sampling region, where each boundary subregion of R_n is replaced by a cubic subregion. Alternatively, for each of the boundary subregions, we may only make use of the observations that lie on a *congruent* subset of a resampled block. This would define a copy of the Z-process exactly over the sampling region R_n.

To get a better insight, consider the one dimensional case where $R_n = [1, n]$ and the Z-process is observed at locations $\{1, 2, \ldots, n\}$. With $a_n = \ell$ denoting the block size, let b be the largest integer not exceeding n/ℓ. If $n > b\ell$, then the interior subregions are given by $\{1, \ldots, \ell\}, \ldots, \{(b-1)\ell + 1, \ldots, b\ell\}$ and the boundary subregion by $\{b\ell + 1, \ldots, n\}$, with $K_1 = b$ and $K_2 = 1$. We, therefore, resample a total of $b + 1$ blocks and use the first b blocks to reconstruct the process over the union of the interior subregions, given by $\{1, \ldots, b\ell\}$. For the boundary subregion, under the first version, *all* of the last resampled block is added to the collection $\{1, \ldots, b\ell\}$, yielding bootstrap observations at each point of the *enlarged* set $\{1, \ldots, (b+1)\ell\}$. On the other hand, in the second version, only the first $n - b\ell$ observations from the last resampled block are retained and we get bootstrap observations over $\{1, \ldots, n\}$. Note that for regularly spaced sampling sites, the first version may have a larger resample size than the original sample size. In comparison, for irregularly spaced sampling sites, the resample size may be different from the original sample size under both versions of the block bootstrap. For sampling regions satisfying some mild boundary conditions, the difference between the two versions of the block bootstraps can be shown to be asymptotically negligible.

Lahiri (2003c) establishes validity of this spatial bootstrap method for irregularly spaced spatial data where the locations of the sampling sites are generated by a stochastic design allowing a nonuniform sampling density. Extension of the method to M-estimation under a spatial regression model is recently given by Lahiri and Zhu (2005). For spatial data observed on a regular grid, Zhu and Lahiri (2005) establish validity of a variant of the spatial block bootstrap method for the empirical process. Lee and Lahiri (2002) develop a resampling based method of estimating covariance parameters of a spatial process. For formulations and applications of resampling methods in spatial prediction problems, see Lahiri (1999b), Lahiri *et al.* (1999), Sjöstedt-DeLuna and Young (2003), Chapter 12 of Lahiri (2003a) and the references therein. For related work on other resampling methods for spatial data, see Sherman and Carlstein (1994), Politis, Paparoditis, and Romano (1998), Politis, Romano and Wolf (1999), Nordman and Lahiri (2004), and the references therein.

Acknowledgements

This work is partially supported by NSF grant no. DMS 0306574. The author thanks Professor H. L. Koul for his careful reading and comments on an earlier draft of the article.

References

Anderson, T. W. (1971). *The statistical analysis of time series.* Wiley, New York.

Arcones, M. A. and Giné, E. (1989). The bootstrap of the mean with arbitrary bootstrap sample size. *Annales de l'Institut Henri Poincare, Section B, Calcul des Probabilities et Statistique* **25** 457-481. (Corrections: 1991, Vol. 27, p583-595).

Athreya, K. B. , and Fuh, C. D. (1992). Bootstrapping Markov chains: Countable case. *Journal of Statistical Planning and Inference* **33** 311-331.

Athreya, K. B. and Ney, P. (1978). A new approach to the limit theory of recurrent Markov chains. *Transactions of the American Mathematical Society* **245**, 493–501.

Babu, G. J. and Singh, K. (1984). On one term Edgeworth correction by Efron's bootstrap. *Sankhya, Series A* **46**, 219-232.

Bickel, P. J. and Freedman, D. A. (1981). Some asymptotic theory for the bootstrap. *Annals of Statistics* **9**, 1196-1217.

Bickel, P. J., Götze, F., and van Zwet, W. R. (1997). Resampling fewer than n observations: Gains, losses, and remedies for losses. *Statistica Sinica* **7**, 1-32.

Bose, A. (1988). Edgeworth correction by bootstrap in autoregressions. *Annals of Statistics* **16**, 1709-1722.

Brockwell, P. J. , and Davis, R. A. (1996). *Introduction to time series and forecasting.* Springer-Verlag Inc. New York.

Bühlmann, P. (1997). Sieve bootstrap for time series. *Bernoulli* **3**, 123–148.

Bühlmann, P. (2002). Sieve bootstrap with variable-length Markov chains for stationary categorical time series. With comments and a rejoinder by the author. *Journal of the American Statistical Association* **97**, 458, 443–471.

Bühlmann, P. and Künsch, H. R. (1999). Block length selection in the bootstrap for time series. *Computational Statistics and Data Analysis* **31**, 295-310.

Carlstein, E. (1986). The use of subseries values for estimating the variance of a general statistic from a stationary sequence. *Annals of Statistics* **14**, 1171-1179.

Carlstein, E., Do, K-A., Hall, P., Hesterberg, T., and Künsch, H. R. (1998). Matched-block bootstrap for dependent data. *Bernoulli* **4**, 305-328

Chatterjee, S. and Bose, A. (2002). Dimension asymptotics for generalised bootstrap in linear regression. *Annals of the Institute of Statistical Mathematics* **54** 367–381.

Choi, E. and Hall, P. (2000). Bootstrap confidence regions computed from autoregressions of arbitrary order. *Journal of the Royal Statstical Society, Series B* **62**, 461–477.

Dahlhaus, R. and Janas, D. (1996). A frequency domain bootstrap for ratio statistics in time series analysis. *Annals of Statistics* **24**, 1934–1963.

Datta, S. (1995). Limit theory and bootstrap for explosive and partially explosive autoregression. *Stochastic Processes and their Applications* **57**, 285-304.

Datta, S. (1996). On asymptotic properties of bootstrap for AR(1) processes. *Journal of Statistical Planning and Inference* **53**, 361-374.

Datta, S. and McCormick, W. P. (1995a). Bootstrap inference for a first-order autoregression with positive innovations. *Journal of the American Statistical Association* **90**, 1289-1300.

Datta, S. and McCormick, W. P. (1995b). Some continuous Edgeworth expansions for Markov chains with applications to bootstrap. *Journal of Multivariate Analysis* **52**, 83-106.

Davison, A. C. and Hinkley, D. V. (1997). *Bootstrap methods and their application*. Cambridge University Press, Cambridge, UK.

Dobrushin, R. L. , and Major, P. (1979). Non-central limit theorems for non-linear functionals of Gaussian fields. *Zeitschrift für Wahrscheinlichkeitstheorie und Verwandte Gebiete* **50**, 27-52.

Efron, B. (1979). Bootstrap methods: Another look at the jackknife. *Annals of Statistics* **7**, 1-26.

Efron, B. (1987). Better bootstrap confidence intervals. (With comments and a rejoinder by the author). *Journal of the American Statistical Association* **82**, 171–200.

Efron, B. (1992). Jackknife-after-bootstrap standard errors and influence functions (Disc: p111-127). *Journal of the Royal Statistical Society, Series B* **54**, 83-111.

Efron, B. and Tibshirani, R. J. (1986). Bootstrap methods for standard errors, confidence intervals, and other measures of statistical accuracy (C/R: p75-96). *Statistical Science* **1**, 54-75.

Efron, B. and Tibshirani, R. J. (1993). *An introduction to the bootstrap*. Chapman & Hall Ltd, New York.

Fukuchi, J. I. (1994). Bootstrapping extremes of random variables. *Ph.D. Dissertation*. Iowa State University, Ames, IA.

Falk, M. and Reiss, R.-D. (1989). Weak convergence of smoothed and non-smoothed bootstrap quantile estimates. *Annals of Probability* **17** 362–371.

Ferretti, N. and Romo, J. (1996). Unit root bootstrap tests for AR(1) models. *Biometrika* **83** 849-860.

Franke, J.and Härdle, W. (1992) On bootstrapping kernel spectral estimates. *Annals of Statistics* **20**, 121–145.

Freedman, D. A. (1981). Bootstrapping regression models. *Annals of Statistics* **9**, 1218-1228.

Freedman, D. A. and Peters, S. C. (1984). Bootstrapping a regression equation: Some empirical results. *Journal of the American Statistical Association* **79**, 97-106.

Ghosh, M., Parr, W.C., Singh, K. and Babu, G. J. (1984). A Note on Bootstrapping the Sample Median. *Annals of Statistics* **12**, 1130-1135.

Giné, E. and Zinn, J. (1989). Necessary conditions for the bootstrap of the mean. *Annals of Statistics* **17**, 684-691.

Götze, F. and Künsch, H. R. (1996). Second-order correctness of the blockwise bootstrap for stationary observations. *Annals of Statistics* **24**, 1914–1933.

Heimann, G. and Kreiss, J-P. (1996). Bootstrapping general first order autoregression. *Statistics & Probability Letters* **30** 87-98.

Hall, P. (1988). Theoretical comparison of bootstrap confidence intervals (C/R: p953-985). *Annals of Statistics* **16**, 927-953.

Hall, P. (1992). *The bootstrap and Edgeworth expansion*. Springer-Verlag Inc. New York.

Hall, P., DiCiccio, T. J. and Romano, J. P. (1989). On smoothing and the bootstrap. *Annals of Statistics* **17**, 692–704.

Hall, P., Horowitz, J. L., and Jing, B-Y. (1995). On blocking rules for the bootstrap with dependent data. *Biometrika* **82**, 561-574.

Hall, P., Jing, B.-Y., and Lahiri, S. N.(1998). On the sampling window method for long-range dependent data. *Statistica Sinica* **8**, 1189-1204.

Helmers, R. (1991). On the Edgeworth expansion and the bootstrap approximation for a studentized *U*-statistic. *Annals of Statistics* **19**, 470-484.

Ho, Y. H. S. and Lee, S. M. S. (2005). Iterated smoothed bootstrap confidence intervals for population quantiles. *Annals of Statistics* **33** (In press).

Hurvich C. M. and Zeger, S. L. (1987). Frequency domain bootstap methods for time series. *Statistics and Operations Reserach working paper*. New York university, New York.

Koul, H. L. and Lahiri, S. N. (1994). On bootstrapping M-estimated residual processes in multiple linear regression models. *Journal of Multivariate Analysis* **49**, 255-265.

Kreiss, J-P. and Franke, J. (1992). Bootstrapping stationary autoregressive moving-average models. *Journal of Time Series Analysis* **13**, 297-317.

Kreiss, J-P. and Paparoditis, E. (2003). Autoregressive-aided periodogram bootstrap for time series. *Annals of Statistics* **31**, 1923–1955.

Kulperger, R. J. and Prakasa Rao, B. L. S. (1989). Bootstrapping a finite state Markov chain. *Sankhya, Series A* **51** 178-191.

Lahiri, S. N. (1991). Second order optimality of stationary bootstrap. *Statistics and Probability Letters* **11**, 335-341.

Lahiri, S. N. (1992a). Bootstrapping M-estimators of a multiple regression parameter. *Annals of Statistics* **20**, 1548-1570.

Lahiri, S. N. (1992b). Edgeworth correction by 'moving block bootstrap' for stationary and nonstationary data. In *Exploring the limits of bootstrap*. - Eds. R. Lepage and L. Billard, Wiley, NY, 183-214.

Lahiri, S. N. (1994a). Rates of bootstrap approximation for the mean of lattice variables. *Sankhya, Ser. A.* **56**, 77-89.

Lahiri, S. N. (1994b). Two term Edgeworth expansion and bootstrap approximation for multivariate studentized M-estimators. *Sankhya, Ser. A.* **56**, 201-226.

Lahiri, S. N. (1996). On Edgeworth expansions and the moving block bootstrap for studentized M-estimators in multiple linear regression models. *Journal of Multivariate Analysis* **56**, 42-59.

Lahiri, S. N. (1999a). Theoretical comparison of block bootstrap methods. *Annals of Statistics* **27**, 386-404.

Lahiri, S.N. (1999b). Asymptotic distribution of the empirical spatial cumulative distribution function predictor and prediction bands based on a subsampling Method. *Probability Theory and Related Fields* **114**, 55-84.

Lahiri, S. N. (2002). On the Jackknife after Bootstrap method for dependent data and its consistency properties. *Econometric Theory* **18**, 79-98.

Lahiri, S. N. (2003a). *Resampling methods for dependent data.* Springer-Verlag Inc., New York.

Lahiri, S. N. (2003b). A necessary and sufficient condition for asymptotic independence of discrete Fourier transforms under short- and long-range dependence. *Annals of Statistics* **31**, 613-641.

Lahiri, S.N. (2003c). Validity of a block bootstrap method for irregularly spaced spatial data under nonuniform stochastic designs. *Preprint*, Department of Statistics, Iowa State University, Ames, IA.

Lahiri, S. N., Furukawa, K., and Lee, Y-D. (2003). A nonparametric plug-in method for selecting the optimal block length. *Preprint*, Department of Statistics, Iowa State University, Ames, IA.

Lahiri, S. N., Kaiser, M. S., Cressie, N., and Hsu, N-J. (1999). Prediction of spatial cumulative distribution functions using subsampling (with discussion.) *Journal of the American Statistical Association* **94**, 86 -110.

Lahiri, S. N. (2005). A note on the subsampling method under long range dependence. *Preprint*, Department of Statistics, Iowa State University, Ames, IA.

Lahiri, S.N. and Zhu, J. (2005). Resampling methods for spatial regression models under a class of stochastic designs. *Preprint*, Department of Statistics, Iowa State University, Ames, IA.

Lee, Y-D. and Lahiri, S.N. (2002). Least squares variogram fitting by spatial subsampling. *Journal of the Royal Statistical Society, Series B* **64**, 837-854.

Lee, S. M. S. and Young, G. A. (1995). Asymptotic iterated bootstrap confidence intervals. *Annals of Statistics* **23**, 1301–1330.

Liu, R. Y. , and Singh, K. (1992a). Moving blocks jackknife and bootstrap capture weak dependence. In *Exploring the Limits of Bootstrap* Ed. R.Lepage and L. Billard. 225-248.

Liu, R. Y. , and Singh, K. (1992b). Efficiency and robustness in resampling. *Annals of Statistics* **20**, 370-384.

Loh, W-Y. (1987). Calibrating confidence coefficients. *Journal of the American Statistical Association* **82**, 155-162.

Nordman, D. and Lahiri, S. N. (2004). On optimal spatial subsample size for variance estimation. *Annals of Statistics.* **32** 1981-2027.

Nordman, D., Lahiri, S.N., and Sibbertsen, P. (2004). Empirical likelihood confidence intervals for the mean of a long range dependent process. *Preprint*, Department of Statistics, Iowa State University, Ames, IA.

Paparoditis, E. and Politis, D. N. (2001a). Tapered block bootstrap. *Biometrika* **88** 1105-1119.

Paparoditis, E., and Politis, D. N. (2001b). A Markovian local resampling scheme for nonparametric estimators in time series analysis. *Econometric Theory* **17**, 540-566.

Percival, D. B., Sardy, S., and Davison, A. C. (2000). Wavestrapping time series: adaptive wavelet-based bootstrapping. In *Nonlinear and nonstationary signal processing (Cambridge, 1998).* p. 442–471, Cambridge Univ. Press, Cambridge.

Politis, D. N., Paparoditis, E., and Romano, J. P. (1998). Large sample inference for irregularly spaced dependent observations based on subsampling. *Sankhya, Series A* **60**, 274-292.

Politis, D. N., Paparoditis, E., and Romano, J. P. (1999). Resampling marked point processes. In *Multivariate Analysis, Design of Experiments, and Survey Sampling: A tribute to J.N. Srivastava* S. Ghosh (Ed.). Mercel Dekker, New York. P. 163-185.

Politis, D. N. and Romano, J. P. (1992). A circular block-resampling procedure for stationary data. In *Exploring the Limits of Bootstrap*. 263-270.

Politis, D. N. and Romano, J. P. (1993). Nonparametric resampling for homogeneous strong mixing random fields. *Journal of Multivariate Analysis*. **47**, 301-328.

Politis, D. N. , and Romano, J. P. (1994). The stationary bootstrap. *Journal of the American Statistical Association* **89**, 1303-1313.

Politis, D.N. and White, H. (2003). Automatic block-length selection for the dependent bootstrap. *Preprint* Department of Mathematics, University of California at San Diego, LaJolla, CA.

Politis, D. N., Romano, J. P., and Wolf, M. (1999). *Subsampling*. Springer-Verlag Inc., New York.

Sakov, A. and Bickel, P. J. (1999). Choosing m in the m out of n bootstrap. *ASA Proceedings of the Section on Bayesian Statistical Science* 125-128.

Shao, J. and Tu, D. (1995). The jackknife and bootstrap. Springer-Verlag Inc. New York.

Shorack, G. R. (1982). Bootstrapping robust regression. *Communications in Statistics, Part A – Theory and Methods* **11**, 961-972

Sherman, M. and Carlstein, E. (1994). Nonparametric estimation of the moments of a general statistic computed from spatial data. *Journal of the American Statistical Association* **89**, 496-500.

Singh, K. (1981). On the asymptotic accuracy of Efron's bootstrap. *Annals of Statistics* **9**, 1187-1195.

Sjöstedt-DeLuna, S. and Young, G. A. (2003). The bootstrap and Kriging prediction intervals. *Scandinavian Journal of Statistics* **30**, 175-192.

Swanepoel, J. W. H. , and van Wyk, J. W. J. (1986). The bootstrap applied to power spectral density function estimation. *Biometrika* **73**, 135-141.

Taqqu, M. S. (1975). Weak convergence to fractional Brownian motion and to the Rosenblatt process. *Zeitschrift für Wahrscheinlichkeitstheorie und Verwandte Gebiete* **31**, 287-302.

Taqqu, M. S. (1979). Convergence of integrated processes of arbitrary hermite rank. *Zeitschrift für Wahrscheinlichkeitstheorie und Verwandte Gebiete* **50**, 53-83.

Zhu, J. and Lahiri, S. N. (2005). Weak convergence of the spatial empirical process and its bootstrap version. *Statistical Inference from Stochastic Processes*. (In press).

CHAPTER 12

An Expansion for a Discrete Non-Lattice Distribution

Friedrich Götze and Willem R. van Zwet

Fakultat fur Mathematik
University of Bielefeld
Postfach 100131, 33501 Bielefeld Germany
goetze@mathematik.uni-bielfeld.de

Mathematical Institute
University of Leiden
P.O.Box 9512, 2300 RA Leidenm, The Netherlands
vanzwet@math.leidenuniv.nl

Much is known about asymptotic expansions for asymptotically normal distributions if these distributions are either absolutely continuous or pure lattice distributions. In this paper we begin an investigation of the discrete but non-lattice case. We tackle one of the simplest examples imaginable and find that curious phenomena occur. Clearly more work is needed.

12.1. Introduction

There is a voluminous literature on second order analysis of distribution functions $F_N(z) = P(Z_N \leq z)$ of statistics $Z_N = \zeta_N(X_1, X_2, \ldots, X_N)$ that are functions of i.i.d. random variables $X_1, X_2, \ldots$. The results obtained are generally refinements of the central limit theorem. Suppose that Z_N is asymptotically standard normal, that is $\sup_z |F_N(z) - \Phi(z)| \to 0$ as $N \to \infty$, where Φ denotes the standard normal distribution function. Then second order results are concerned with the speed of this convergence, or with attempts to increase this speed by replacing the limit Φ by a series expansion Ψ_N that provides a better approximation. Results of the first kind are called theorems of Berry-Esseen type and assert that for all N,

$$\sup_z |F_N(z) - \Phi(z)| \leq CN^{-\frac{1}{2}},$$

where C is a constant that depends on the particular statistic Z_N and the distribution of the variables X_i, but not on N. Such results are often valid under mild restrictions such as the existence of a third moment of X_i. The original Berry-Esseen theorem dealt with the case where Z_N is a sum of i.i.d. random variables, Esseen (1942), Berry (1941). For a more general version see van Zwet (1984).

Results of the second kind concern so-called Edgeworth expansions. These are series expansions such as

$$\Psi_{N,1}(z) = \Phi(z) + \varphi(z)N^{-\frac{1}{2}}Q_1(z), \quad \text{or}$$
$$\Psi_{N,2}(z) = \Phi(z) + \varphi(z)\left[N^{-\frac{1}{2}}Q_1(z) + N^{-1}Q_2(z)\right], \tag{12.1}$$

where φ is the standard normal density and Q_1 and Q_2 are polynomials depending on low moments of X_i . One then shows that

$$\sup_z |F_N(z) - \Psi_{N,1}(z)| \le CN^{-1}, \quad \text{or}$$
$$\sup_z |F_N(z) - \Psi_{N,2}(z)| \le CN^{-\frac{3}{2}}. \tag{12.2}$$

For this type of result the restrictions are more severe. Apart from moment assumptions one typically assumes that Z_N is not a lattice random variable. For the case where Z_N is a sum of i.i.d. random variables a good reference is Feller (1965, Chapter XVI). There are numerous papers devoted to special types of statistics. For a somewhat more general result we refer to Bickel, Götze and van Zwet (1986) and Bentkus, Götze, and van Zwet (1997).

For the case where Z_N assumes its values on a lattice, say the integers, an alternative approach is to generalize the local central limit theorem and provide an expansion for the point probabilities $P(Z_N = z)$ for values of z belonging to the lattice. A typical case is the binomial distribution for which local expansions are well known. It is obvious that for the binomial distribution one can not obtain Edgeworth expansions as given in (12.1) for which (12.2) holds. The reason is that out of the N possible values for a binomial (N, p) random variable, only $cN^{\frac{1}{2}}$ values around the mean Np really count and each of these has probability of order $N^{-\frac{1}{2}}$. Hence the distribution function has jumps of order $N^{-\frac{1}{2}}$ and can therefore not be approximated by a continuous function such as given in (12.1) with an error of smaller order than $N^{-\frac{1}{2}}$.

In a sense the binomial example is an extreme case where the ease of the approach through local expansions for $P(Z_N = z)$ precludes the one through expansions of Edgeworth type for $P(Z_N \le z)$. In Albers, Bickel and van Zwet (1976) the authors found somewhat to their surprise that for the Wilcoxon statistic which is a pure lattice statistic, an Edgeworth expansion with remainder $O(N^{-\frac{3}{2}})$ for the distribution function is perfectly possible. In this case the statistic ranges over N^2 possible integer values, of which the central $N^{\frac{3}{2}}$ values have probabilities of order $N^{-\frac{3}{2}}$ so that one can approximate a distribution function with such jumps by a continuous one with error $O(N^{-\frac{3}{2}})$.

On the basis of these examples one might guess that the existence of an Edgeworth expansion with error $O(N^{-p})$ for the distribution function $F_N(z) = P(Z_N \leq z)$ would merely depend on the existence of some moments of Z_N combined with the requirement that F_N does not exhibit jumps of large order than N^{-p}. But one can envisage a more subtle problem if F_N would assign a probability of larger order than N^{-p} to an interval of length N^{-p}. Since Edgeworth expansions have bounded derivative, this would also preclude the existence of such an expansion with error $O(N^{-p})$.

Little seems to be known about the case where Z_N has a discrete but non-lattice distribution. Examples abound if one considers a lattice random variable with expectation 0 and standardized by dividing by its sample standard deviation. As a simple example, one could for instance consider Student's t-statistic $\tau_N = N^{-1/2} \sum_i X_i / \sqrt{\sum_i \left(X_i - m\right)^2 / (N-1)}$ with $m = \sum_i X_i / N$ and $X_1, X_2, \ldots$ i.i.d. random variables with a lattice distribution. Since we are not interested in any particular statistic, but merely in exploring what goes on in a case like this, we shall simplify even further by deleting the sample mean m and considering the statistic

$$W_N = \sum_{i=1}^{N} \frac{X_i}{\sqrt{\sum_{i=1}^{N} X_i^2}}, \tag{12.3}$$

with $X_1, X_2, \ldots$ i.i.d. with $P(X_i = -1) = P(X_i = 0) = P(X_i = 1) = \frac{1}{3}$.

We should perhaps point out that for $w > 0$

$$P(0 < \tau_N \leq w) = P\left(0 < W_N \leq \frac{\sqrt{N/(N-1)}x}{1 + x^2/(N-1)}\right)$$
$$= P(0 < W_N \leq x + O(x(1+x^2)/N)),$$

and since both τ_N and W_N have distributions that are symmetric about the origin, Theorem 12.1 ensures that under this model, the expansions for the distributions of τ_N and W_N are identical up to order $O(N^{-1})$. Hence the fact that we discuss an expansion for W_N rather than for Student's statistic τ_N is not an essential difference, but merely a matter of convenience.

Notice that in (12.3), $\sum X_i^2$ equals the number of non-zero X_i and that each of these equals -1 or $+1$ with probability $\frac{1}{2}$. Hence we may also describe the model given by (12.3) as follows.

For $N = 1, 2, \cdots$, T_N has a binomial distribution with parameters N and $\frac{2}{3}$. Given T_N, S_N has a binomial distribution with parameters T_N and $\frac{1}{2}$. Define $D_N = 2S_N - T_N$ and notice that D_N and T_N are either both even or both odd. We consider the statistic

$$W_N = \begin{cases} 0 & \text{if } T_N = 0 \\ \frac{D_N}{\sqrt{T_N}} & \text{otherwise.} \end{cases} \tag{12.4}$$

Notice that unconditionally S_N has a binomial distribution with parameters N and $\frac{1}{3}$.

Let F_N be the distribution function of W_N. Obviously W_N has expected value $EW_N = 0$ and variance $\sigma^2(W_N) = 1$, and is asymptotically normal. By an appropriate Berry-Esseen theorem $\sup_x |F_N(x) - \Phi(x)| = O(N^{-\frac{1}{2}})$, where Φ denotes the standard normal distribution function as before, see e.g. Bentkus, Götze, and Tikhomirov (1997). A curious thing about the distribution of W_N is that $P(W_N = 0) = P(D_N = 0) = P(S_N = \frac{1}{2}T_N)$ which is obviously of exact order $N^{-\frac{1}{2}}$, but all other point probabilities $P(W_N = w)$ for $w \neq 0$ are clearly of smaller order, and we shall see that these are actually $O(N^{-1})$. Hence the following question arises: if we remove the point probability at the origin, to what order of magnitude can we approximate the distribution function of W_N by an expansion?

Let $\text{frac}(x)$ denote the fractional part of a number $x \geq 0$, i.e. $\text{frac}(x) = x - [x]$ if $[x]$ is the largest integer smaller than or equal to x. For $N = 1, 2, \ldots$, define a function Ψ_N on $(-\infty, \infty)$ as follows. For $w \geq 0$,

$$\Psi_N(w) = \Phi(w) + N^{-1/2}\Lambda_N(w), \quad \text{with} \tag{12.5}$$

$$\Lambda_N(w) = -\sqrt{\frac{3}{2}}\,\varphi(w) \sum_{0 \leq n \leq N} \frac{3}{\sqrt{\pi N}} e^{-\frac{9}{N}\left(n - \frac{N}{3}\right)^2} \left[\text{frac}\left(w\sqrt{2n}\right) - \frac{1}{2}\right],$$

$$\Psi_N(-w) = 1 - \Psi_N(w-), \quad w < 0.$$

Here the argument $w-$ denotes a limit from the left at w. Ψ_N is of bounded variation and for sufficiently large N it is a probability distribution function. It has upward jump discontinuities of order $O(N^{-1})$ at points w where $w\sqrt{2n}$ assumes an integer value $k \neq 0$. At the origin it has a jump discontinuity of magnitude

$$\Psi_N(0) - \Psi_N(0-) \sim \sqrt{\frac{3}{2N}}\varphi(0) = \sqrt{\frac{3}{4\pi N}}. \tag{12.6}$$

Theorem 12.1: *As $N \to \infty$,*

$$P(W_N = 0) \sim \sqrt{\frac{3}{(4\pi N)}} = O\left(\frac{1}{\sqrt{N}}\right),$$

$$P(W_N = w) = O\left(\frac{1}{N}\right), \quad w \neq 0,$$

$$\sup_w |F_N(w) - \Psi_N(w)| = O\left(\frac{1}{N}\right).$$

Moreover the distribution Ψ_N — and hence F_N — assigns probability $O(N^{-1})$ to any closed interval of length $O(N^{-1})$ that does not contain the origin.

The reader may want to compare this result with the expansion obtained in Brown, Cai and DasGupta (2002) for the distribution of a normalized binomial variable $(Y_n - np)/\sqrt{np(1-p)}$, where Y_n has a binomial distribution with parameters n and p. For $p = 1/2$ this distribution coincides with the conditional distribution of W_N given $N = n$, but of course this is a pure lattice distribution.

Since $\sum_n 3(\pi N)^{-1/2} \exp\{-(9/N)(n - N/3)^2\}$ is the sum of the density of a normal $N(N/3, N/18)$ distribution taken at the integer values $n \in (-\infty, \infty)$, it is asymptotic to 1 as $N \to \infty$, and hence bounded for all N. Because $|\mathrm{frac}(x) - 1/2| \leq 1/2$ for all x, $\Lambda_N(w)$ is bounded and $N^{-1/2}\Lambda_N(w) = O(N^{-1/2})$ uniformly in w. At first sight there is a striking similarity between the expansion Ψ_N in Theorem 12.1 and the two term Edgeworth expansion $\Psi_{N,1}$ in (12.1). However, the term $\phi(z)N^{-1/2}Q_1(z)$ of order $O(N^{-1/2})$ in the Edgeworth expansion is a skewness correction that vanishes for a symmetric distribution F_N. As we are dealing with a symmetric case, such a term is not present and for the continuous case the Edgeworth expansion with remainder $O(N^{-1})$ is simply $\Phi(z)$. The origin of the term $N^{-1/2}\Lambda_N(w)$ is quite different. It arises from the fact that we are approximating a discrete distribution function by a continuous one, and as such it is akin to the classical continuity correction.

To make sure that the term $N^{-1/2}\Lambda_N(w)$ is not of smaller order than $N^{-1/2}$, we shall bound $|\Lambda_N(w)|$ from below by the absolute value of the following series. Assume that N is divisible by 3 and let

$$\lambda_N(w) = \sqrt{\frac{3}{2}}\varphi(w) \sum_{k=1}^{M} \frac{1}{\pi k} f_{N,k} \exp\left(-\frac{\pi^2}{6}k^2 w^2\right) \sin\left(2\pi k w \sqrt{\frac{2N}{3}}\right)$$

$$+ O\left(N^{-1/2}(\log N)^5\right), \quad M = [\log N], \tag{12.7}$$

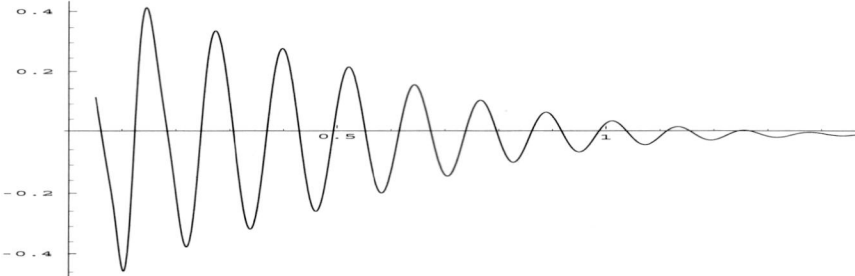

Figure 12.1. $\lambda_{100}(w)$: $0.05 \leq w \leq 2.34$, $M = 10$, $f_{100,k} = \exp[-(k/M)^{2/3}]$

where $f_{N,k} = 1 + O((k/M)^2)$ is defined in (12.22). Thus $\lambda_N(w)$ is a rapidly converging Fourier series, (illustrated in Figure 12.1 above) the modulus of which is larger than a positive constant $c(w) > 0$, provided that $4w\sqrt{\frac{2N}{3}}$ is an odd integer. Hence, we shall prove

Theorem 12.2: *For any N divisible by 3, we have*

$$\sup_{w>0} |F_N(w) - \Phi(w)| \geq \sup_{w \geq 1} N^{-\frac{1}{2}} |\lambda_N(w)| + O\left(N^{-1}(\log N)^5\right) > \frac{c}{\sqrt{N}},$$
(12.8)

for some absolute constant $c > 0$.

The proof of Theorem 12.1 is given in the next section. In Section 3 we investigate the oscillatory part of Ψ_N in (12.5), relating it to the Fourier series $\lambda_N(w)$ above and thus proving Theorem 12.2.

12.2. Proof of Theorem 12.1

The event $W_N = 0$ occurs if $D_N = 0$. Let $Z_1, Z_2, \ldots, Z_N$ be i.i.d. random variables assuming the values 0, -1 and $+1$, each with probability $\frac{1}{3}$. Then D_N is distributed as $\sum Z_i$, which has mean 0 and variance $\frac{2N}{3}$. By the local central limit theorem $P(\sum Z_i = 0) \sim (2\pi)^{-\frac{1}{2}} \left(\frac{2N}{3}\right)^{-\frac{1}{2}} = \sqrt{\frac{3}{4\pi N}}$ which proves the first statement of Theorem 12.1. Because the distribution of W_N is symmetric about the origin, this implies that in the remainder of the proof we only need to consider positive values of W_N. Hence we suppose that $w > 0$ throughout and this implies that we need only be concerned with positive values of D_N also.

Hoeffding's inequality ensures that for all $N \geq 2$,

$$P\left(|D_N| \geq \sqrt{6N \log N}\right) \leq \frac{2}{N^3}$$

and

$$P\left(|T_N - 2N/3| \geq \sqrt{2N \log N}\right) \leq \frac{2}{N^3}.$$

Since the joint distribution of T_N and D_N assigns positive probability to at most N^2 points and events with probability $O(N^{-1})$ are irrelevant for the remainder of the proof, we may at any point restrict attention to values $D_N = d$ and $T_N = t$ with $|d| \leq t$ and satisfying

$$|d| < \sqrt{6N \log N} \quad \text{and} \quad \left|t - \frac{2N}{3}\right| < \sqrt{2N \log N}. \tag{12.9}$$

For a positive integer $m \leq n$ we have

$$P(D_N = 2m, T_N = 2n) = P(S_N = m + n, T_N = 2n)$$
$$= \frac{N!}{3^N (n+m)!(n-m)!(N-2n)!}.$$

If $d = 2m$ and $t = 2n$ satisfy (12.9), then $(n + m)$, $(n - m)$ and $(N - 2n)$ are of exact order N and we may apply Stirling's formula to see that

$$P(D_N = 2m, T_N = 2n)$$

$$= \frac{N^{N+\frac{1}{2}}\left(1 + O\left(\frac{1}{N}\right)\right)}{2\pi 3^N (n+m)^{(n+m+\frac{1}{2})}(n-m)^{(n-m+\frac{1}{2})}(N-2n)^{(N-2n+\frac{1}{2})}}$$

$$= \frac{3^{\frac{3}{2}}\left(1 + O\left(\frac{1}{N}\right)\right)}{2\pi N \left(\frac{3(n+m)}{N}\right)^{(n+m+\frac{1}{2})}\left(\frac{3(n-m)}{N}\right)^{(n-m+\frac{1}{2})}\left(\frac{3(N-2n)}{N}\right)^{(N-2n+\frac{1}{2})}}$$

$$= \frac{3^{\frac{3}{2}}\left(1 + O\left(\frac{1}{N}\right)\right)}{2\pi N} \exp\left\{-\left(n+m+\frac{1}{2}\right)\log\left(1 + \frac{3}{N}\left(n+m-\frac{N}{3}\right)\right)\right.$$

$$-\left(n-m+\frac{1}{2}\right)\log\left(1 + \frac{3}{N}\left(n-m-\frac{N}{3}\right)\right)$$

$$\left.-\left(N-2n+\frac{1}{2}\right)\log\left(1 + \frac{3}{N}\left(\frac{2N}{3}-2n\right)\right)\right\}.$$

Next we expand the logarithms in the exponent. For the first order terms we obtain

$$-\frac{3}{N}\left[\left(n+m+\frac{1}{2}\right)\left(n+m-\frac{N}{3}\right)+\left(n-m+\frac{1}{2}\right)\left(n-m-\frac{N}{3}\right)\right.$$
$$\left.+\left(N-2n+\frac{1}{2}\right)\left(\frac{2N}{3}-2n\right)\right]$$
$$=-\frac{3}{N}\left[\left(n+m-\frac{N}{3}\right)^2+\left(n-m-\frac{N}{3}\right)^2+\left(\frac{2N}{3}-2n\right)^2\right]$$
$$=-\frac{3}{N}\left(6\tilde{n}^2+2m^2\right),$$

where $\tilde{n}=\left(n-\frac{N}{3}\right)$. The second order terms yield

$$\frac{1}{2}\left(\frac{3}{N}\right)^2\left[\left(n+m+\frac{1}{2}\right)\left(n+m-\frac{N}{3}\right)^2\right.$$
$$\left.+\left(n-m+\frac{1}{2}\right)\left(n-m-\frac{N}{3}\right)^2+\left(N-2n+\frac{1}{2}\right)\left(\frac{2N}{3}-2n\right)^2\right]$$
$$=\frac{1}{2}\left(\frac{3}{N}\right)^2\left[-6\tilde{n}^3+(2N+3)\tilde{n}^2+6\tilde{n}m^2+\left(\frac{2N}{3}+1\right)m^2\right]$$
$$=\frac{3}{N}\left(3\tilde{n}^2+m^2\right)+\frac{27}{N^2}\left(-\tilde{n}^3+\tilde{n}m^2\right)+O\left(\frac{\tilde{n}^2+m^2}{N^2}\right).$$

The third order terms contribute

$$-\frac{1}{3}\left(\frac{3}{N}\right)^3\left[\left(n+m+\frac{1}{2}\right)\left(n+m-\frac{N}{3}\right)^3\right.$$
$$\left.+\left(n-m+\frac{1}{2}\right)\left(n-m-\frac{N}{3}\right)^3+\left(N-2n+\frac{1}{2}\right)\left(\frac{2N}{3}-2n\right)^3\right]$$
$$=\frac{18(\tilde{n}^3-\tilde{n}m^2)}{N^2}+O\left(\frac{\tilde{n}^4+m^4}{N^3}\right).$$

As $d=2m$ and $t=2n$ satisfy (12.9), the contribution of the remaining terms is dominated by that of the fourth order terms and equals

$$O\left(\frac{\tilde{n}^4+m^4}{N^3}\right).$$

Collecting the results of these computations we arrive at

$$P(D_N=2m,T_N=2n) \tag{12.10}$$
$$=\frac{3^{\frac{3}{2}}}{2\pi N}\exp\left\{-\frac{3(3\tilde{n}^2+m^2)}{N}-\frac{9(\tilde{n}^3-\tilde{n}m^2)}{N^2}+O\left(\frac{1}{N}+\frac{\tilde{n}^4+m^4}{N^3}\right)\right\},$$

provided $m\le n$ are integers between 1 and $\frac{1}{2}N$ satisfying $m<\sqrt{2N\log N}$ and $|\tilde{n}|=|n-\frac{N}{3}|<\sqrt{N\log N}$. However, we shall also use (12.10) if these inequalities do not hold, since in that case both left- and right-hand members of (12.10) are negligible for our purposes.

By Taylor expansion of the integrand about $x = m$, we find that for integer $0 < m \leq n$ with $m < \sqrt{2N \log N}$ and $|\tilde{n}| = \left|n - \frac{N}{3}\right| < \sqrt{N \log N}$,

$$\int_{[m-\frac{1}{2}, m+\frac{1}{2})} \exp\left\{-\frac{3}{N}x^2 + \frac{9}{N^2}\tilde{n}x^2 + O\left(\frac{1}{N} + \frac{1}{N^3}\left(\tilde{n}^4 + x^4\right)\right)\right\} dx$$

$$= \exp\left\{-\frac{3}{N}m^2 + \frac{9}{N^2}\tilde{n}m^2 + O\left(\frac{1}{N} + \frac{1}{N^3}\left(\tilde{n}^4 + m^4\right)\right)\right\}.$$

It follows that for integers $0 < m \leq n$ with $m < \sqrt{2N \log N}$ and $|\tilde{n}| = \left|n - \frac{N}{3}\right| < \sqrt{N \log N}$,

$$P(2 \leq D_N \leq 2m, T_N = 2n)$$

$$= \frac{3^{\frac{3}{2}}}{2\pi N} \int_{[\frac{1}{2}, m+\frac{1}{2})} e^{-\left\{\frac{3}{N}x^2 - \frac{9}{N^2}\tilde{n}x^2 + O\left(\frac{1}{N^3}x^4\right)\right\}} dx$$

$$\times e^{-\left\{\frac{9}{N}\tilde{n}^2 + \frac{9}{N^2}\tilde{n}^3 + O\left(\frac{1}{N} + \frac{1}{N^3}\tilde{n}^4\right)\right\}}$$

$$= \frac{3^{\frac{3}{2}}}{2\pi N} \int_{[\frac{1}{2}, m+\frac{1}{2})} e^{-\left\{\frac{3}{N}x^2 - \frac{9}{N^2}\tilde{n}x^2\right\}} dx\, e^{-\left\{\frac{9}{N}\tilde{n}^2 + \frac{9}{N^2}\tilde{n}^3 + O\left(\frac{1}{N} + \frac{1}{N^3}\tilde{n}^4\right)\right\}},$$

and again we may use this for all integers m and n with $0 < m \leq n \leq \frac{1}{2}N$ with impunity.

For real $r > \frac{1}{2}$ we write $r = m + \theta$ where $m = \lfloor r \rfloor$ and $\theta = \text{frac}(r) = r - \lfloor r \rfloor \in [0, 1)$ denote the integer and fractional parts of r respectively. Then for $r < \sqrt{2N \log N}$ and $|\tilde{n}| = \left|n - \frac{N}{3}\right| < \sqrt{N \log N}$,

$$P(2 \leq D_N \leq 2r, T_N = 2n) = P(2 \leq D_N \leq 2m, T_N = 2n)$$

$$= \frac{3^{\frac{3}{2}}}{2\pi N} e^{-\left\{\frac{9\tilde{n}^2}{N} + \frac{9\tilde{n}^3}{N^2} + O\left(\frac{1}{N} + \frac{\tilde{n}^4}{N^3}\right)\right\}} \left[\int_{[\frac{1}{2}, r)} e^{\frac{-3x^2}{N} + \frac{9\tilde{n}x^2}{N^2}} dx \right.$$

$$\left. + \int_{[r, m+\frac{1}{2})} e^{\frac{-3x^2}{N} + \frac{9\tilde{n}x^2}{N^2}} dx\right].$$

Evaluating the second integral by expanding the integrand about the point

$x = m + \frac{1}{2}$, we arrive at

$$P(2 \le D_N \le 2r, T_N = 2n) = \frac{3^{\frac{3}{2}}}{2\pi N} e^{-\left\{\frac{9}{N}\tilde{n}^2 + \frac{9}{N^2}\tilde{n}^3 + O\left(\frac{1}{N} + \frac{1}{N^3}\tilde{n}^4\right)\right\}}$$
$$\times \left[\int_{[\frac{1}{2},r)} e^{-\frac{3}{N}x^2 + \frac{9}{N^2}\tilde{n}x^2} dx - e^{-\frac{3}{N}r^2 + \frac{9}{N^2}\tilde{n}r^2} \left(\text{frac}(r) - \frac{1}{2} + O\left(\frac{r}{N}\right)\right) \right].$$

Again we may use this for all $r > 0$ and integer $n \le \frac{1}{2}N$.

Choose $w > 0$ and $r = w\sqrt{\frac{n}{2}}$. We have

$$P(0 < W_N \le w, T_N \text{ is even }) = \sum_{1 \le n \le \frac{N}{2}} P(2 \le D_N \le 2r, T_N = 2n)$$

$$= \sum_{1 \le n \le \frac{N}{2}} \frac{3^{\frac{3}{2}}}{2\pi N} e^{-\left\{\frac{9}{N}\tilde{n}^2 + \frac{9}{N^2}\tilde{n}^3 + O\left(\frac{1}{N} + \frac{1}{N^3}\tilde{n}^4\right)\right\}}$$
$$\times \left[\int_{[\frac{1}{2},r)} e^{-\frac{3}{N}x^2 + \frac{9}{N^2}\tilde{n}x^2} dx - e^{-\frac{3}{N}r^2 + \frac{9}{N^2}\tilde{n}r^2} \left(\text{frac}(r) - \frac{1}{2} + O\left(\frac{r}{N}\right)\right) \right].$$

As $|\tilde{n}| = |n - \frac{N}{3}| \le \frac{N}{3}$, the expression between square brackets is of order $\sqrt{N}$. Next, comparison with the normal $N(\frac{N}{3}, \frac{N}{18})$ distribution with mean $\frac{N}{3}$ and variance $\frac{N}{18}$ shows that

$$\frac{3}{\sqrt{\pi N}} \sum_{1 \le n \le \frac{N}{2}} e^{-\left\{\frac{9}{N}\tilde{n}^2 + \frac{9}{N^2}\tilde{n}^3 + O\left(\frac{1}{N} + \frac{1}{N^3}\tilde{n}^4\right)\right\}} = \frac{3}{\sqrt{\pi N}} \sum_{1 \le n \le \frac{N}{2}} e^{-\frac{9}{N}\tilde{n}^2} + O\left(\frac{1}{N}\right),$$

and hence

$$P(0 < W_N \le w, T_N \text{ is even}) = O\left(\frac{1}{N}\right) + \frac{1}{2} \sum_{1 \le n \le \frac{N}{2}} \frac{3}{\sqrt{\pi N}} e^{-\frac{9}{N}\tilde{n}^2} \sqrt{\frac{3}{\pi N}}$$

$$\times \left[\int_{[\frac{1}{2},r)} e^{-\frac{3}{N}x^2 + \frac{9}{N^2}\tilde{n}x^2} dx - e^{-\frac{3}{N}r^2 + \frac{9}{N^2}\tilde{n}r^2} \left(\text{frac}(r) - \frac{1}{2} + O\left(\frac{r}{N}\right)\right) \right].$$

$$(12.11)$$

Since $\sqrt{\frac{3}{\pi N}} e^{-\frac{3}{N}x^2}$ is the density of the $N\left(0, \frac{N}{6}\right)$ distribution, we see that

$$\sqrt{\frac{3}{\pi N}} \int_{[\frac{1}{2},r)} e^{-\frac{3}{N}x^2 + \frac{9}{N^2}\tilde{n}x^2} dx = \sqrt{\frac{3}{\pi N}} \int_{[\frac{1}{2},r)} e^{-\frac{3}{N}x^2} dx + \frac{3\tilde{n}g(r)}{2N} + O\left(\frac{1}{N}\right),$$

where

$$g(r) = \frac{1}{\sqrt{2\pi}} \int_B x^2 e^{-\frac{1}{2}x^2} \, dx,$$

with $B = \left(\frac{1}{2}\sqrt{\frac{6}{N}}, r\sqrt{\frac{6}{N}} \right)$. Now

$$r\sqrt{\frac{6}{N}} = w\sqrt{\frac{3n}{N}} = w + \frac{3w\tilde{n}}{2N} + O\left(w\left(\frac{\tilde{n}}{N}\right)^2 \right),$$

and splitting the integral in one over $\left(\frac{1}{2}\sqrt{\frac{6}{N}}, w \right)$ and one over $\left(w, r\sqrt{\frac{6}{N}} \right)$ and expanding the latter around the point w, we obtain

$$g(r) = h(w, N) + O\left(\frac{\tilde{n}}{N}\right),$$

for a bounded function h. It follows that

$$\sqrt{\frac{3}{\pi N}} \int_{[\frac{1}{2}, r)} e^{-\frac{3}{N}x^2 + \frac{9}{N^2}\tilde{n}x^2} \, dx$$

$$= \sqrt{\frac{3}{\pi N}} \int_{[\frac{1}{2}, r)} e^{-\frac{3}{N}x^2} \, dx + \frac{3\tilde{n}\, h(w, N)}{2N} + O\left(\frac{1}{N} + \left(\frac{\tilde{n}}{N}\right)^2 \right).$$

Substituting this in (12.11) and comparing the distribution of n once more with $N\left(\frac{N}{3}, \frac{N}{18} \right)$, we see that the last two terms above contribute only $O(N^{-1})$. Finally the term containing $N^{-\frac{1}{2}} e^{-\frac{3}{N}r^2} \cdot O\left(\frac{r}{N}\right)$ is clearly $O(N^{-1})$. Hence we have reduced (12.11) to

$$P(0 < W_N \le w, T_N \text{ is even}) = \frac{1}{2} \sum_{1 \le n \le \frac{N}{2}} \frac{3}{\sqrt{\pi N}} e^{-\frac{9}{N}\tilde{n}^2} \sqrt{\frac{3}{\pi N}}$$

$$\times \left[\int_{[\frac{1}{2}, r)} e^{-\frac{3}{N}x^2} \, dx - e^{-\frac{3}{N}r^2 + \frac{9}{N^2}\tilde{n}r^2} \left(\text{frac}(r) - \frac{1}{2} \right) \right] + O\left(\frac{1}{N}\right). \tag{12.12}$$

Now

$$\frac{1}{\sqrt{N}} e^{-\frac{3}{N}r^2 + \frac{9}{N^2}\tilde{n}r^2} = \frac{1}{\sqrt{N}} e^{-\frac{1}{2}w^2 n\left(\frac{3}{N} - \frac{9}{N^2}\tilde{n} \right)} = \frac{1}{\sqrt{N}} e^{-\frac{1}{2}w^2\left(1 - \frac{9}{N^2}\tilde{n}^2 \right)}$$

$$= \frac{1}{\sqrt{N}} e^{-\frac{1}{2}w^2} \left(1 + O\left(\frac{9}{N^2}\tilde{n}^2\right) \right)$$

and the remainder term will give rise to another $O(N^{-1})$ term in (12.12). We obtain

$$P(0 < W_N \le w, T_N \text{ is even}) = O\left(\frac{1}{N}\right) +$$

$$\frac{1}{2}\sum_{1\le n\le\frac{N}{2}}\frac{3}{\sqrt{\pi N}}e^{-\frac{9}{N}\tilde{n}^2}\sqrt{\frac{3}{\pi N}}\Big[\int_{[\frac{1}{2},r)}e^{-\frac{3}{N}x^2}dx-e^{-\frac{1}{2}w^2}\Big\{\mathrm{frac}\Big(w\sqrt{\frac{n}{2}}\Big)-\frac{1}{2}\Big\}\Big].$$

As $r=w\sqrt{n/2}$,we have

$$\sqrt{\frac{3}{\pi N}}\int_{[\frac{1}{2},r)}e^{-\frac{3}{N}x^2}dx=\Phi\Big(w\sqrt{\frac{3n}{N}}\Big)-\Phi\Big(\sqrt{\frac{3}{2N}}\Big)$$

$$=\Phi(w)-\frac{1}{2}-\sqrt{\frac{3}{4\pi N}}+\frac{3\tilde{n}w\varphi(w)}{2N}+O\Big(\Big(\frac{\tilde{n}}{N}\Big)^2\Big),$$

where Φ and φ denote the standard normal distribution function and its density. Obviously, the linear term containing $\tilde{n}$ as well as the remainder term contribute $O(N^{-1})$ to (12.12). Since

$$\sum_{1\le n\le\frac{N}{2}}\frac{3}{\sqrt{\pi N}}e^{-\frac{9}{N}\tilde{n}^2}=1+O\Big(\frac{1}{N}\Big),$$

we find

$$P(0<W_N\le w,T_N\text{ is even})=\frac{1}{2}\Big(\Phi(w)-\frac{1}{2}-\sqrt{\frac{3}{4\pi N}}\Big)$$

$$-\sqrt{\frac{3}{2N}}\varphi(w)\sum_{1\le n\le\frac{N}{2}}\frac{3}{\sqrt{\pi N}}e^{-\frac{9}{N}\tilde{n}^2}\Big\{\mathrm{frac}\Big(w\sqrt{\frac{n}{2}}\Big)-\frac{1}{2}\Big\}+O\Big(\frac{1}{N}\Big).$$

$$(12.13)$$

An almost identical computation produces an asymptotic expression for the case where T_N is odd. We have

$$P(0<W_N\le w,T_N\text{ is odd})\qquad\qquad\qquad\qquad(12.14)$$

$$=\frac{1}{2}\Big(\Phi(w)-\frac{1}{2}\Big)$$

$$-\sqrt{\frac{3}{2N}}\varphi(w)\sum_{n=0}^{(N-1)/2}\frac{3}{\sqrt{\pi N}}e^{-\frac{9}{N}(n+\frac{1}{2}-\frac{N}{3})^2}\Big\{\mathrm{frac}\Big(\frac{1}{2}w\sqrt{2n+1}+\frac{1}{2}\Big)-\frac{1}{2}\Big\}$$

$$+O\Big(\frac{1}{N}\Big).$$

A few minor modifications of (12.13) and (12.14) are in order. In both formulas we may replace the sum by a sum over $0\le n\le N$ since the terms added are exponentially small in N. Next, in (12.14), we may change the factor $\Big(n+\frac{1}{2}-\frac{N}{3}\Big)^2$ in the exponent to $\tilde{n}^2=(n-\frac{N}{3})^2$ because the additional factor $e^{O\left((n-\frac{N}{3})/N\right)} = 1 + O\left(\tilde{n}/N\right)$ only produces an additional remainder term of order N^{-1}. Also in (12.14), we may replace $\operatorname{frac}\left(\frac{1}{2}w\sqrt{2n+1}+\frac{1}{2}\right)$ by $\operatorname{frac}\left(\frac{1}{2}w\sqrt{2n}+\frac{1}{2}\right) = \operatorname{frac}\left(w\sqrt{\frac{n}{2}}+\frac{1}{2}\right)$. This is because $(\frac{1}{2}w\sqrt{2n+1}+\frac{1}{2})-(\frac{1}{2}w\sqrt{2n}+\frac{1}{2})$ is of exact order $\frac{w}{\sqrt{n}}$ and hence $\frac{w}{\sqrt{N}}$, so $\operatorname{frac}\left(\frac{1}{2}w\sqrt{2n+1}+\frac{1}{2}\right) - \operatorname{frac}\left(\frac{1}{2}w\sqrt{2n}+\frac{1}{2}\right)$ is roughly equal to -1 for only one out of every $\frac{c\sqrt{N}}{w}$ values of n, i.e. for $O\left(w\sqrt{N}\right)$ values of n at distances of exact order $\sqrt{N}/w$. For all other n, this difference is $O(N^{-\frac{1}{2}})$ and this implies that the substitution only adds another $O(N^{-1})$ remainder term. Hence

$$P\left(0 < W_N \le w, T_N \text{ is odd}\right) = \frac{1}{2}\left(\Phi(w) - \frac{1}{2}\right) - \sqrt{\frac{3}{2N}}\varphi(w)$$

$$\times \sum_{0 \le n \le N} \frac{3}{\sqrt{\pi N}} e^{-\frac{9}{N}\tilde{n}^2}\left[\operatorname{frac}\left(w\sqrt{\frac{n}{2}}+\frac{1}{2}\right) - \frac{1}{2}\right] + O\left(\frac{1}{N}\right). \tag{12.15}$$

If $0 \le \operatorname{frac}\left(w\sqrt{\frac{n}{2}}\right) < \frac{1}{2}$, then $\operatorname{frac}\left(w\sqrt{\frac{n}{2}}\right) + \operatorname{frac}\left(w\sqrt{\frac{n}{2}}+\frac{1}{2}\right) = 2\operatorname{frac}\left(w\sqrt{\frac{n}{2}}\right) + \frac{1}{2} = \operatorname{frac}\left(2w\sqrt{\frac{n}{2}}\right) + \frac{1}{2}$. On the other hand, if $\frac{1}{2} \le \operatorname{frac}\left(w\sqrt{\frac{n}{2}}\right) < 1$, then $\operatorname{frac}\left(w\sqrt{\frac{n}{2}}\right) + \operatorname{frac}\left(w\sqrt{\frac{n}{2}}+\frac{1}{2}\right) = 2\operatorname{frac}\left(w\sqrt{\frac{n}{2}}\right) - \frac{1}{2} = \operatorname{frac}\left(2w\sqrt{\frac{n}{2}}\right) + \frac{1}{2}$. Hence $\operatorname{frac}\left(w\sqrt{\frac{n}{2}}\right) + \operatorname{frac}\left(w\sqrt{\frac{n}{2}}+\frac{1}{2}\right) = \operatorname{frac}\left(2w\sqrt{\frac{n}{2}}\right) + \frac{1}{2} = \operatorname{frac}\left(w\sqrt{2n}\right) + \frac{1}{2}$ in both cases. Hence we may combine (12.13) and (12.15) to obtain for $w > 0$,

$$P\left(0 < W_N \le w\right) = \Phi(w) - \frac{1}{2} - \frac{1}{2}\sqrt{\frac{3}{4\pi N}} - \sqrt{\frac{3}{2N}}\varphi(w)$$

$$\times \sum_{0 \le n \le N} \frac{3}{\sqrt{\pi N}} e^{-\frac{9}{N}\left(n-\frac{N}{3}\right)^2}\left[\operatorname{frac}\left(w\sqrt{2n}\right) - \frac{1}{2}\right] + O\left(\frac{1}{N}\right). \tag{12.16}$$

Because the distribution of W_N is symmetric about the origin and $P(W_N = 0) = \sqrt{\frac{3}{4\pi N}} + O(N^{-1})$, this determines the expansion for the distribution function F_N of W_N. The expansion is uniform in $w > 0$. Since it is identical to (12.6), this proves the third statement of the theorem.

It remains to prove that any closed interval of length $O(N^{-1})$ that does not contain the origin has probability $O(N^{-1})$ under Ψ_N and hence F_N. Clearly, this will imply the second statement of the theorem. Obviously, the only term in (2.8) that we need to consider is

$$R(w) = \frac{\Lambda_N(w)}{\sqrt{N}} \tag{12.17}$$

$$= -\sqrt{\frac{3}{2N}}\varphi(w) \sum_{0 \le n \le N} \frac{3}{\sqrt{\pi N}} e^{-\frac{9}{N}\left(n - \frac{N}{3}\right)^2} \left(\mathrm{frac}\left(w\sqrt{2n}\right) - \frac{1}{2}\right),$$

as the remainder of the expansion obviously has bounded derivative.

We begin by noting that if for a given $w > 0$, $w\sqrt{2n}$ is an integer for some $1 \le n \le N$, then $\mathrm{frac}\left(w\sqrt{2n}\right)$ and hence R has a jump discontinuity at this value of w. In the range where $|n - \frac{N}{3}| = x\sqrt{N}$ for $|x| \le y$, there can be a most wy such integer values of n. To see this, simply note that if $w\sqrt{2n} = k$ and $w\sqrt{2n'} = k + 1$, then $|n' - n| \ge \frac{2\sqrt{N}}{w}$, so there can be only $\frac{2y\sqrt{N}}{\frac{2\sqrt{N}}{w}} = wy$ values of n in the required interval. Such a value of n contributes an amount $O\left(N^{-1}\varphi(w)e^{-9x^2}\right)$ to the jump discontinuity at w, and hence $R(w) - R(w - 0) = O(N^{-1})$ at such a point w. Incidentally, this proves the second part of Theorem 12.1.

Choose $\epsilon > 0$ and consider two such jump points $w \ne w'$ in $[\epsilon, \infty)$ with $w\sqrt{2n} = k$ and $w'\sqrt{2n'} = k'$ for integers k, k', n and n' with $(n - \frac{N}{3}) = x\sqrt{N}$, $\left(n' - \frac{N}{3}\right) = x'\sqrt{N}$ and $|x| \vee |x'| \le y$. Suppose that $(w' - w) = O(N^{-1})$ and hence $\frac{w' - w}{w} = O(N^{-1})$ since $w \ge \epsilon$. For given w, n and k, we ask how many integer values of n' satisfy these conditions.

First we note that, for some positive c there are only at most $cw(y + 1)$ possible choices for k' since $\sqrt{2n} = \sqrt{2\frac{N}{3} + 2x\sqrt{N}} = \sqrt{\frac{2N}{3}} + \sqrt{\frac{3}{2}}x + O\left(\frac{y^2}{\sqrt{N}}\right)$, $\sqrt{2n'} = \sqrt{\frac{2}{3}N} + \sqrt{\frac{3}{2}}x' + O\left(\frac{y^2}{\sqrt{N}}\right)$ and hence $|k' - k| \le 2wy + O\left(w\frac{y^2}{\sqrt{N}} + |w' - w|\sqrt{N}\right) \le \left(\frac{c}{2}\right)w(y + 1)$. For each choice of k', the corresponding n' satisfies $n' = \frac{1}{2}\left(\frac{k'}{w'}\right)^2$ for some admissible w', and since $w, w' \ge \epsilon$ and $(w' - w) = O(N^{-1})$, this leaves a range of order $O\left(\left(\frac{k'}{w'}\right)^2 N^{-1}\right) = O(1)$ for n'. Hence, for some $C > 0$, there are at most $Cw(y + 1)$ possible values of n' for which there exists an integer k' with $(w' - w) = O(N^{-1})$. By the same argument as above, the total contribution of discontinuities to $|R(w') - R(w)|$ is $O(N^{-1})$ as long as $|w - w'| = O(N^{-1})$. As any closed interval of length $O(N^{-1})$ that does not contain the origin is bounded away from 0, this holds for the sum of the discontinuities in such an interval.

At all other points $w > 0$, R is differentiable and the derivative of $\text{frac}\big(w\sqrt{2n}\big)$ equals $\sqrt{2n}$. Hence the derivative of R is $O(1)$ and its differentiable part contributes at most $O(N^{-1})$ to the probability of any interval of length $O(N^{-1})$. This completes the proof of the Theorem 12.1.

12.3. Evaluation of the Oscillatory Term

Let W denotes a r.v. with non negative c.f. $\psi(t) \geq 0$ of support contained in $[-1, 1]$ and exponential decay of density of type $\exp\{-|x|^{2/3}\}$, $x \to \infty$, see e.g. Bhattacharya and Ranga Rao (1986). Introduce r.v. $w_N = w + N^{-1/2}(\log N)^{-1} W$, $w > 0$ and let $c > 0$ denote an positive absolute constant. Then we may bound the normal approximation error in (12.5) using similar arguments as in the proof of the well-known smoothing inequality, (see Lemma 12.1 of Bhattacharya and Rao), obtaining, for $w \geq 1$,

$$
\begin{aligned}
N^{-1/2}\big|\mathbf{E}\Lambda_N(w_N)\big| &\leq \big|\mathbf{E}\big(F_N(w_N) - \Phi(w_N)\big)\big| + cN^{-1} \\
&\leq \sup_{x \in [w-1/2,\, w+1/2]} \big|F_N(x) - \Phi(x)\big| + cN^{-1},
\end{aligned}
$$

where

$$
\Lambda_N(w) = -\varphi(w) \sum_{1 \leq n \leq N} \frac{3^{3/2}}{\sqrt{2\pi N}} \exp\{-\frac{9}{N}(n - \frac{N}{3})^2\}(\text{frac}(w\sqrt{2n}) - 1/2).
$$

We start with the following Fourier series expansion

$$
\tau(x) = \text{frac}(x) - 1/2 = -\sum_{k=1}^{\infty} 2\frac{\sin(2\pi k x)}{2k\pi},
$$

which holds for all nonintegral x.

Note that by the properties of W (i.e. the vanishing of Fourier coefficients)

$$
\mathbf{E}\tau(w_N\sqrt{2n}) = -\sum_{k=1}^{M_n} \mathbf{E}\frac{\sin(2\pi k \sqrt{2n}(w + N^{-1/2}(\log N)^{-1} W))}{k\pi},
$$

where $M_n = [\sqrt{N}\log N/(2\pi\sqrt{2n})] + 1$, i.e. $M_n = O(\log N)$ for $|n - N/3| < \sqrt{N}\log N$.

Rewriting $\Lambda_N(w)$ in the form

$$
\Lambda_N(w) = -\frac{3^{3/2}}{(2\pi N)^{1/2}}\varphi(w) \sum_{n=1}^{N} \exp\{-9(\tilde{n}^2/N)\}\tau\big(w(2n)^{1/2}\big),
$$

where $\tilde{n} = n - N/3$, we get

$$\mathbf{E}\Lambda_N(w_N) = \sqrt{\frac{3}{2}}\pi^{-1}\sum_{k=1}^{M}\frac{1}{k}\lambda_{N,k} + O(N^{-3}), \text{ where } M = [\log N] \text{ and}$$

$$\lambda_{N,k} = \frac{3}{\sqrt{\pi N}}\mathbf{E}\varphi(w_N)\sum_{n=1}^{N}\exp\{-9\tilde{n}^2/N\}\sin(2\pi k w_N \sqrt{2n}).$$

$$(12.18)$$

In the arguments of the sin function we use a Taylor expansion, for $|n - N/3| < \sqrt{N}\log N$,

$$\sqrt{n} = \sqrt{N/3} + \sqrt{3}\tilde{n}/(2\sqrt{N}) + O(\tilde{n}^2/N^{3/2}).$$

Thus, for $|\tilde{n}| < \sqrt{N}\log N$,

$$\sin(2\pi k w_N \sqrt{2n}) = \sin(d_0 + \pi d_1\tilde{n}) + O(k w_N N^{-3/2}\tilde{n}^2),$$

where $d_0 = 2\pi k w_N (\frac{2}{3})^{1/2}\sqrt{N}$, $d_1 = k w_N (\frac{3}{2})^{1/2}/\sqrt{N}$. Hence we may write

$$\lambda_{N,k} = \frac{3}{\sqrt{\pi N}}\mathbf{E}\varphi(w_N)\sum_{n\in\mathbf{Z}}\exp\{-9\tilde{n}^2/N\}\sin(d_0 + 2\pi d_1\tilde{n})$$

$$+ O(kN^{-1/2}\log N). \qquad (12.19)$$

We shall now evaluate the theta sum on the left hand side using Poisson's formula, see e.g. Mumford (1983, p. 189, Section (5.1)).

$$\sum_{m\in\mathbf{Z}}\exp\{-zm^2 + i2\pi mb\} = \pi^{1/2}z^{-1/2}\sum_{l\in\mathbf{Z}}\exp\{-\pi^2 z^{-1}(l-b)^2\},$$

$$(12.20)$$

where $b \in \mathbb{R}$, $\text{Re}\,z > 0$ and $z^{1/2}$ denotes the branch with positive real part. Writing $\sin(x) = (\exp[ix] - \exp[-ix])/2$ in (12.19) and assuming for simplicity $N/3 \in \mathbf{Z}$ we may replace summation over n by summation over $m = \tilde{n} = n - N/3 \in \mathbf{Z}$ in (12.19). Applying now (12.20) we have to bound the imaginary part of expectations of theta functions of type

$$I_k = \frac{3}{\sqrt{\pi N}}\mathbf{E}\varphi(w_N)\exp\{id_0\}\sum_{m\in\mathbf{Z}}\exp\{-9m^2 N^{-1} + i2\pi d_1 m\}. \quad (12.21)$$

We obtain for $k \leq M = [\log N]$ that $|d_1| \leq 2N^{-1/2}(\log N)|w_N| \leq 4N^{-1/2}(\log N)^2$ with probability $1 - O(N^{-3/2})$ by the assumption $w \leq \log N$. Hence the dominant term in (12.21) below is the term with $l = 0$

and we obtain with $c_{N,k} = \exp\{2\pi i\, k\, w_N\, (\frac{2}{3})^{1/2} \sqrt{N}\}$

$$
\begin{aligned}
I_k &= \mathbf{E} c_{N,k} \varphi(w_N) \sum_{l \in \mathbb{Z}} \exp\{-N\,(l - d_1)^2\, \pi^2/9\} \\
&= \mathbf{E} c_{N,k}\, \varphi(w_N) \exp\{-N\, d_1^2\, \pi^2/9\} + O\big(N^{-3/2}\big) \\
&= f_{N,k}\, \varphi(w) \exp\{-\pi^2 k^2\, w^2/6 + i\, 2\pi k\, w\, (\frac{2}{3})^{1/2} \sqrt{N}\} \\
&\quad + O\big(N^{-1/2} (\log N)^4\big),
\end{aligned}
\tag{12.22}
$$

where $f_{N,k} = \psi\big(2\pi\,(\frac{2}{3})^{1/2}\, \frac{k}{\log N}\big) = 1 + O\big((k/\log N)^2\big)$. Using this equation in (12.18) we get

$$
\begin{aligned}
\mathbf{E}\Lambda_N(w_N) &= \sqrt{\frac{3}{2}}\, \varphi(w) \, \mathrm{Im} \sum_{k=1}^{M} \frac{f_{N,k}}{k\pi} \exp\Big\{-\frac{\pi^2}{6} k^2\, w^2 + 2\pi i\, k\, w\, \sqrt{\frac{2N}{3}}\Big\} \\
&\quad + O\big(N^{-1/2} (\log N)^5\big).
\end{aligned}
$$

Hence, there exists a constant $c_0(w) > 0$ such that $|\mathbf{E}\Lambda_N(w_N)| > c_0(w) > 0$, provided that $4\,w\,\sqrt{\frac{2N}{3}}$ is an odd integer, which proves Theorem 12.2.

Acknowledgment

The authors would like to thank G.Chistyakov for a careful reading of the manuscript and J. Landes for his help with the tex version of this paper.

References

Albers, W., Bickel, P. J., and van Zwet, W. R. (1976). Asymptotic expansions for the power of distribution free tests in the one-sample problem. *Ann. Statist.* **4**, no. 1, 108–156.

Bentkus, V., Götze, F., and Tikhomirov, A. (1997). Berry-Esseen bounds for statistics of weakly dependent samples. *Bernoulli*, **3**, no. 3, 329–349.

Bentkus, V., Götze, F., and van Zwet, W. R. (1997). An Edgeworth expansion for symmetric statistics. *Ann. Statist.* **25**, no. 2, 851–896.

Berry, Andrew C. (1941). The accuracy of the Gaussian approximation to the sum of independent variates. *Trans. Amer. Math. Soc.* **49**, 122–136.

Bhattacharya, R. N. and Ranga Rao, R. *Normal approximation and asymptotic expansions*. Reprint of the 1976 original. Robert E. Krieger Publishing Co., Inc., Melbourne, FL, 1986.

Bickel, P. J., Götze, F. and van Zwet, W. R. (1986). The Edgeworth expansion for U-statistics of degree two. *Ann. Statist.* **14**, no. 4, 1463–1484.

Brown, L. D.; Cai, T. Tony; DasGupta, A. (2002). Confidence intervals for a binomial proportion and asymptotic expansions. *Ann. Statist.* **30**, no. 1, 160–201.

Feller, W. *An introduction to probability theory and its applications.* Vol. II. John Wiley & Sons, Inc., New York-London-Sydney 1966.

Mumford, D. Tata lectures on theta. I. With the assistance of C. Musili, M. Nori, E. Previato and M. Stillman. *Progress in Mathematics,* **28**. Birkhuser Boston, Inc., Boston, MA, 1983.

van Zwet, W. R. (1984). A Berry-Esseen bound for symmetric statistics. *Z. Wahrsch. Verw. Gebiete* **66**, no. 3, 425–440.

Part IV

Longitudinal Data Analysis

CHAPTER 13

An Overview on Nonparametric and Semiparametric Techniques for Longitudinal Data

Jianqing Fan and Runze Li

Department of Operation Research and Financial Engineering
Princeton University, Princeton, NJ 08544
jqfan@princeton.edu

Department of Statistics and the Methodology Center
Pennsylvania State University
University Park, PA 16802-2111
rli@stat.psu.edu

In the last two decades, there has been considerable literature on the topic of longitudinal data analysis. In particular, many authors have made much effort on developing diverse nonparametric and semiparametric models, along with their inference procedures, for longitudinal data. This chapter presents a review on recent development on this topic.

13.1. Introduction

Longitudinal data are often highly unbalanced because data were collected at irregular and possibly subject-specific time points. It is difficult to directly apply traditional multivariate regression techniques for analyzing such highly unbalanced collected data. This has led biostatisticians and statisticians to develop various modeling procedures for longitudinal data.

Parametric regression models have been extended to longitudinal data analysis (Diggle, Heagerty, Liang, and Zeger (2002)). They are very useful for analyzing longitudinal data and for providing a parsimonious description of the relationship between the response variable and its covariates. However, the parametric assumption likely introduce modeling biases. To relax the assumptions on parametric forms, various nonparametric models have been proposed for longitudinal data analysis. Earlier works on nonparametric regression analysis for longitudinal data were summarized in Müller (1988). Kernel regression was applied to repeated measurements

277

data with continuous responses in Hart and Wehrly (1986), and for data with time-series errors in Altman (1990) and Hart (1991). Time-varying coefficient models and varying-coefficient models for continuous responses were proposed for longitudinal data in Faraway (1997), Faraway (1999), Hoover, Rice, Wu, and Yang (1998), Wu, Chiang and Hoover (1998), Fan and Zhang (2000), Wu and Chiang (2000), Chiang, Rice and Wu (2001), and Huang, Wu and Zhou (2002). Nonparametric regression with a single covariate has also been extended for longitudinal data in the setting of generalized linear models in Lin and Carroll (2000) and Wang (2003). Qu and Li (2005) studied time-varying coefficient models under the generalized linear model framework using the quadratic inference function approach (Qu, Lindsay, and Li (2000)). More references will be given in Sections 2 and 4.

Although parametric models may be restrictive for some applications, they are more parsimoneous than nonparametric models may be too flexible to make concise conclusions. Semiparametric models are good compromises and retain nice features of both the parametric and nonparametric models. Thus, various semiparametric models have been extended for longitudinal data. Zeger and Diggle (1994) and Moyeed and Diggle (1994) extended partially linear models for longitudinal data. There are many papers on semiparametric modeling for longitudinal data published in the recent literature (Martinussen and Scheike (1999), Lin and Carroll (2001a), Lin and Carroll (2001b), Lin and Ying (2001), Fan and Li (2004), Wang, Carroll and Lin (2005)).

This chapter aims to present a selective overview of recent developments on the topic of nonparametric and semiparametric regression modeling for longitudinal data. A complete, detailed review on this topic is impossible due to the limited space. The rest of this chapter is organized as follows. Section 2 provides a review of nonparametric smoothing procedures for longitudinal data with a single covariate. In Section 3, we summarize recent developments on partially linear models for longitudinal data. A review on time-varying coefficient models and functional linear models is given in Section 4. An illustration is presented in Section 5. Some generalizations of models introduced in Sections 2, 3 and 4 are given in Section 6. We will briefly review the recent developments in estimation of covariance functions for the analysis of longitudinal data.

13.2. Nonparametric Model with a Single Covariate

Suppose that $\{(x_{ij}, y_{ij}), j = 1, \cdots, J_i\}$ is a random sample collected from the i-th subject or cluster, $i = 1, \cdots, n$. In this chapter, we assume that J_i is finite. Denote the conditional mean and variance as $\mu_{ij} = E(y_{ij}|x_{ij})$ and $\sigma_{ij}^2 = \mathrm{Var}(y_{ij}|x_{ij})$, respectively. Here it is assumed that the regression function $E(y_{ij}|x_{ij})$ is a nonparametric smooth function of x_{ij}. For longitudinal data and clustered data, it is known that samples collected within a subject are correlated and samples between subjects are often independent. It has been interesting to study how to incorporate within-subject correlation information into estimation of the mean function in the literature. Several statistical models have been proposed in existing works. For example, Ruckstuhl, Welsh and Carroll (2000) and Wu and Zhang (2002) studied the case in which the response variable is normally distributed. Severini and Staniswalis (1994), Lin and Carroll (2000) and Wang (2003) proposed estimation procedures for the marginal mean function under the framework of generalized linear models.

Let us begin with a summary of the work by Lin and Carroll (2000), in which it is assumed that $\sigma_{ij}^2 = \phi_j w_{ij} v(\mu_{ij})$, where ϕ_j is a scale parameter, w_{ij} is a known weight and $v(\cdot)$ is a variance function. Under the framework of generalized linear models, assume that the marginal mean μ_{ij} depends on x_{ij} through a known canonical link function $\mu(\cdot)$,

$$\mu_{ij} = \mu\{\theta(x_{ij})\},$$

where $\theta(\cdot)$ is an unknown smooth function. The local likelihood approach (Fan, Heckman, and Wand (1995)) is employed to estimate $\theta(\cdot)$. The use of canonical link guarantees that the associated optimization problem is either a convex minimization or concave maximization.

Here we focus on the local linear method for simplicity of notation. The idea is applicable for local polynomial methods (Lin and Carroll (2000)). To motivate the kernel generalized estimation equations (GEE) approach, let us temporarily assume that $\theta(x)$ is a linear function of x:

$$\theta(x) = \beta_0 + \beta_1 x \equiv \mathbf{g}(x)^T \boldsymbol{\beta},$$

where $\boldsymbol{\beta} = (\beta_0, \beta_1)^T$, and $\mathbf{g}(x) = (1, x)^T$. Let $\mathbf{y}_i = (y_{i1}, \cdots, y_{iJ_i})^T$ and $\boldsymbol{\mu}_i = E(\mathbf{y}_i) = [\mu\{\mathbf{g}(x_{i1})^T \boldsymbol{\beta}\}, \cdots, \mu\{\mathbf{g}(x_{iJ_i})^T \boldsymbol{\beta}\}]^T$. The conventional GEE (Liang and Zeger (1986)) approach estimates $\boldsymbol{\beta}$ by solving the following equations:

$$\sum_{i=1}^{n} \mathbf{G}_i^T \Delta_i \mathbf{V}_i^{-1} (\mathbf{y}_i - \boldsymbol{\mu}_i) = 0,$$

where $\mathbf{G}_i = (\mathbf{g}(x_{i1}), \cdots, \mathbf{g}(x_{iJ_i}))^T$, $\Delta_i = \mathrm{diag}\{\mu'\{\mathbf{g}^T(x_{ij})\boldsymbol{\beta}\}\}$, $\mu'(\cdot)$ is the first order derivative of $\mu(\cdot)$, $\mathbf{V}_i = \mathbf{S}_i^{1/2}\mathbf{R}_i(\boldsymbol{\delta})\mathbf{S}_i^{1/2}$, $\mathbf{S}_i = \mathrm{diag}[\phi_j w_{ij} v(\mu\{\mathbf{g}^T(x_{ij})\boldsymbol{\beta}\})]$ and $\mathbf{R}_i$ is an invertible working correlation matrix, which possibly depends on a parameter vector $\boldsymbol{\delta}$. The purpose of including the working correlation matrix $\mathbf{R}_i$ is to improve efficiency of $\boldsymbol{\beta}$. When the true correlation matrix is known, one would use it as the working correlation matrix. When there is no knowledge available on the within-subject correlation, the identity matrix is a convenient choice for working correlation matrix. As shown in Liang and Zeger (1986), if the mean function is correctly specified, the choice of $\mathbf{R}_i$ affects only the efficiency, but not the root n consistency of the resulting estimate $\widehat{\boldsymbol{\beta}}$. It has been shown that the resulting estimate is most efficient when $\mathbf{R}_i$ equals the true correlation matrix. The GEE strategy has been extended for a nonparametric smooth function $\theta(x)$ in Lin and Carroll (2000).

The local linear method is to locally, linearly approximate $\theta(z)$ at a neighborhood of x by

$$\theta(z) = \theta(x) + \theta'(x)(z - x) \equiv \boldsymbol{\beta}^T\mathbf{g}(z - x). \tag{13.1}$$

Here we slightly abuse the notation of $\boldsymbol{\beta}$, which should depend on the given point x. In particular, in this local modeling the first component β_1 of $\boldsymbol{\beta}$ represents $\theta(x)$. Let $K(x)$ be a symmetric kernel density function and h be a bandwidth. Denote $K_h(x) = h^{-1}K(x/h)$. As direct extension of the conventional GEE approach, Lin and Carroll (2000) suggested two ways incorporating kernel weight function in generalized estimation equation. The two ways lead to two sets of kernel GEE's for $\boldsymbol{\beta}$:

$$\sum_{i=1}^n \mathbf{G}_i^T(x)\Delta_i(x)\mathbf{V}_i^{-1}(x)\mathbf{K}_{ih}(x)\{\mathbf{y}_i - \boldsymbol{\mu}_i(x)\} = 0, \tag{13.2}$$

and

$$\sum_{i=1}^n \mathbf{G}_i^T(x)\Delta_i(x)\mathbf{K}_{ih}^{1/2}(x)\mathbf{V}_i^{-1}(x)\mathbf{K}_{ih}^{1/2}(x)\{\mathbf{y}_i - \boldsymbol{\mu}_i(x)\} = 0, \tag{13.3}$$

where $\mathbf{K}_{ih}(x) = \mathrm{diag}\{K_h(x_{ij} - x)\}$, and $\{\boldsymbol{\mu}_i(x), \Delta_i(x), \mathbf{V}_i(x), \mathbf{S}_i(x)\}$ are the same as those in the conventional GEE except that they are evaluated at $\mu_{ij} = \mu\{\mathbf{g}(x_{ij} - x)^T\boldsymbol{\beta}\}$. Solving equation (13.2) or (13.3) yields an estimate of $\widehat{\boldsymbol{\beta}}$. Having estimated $\boldsymbol{\beta}$ at x, let $\widehat{\theta}(x) = \widehat{\beta}_1$, the first component of $\boldsymbol{\beta}$. The standard error for $\widehat{\theta}(x)$ can be estimated by a sandwich formula, which is a conventional technique in GEE.

Severini and Staniswalis (1994) suggested letting $\mathbf{R}_i$ be an estimator of the actual correlation matrix.Lin and Carroll (2000) showed that it is generally the best strategy to ignore entirely the correlation structure within each subject/cluster and to instead pretend that all observations are independent. This implies that the behavior of kernel GEE is quite different from that of the parametric GEE. The intuition is that in a local neighborhood around x, there is unlikely to be more than one data point contributed from a subject so that the effective data points used in fitting (13.2) and (13.3) are nearly independent.

Lin and Carroll (2001b) extended the kernel GEE to generalized partially linear model for longitudinal data. Their works inspired other authors to study how to incorporate correlation information into nonparametric regression with longitudinal data. Peterson, Zhao and Eapen (2003) proposed a simple extension of the local polynomial regression smoother that retains the asymptotic properties of the working independence estimator, while typically reducing both the conditional bias and variance for practical sample sizes. Xiao, Linton, Carroll, and Mammen (2003) proposed a modification of local polynomial time series regression estimators that improves efficiency when innovation process is autocorrelated. Wu and Zhang (2002) considered local polynomial mixed effects (LLME) models for longitudinal data with continuous response. They proposed an estimation procedure for the LLME models using local polynomial regression. Their procedure incorporates correlation information by introducing a subjectwise random intercept function. They showed that the asymptotic bias and variance essentially the same as those of the kernel GEE. For finite sample performance, they empirically demonstrated that their procedure is more efficient than the kernel GEE approach.

Wang (2003) illustrated how the kernel GEE methods account within-subject correlation and found that the kernel GEE method uses kernel weights to control biases. Unfortunately, the kernel weight eliminates biases but also eliminates input from correlated elements in the same subject. It is challenging to control bias and to reduce the variation simultaneously. Wang (2003) proposed the marginal kernel method for accomplishing both tasks. The idea is closely related to the method of using control variable for variance reduction in the simulation literature. See, for example, Ross (1997). Suppose that we want to evaluate $\mu_f = Ef(X)$ and it is known from the side information that $E\mathbf{g}(X) = 0$ for a vector of functions $\mathbf{g}$. Then, for any given constant vector $\mathbf{a}$, $\mu_f = E\{f(X) - \mathbf{a}^T\mathbf{g}(X)\}$ can be estimated by its sample average from the simulation. The optimal choice of **a** is to minimize $\text{Var}\{f(X) - \mathbf{a}^T\mathbf{g}(X)\}$ with respect to **a**. This results in the optimal choice $\mathbf{a} = \text{Var}(\mathbf{g})^{-1}\text{cov}(\mathbf{g}, f)$.

The innovation of Wang (2003) is to create the side information via a preliminary estimate and to subtract it from the observed data so that the observed data have approximate mean zero, which plays the same role as $E\mathbf{g}(X) = 0$. Let $\breve{\theta}(x)$ be a consistent estimator of $\theta(x)$. For example, $\breve{\theta}(x)$ might be taken to be the resulting estimate of the kernel GEE with working independent correlation matrix. Let $\mathbf{G}_{*j}^i$ be an $J_i \times 2$ matrix with the first column being e_j and the second column being $e_j(x - x_{ij})/h$, where e_j denotes the indicator vector with jth entry equal to 1. Wang (2003) proposed solving the following kernel-weighted estimation equation with respect to β_0 and β_1:

$$0 = \frac{1}{n}\sum_{i=1}^n\sum_{j=1}^{J_i} K_h(x - x_{ij})\mu'\{\beta_0 + \beta_1(x - x_{ij})\}(\mathbf{G}_{*j}^i)^T\mathbf{V}_i^{-1}(x)(\mathbf{y}_i - \boldsymbol{\mu}_{*j}),$$

where the l-th element of $\boldsymbol{\mu}_{*j}$ is $\mu\{\beta_0 + \beta_1(x - x_{il})/h\}$, when $l = j$, and is $\mu\{\breve{\theta}(x_{ij})\}$, when $l \neq j$. The basic idea behind the equations is as follows: once a data point, (x_{ij}, y_{ij}) say, within a cluster has its x-value within a local neighborhood of x and is used to estimate $\theta(x)$, all data points in that cluster are used. To improve the efficiency, as in the simulation literature, the contributions of all points but point (x_{ij}, y_{ij}) within cluster to the local estimate of $\theta(x)$ are through their residuals $(\mathbf{y}_i - \boldsymbol{\mu}_{*j})$, namely the side information $E(\mathbf{y}_i - \boldsymbol{\mu}_{*j}) \approx 0$ is used. Denote the solution to be $(\widehat{\beta}_0, \widehat{\beta}_1)^T$. Then $\widehat{\theta}(x) = \widehat{\beta}_0$. As shown in Wang (2003), the smallest variance of the $\widehat{\theta}(x)$ is achieved when the true correlation is employed. Asymptotically, the smallest variance is uniformly smaller than that of the most efficient estimate of kernel GEE.

It is well known that for independent data, kernel regression and spline smoothing are asymptotically equivalent for nonparametric model with a single covariate (Silverman (1984)). Welsh, Lin, and Carroll (2002) shows this is not the case for longitudinal/clustered data. Splines and conventional kernels are different in localness and ability to account for within-cluster correlation. Lin, Wang, Welsh, and Carroll (2004) showed that a smoothing spline estimator is asymptotically equivalent to the marginal kernel method proposed in Wang (2003). They further showed that both the marginal kernel method and the smoothing spline estimator are nonlocal unless working independence is assumed but have asymptotically negligible bias.

Carroll, Hall, Apanasovivh and Lin (2004) proposed histospline method for nonparametric regression models for clustered longitudinal data. This provides a simple approach to deal with the difficulty which arises in situations in which aspects of the regression function are known parametrically, or the sampling scheme has significant structure. The histospline technique converts a problem in the continuum to one that is governed by only a finite number of parameters.

13.3. Partially Linear Models

A natural extension of the nonparametric model with a single covariate and the ordinary linear regression model with multiple covariates is the partially linear model, which has been well studied for independent data in the literature (Wahba (1984) Engle,Granger, Rice, and Weiss (1986), Heckman (1986), Speckman (1988), Härdle, Liang, and Gao (1999). This accommodates multiple covariates while retaining a nonparametric baseline function. Let y be a response variable and u and $\mathbf{x}$ be covariates. A partially linear model is defined as

$$y = \alpha(u) + \mathbf{x}^T \boldsymbol{\beta} + \varepsilon, \tag{13.4}$$

where $\alpha(\cdot)$ is a nonparametric smooth baseline function, $\boldsymbol{\beta}$ is an unknown regression coefficient vector and ε is a random error with $E(\varepsilon|u, \mathbf{x}) = 0$. For independent data, Heckman (1986) parameterized $\alpha(u)$ by splines and used the smoothing spline approach to estimate both α and $\boldsymbol{\beta}$. Note that

$$E(y|u) = \alpha(u) + E(\mathbf{x}|u)^T \boldsymbol{\beta}.$$

Then

$$y - E(y|u) = \{\mathbf{x} - E(\mathbf{x}|u)\}^T \boldsymbol{\beta} + \varepsilon. \tag{13.5}$$

Speckman (1988) suggested smoothing y and $\mathbf{x}$ over u and substituting the estimates of $E(y|u)$ and $E(\mathbf{x}|u)$ into (13.5). Thus, we can estimate $\boldsymbol{\beta}$ easily. This approach is referred to as a partial residual approach or more generally a profile least squares method. Plugging-in the estimate of $\boldsymbol{\beta}$ into (13.4), we can further estimate $\alpha(u)$ using one-dimensional smoothing techniques. We next present some existing estimation procedures for partially linear models with longitudinal data.

13.3.1. *Estimation procedures*

Suppose that we have a sample of n subjects. For the i-th subject, the response variable $y_i(t)$ and the covariate vector $\mathbf{x}_i(t)$, are collected at time points $t = t_{i1}, \cdots, t_{iJ_i}$, where J_i is the total number of observations on the i-th subject. A partially linear model for longitudinal data has the following form:

$$y_i(t_{ij}) = \alpha(t_{ij}) + \boldsymbol{\beta}^T \mathbf{x}_i(t_{ij}) + \varepsilon_i(t_{ij}) \tag{13.6}$$

for $i = 1, \cdots, n$, and $j = 1, \cdots, J_i$. Zeger and Diggle (1994) suggested using a backfitting algorithm to find an estimate for $\alpha(u)$ and $\boldsymbol{\beta}$. Specifically, starting with an initial value of $\boldsymbol{\beta}$, denoted by $\boldsymbol{\beta}^{(0)}$, we smooth the residual $y_i(t_{ij}) - \mathbf{x}_i^T(t_{ij})\boldsymbol{\beta}^{(0)}$ over t_{ij} to estimate $\alpha(\cdot)$. Having an estimate for $\alpha(\cdot)$, denoted by $\widehat{\alpha}(\cdot)$, we conduct linear regression of $y_i(t_{ij}) - \widehat{\alpha}(t_{ij})$ on $\mathbf{x}_i(t_{ij})$. Iterate this procedure until it converges. This is basically the same as the Gauss-Seidal algorithm to compute the profile least squares estimate. Moyeed and Diggle (1994) proposed an improved version of the backfitting algorithm based on a partial residual approach.

Lin and Ying (2001) introduced the counting process technique to the estimation scheme. The time points where the observations on the i-th subject are made are characterized by the counting process: $N_i(t) \equiv \sum_{j=1}^{J_i} I(t_{ij} \leq t)$, where $I(\cdot)$ is the indicator function. Both $y(t)$ and time-varying covariates $\mathbf{x}(t)$ were observed at the jump points of $N_i(t)$. The observation times are regarded as realizations from an arbitrary counting process that is censored at the end of follow-up. Specifically, $N_i(t) = N_i^*(t \wedge c_i)$, where $N_i^*(t)$ is a counting process in discrete or continuous time, c_i is the follow-up or censoring time, and $a \wedge b = \min(a, b)$. The censoring time c_i is allowed to depend on the vector of covariates $\mathbf{x}_i(\cdot)$ in an arbitrary manner. It is assumed that the censoring mechanism is noninformative in the sense that $E\{y_i(t)|\mathbf{x}_i(t), c_i \geq t\} = E\{y_i(t)|\mathbf{x}_i(t)\}$. Lin and Ying (2001) proposed minimizing the following least squares function with counting process notation,

$$\sum_{i=1}^{n} \int_0^{+\infty} w(t)\{y_i(t) - \alpha(t) - \boldsymbol{\beta}^T \mathbf{x}_i(t)\}^2 \, dN_i(t), \tag{13.7}$$

where $w(t)$ is a possibly data-dependent weight function.

Lin and Ying (2001) allow that the potential observation times to depend on the covariates and assume that

$$E\{dN_i^*(t)|\mathbf{x}_i(t), y_i(t), c_i \geq t\} = \exp\{\boldsymbol{\gamma}' \mathbf{x}_i(t)\} \, d\Lambda(t), \quad i = 1, \cdots, n, \tag{13.8}$$

where $\boldsymbol{\gamma}$ is a vector of unknown parameters and $\Lambda(\cdot)$ is an arbitrary nondecreasing function. Denote

$$\bar{\mathbf{x}}(t, \boldsymbol{\gamma}) = \frac{\sum_{i=1}^{n} \xi_i(t) \exp\{\boldsymbol{\gamma}^T \mathbf{x}_i(t)\}\mathbf{x}_i(t)}{\sum_{i=1}^{n} \xi_i(t) \exp\{\boldsymbol{\gamma}^T \mathbf{x}_i(t)\}},$$

and

$$\bar{y}(t, \boldsymbol{\gamma}) = \frac{\sum_{i=1}^n \xi_i(t) \exp\{\boldsymbol{\gamma}^T \mathbf{x}_i(t)\} y_i(t)}{\sum_{i=1}^n \xi_i(t) \exp\{\boldsymbol{\gamma}^T \mathbf{x}_i(t)\}},$$

where $\xi_i(t) = I(c_i \geq t)$. Replacing $dN_i(t)$ in (13.7) by its expectation (13.8), we obtain

$$\sum_{i=1}^n \int_0^\infty w(t)\{y_i(t) - \alpha(t) - \boldsymbol{\beta}^T \mathbf{x}_i(t)\}^2 \xi_i(t) \exp\{\boldsymbol{\gamma}^T \mathbf{x}_i(t)\} \, d\Lambda(t).$$

For each given $\boldsymbol{\beta}$ and $\boldsymbol{\gamma}$, minimizing the above criterion function with respect to function $\alpha(t)$ is equivalent to minimizing it at each given time t. This results in estimating the baseline function by

$$\widehat{\alpha}(t; \boldsymbol{\beta}, \boldsymbol{\gamma}) = \bar{y}(t, \boldsymbol{\gamma}) - \boldsymbol{\beta}^T \bar{\mathbf{x}}(t, \boldsymbol{\gamma}). \tag{13.9}$$

Substituting $\alpha(t)$ with $\widehat{\alpha}(t; \boldsymbol{\beta}, \boldsymbol{\gamma})$ in (13.7) yields

$$\ell(\boldsymbol{\beta}, \boldsymbol{\gamma}) = \sum_{i=1}^n \int_0^{+\infty} w(t)[\{y_i(t) - \bar{y}(t, \boldsymbol{\gamma})\} - \boldsymbol{\beta}^T \{\mathbf{x}_i(t) - \bar{\mathbf{x}}(t, \boldsymbol{\gamma})\}]^2 \, dN_i(t).$$

$$\tag{13.10}$$

The parameter $\boldsymbol{\gamma}$ can be consistently estimated by its moment estimator $\widehat{\boldsymbol{\gamma}}$, the solution to

$$\sum_{i=1}^n \int_0^{+\infty} \{\mathbf{x}_i(t) - \bar{\mathbf{x}}(t, \boldsymbol{\gamma})\} \, dN_i(t) = 0.$$

Substituting $\widehat{\boldsymbol{\gamma}}$ for $\boldsymbol{\gamma}$ in (13.10), an explicit form for $\widehat{\boldsymbol{\beta}}$ can be derived.

The weighted least squares problem (13.10) requires that the processes $y_i(t)$ and $\mathbf{x}_i(t)$ are fully observable until the censoring time c_i. This is an unrealistic assumption. Thus, Lin and Ying (2001) replaced the processes by their corresponding values at the nearest time where their values are observed. While this helps in practical implementations of the procedure, it introduces biases due to the nearest neighborhood approximation. See Figure 1 for the within subject approximations. Furthermore, since for each subject the spaces among observation times $\{t_{ij}, j = 1, \cdots, J_i\}$ do not tend to zero even when the sample size n tends to infinity, the approximation biases cannot always be negligible in practice.

To improve efficiency and avoid nonnegligible biases due to approximations, Fan and Li (2004) proposed two estimators: a difference-based estimator and a profile least squares estimator. Dropping the subscript j, the observed data

$$\{(t_{ij}, \mathbf{x}(t_{ij})^T, \mathbf{y}(t_{ij})), j = 1, \cdots, J_i, i = 1, \cdots, n\},$$

can be expressed in the vector notation as

$$\{(t_i, \mathbf{x}_i^T, \mathbf{y}_i), i = 1, \cdots, n^*\}, \quad \text{with} \quad n^* = \sum_{i=1}^{n} J_i,$$

ordered according to the time $\{t_{ij}\}$. By the marginal model (13.6), it follows that

$$y_i = \alpha(t_i) + \boldsymbol{\beta}^T \mathbf{x}_i + \varepsilon_i, \quad \text{with} \quad E(\varepsilon_i|\mathbf{x}_i) = 0. \qquad (13.11)$$

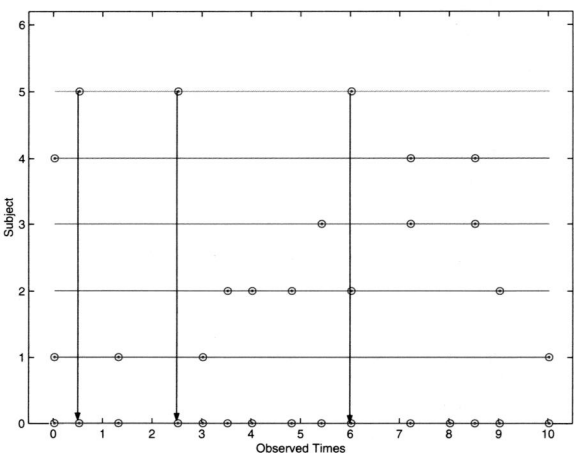

Figure 13.1.   Projection of observed time points

As illustrated in Figure 13.1, all observed times across all subjects may be dense and $t_{i+1} - t_i$ may be very small, although observed times for an individual subject may be very sparse. Observe that

$$y_{i+1} - y_i = \alpha(t_{i+1}) - \alpha(t_i) + \boldsymbol{\beta}^T(\mathbf{x}_{i+1} - \mathbf{x}_i) + e_i, \quad i = 1, \cdots, n^* - 1, \quad (13.12)$$

where stochastic error $e_i = \varepsilon_{i+1} - \varepsilon_i$. Under some mild conditions, the spacing between t_i and t_{i+1} is of order $O(1/n)$. Hence, the term $\alpha(t_{i+1}) - \alpha(t_i)$ in (13.12) is negligible. The least squares approach can be employed to estimate the parameter $\boldsymbol{\beta}$. The method can be further improved by fitting the following linear model

$$y_{i+1} - y_i = \alpha_0 + \alpha_1(t_{i+1} - t_i) + \boldsymbol{\beta}^T(\mathbf{x}_{i+1} - \mathbf{x}_i) + e_i, \quad i = 1, \cdots, n^* - 1. \quad (13.13)$$

Fitting model (13.13) yields an estimate of $\boldsymbol{\beta}$. For simplicity, we will call this method the Difference Based Estimator (DBE). From simulation comparisons in Fan and Li (2004), the DBE outperforms Lin and Ying's approach. This is mainly due to the fact that within-subject nearest neighborhood approximations are much rougher than those in the pooled samples since the former has much wider time gaps (See Figure 13.1). We next present the profile least squares approach.

For a given $\boldsymbol{\beta}$, let $y^*(t) \equiv y(t) - \boldsymbol{\beta}^T \mathbf{x}(t)$. Then partially linear model (13.6) can be written as

$$y^*(t) = \alpha(t) + \varepsilon(t). \tag{13.14}$$

This is a nonparametric regression problem. Thus, one can use a nonparametric regression technique to estimate $\alpha(t)$. We will focus only on the local linear regression technique (Fan and Gijbels, 1996). For t in a neighborhood of t_0, it follows by the Taylor expansion that

$$\alpha(t) \approx \alpha(t_0) + \alpha'(t_0)(t - t_0) \equiv a_0 + a_1(t - t_0).$$

Let $K(\cdot)$ be a kernel function and h be a bandwidth. The local linear fit is to find $(\widehat{a}, \widehat{b})$ minimizing

$$\sum_{i=1}^{n} \sum_{j=1}^{J_i} \{y_i^*(t_{ij}) - a_0 - a_1(t_{ij} - t_0)\}^2 w(t_{ij}) K_h(t_{ij} - t_0). \tag{13.15}$$

Here the weight function, $w(t_{ij})$, serves a similar purpose to that in (13.7). The local linear estimate is simply $\widehat{\alpha}(t_0; \boldsymbol{\beta}) = \widehat{a}_0$.

We may derive succinct expression of the profile least squares estimator using matrix notation. Denote by $\mathbf{y}_i = (y_i(t_{i1}), \cdots, y_i(t_{iJ_i}))^T$, $\mathbf{y} = (\mathbf{y}_1^T, \cdots, \mathbf{y}_n^T)$, $\mathbf{X} = (\mathbf{X}_1^T, \cdots, \mathbf{X}_n^T)^T$ $\mathbf{X}_i = (\mathbf{x}_i(t_{i1}), \cdots, \mathbf{x}_i(t_{iJ_i}))^T$, $\boldsymbol{\alpha}_i = (\alpha(t_{i1}), \cdots, \alpha(t_{iJ_i}))^T$, and $\boldsymbol{\alpha} = (\boldsymbol{\alpha}_1^T, \cdots, \boldsymbol{\alpha}_n^T)^T$. Then, model (13.11) can be written as

$$\mathbf{y} = \boldsymbol{\alpha} + \mathbf{X}\boldsymbol{\beta} + \boldsymbol{\varepsilon}, \tag{13.16}$$

where $\boldsymbol{\varepsilon}$ is the vector of stochastic errors. It is well known that the local linear fit is linear in $y_i^*(t_{ij})$ (Fan and Gijbels (1996)). Thus, the estimate of $\alpha(t)$ is linear in $\mathbf{y} - \mathbf{X}\boldsymbol{\beta}$. Hence, the estimate for the vector $\boldsymbol{\alpha}$ can be expressed as $\widehat{\boldsymbol{\alpha}} = \mathbf{S}(\mathbf{y} - \mathbf{X}\boldsymbol{\beta})$. The matrix $\mathbf{S}$ is usually called a smoothing matrix of the local linear smoother. It depends only on the observation times $\{t_{ij}, i = 1, \cdots, n, j = 1, \cdots, J_i\}$ and the amount of smoothing h. Substituting $\widehat{\boldsymbol{\alpha}}$ into (13.16), we obtain

$$(\mathbf{I} - \mathbf{S})\mathbf{y} = (\mathbf{I} - \mathbf{S})\mathbf{X}\boldsymbol{\beta} + \boldsymbol{\varepsilon}, \tag{13.17}$$

where $\mathbf{I}$ is the identity matrix of order $n^* = \sum_i n_i$. Applying weighted least squares to the linear model (13.17), we obtain

$$\widehat{\boldsymbol{\beta}} = \{\mathbf{X}^T(\mathbf{I} - \mathbf{S})^T\mathbf{W}(\mathbf{I} - \mathbf{S})\mathbf{X}\}^{-1}\mathbf{X}^T(\mathbf{I} - \mathbf{S})^T\mathbf{W}(\mathbf{I} - \mathbf{S})\mathbf{y}, \qquad (13.18)$$

where $\mathbf{W}$ is the weight matrix in the general least squares, which can incorporate the within subject correlation. Working independence is also allowed, in which $\mathbf{W}$ is a diagonal matrix. The estimator in (13.18) is called the profile least squares estimator. The profile least squares estimator for the nonparametric component is simply $\alpha(\cdot; \widehat{\boldsymbol{\beta}})$. Fan and Li (2004) derived an estimate for the covariance matrix using a sandwich formula and discussed the issue of bandwidth selection. They suggested the following procedure for selecting a bandwidth.

Using (13.18) and noting that $\widehat{\alpha} = \mathbf{S}(\mathbf{y} - \mathbf{X}\widehat{\boldsymbol{\beta}})$, $\widehat{\alpha}$ is linear in $\mathbf{y}$. Data driven methods, such as cross-validation (CV), or generalized cross-validation (GCV), can be used to select the bandwidth. However, this can computationally expensive. To avoid expensive computations and possibly unstable numerical implementations, our practical choice of bandwidth is as follows. Use the DBE to get an estimate $\widehat{\boldsymbol{\beta}}_{DBE}$. Substituting it into (13.14), we have a univariate nonparametric regression problem. Let $\widehat{h}$ be the bandwidth that is appropriate for this problem. This can be obtained either by a subjective choice via visualization, or by a data-driven procedure, such as substitution methods or cross-validation methods. Use this $\widehat{h}$ for the profile least squares estimate. From nonparametric theory, this optimal choice of bandwidth is of order $h_n = bn^{-1/5}$. Fan and Li (2004) established the asymptotic normality of the profile least squares estimators. The performance of $\widehat{\boldsymbol{\beta}}$ is not very sensitive to the choice of h, namely, for a wide range of choice of bandwidth, the performance of $\widehat{\boldsymbol{\beta}}$ remains approximately the same. Liang, Wang, Robins, and Carroll (2004) consider estimation for the partially linear model with missing covariates.

13.3.2. *Variable selection*

Like parametric regression models, variable selection is important in the semiparametric model (13.6). The number of variables in (13.6) can easily be large when nonlinear terms and interactions between covariates are introduced to reduce possible modeling biases. It is common in practice to include only important variables in the model to enhance predictability and to give a parsimonious description of the relationship between the response and the covariates. Fan and Li (2004) proposed a class of variable selection procedure via the nonconvex penalized quadratic loss

$$\frac{1}{2}\sum_{i=1}^{n}(\mathbf{y}_i - \boldsymbol{\alpha}_i - \mathbf{X}_i\boldsymbol{\beta})^T W_i(\mathbf{y}_i - \boldsymbol{\alpha}_i - \mathbf{X}_i\boldsymbol{\beta}) + n\sum_{j=1}^{d}\lambda_j p_j(|\beta_j|),$$

where the $p_j(\cdot)$'s are penalty functions, and the λ_j's are tuning parameters, which control the model complexity and can be selected by some data-driven methods, such as cross validation or generalized cross validation.

After eliminating the nonparametric function $\alpha(\cdot)$ using the profiling technique [see (13.17)], we obtain the following penalized least squares:

$$\frac{1}{2}(\mathbf{y} - \mathbf{X}\boldsymbol{\beta})^T(\mathbf{I} - \mathbf{S})^T\mathbf{W}(\mathbf{I} - \mathbf{S})(\mathbf{y} - \mathbf{X}\boldsymbol{\beta}) + n\sum_{j=1}^{d}\lambda_j p_j(|\beta_j|). \qquad (13.19)$$

By minimizing (13.19), with a special construction of the penalty function, some coefficients are estimated as zero, which deletes the corresponding variables, while others are not. Thus, the procedure selects variables and estimates coefficients simultaneously by minimizing (13.19). The resulting estimate is called a penalized least squares estimate.

The penalty functions $p_j(\cdot)$ and the regularization parameters λ_j are not necessarily the same for all j. This allows us to incorporate prior information for the unknown coefficients by using different penalty functions or taking different values of λ_j. For instance, we may wish to keep important predictors in linear regression models and hence do not want to penalize their coefficients; so we take their $\lambda_j's$ to be zero. For ease of presentation, we denote $\lambda_j p_j(\cdot)$ by $p_{\lambda_j}(\cdot)$.

Many penalty functions, such as the family of L_q-penalty ($q \geq 0$), have been used for penalized least squares and penalized likelihood in various parametric models. Fan and Li (2001) provided various insights into how a penalty function should be chosen. and suggested the use of the smoothly clipped absolute deviation (SCAD) penalty. Its first derivative is defined by

$$p'_\lambda(\beta) = \lambda\left\{I(\beta \leq \lambda) + \frac{(a\lambda - \beta)_+}{(a-1)\lambda}I(\beta > \lambda)\right\}, \text{ for some } a > 2 \text{ and } \beta > 0,$$

and $p_\lambda(0) = 0$. The SCAD penalty involves two unknown parameters, λ and a. Fan and Li (2001) suggested using $a = 3.7$ from a Bayesian point of view.

Figure 13.2 depicts the plots of the SCAD, $L_{0.5}$ and L_1 penalty functions. As shown in Figure 13.2, the three penalty functions all are singular at the origin. This is a necessary condition for sparsity in variable selection: the resulting estimator automatically sets some small coefficients to be zero

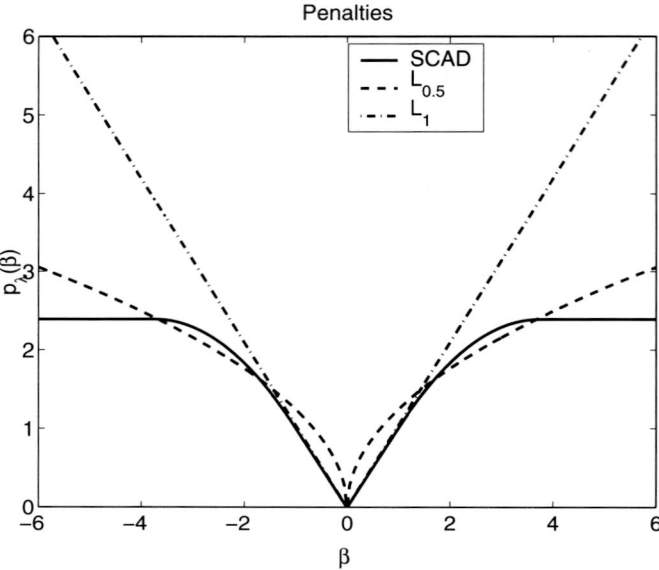

Figure 13.2.   Plot of Penalty Functions

(Antoniadis and Fan (2001)). Furthermore, the SCAD and $L_{0.5}$ penalties are nonconvex over $(0, +\infty)$ in order to reduce estimation bias. We refer to penalized least squares with the nonconvex penalties over $(0, \infty)$ as *non-convex penalized least-squares* in order to distinguish from the L_2 penalty, which yields a ridge regression. The SCAD is an improvement over the L_0-penalty in two aspects: saving computational cost and resulting in a continuous solution to avoid unnecessary modeling variation. Furthermore, the SCAD improves bridge regression (using L_p-penalty) by reducing modeling variation in model prediction. Although similar in spirit to the L_1-penalty, the SCAD may improve the L_1-penalty by avoiding excessive estimation bias because the solution of the L_1-penalty always shrinks regression coefficients by a constant, for instance, the soft thresholding rule (Donoho and Johnstone (1994) and Tibshirani (1996)). In contrast, the SCAD does not excessively over-penalize large coefficients.

Fan and Li (2001) suggested using local quadratic approximation for the nonconvex penalty functions, such as the SCAD penalty. With the aid of local quadratic approximation, Fan and Li (2004) proposed an iterative ridge regression algorithm for the penalized least squares (13.19). They further studied the sampling properties of the proposed variable selection procedures. They demonstrated that with a proper choice of regularization parameters and penalty functions, the proposed variable selection procedures perform as well as an oracle estimator.

13.3.3. *Extensions*

As an extension of partially linear models, Severini and Staniswalis (1994) considered generalized partially linear models. In the longitudinal data analysis, the following generalized partially linear models are usually considered

$$E\{y_i(t_{ij})|\mathbf{x}_i(t_{ij}\} = \mu\{\alpha(t_{ij}) + \mathbf{x}_i^T(t_{ij})\boldsymbol{\beta}\}. \tag{13.20}$$

Severini and Staniswalis (1994) proposed a profile quasi-likelihood estimation procedure for model (13.20). Lin and Carroll (2001a) and Lin and Carroll (2001b) developed estimation procedures for model (13.20) using profile-kernel estimation equation method. These authors have also studied the sampling properties of their proposed estimation procedures.

He, Zhu and Fung (2002) extended M-estimators for model (13.6). They approximate the baseline function $\alpha(t)$ by a regression spline. Thus, any M-estimation algorithm for the usual linear models can be used to obtain estimators of the model.

13.4. Varying-Coefficient Models

With the same notation in (13.6), the varying-coefficient model is defined as

$$y_i(t_{ij}) = \mathbf{x}_i^T(t_{ij})\boldsymbol{\beta}(t_{ij}) + \varepsilon_i(t_{ij}), \tag{13.21}$$

where we set the first elements of $\mathbf{x}_i(t_{ij})$ to be 1 to include an intercept function. Compared with model (13.6), the varying-coefficient model allows all regression coefficients to vary over time. Hence, it is also referred to as a time-varying coefficient model, which was introduced in Hastie and Tibshirani (1993). The main distinction between the functional data and longitudinal data setting is that the functional data in the current setting are observed at a much more frequently as if the whole functions were observed. Faraway (1997) considered model (13.21) with time-invariant covariate for functional data and proposed a smoothing splines procedure for the coefficients. Hoover, Rice, Wu, and Yang (1998) and Wu, Chiang and Hoover (1998) are among the first to introduce the varying-coefficient models for analysis of longitudinal data.

Hoover, Rice, Wu, and Yang (1998) proposed two estimation procedures for model (13.21). One is to globally approximate $\boldsymbol{\beta}(t)$ by splines and then estimate $\boldsymbol{\beta}(t)$ by smoothing spline. The other one is to locally approximate $\boldsymbol{\beta}(t)$ and to use local polynomial regression techniques to derive an estimator for $\boldsymbol{\beta}(t)$. Wu, Chiang and Hoover (1998) carefully studied the asymptotic properties of the local polynomial regression estimator for $\boldsymbol{\beta}(t)$ and derived asymptotic confidence region for $\boldsymbol{\beta}(t)$. Faraway (1999) proposed a graphical method of exploring the mean structure using model (13.21). Wu and Chiang (2000) proposed a cross-validation criterion for selecting data-driven bandwidth and a bootstrap procedure for constructing confidence intervals.

Fan and Zhang (2000) proposed a two-step estimation procedure for model (13.21) for functional data. For simplicity of description, assume that (13.21) is observed at the same time and dense points $\{t_j, j = 1, \cdots, J\}$. This can be achieved by binning the functional data with respect to observed times t_{ij} if necessary. They suggested that in the first step, we estimate $\boldsymbol{\beta}(t_j)$ by linear regression using n data points collected at time t_j. Having estimated $\widehat{\boldsymbol{\beta}}(t)$ over $\{t_1, \cdots, t_J\}$, in the second step, they use the local polynomial regression to smooth $(\{(t_j, \widehat{\boldsymbol{\beta}}(t_j)), j = 1, \cdots, J\}$ componentwise. This procedure can be easily implemented and allows different coefficients to have different degrees of smoothness. Chiang, Rice and Wu (2001) applied the two-step estimation procedure to model (13.21) with time-invariant covariate and use smoothing splines in the second step.

Statistical inference on model (13.21) is still an active research area. Huang, Wu and Zhou (2002) and Huang, Wu and Zhou (2004) proposed regression splines for varying coefficient models and studied the asymptotic properties of the resulting estimate. Eubank, Huang, Maldonado, Wang, Wang, and Buchanan (2004) proposed smoothing spline estimators for inference in model (13.21) and developed Bayesian confidence intervals for the regression coefficient functions.

We can extend model (13.21) in the fashion of generalized linear models. This yields a generalized varying-coefficient model

$$E\{y_i(t_{ij})|\mathbf{x}_i(t_{ij})\} = \mu\{\mathbf{x}_i^T(t_{ij})\boldsymbol{\beta}(t)\}. \tag{13.22}$$

Cai, Fan and Li (2000) proposed efficient statistical inference procedures for the generalized varying coefficient model. Fan, Yao and Cai (2003) proposed adaptive varying coefficient model by allowing the link function to be unknown. Kauermann (2000) used model (13.22) to fit ordinal response longitudinal data. Qu and Li (2005) proposed a quadratic inference function approach (Qu, Lindsay, and Li (2000)) to include within-subject correlation information into statistical inference on $\boldsymbol{\beta}(t)$.

Another extension of model (13.21) is the varying-coefficient mixed model

$$y_i(t_{ij}) = \mathbf{x}_i^T(t_{ij})\boldsymbol{\beta}(t_{ij}) + \mathbf{z}_i^T(t_{ij})\mathbf{b}_i(t_{ij}) + \varepsilon_i(t_{ij}), \qquad (13.23)$$

where $\mathbf{b}_i(t)$ is random effect and both $\mathbf{x}_i$ and $\mathbf{z}_i$ are covariate vectors. The model considered in Wu and Zhang (2002) and Rice and Wu (2003) is a specific case with $\mathbf{x}_i(t_{ij}) = 1$ and $\mathbf{z}_i(t_{ij}) = 1$. Wu and Liang (2004) proposed an estimation procedure for model (13.23) using local polynomial regression techniques. Li, Root, and Shiffman (2005) applied model (13.23) for analysis of intensively correlated longitudinal data from a study of the subjective sensation of nicotine withdrawal. Zhang (2004) included a constant random effects in model (13.22) and obtained a generalized linear mixed effects model with time-varying coefficients. He further proposed an estimation procedure for his model using smoothing splines.

13.5. An Illustration

We now illustrate the proposed procedures in Sections 2 and 3 via an analysis of a subset of data from the Multi-Center AIDS Cohort study. The data set contains the HIV status of 283 homosexual men who became infected with HIV during the follow-up period between 1984 and 1991. Details about the design, methods and medical implications of the study can be found in Kaslow, Ostrow, Detels, Phair, Polk and Rinaldo (1987). During this study, all participants were scheduled to have their measurements taken during semi-annual visits. But, because many participants missed some of their scheduled visits, and the HIV infections happened randomly during the study, there are unequal numbers of repeated measurements and different measurement times per individual. Fan and Zhang (2000), and Huang, Wu and Zhou (2002) analyzed the same data set by using varying-coefficient models.

Take x_1 to be the smoking status: 1 for a smoker and 0 for a non-smoker, $x_2(t)$ to be the standardized variable for age, and x_3 to be the standardized variable for PreCD4, the baseline CD4 percentage before HIV infection. It is of interest to examine whether there are any interaction effects and quadratic effects from these covariates. Based on the analysis of Huang, Wu and Zhou (2002), Fan and Li (2004) considered the following semiparametric model:

$$y(t) = \alpha(t) + \beta_1 x_1 + \beta_2 x_2(t) + \beta_3 x_3 + \beta_4 x_2^2(t) + \beta_5 x_3^2$$
$$+ \beta_6 x_1 x_2(t) + \beta_7 x_1 x_3 + \beta_8 x_2(t) x_3 + \varepsilon(t). \qquad (13.24)$$

Table 13.1. Estimated Coefficients for Model (13.24), adapted from Fan and Li (2004)

Variable	Profile LS $\widehat{\beta}(\mathrm{se}(\widehat{\beta}))$	SCAD $\widehat{\beta}(\mathrm{se}(\widehat{\beta}))$
Smoking	0.5333(1.0972)	0(0)
Age	-0.1010(0.9167)	0(0)
PreCD4	2.8252(0.8244)	3.1993(0.5699)
Age^2	0.1171(0.4558)	0(0)
PreCD4^2	-0.0333(0.3269))	0(0)
Smoking*Age	-1.7084(1.1192)	-1.0581(0.5221)
Smoking*PreCD4	1.3277(1.3125)	0(0)
Age*PreCD4	-0.1360(0.5413)	0(0)

The DBE estimate for $\boldsymbol{\beta}$ was computed to obtain the partial residuals for $\alpha(\cdot)$, and then the bandwidth $h = 0.5912$ was selected by the plug-in method proposed in Ruppert, Sheather and Wand (1995). After that, the profile least squares method with weight $w(t) \equiv 1$ was applied to this model. The resulting estimates and standard errors are depicted in Table 13.1. Figure 13.3 depicts the estimated baseline function $\alpha(t)$ along with its 95% pointwise confidence interval without taking account into the bias of the nonparametric fit. A decreasing trend can easily be seen, as the CD4 percentage depletes over time.

Fan and Li (2004) further applied the penalized profile least squares approach to select significant variables. The generalized cross validation is used to select the tuning parameter and the selected $\lambda = 0.7213$ for the SCAD penalty. The results are also shown in Table 13.1. From Table 13.1, the result is in the line with that of Huang, Wu and Zhou (2002), but indicates possible interactions between Smoking status and Age.

13.6. Generalizations

A natural extension of varying-coefficient models and partially linear models is the semiparametric varying-coefficient model:

$$y_i(t_{ij}) = \mathbf{x}_i(t_{ij})^T \boldsymbol{\alpha}(t_{ij}) + \mathbf{z}_i(t_{ij})^T \boldsymbol{\beta} + \varepsilon_i(t_{ij}), \qquad (13.25)$$

where both $\mathbf{x}_i(t)$ and $\mathbf{z}_i(t)$ are covariate vectors, $\boldsymbol{\alpha}(t)$ consists of p unknown smooth functions, $\boldsymbol{\beta}$ is a q-dimensional unknown parameter vector, and

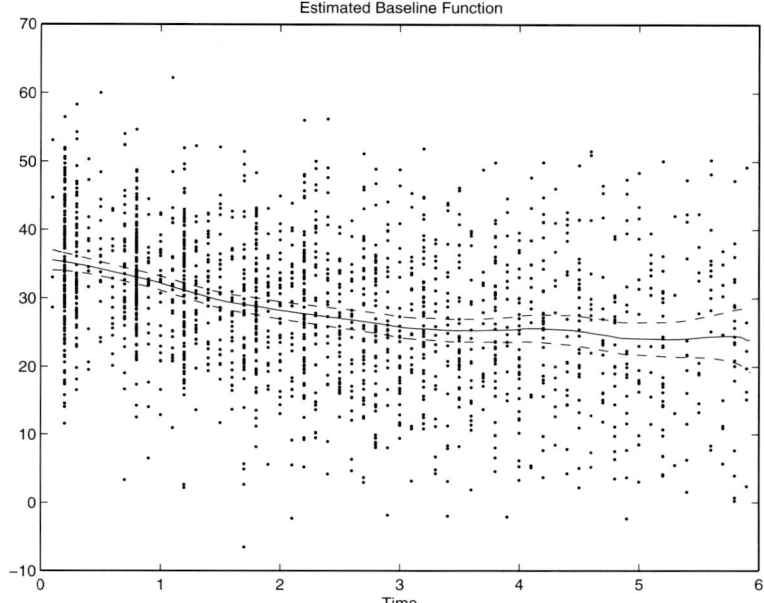

Figure 13.3.   Estimated Baseline Function. The solid line stands for the estimated baseline function, the dash lines are the estimated baseline function plus/minus twice standard errors. The dots are the residual on parametric part $r(t) = y(t) - \widehat{\boldsymbol{\beta}}^T \mathbf{x}(t)$. Taken from Fan and Li (2004).

$E\{\varepsilon_i(t_{ij})|\mathbf{x}_i(t_{ij}), \mathbf{z}_i(t_{ij})\} = 0$. Martinussen and Scheike (1999) proposed an estimation procedure for model (13.25) using the notion of a counting process. Sun and Wu (2005) extended the estimation procedure of Lin and Ying for partially linear models to model (13.25). Fan, Huang and Li (2005) extended the profile least squares approach for model (13.25), and further proposed semiparametric modeling strategy for the covariance function of random error process $\varepsilon(t)$.

A generalization of model (13.25) is the semiparametric varying-coefficient mixed effects model:

$$y_i(t_{ij}) = \mathbf{x}_{i1}^T(t_{ij})\boldsymbol{\alpha}(t) + \mathbf{z}_{i1}^T(t_{ij})\boldsymbol{\beta} + \mathbf{x}_{i2}^T(t_{ij})\mathbf{a}_i^T(t_{ij}) + \mathbf{b}_i^T \mathbf{z}_{i2}(t_{ij}) + \varepsilon_i(t_{ij}),$$
$$(13.26)$$

where $\boldsymbol{\alpha}(t)$ consists of time-dependent fixed effects, $\boldsymbol{\beta}$ is time-independent fixed effects, $\mathbf{a}_i(t)$ is time-dependent random effects, and $\mathbf{b}_i$ is time-independent random effects. Fung, Zhu, Wei, and He (2002) considered partially linear mixed effects model which coincides with model (13.26) with $\mathbf{x}_{i1}(t_{ij}) = 1$ and $\mathbf{x}_{i2}(t_{ij}) = 1$.

In situations where the response variable $y(t)$ is discrete (e.g., binary, categorical, count), models (13.25) and (13.26) may not be appropriate. For such cases, one may consider the following further generalization

$$E\{y_i(t_{ij})|\mathbf{x}_{i1}(t_{ij}), \mathbf{x}_{i2}(t_{ij}), \mathbf{z}_{i1}(t_{ij}), , \mathbf{z}_{i2}(t_{ij})\}$$
$$= \mu\{\mathbf{x}_{i1}^T(t_{ij})\boldsymbol{\alpha}(t) + \mathbf{z}_{i1}^T(t_{ij})\boldsymbol{\beta} + \mathbf{x}_{i2}^T(t_{ij})\boldsymbol{\alpha}_i^T(t_{ij}) + \boldsymbol{\beta}_i^T\mathbf{z}_{i2}(t_{ij})\}, \qquad (13.27)$$

where $\mu(\cdot)$ is a known link function. It is of interest to develop statistical inference procedures for models (13.26) and (13.27) and their associated theory. Further research is needed in this area.

Other extensions of aforementioned models are semiparametric additive mixed models (Zhang, Lin, Raz, and Sowers (1998), Zhang and Lin (2003), generalized additive models for longitudinal data (Berhane and Tibshirani (1998)) and generalized additive mixed models (Lin and Zhang (1999), Zhang and Davidian (2004)).

13.7. Estimation of Covariance Matrix

Estimation of covariance functions is an important issue in the analysis of longitudinal data. It features prominently in forecasting the trajectory of an individual response over time and is closely related with improving the efficiency of estimated regression coefficients. Challenges arise in estimating covariance functions due to the fact that longitudinal data are frequently collected at irregular and possibly subject-specific time points. Interest in issue has surged in the recent literature.

For parametric regression model in the analysis of longitudinal data, various efforts have been made to improve efficiency for estimating the regression coefficients. For instance, Liang and Zeger (1986) discussed how to incorporate correlation structure into the GEE framework, and showed that the resulting estimate is the most efficient when the working correlation equals the inverse of the actual correlation. Much attention has been paid to nonparametric regression analysis for longitudinal data in the recent literature. Wang (2003) proposed a marginal kernel GEE and showed that when the working correlation matrix equals the inverse of the true one, the resulting estimate is the most efficient. The marginal kernel GEE approach is extended to the generalized partial linear model in the seminal paper by Welsh, Lin, and Carroll (2002). All of these works indicate that the estimation of covariance function plays an important role in the analysis of longitudinal data.

Some estimation procedures for large covariance matrices have been proposed in the literature. Daniels and Kass (2001) proposed a Bayesian approach which places priors on a covariance matrix so as to shrink it toward some structures. Using an unconstrained and statistically meaningful reparametrization of the covariance matrix, Daniels and Pourahmadi (2002) further introduce more flexible priors with many parameters to control shrinkage. Pourahmadi (1999) and Pourahmadi (2000) proposed a flexible, data based parametric approach to formulating models for covariance matrices. Wu and Pourahmadi (2003) further proposed nonparametric estimation of large covariance matrices for balanced or nearly balanced longitudinal data, based on Cholesky decomposition and using two-step estimation procedure (Fan and Zhang (2000)). The key idea of the series work by Pourahmadi and his coauthors is that the covariance matrix, denoted by Σ, of a zero mean random vector $\mathbf{z} = (z_1, \cdots, z_m)^T$ can be diagonalized by a lower triangular matrix constructed from the regression coefficients when z_t is regressed on its predecessors $z_1, \cdots, z_t$. Specifically, for $t = 2, \cdots, m$,

$$y_t = \sum_{j=1}^{t-1} \phi_{t,t-j} y_{t-j} + \varepsilon_t, \quad L\Sigma L^T = D, \qquad (13.28)$$

where L and D are unique, L is a unit lower triangular matrix having ones on its diagonal and $-\phi_{ij}$ at its (i,j)th element for $j \leq t$, and D is diagonal with $\sigma_t^2 = \text{Var}(\varepsilon_t)$ as its diagonal entries. The Cholesky decomposition (13.28) converts the constraint entries of Σ into two groups of unconstrainted regression and variance parameters given by $\{\phi_{tj}, t = 2, \cdots, m; j = 1, \cdots, t-1\}$ and $\{\log \sigma_1^2, \cdots, \log \sigma_m^2\}$, respectively. Let

$$\beta_{j,m}(t/m) = \phi_{t,t-j}, \quad \sigma_m(t/m) = \sigma_t.$$

Then the decomposition (13.28) yields

$$y_t = \sum_{j=1}^{t-1} \beta_{j,m}(t/m) y_{t-j} + \sigma_m(t/m)\varepsilon_t, \quad (t = 0, 1, ...), \qquad (13.29)$$

which can be viewed as a time-varying coefficient model in which $\beta_{j,m}(\cdot)$ and $\sigma_m(\cdot)$ are assumed to be smooth function. This allows us to smooth along the subdiagonals of L. Wu and Pourahmadi (2003) directly applied two-step estimation for time-varying coefficient model (Fan and Zhang (2000), also see Section 4 for a brief introduction) to (13.29). Using the Cholesky decomposition, Huang, Liu and Pourahmadi (2005) further introduced a penalized likelihood method for estimating a large covariance matrix. Yao, Müller, and Wang (2005a) and Yao, Mülluer, and Wang (2005b) proposed other new estimation procedures for the covariance function of functional data.

Diggle and Verbyla (1998) proposed using local linear regression techniques to estimate covariance structure for longitudinal data. Their approach is to estimate variance function and variogram by smoothing the squared residual and the variogram cloud of the squares residuals. Then using the relationship between variogram and covariance function, one may derive an estimation for covariance function. However, the resulting estimate for covariance function may not be positive definite. Fan, Huang and Li (2005) gave a method for parsimonious modeling of the covariance function of random error process $\varepsilon(t)$ for the analysis of longitudinal data when they are collected at irregular and possibly subject-specific time points. They approach this by assuming that $\text{Var}\{\varepsilon(t)|\mathbf{x}(t), \mathbf{z}(t)\} = \sigma^2(t)$, which is a nonparametric smoothing function, but the correlation function between $\varepsilon(s)$ and $\varepsilon(t)$ has a parametric form $\text{corr}\{\varepsilon(s), \varepsilon(t)\} = \rho(t, s, \boldsymbol{\theta})$, where $\rho(s, t, \boldsymbol{\theta})$ is a positive definite function of s and t, and $\boldsymbol{\theta}$ is an unknown parameter vector. Specification of the correlation function may be motivated from the random error structure in hierarchical linear models and mixed effects models, or by specifying a working correlation function as in GEE (Liang and Zeger (1986)). For instance, an AR(1)-type correlation structure yields $\rho(s, t, \theta) = \exp(-\theta|s - t|)$ with $\theta > 0$, and a compound symmetric correlation structure results in $\rho(s, t, \theta) = \theta$ with $0 < \theta < 1$, which is the same as the correlation structure of random effects model in the presence of only random intercept. More complicated correlation structures can be introduced by using ARMA models or adopting a hierarchical linear model including various levels of random effects.

In Fan, Huang and Li (2005), the covariance function is fitted by a semiparametric model. The semiparametric model allows random error process $\varepsilon(t)$ to be nonstationary as its variance function $\sigma^2(t)$ may be time-dependent. The semiparametric model to the covariance function guarantees the positive definite property of the resulting estimate and retains the flexibility of nonparametric modeling and the parsimony of parametric modeling. Furthermore, this semiparametric model allows one to incorporate easily prior information about the correlation structure. It can be used to improve the efficiency of regression coefficient $\boldsymbol{\beta}$ even if the correlation matrix is misspecified. For example, let $\rho_0(s, t)$ be a working correlation function (e.g. working independence) and $\rho(s, t, \boldsymbol{\theta})$ be a family of correlation functions that contains ρ_0, the semiparametric model allows one to choose an appropriate $\boldsymbol{\theta}$ to improve the efficiency of the estimator of $\boldsymbol{\beta}$. Obviously, to improve the efficiency, the family of correlation functions $\{\rho(s, t, \boldsymbol{\theta})\}$ does not need to contain the true correlation structure even if the correlation matrix is misspecified. Fan, Huang and Li (2005) suggested estimating the variance function by using a kernel smoothing over the squares of residuals. They further proposed two estimation procedures for $\boldsymbol{\theta}$. One is motivated by maximizing the likelihood of the data, and the other is motivated by minimizing the volume of confidence ellipsoid of regression coefficients.

Acknowledgements

Fan's research was partially supported by DMS-0354223 and NIH grants R01-HL69720, R01-GM072611. Li's research was supported by a National Institute on Drug Abuse (NIDA) grant P50 DA10075 and a NSF grant DMS-0348869.

References

Altman, N. S. (1990). Kernel smoothing of data with correlated errors. *Journal of American Statistical Association*, **85**, 749-759.

Antoniadis, A. and Fan, J. (2001). Regularization of wavelets approximations (with discussions). *Journal of American Statistical Association*, **96**, 939-967.

Berhane, K. and Tibshirani, R. J. (1998). Generalized additive models for longitudinal data. *Canadian Journal of Statistics*, **26**, 517-535.

Cai, Z, Fan, J. and Li, R. (2000). Efficient estimation and inferences for varying-coefficient models. *Journal of American Statistical Association*, **95**, 888-902.

Carroll, R. J., Hall, P., Apanasovivh, T. V. and Lin, X. (2004). Histospline method in nonparametric regression models with application to clustered longitudinal data. *Statistica Sinica*, **14**, 649-674.

Chiang, C.-T., Rice, J. A. and Wu, C. O. (2001). Smoothing spline estimation for varying coefficient models with repeatedly measured dependent variables. *Journal of American Statistical Association*, **96**, 605-619.

Daniels, M. J. and Kass, R. E. (2001). Shrinkage estimators for covariance matrices. *Biometrics*, **57**, 1173-1184.

Daniels, M. J. and Pourahmadi, M. (2002). Bayesian analysis of covariance matrices and dynamic models for longitudinal data. *Biometrika*, **89**, 553-566.

Diggle, P. J., Heagerty, P., Liang, K. Y., Zeger, S. L. (2002). *Analysis of Longitudinal Data*, 2nd Edition, Oxford, U. K. Oxford University Press.

Diggle, P. J. and Verbyla, A. P. (1998). Nonparametric estimation of covariance structure of longitudinal data. *Biometrics*, **54**, 401-415.

Donoho, D. L. and Johnstone, I. M. (1994). Ideal spatial adaptation by wavelets shrinkage. *Biometrika* **81**, 425-455.

Engle, R. F., Granger,C. W. J., Rice, J. A. and Weiss, A. (1986). Semiparametric estimates of the relation between weather and electricity sales. *Journal of American Statistical Association*, **81**, 310-320.

Eubank, R. L., Huang, C. F., Maldonado, Y. M., Wang, N., Wang, S. and Buchanan, R. J. (2004). Smoothing spline estimation in varying-coefficient models *Journal of Royal Statistical Society, Series B*, **66**, 653-667.

Fan, J. and Gijbels, I. (1996). *Local Polynomial Modelling and Its Applications*, Chapman and Hall, London.

Fan, J., Heckman, N. E. and Wand, M. P. (1995). Local polynomial kernel regression for generalized linear models and quasi-likelihood functions. *Journal of American Statistical Association*, **90**, 141–150.

Fan, J., Huang, T. and Li, R. (2005). Analysis of longitudinal data with semi-parametric estimation of covariance function. Manuscript.

Fan, J. and Li, R. (2001). Variable selection via nonconcave penalized likelihood and its oracle properties. *Journal of American Statistical Association*, **96**, 1348-1360.

Fan, J and Li, R. (2004). New estimation and model selection procedures for semi-parametric modeling in longitudinal data analysis, *Journal of American Statistical Association*, **99**, 710-72.

Fan, J. and Zhang, J. (2000). Two-step estimation of functional linear models with applications to longitudinal data. *Journal of Royal Statistical Society, Series B*, **62**, 303-322.

Fan, J., Yao, Q., and Cai, Z. (2003). Adaptive varying-coefficient linear models, *Journal of Royal Statistical Society, Series B*, **65**, 57-80.

Faraway, J. J. (1997). Regression analysis for a functional response. *Technometrics*, **39**, 254-261.

Faraway, J.J. (1999). A graphical method of exploring the mean structure in longitudinal data analysis. *Journal of Computational and Graphical Statistics*, **8**, 60-68.

Fung, W. K., Zhu, Z. Y., Wei, B. C. and He, X. (2002). Influence diagnostics and outlier tests for semiparametric mixed models. *Journal of Royal Statistical Society, Series B*, **64** 565-579.

Härdle, W. Liang, H. and Gao, J. (1999). *Partially Linear Models*, Springer-Verlag, New York.

Hart, J.D. (1991). Kernel regression estiamtion with time-series errors. *Journal of Royal Statistical Society, Series B*, **53**, 173-187.

Hart, J. D. and Wehrly, T. E. (1986). Kernel regression estimation using repeated measurements data. *Journal of American Statistical Association*, **81**, 1080-1088.

Hastie, T. and Tibshirani, R. J. (1993). Varying-coefficient models (with discussion). *Journal of Royal Statistical Society, Series B*, **55**, 757-796.

He, X., Zhu, Z. Y., and Fung, W. K. (2002). Estimation in a semiparametric model for longitudinal data with unspecified dependence structure. *Biometrika*, **89**, 579-590.

Heckman, N. E. (1986). Spline smoothing in partly linear models, *Journal of Royal Statistical Society, Series B*, **48**, 244-248.

Hoover, D. R., Rice, J. A., Wu, C. O. and Yang, L. P. (1998). Nonparametric smoothing estimates of time-varying coefficient models with longitudinal data. *Biometrika*, **85**, 809–822.

Huang, J. Z., Liu, N. and Pourahmadi, M. (2005). Covariance selection and estimation via penalized normal likelihood. *Biometrika*. To appear.

Huang, J. Z., Wu, C. O. and Zhou, L. (2002). Varying-coefficient models and basis function approximations for the analysis of repeated measurements. *Biometrika*, **89**, 111-128.

Huang, J. Z, Wu, C.O. and Zhou, L. (2004). Polynomial spline estimation and inference for varying coefficient models with longitudinal data. *Statistica Sinica*, **14**, 763-788.

Kaslow, R.A., Ostrow, D.G., Detels, R. Phair, J.P., Polk, B.F. and Rinaldo, C.R. (1987). The Multicenter AIDS Cohort Study: rationale, organization and selected characteristics of the participants. *Am. J. Epidem.*, **126**, 310-318.

Kauermann G. (2000). Modeling longitudinal data with ordinal response by varying coefficients, *Biometrics*, **56**, 692-698.

Li, R., Root, T. and Shiffman, S. (2005). A local linear estimation procedure for functional multilevel modeling. In *Models for Intensively Longitudinal Data*, (T. Walls and J. Schafer eds), 63-83. Oxford University Press.

Liang, H., Wang, S.J., Robins, J.M. and Carroll, R.J. (2004). Estimation in partially linear models with missing covariates *Journal of American Statistical Association*, **99**, 357-367.

Liang, K. Y. and Zeger, S. L. (1986). Longitudinal data analysis using generalized linear models. *Biometrika*, **73**, 13-22

Lin, D.Y. and Ying, Z. (2001). Semiparametric and nonparametric regression analysis of longitudinal data (with discussion). *Journal of American Statistical Association*, **96**, 103-126.

Lin, X. and Carroll, R. J. (2000). Nonparametric function estimation for clustered data when the predictor is measured without/with error. *Journal of American Statistical Association*, **95**, 520-534.

Lin, X. and Carroll, R.J. (2001a). Semiparametric regression for clustered data. *Biometrika*, **88**, 1179-1185.

Lin, X, and Carroll, R.J. (2001b). Semiparametric regression for clustered data using generalized estimating equations. *Journal of American Statistical Association*, **96**, 1045-1056.

Lin, X., Wang, N. Y., Welsh, A. H, and Carroll, R.J. (2004). Equivalent kernels of smoothing splines in nonparametric regression for clustered/longitudinal data. *Biometrika*, **91**, 177-193.

Lin, X. and Zhang, D. (1999) Inference in generalized additive mixed models by using smoothing splines. *Journal of Royal Statistical Society, Series B*, **61**, 381-400.

Martinussen, T. and Scheike, T.H. (1999). A semiparametric additive regression model for longitudinal data. *Biometrika*, **86**, 691-702.

Moyeed, R.A. and Diggle, P.J. (1994). Rates of convergence in semiparametric modeling of longitudinal data. *Austr. Jour. Statist.*, **36**, 75-93.

Müller, H. G. (1988). *Nonparametric Regression Analysis of Longitudinal Data.* Springer-Verlag, New York.

Pourahmadi, M. (1999). Joint mean-covariance models with applications to longitudinal data: unconstrained parameterization. *Biometrika,* **86**, 677-690.

Pourahmadi, M. (2000). Maximum likelihood estimation of generalized linear models for multivariate normal covariance matrix. *Biometrika,* **87**, 425-435.

Peterson, D.R., Zhao, H. and Eapen, S. (2003). Using local correlation in kernel-based smoothers for dependent data. *Biometrics,* **59**, 984-991.

Qu, A. and Li, R. (2005). Quadratic inference functions for varying coefficient models with longitudinal data, *Biometrics.* To appear.

Qu, A., Lindsay, B. G., and Li, B. (2000). Improving generalised estimating equations using quadratic inference functions. *Biometrika,* **87**, 823-836.

Rice, J. A. and Wu, C. O. (2003). Nonparametric mixed effects models for unequally sampled noisy curves. *Biometrics,* **57**, 253-259.

Ross, S. M. (1997). *Simulation.* Second Edition. Academic Press, Inc., San Diego, CA.

Ruckstuhl, A.,Welsh, A. H. and Carroll, R.J. (2000). Nonparametric function estimation of the relationship between two repeatedly measured variables. *Statistica Sinica,* **10**, 51-71.

Ruppert, D., Sheather, S.J. and Wand, M.P. (1995). An effective bandwidth selector for local least squares regression. *Journal of American Statistical Association,* **90**, 1257-1270.

Severini, T.A. and Staniswalis, J.G. (1994). Quasi-likelihood estimation in semiparametric models. *Journal of American Statistical Association,* **89**, 501–511.

Silverman, B. W. (1984). Spline smoothing: the equivalent variable kernel method. *Annals of Statistics,* **12**, 501-511.

Speckman, P. (1988). Kernel smoothing in partial linear models. *Journal of Royal Statistical Society, Series B,* **50**, 413–436.

Sun, Y. and Wu, H. (2005). Semiparametric time-varying coefficients regression model for longitudinal data. *Scandinavian Journal of Statistics,* **32**, 21 – 47.

Tibshirani, R. J. (1996). Regression shrinkage and selection via the LASSO. *Journal of Royal Statistical Society, Series B,* **58**, 267-288.

Wahba, G. (1984). Partial spline models for semiparametric estimation of functions of several variables, in *Statist. Analysis of Time Ser.,* Proceedings of the Japan U.S. Joint Seminar, Tokyo, 319–329.

Wang, N. (2003). Marginal nonparametric kernel regression accounting for within-subject correlation. *Biometrika,* **90**, 43-52.

Wang, N., Carroll, R. J. and Lin, X. (2005). Efficient semiparametric marginal estimation for longitudinal/clustered data. *Journal of American Statistical Association,* **100**, 147-157.

Welsh, A. H., Lin, X. and Carroll, R.J. (2002). Marginal longitudinal nonparametric regression: Locality and efficiency of spline and kernel methods. *Journal of American Statistical Association,* **97**, 482-493.

Wu, C. O. and Chiang, C.-T. (2000). Kernel smoothing on varying coefficient models with longitudinal dependent variable. *Statistica Sinica*, **10**, 433-456.

Wu, C. O., Chiang, C.-T. and Hoover, D. R. (1998). Asymptotic confidence regions for kernel smoothing of a varying-coefficient model with longitudinal data. *Journal of American Statistical Association*, **93**, 1388-1402.

Wu, H. and Liang, H. (2004). Random Varying-Coefficient Models with Smoothing Covariates, Applications to an AIDS Clinical Study. *Scan. J. Statist.*, **31**, 3-19.

Wu, F. and Zhang, J. (2002). Local polynomial mixed-effects models for longitudinal data. *Journal of American Statistical Association*, **97** , 883-897.

Wu, W. B. and Pourahmadi, M. (2003). Nonparametric estimation of large covariance matrices of longitudinal data. *Biometrika*, **90**, 831–844.

Xiao, Z., Linton, O.B., Carroll, R.J, and Mammen, E. (2003). More efficient local polynomial estimation in nonparametric regression with autocorrelated errors. *Journal of American Statistical Association*, **98**, 980-992.

Yao, F., Müller, H.G. and Wang, J.-L. (2005a). Functional data analysis for sparse longitudinal data. *Journal of the American Statistical Association*, **100**, 577-590.

Yao, F., Müller, H.G., Wang, J. L. (2005b). Functional Regression Analysis for Longitudinal Data. *The Annals of Statistics*, in press.

Zeger, S. L. and Diggle, P. J. (1994). Semiparametric models for longitudinal data with application to CD4 cell numbers in HIV seroconverters. *Biometrics*, **50**, 689-699.

Zhang, D. (2004). Generalized linear mixed models with varying coefficients for longitudinal data. *Biometrics*, **60**, 8-15.

Zhang, D. and Davidian, M. (2004). Likelihood and conditional likelihood inference for generalized additive mixed models for clustered data. *Journal of Multivariate Analysis*, **91**, 90-106.

Zhang, D. and Lin, X. (2003). Hypothesis testing in semiparametric additive mixed models. *Biostatistics*, **4**, 57-74.

Zhang, D., Lin, X., Raz, J., and Sowers, M. F. (1998). Semiparametric stochastic mixed models for longitudinal data. *Journal of American Statistical Association*, **93**, 710-719.

Zhang, D., Lin, X. and Sowers, M.F. (2000). Semiparametric regression for periodic longitudinal hormone data from multiple menstrual cycles. *Biometrics*, **56**, 31-39.

CHAPTER 14

Regressing Longitudinal Response Trajectories on a Covariate

Hans-Georg Müller and Fang Yao

Department of Statistics
University of California
Davis, Davis, CA 95616
mueller@wald.ucdavis.edu

Department of Statistics
Colorado State University
Fort Collins, CO 80523
fyao@stat.colostate.edu

Working within the framework of functional data analysis, we consider the analysis of longitudinal responses which are considered random trajectories and are sampled on a sparse and irregular design. In this functional response setting, the nonlinear influence of a one-dimensional covariate on the response curves is often of interest. We propose a simple two-dimensional smoothing approach and study its consistency. Functional principal component analysis for sparse data, using the PACE (principal analysis through conditional expectation) algorithm has been proposed recently for dimension reduction of irregularly sampled longitudinal data, and is shown to provide useful tools for the study of the variation of the observed data around the predicted trajectories. For example, the functional principal component scores of the differences between longitudinally-predicted and covariate-predicted trajectories can be used to study goodness-of-fit. We illustrate our methods with longitudinal CD4 count data. Brief reviews of functional regression models and of PACE are also included.

14.1. Introduction and Review

Functional data analysis is a field that is still in rapid development. It is concerned with data that include functional components, i.e., components that can be considered to be (random) functions. Usually these functions are assumed to be observed either as continuous random trajectories, or on a dense grid with added measurement errors. Recent developments in functional data analysis include extensions of various functional regression models. A characteristic feature of functional regression models is that one of the predictors or responses is a function. The functional components are typically assumed to correspond to the realization of a stochastic process. Overviews of various aspects of functional regression can be found for example in the books by Ramsay and Silverman (2002) and Ramsay and Silverman (2005), where many other approaches in addition to those discussed below can be found, and in a review article by Rice (2004). A more detailed review of some of the approaches that are only briefly mentioned in this article can be found in Müller (2005). In the following we give a brief overview of the various functional regression models that have been considered in the literature to date.

14.1.1. *Brief review of regression models with functional predictors*

The classical linear functional regression models include a functional component as a predictor and are direct extensions of the corresponding multivariate counterparts (multiple regression or multivariate regression). Frequently considered models are the functional linear regression models with predictor process X,

$$E(Y|X) = \mu + \int X(s)\beta(s)\, ds$$

for a scalar response Y, and

$$E(Y(t)|X) = \mu(t) + \int X(s)\beta(s,t)\, ds$$

for a functional response $Y(t)$. The functions β are the regression parameter functions, while μ or $\mu(t)$ is the mean of the responses. Basic ideas for such models go back to Grenander (1950) while statistical features and analysis are the focus of papers by Ramsay and Dalzell (1991), Cardot, Ferraty and Sarda (1999, 2003), and Cardot, Ferraty, Mas, and Sarda (2003), among others. Various estimation approaches and optimal theoretical results for the case of a scalar response have been recently investigated by Hall and Horowitz (2005). Crucial differences in regard to the asymptotics between estimation of the regression function β and predicting observations Y are highlighted in Cai and Hall (2005).

One basic problem for statistical estimation in functional regression is that the estimation of the regression parameter function β is an inverse problem, while the corresponding operators are not directly invertible. Uniqueness and existence of solutions for functional regression problems are therefore not guaranteed in general. Under mild conditions, a unique solution exists in the image space of the operator. The inversion problem in function space has been discussed in He, Müller and Wang (2000). When estimating the regression parameter function or other characteristics from actual data, regularization is a must. General theory for regularizing estimated operators was recently developed in Hoffmann and Reiss (2005) and such developments may lead to a more comprehensive theory of regularization in functional regression.

Two main approaches have emerged with regard to implementing the regularization step: The first approach is to project on a functional basis. For the functional basis, especially the eigenbasis of the auto-covariance operator of the predictor processes X has been studied, and the projection is obtained by truncating the number of included basis functions (Rice and Silverman (1991)). This approach has also been implemented with other orthogonal bases of the function space L^2 of square differentiable functions, ranging from the traditional Fourier basis to other orthogonal function systems, such as the wavelet basis (Morris, Vannucci, Brown and Carroll (2003)) or a suitably restricted family of B-splines, for example with a fixed number of knots (Shi, Weiss and Taylor (1996); Rice and Wu (2000)). A second common approach to regularization is based on penalized likelihood or penalized least squares. To date, this approach has been implemented via penalized smoothing splines (Cardot, Crambes, and Sarda (2005)), ridge regression (Hall and Horowitz (2005)), and P-splines (Yao and Lee (2005)). Other forms of regularization are conceivable and will present interesting avenues of research.

Several variants have been proposed to extend the basic functional regression models with functional predictor to more complex models. These variants are motivated by the corresponding extensions of the classical multivariate regression model towards more general regression models, and include:

(1) The extension from one single index on the right hand side of the functional linear model to the case of several single indices, or projections. This is an extension analogous to projection pursuit regression (Friedman and Stuetzle (1981); Lingjærde and Liestøl (1998)) and the multiple-index model (Chiou and Müller (2004)) and was proposed by James and Silver-

man (2005), who named it Functional Adaptive Model Estimation.

(2) Another variant is the generalized functional linear model (James (2002); Cardot and Sarda (2005); Müller and Stadtmüller (2005)). Here the response is a scalar Y which belongs to a general distribution family such as the exponential family. This extension corresponds to extending the classical linear regression model to a generalized linear model for the case of functional data. The generalized functional linear model can be described as

$$E(Y|X) = g\left(\alpha + \int \beta(t)X(t)\,dt\right),$$

where α is an intercept parameter in the functional linear predictor and g a suitably chosen link function. The variance of the responses typically is a function of the mean, either through an exponential family assumption as in the generalized linear model, or through an otherwise postulated variance function via quasi-likelihood, in which case no exponential family distribution needs to be specified. A nonparametric quasi-likelihood version where both link function and variance function are estimated nonparametrically is included in the approach of Müller and Stadtmüller (2005).

(3) The case of sparse and irregular noisy data as they are encountered in longitudinal studies; more about this below. For the model with sparse functional predictors and sparse functional responses such an extension was discussed in Yao, Müller, and Wang (2005a) and for the case of the generalized functional linear model with sparse predictors in Müller (2005).

(4) Varying-coefficient functional models, for example as the domains of the prediction functions increase, have been considered in the form of a "historical linear functional model" as proposed by Malfait and Ramsay (2003) or of a varying-coefficient regression model discussed in Müller and Zhang (2005), where the goal was the nonparametric analysis for mean residual lifetime as outcome Y. The latter paper also contains a proposal for functional quantile regression based on functional principal components analysis. An alternative approach for functional quantile regression based on penalized splines is discussed in Cardot, Crambes, and Sarda (2005).

14.1.2. *Brief review of regression models with functional response*

Another class of functional regression models that is of particular interest in many medical, biological or social sciences applications concerns the case where the response is a random function and the predictor a scalar or vector. This is often the situation in longitudinal studies where individuals are followed over time and repeated measurements are made of a quantity of interest. The goal then is to study the relationship between the functional responses and some subject-specific covariate or measurement. Such models have various applications (Faraway (1997)) and have been referred to as Functional Response Models in Chiou, Müller and Wang (2004), which also includes a review of models with functional response.

We assume in the following the covariate X is scalar. Since the models of interest are generally nonparametric, in order to avoid the curse of dimension, it is necessary to include a dimension reduction step if the covariate is multivariate. Dimension reduction can be achieved through an additive model, a single index model or other methods; a model for response curves based on multiple single indices with unknown link functions and quasi-likelihood was proposed in Chiou, Müller and Wang (2003). A special case of a functional response model is the product surface model $E(Y(t)|X = x) = \mu(t)\rho(x)$ for a function ρ such that $E(\rho(X)) = 1$, so that $E(Y(t)) = E[E(Y(t)|X)] = \mu(t)$. This simple approach was found to be useful for applications to medfly data in Cardot, Ferraty, Mas, and Sarda (2003), and has been referred to as functional multiplicative effects model.

Given a sample of response functions $Y(t)$, let $\mu(t) = E(Y(t))$ and define the auto-covariance operator $A(f)(t) = \int f(s)\text{cov}\{X(s), X(t)\}\,ds$. We assume existence of the quantities discussed. The operator A has orthonormal eigenfunctions ψ_k and ordered eigenvalues $\lambda_1 \geq \lambda_2 \geq \ldots$. Then the process Y can be represented through the Karhunen-Loève expansion (Karhunen, 1946)

$$Y(t) = \mu(t) + \sum_{k=1}^{\infty} A_k \, \psi_k(t), \tag{14.1}$$

where the expansion coefficients A_k are uncorrelated random variables (r.v.s), known as functional principal component scores. They satisfy the properties $E(A_k) = 0$ and $\text{Var}(A_k) = \lambda_k$, such that $\sum_k \lambda_k < \infty$ and

$$A_k = \int (Y(t) - \mu(t))\psi_k(t)\,dt. \tag{14.2}$$

These developments are a direct extension of the corresponding spectral decomposition of a multivariate random vector. A crucial difference between multivariate and functional data is that only for the latter, order and neighborhood relations matter, while for the former, permutation of the components of a sample of random vectors does not affect the results of the analysis. This crucial difference makes it possible and indeed necessary to use smoothing methods when dealing with functional data.

With a covariate X, our focus is the regression

$$E(Y(t)|X = x) = \mu(t) + \sum_{k=1}^{\infty} E(A_k|X = x)\, \psi_k(t), \qquad (14.3)$$

which is a bivariate function of t and x. For regularization of the functional data, we consider here projection onto the truncated eigenbase, where we would include K components in the sum on the right hand side. When one deals with densely sampled data with errors or the case where entire trajectories are available, several methods have been described to obtain the functional principal components of $Y(t)$; given a sample $Y_i(t)\, i = 1, \ldots, n$, of fully observed smooth trajectories, one can simply obtain

$$\widehat{\mu}(t) = \frac{1}{n}\sum_{i=1}^{n} Y_i(t), \quad \widehat{\operatorname{cov}}(Y(t), Y(s)) = \frac{1}{n}\sum_{i=1}^{n}(Y_i(t) - \widehat{\mu}(t))(Y_i(s) - \widehat{\mu}(s))$$

and these estimates are easily seen to converge at the $n^{-1/2}$ rate. By using perturbation methods for linear operators in Hilbert space, these results can be extended to the convergence of the resulting eigenfunction and eigenvalue estimates.

Then one usually discretizes the estimated covariance surface $\widehat{\operatorname{cov}}(Y(t), Y(s))$ to obtain the corresponding multivariate spectral decomposition. The resulting eigenvectors are smoothed or interpolated to give the eigenfunction estimates $\widehat{\psi}_k$ along with the eigenvalue estimates $\widehat{\lambda}_k$. The number K of components to include is often obtained in the same way as in multivariate analysis, by the scree plot or by the fraction of variance explained which is $X(t) = \mu(t) + \sum_{k=1}^{K} \tilde{\xi}_k\, \phi_k(t)$ (Mardia, Kent and Bibby, 1979). Similar results are possible if the observations are subject to additional independent errors. Regarding inference, only very few results are currently available. Besides the bootstrap which is based on resampling entire trajectories, a promising approach is Fan and Lin (1998).

14.1.3. *Overview*

In Section 2, the focus is on the functional response model where the data are sparse and irregular, as is common in longitudinal studies. An important situation in applications is the presence of a non-functional covariate. The principal analysis through conditional expectation (PACE) method for sparse longitudinal data is briefly reviewed. A simple smoothing procedure for regressing longitudinal responses on a covariate is introduced in Section 3 and its consistency is discussed. The difference between longitudinally predicted and covariate predicted trajectories gives rise to an analogue of residuals, which in this case correspond to residual trajectories. Residual trajectories themselves can be subjected to a functional principal component analysis and can be used for goodness-of-fit checking. A small scale simulation study and an application to longitudinal CD4 measurements can be found in Section 4.

14.2. The Functional Approach to Longitudinal Responses

In longitudinal studies, one does normally not observe complete or densely recorded trajectories and therefore the methods described above will not work. Data from such studies may be viewed as noisy measurements of a subject-specific smooth time course that are typically made at irregular and often sparse time points for a sample of individuals or items. A common approach in longitudinal data analysis is to assume parametric shapes for the trajectories and to fit random coefficient models (Diggle *et al.*, 1994). However, this is often not entirely satisfactory when knowledge about the shape of the individual trajectories is limited, as for example in an exploratory analysis.

More flexible methods that are applicable for such situations can be based on regularized B-splines (providing essentially a flexible parametric model) as proposed by Shi, Weiss and Taylor (1996), Rice and Wu (2000) and James, Hastie, and Sugar (2000) or on regularized eigenfunction expansions (Principal Analysis by Conditional Expectation (PACE); Yao, Müller, and Wang (2005b)). These methods require additional restrictions, typically that the underlying processes are Gaussian. The reason for this additional restriction is that missing information must be imputed; in the principal analysis through conditional expectation (PACE) method this is done explicitly in a conditional expectation step, where it is used that the conditional expectations are linear under Gaussian assumptions.

We briefly review here the PACE method, see Müller (2005) for a more detailed review. The sparseness of the data is reflected by postulating a random number N_i of measurements T_{ij} for the i-th subject where the T_{ij} are i.i.d. The observations of the process Y are then Y_{ij},

$$Y_{ij} = Y(T_{ij}) + \varepsilon_{ij} = \mu(T_{ij}) + \sum_{k=1}^{\infty} A_{ik}\psi_k(T_{ij}) + \varepsilon_{ij}, \qquad (14.4)$$

where the additional measurement errors ε_{ij} are i.i.d. and independent of all other r.v.s, with $E\varepsilon_{ij} = 0$, $\text{Var}(\varepsilon_{ij}) = \sigma^2$. In the PACE method, the mean function $\mu(t)$ is obtained by smoothing the scatterplot $\{(T_{ij}, Y_{ij}) : 1 \leq i \leq n, 1 \leq j \leq N_i\}$, with a local linear smoother to obtain $\widehat{\mu}$; and the covariance surface via $C_{ijk} = (Y_{ij} - \widehat{\mu}(T_{ij}))(Y_{ik} - \widehat{\mu}(T_{ik}))$, $j \neq k$ and two-dimensional smoothing in the scatterplot $\{(T_{ij}, T_{il}; C_{ijk}) : 1 \leq i \leq n, 1 \leq j \neq k \leq N_i\}$, where the diagonal is omitted as it is contaminated by the additional error variance σ^2. This results in covariance estimates $\widehat{\text{cov}}(Y(t), Y(s))$ for all applicable t, s, whence one proceeds as described above for the dense case.

A detailed analysis of the convergence properties of the eigenvalue and eigen-function estimates $(\widehat{\lambda}_k, \widehat{\psi}_k)$ under sparse and less sparse situations can be found in Hall, Müller and Wang (2005). While in the dense case one obtains the rates $n^{-1/2}$ for the eigenfunction estimates, this deteriorates to the nonparametric rate $n^{-2/5}$ for twice differentiable eigenfunctions in the sparse case.

Besides estimating the components (eigenfunctions, eigenvalues and the error variance σ^2) for which consistent estimators are readily available, a major goal when applying FDA to sparse longitudinal data is the prediction of individual trajectories. To achieve such predictions, a sensible idea is to substitute estimates for the terms that appear in the Karhunen-Loève expansion,

$$\widehat{Y}(t) = \widehat{\mu}(t) + \sum_{k=1}^{K} \widehat{A}_k \, \widehat{\psi}_k(t). \tag{14.5}$$

Here the estimates $\widehat{A}_k$ for the functional principal components A_k have not been defined yet. As has been shown in Yao, Müller, and Wang (2005b), substituting estimates based on the integral representation (14.2) for functional principal component scores does not work for sparse data, due to the large approximation error that one incurs in that case. Hence, alternative estimates for A_k are needed.

Such alternative estimates can be found under joint Gaussian assumptions on the random processes Y and the errors ε_{ij}. With $\widetilde{Y}_i = (Y_{i1}, \cdots, Y_{iN_i})^T$, $\mu_i = (\mu(T_{i1}), \cdots, \mu(T_{iN_i}))^T$,

$$\psi_{ik} = (\psi_k(T_{i1}), \cdots, \psi_k(T_{iN_i}))^T$$

and

$$(\Sigma_{Y_i})_{j,l} = \text{cov}(Y_i(T_{ij}), Y_i(T_{il})) + \sigma^2 \delta_{jl},$$

where $\delta_{jl} = 1$ if $j = l$ and $= 0$ otherwise, one can easily calculate

$$E[A_{ik}|Y_{i1}, \ldots, Y_{iN_i}] = \lambda_k \psi_{ik}^T \Sigma_{Y_i}^{-1} (\widetilde{Y}_i - \mu_i). \tag{14.6}$$

Estimates for the quantities on the r.h.s. are available from the procedures described above, leading to estimates for the predicted functional principal component scores,

$$\widehat{E}[A_{ik}|Y_{i1},\ldots,Y_{iN_i}] = \widehat{\lambda}_k \widehat{\psi}_{ik}^T \widehat{\Sigma}_{Y_i}^{-1}(\widetilde{Y}_i - \widehat{\mu}_i). \tag{14.7}$$

When plugging these estimates into (14.5), we refer to the resulting trajectories as the estimated *longitudinally predicted trajectories*, given by

$$\widehat{Y}_i(t) = \widehat{\mu}(t) + \sum_{k=1}^{K} \widehat{E}[A_{ik}|Y_{i1},\ldots,Y_{iN_i}] \, \widehat{\psi}_k(t). \tag{14.8}$$

These estimates target the actual longitudinally predicted trajectories, given by

$$\widetilde{Y}_i(t) = E(Y_i(t)|Y_{i1},\ldots,Y_{iN_i}) \tag{14.9}$$

$$= \mu(t) + \sum_{k=1}^{K} E(A_k|Y_{i1},\ldots,Y_{iN_i}) \, \psi_k(t).$$

Statistical properties of estimates (14.7), (14.8) of longitudinally predicted trajectories and their convergence to the actual longitudinally predicted trajectories (14.9) were investigated in Yao, Müller, and Wang (2005b). Due to the conditioning step, for sparse longitudinal data, the longitudinally predicted trajectories are distinct from the actual individual trajectories $Y_i(t)$. If the N_i (number of measurements per subject) are bounded r.v.s, which reflects the actual situation in many longitudinal studies, longitudinally predicted trajectories will not converge to the true trajectories, which therefore are out of reach. However, convergence to the true individual trajectories will occur if a condition such as $\min_{1 \le i \le n} N_i \to \infty$ holds. It follows from the arguments in Müller (2005), p. 229, that under further regularity conditions one may then obtain $E(A_k|Y_{i1},\ldots,Y_{iN_i}) \to \int Y(t)\psi_k(t)\,dt = A_k$, as $n \to \infty$, which implies that estimated longitudinally predicted trajectories are consistent for the actual individual trajectories as the number of measurements made per subject increases.

14.3. Predicting Longitudinal Trajectories from a Covariate

Estimates for longitudinally predicted trajectories as discussed in the previous section are obtained by conditioning on the sparse measurements obtained for a subject whose trajectory is to be predicted. These estimates are based on the principle of borrowing strength from the entire sample of measurements by pooling the data for all subjects for estimating the mean function, the eigenfunctions and the eigenvalues. A functional data analysis approach, although nonparametric, is therefore very different from an approach where the functions are estimated separately for each individual. Such an approach would not be very efficient.

A covariate, if available, plays no role so far in predicting the longitudinal trajectories. We consider now the problem of predicting a trajectory in the presence of a covariate. We refer to these predictions as *covariate predicted trajectories*, to emphasize their distinction from the *longitudinally predicted trajectories* that were defined previously and for which a covariate is not taken into account.

Analogous to (14.3) and (14.9), the covariate predicted trajectory at covariate level x and time t is defined as

$$\tilde{\mu}(x,t) = E(Y(t)|X = x) = \mu(t) + \sum_{k=1}^{K} E(A_k|X = x)\,\psi_k(t). \quad (14.10)$$

Note that $\tilde{\mu}$ is a surface in x and t. Since in the sparse case we do not have consistent estimates of A_k available, the method proposed in Cardot, Ferraty, Mas, and Sarda (2003) for regressing response functions on a covariate cannot be immediately extended; in particular, substituting $E(A_k|Y_{i1}, \ldots, Y_{iN_i})$ for A_k is not feasible, as in general, $E[E(A_k|Y_{i1}, \ldots, Y_{iN_i})|X] \neq E(A_k|X)$. As an alternative, we propose here to aggregate the sparse longitudinal data with similar covariate levels and then to smooth the resulting aggregated scatterplots.

The available data sample consists of n i.i.d. observations $(X_i, Y_{i1}, \ldots, Y_{iN_i})$, $i = 1, \ldots, n$. Here X_i is a one-dimensional random covariate, and the Y_{ij} are the available sparse measurements for the i-th subject, while N_i is the number of measurements. The N_i are assumed to be i.i.d. with finite moments; an example would be that they are Poisson random variables. The idea for the prediction is simple: Given a covariate level x for which the predicted trajectory is to be calculated, we select a window $[x - h, x + h]$ with bandwidth h, assemble those data for which the associated covariate X_i falls into this neighborhood, and then apply a local linear smoother to the resulting scatterplot which is indexed by time, i.e., $\{(T_{ij}, Y_{ij}) : X_i \in [x-h, x+h]\}$. Essentially, this corresponds to a smoothed surface in t and x, which is implemented by first aggregating the data whose covariate values are near x, followed by a one-dimensional smoothing step in the time direction.

Formally, to fit the weighted local least squares smoother at a fixed point t with bandwidth b (Fan and Gijbels, 1996), one chooses a non-negative symmetric kernel function K and minimizes

$$\sum_{i:|X_i-x|\le h}\sum_{j=1}^{N_i} K\{(t-T_{ij})/b\}\,(Y_{ij}-[a_0+a_1(t-T_{ij})])^2$$

w.r.t. parameters a_0, a_1. The explicit solution to this weighted least squares problem is well known to be a linear estimator in the data that can be written as a weighted average with smoothing weights $w_{ij}(t)$,

$$\widehat{\mu}(x,t) = \sum_{i:|X_i-x|\le h}\sum_{j=1}^{N_i} w_{ij}(t)Y_{ij}$$

$$= \widehat{a}_0(t) = \frac{P_{02}(t)\,P_{10}(t) - P_{01}(t)\,P_{11}(t)}{P_{00}(t)\,P_{02}(t) - P_{01}(t)^2}, \qquad (14.11)$$

where

$$P_{0r}(t) = \sum_{i=1}^{n}\sum_{j=1}^{N_i}(t-T_{ij})^r\,K\{(t-T_{ij})/h\},$$

$$\text{and } P_{1r}(t) = \sum_{i=1}^{n}\sum_{j=1}^{N_i}(t-T_{ij})^r\,Y_{ij}\,K\{(t-T_{ij})/h\},$$

and the $w_{ij}(t)$ are implicitly defined weight functions.

This estimator is linear in the data Y_{ij} and depends on the choice of bandwidths h and b. In practice, one-curve-leave-out cross-validation (Rice and Silverman (1991)) is a good option: For each $1 \le i_0 \le n$, one omits the data $(X_{i_0}, Y_{i_0 1}, \ldots, Y_{i_0 N_{i_0}})$ in turn, and uses the remaining sample to obtain one-curve-leave-out predictions $\widehat{Y}^{(-i_0)}(x,t)$. Then (h,b) are chosen so as to minimize the sum of squared prediction errors (SPE)

$$SPE = \sum_{i_0=1}^{n}\sum_{j=1}^{N_{i_0}}(Y_{i_0 j} - \widehat{Y}^{(-i_0)}(X_{i_0}, T_{i_0 j}))^2.$$

To discuss consistency of this approach, we assume the regularity conditions (A1), (A3), (A4), (A6) and (B2) of Yao, Müller, and Wang (2005b) and additionally assume that the covariate level X_i is independent of number N_i and timing T_{ij} of measurements, furthermore that all joint and marginal densities of (X_i, T_{ij}, Y_{ij}) are twice differentiable and uniformly bounded away from 0 and ∞ on a domain $(x,t) \in S$ where the consistency is desired, furthermore that second moments of all r.v.s exist and are uniformly bounded. We also assume that the target function $\tilde{\mu}(\cdot,\cdot)$ (14.10) is twice continuously differentiable with uniformly bounded derivatives.

Theorem 14.1: Under the above assumptions, if $h \to 0$, $b \to 0$ and $nhb^2 \to \infty$ as $n \to \infty$, it holds that

$$\sup_{(x,t)\in S} |\widehat{\mu}(x,t) - \tilde{\mu}(x,t)| = O_p\left(h^2 + b^2 + \frac{1}{(nh)^{1/2}b}\right).$$

We only give a brief sketch of the proof. Using representation (14.11), one first establishes that $\sum_{i:|X_i-x|\leq h} \sum_{j=1}^{N_i} w_{ij}(t) = 1$ and $\sum_{i:|X_i-x|\leq h} \sum_{j=1}^{N_i} w_{ij}(t)(X_i - x) = O_p(h^2)$, uniformly in t. Then we decompose

$$\sup_{(x,t)\in S} |\widehat{\mu}(x,t) - \tilde{\mu}(x,t)| \leq \sup_{(x,t)\in S} \left| \sum_{i:|X_i-x|\leq h} \sum_{j=1}^{N_i} w_{ij}(t)\{Y_{ij} - \tilde{\mu}(x,t)\} \right|$$

$$\leq \sup_{(x,t)\in S} \left| \sum_{i:|X_i-x|\leq h} \sum_{j=1}^{N_i} w_{ij}(t)\{Y_{ij} - \tilde{\mu}(X_i, T_{ij})\} \right|$$

$$+ \sup_{(x,t)\in S} \left| \sum_{i:|X_i-x|\leq h} \sum_{j=1}^{N_i} w_{ij}(t)\{\tilde{\mu}(X_i, T_{ij}) - \tilde{\mu}(x, T_{ij})\} \right|$$

$$+ \sup_{(x,t)\in S} \left| \sum_{i:|X_i-x|\leq h} \sum_{j=1}^{N_i} w_{ij}(t)\{\tilde{\mu}(x, T_{ij}) - \tilde{\mu}(x,t)\} \right|$$

$$= \sup_{(x,t)\in S} |I| + \sup_{(x,t)\in S} |II| + \sup_{(x,t)\in S} |III|.$$

The analysis is conditional on the T_{ij}. One easily finds $E(I) = 0$, while the terms in the sum which share the same index i are dependent. This situation is analogous to Lemma 1 in Yao, Müller, and Wang (2005b), and adapting the proof leads to $I = O_p((nh)^{-1/2}b^{-1})$. Taylor expansion of $\tilde{\mu}(X_i, T_{ij})$ around $\tilde{\mu}(x, T_{ij})$ yields $II = O_p(h^2)$, and again by a Taylor expansion one finds that $III = O(b^2)$, with bounds that are uniform on S.

Note that the proposed approach for regressing response functions on a covariate is quite simple, amounting to no more than a two-dimensional smoothing step. The smoothing step can be implemented with any available two-dimensional smoother. In the following illustrations, we choose an implementation based on weighted local least-squares.

14.4. Illustrations

The estimated longitudinally predicted trajectories $\widehat{Y}_i(t)$ are closest to the actual trajectories since they incorporate actual measurements from the trajectories, albeit sparse ones. These trajectories can be used to check the goodness-of-fit of the covariate-predicted trajectories $\widehat{\mu}(x, t)$. For this purpose we define the differences

$$R_i(t) = \widehat{Y}_i(t) - \widehat{\mu}(x, t), \qquad (14.12)$$

which assume a role similar to the residuals in ordinary regression analysis. The analysis of these residual trajectories, extending the usual residual analysis, can be done by subjecting the residual trajectories to a functional principal component analysis.

We first describe a small simulation study which confirms that indeed the longitudinally-predicted trajectories are generally closer to actual trajectories than the covariate-predicted trajectories, and then proceed to illustrate the methods with longitudinal CD4 data.

14.4.1. *A simulation comparison of longitudinally-predicted and covariate-predicted trajectories*

The simulation aims at comparing the difference between actual individual trajectories and predicted trajectories that one obtains either with estimated longitudinally predicted trajectories (14.8) or with estimated covariate-predicted trajectories $\widehat{\mu}(X_i, t)$ (14.11), given the covariate level X_i for the i-th subject.

The underlying simulation model for the ith subject is $Y_i(t) = \mu(t) + \sum_{k=1}^{2} A_{ik}\phi_k(t)$, where $\mu(t) = t + \sin(t)$, $\phi_1(t) = -\cos(\pi t/10)/\sqrt{5}$, and $\phi_2(t) = \sin(\pi t/10)/\sqrt{5}$, $0 \leq t \leq 10$. The random coefficients A_{ik} are generated as follows: The covariate $X_i \overset{\text{i.i.d.}}{\sim} N(0, 1)$, $A_{i1} = g_1(X_i) + B_{i1}$ and $A_{i2} = g_2(X_i) + B_{i2}$, where $B_{ik} \overset{\text{i.i.d.}}{\sim} N(0, 3)$. The nonlinear functions $g_1(z) = c_{11}[\sin(z) - c_{12}]$ and $g_2(z) = c_{21}(z^3 - c_{22})$ are used, where the constants $c_{11}, c_{12}, c_{21}, c_{22}$ are adjusted so that $E[g_k(X_i)] = 0$ and $\text{var}(g_k(X_i)) = 1$, $k = 1, 2$. The jth measurement made for the ith subject is given by $Y_{ij} = Y_i(T_{ij}) + \varepsilon_{ij}$, where the additional measurement errors ε_{ij} are also assumed to be normal with mean 0 and constant variance $\sigma^2 = 1$. Each curve is sampled at a random number of points, chosen from a discrete uniform distribution on $\{2, \ldots, 6\}$, and the locations of the measurements were uniformly distributed on $[0, 10]$.

We use the procedures as described in Yao, Müller, and Wang (2005b) to estimate the model components, such as $\mu(t)$, λ_k and $\phi_k(t)$. The number of included components K in the estimates (14.8) for longitudinally predicted trajectories was chosen by AIC, based on a pseudo-likelihood, as described

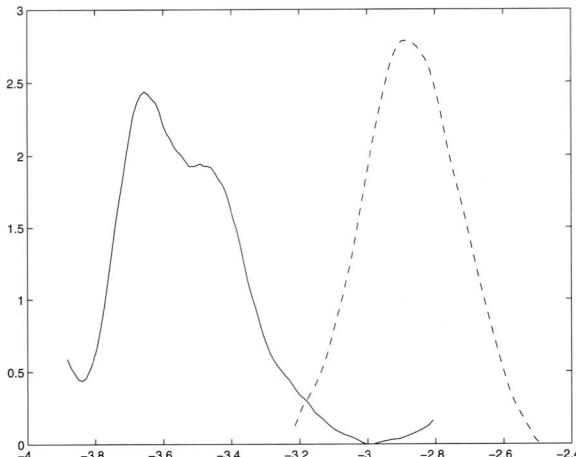

Figure 14.1.   Densities of the distribution of the logarithm of mean integrated relative errors (14.13), obtained for 200 simulated samples, comparing the prediction errors for longitudinally-predicted (14.9) and covariate-predicted (14.11) trajectories. Density estimate for errors of longitudinally-predicted trajectories is the solid curve, while for the covariate-predicted trajectories it is the dashed curve.

in that paper. In the implementation of the estimates for covariate-predicted trajectories $\widehat{\mu}(X_i, t) = \widehat{E}\{Y(t)|X = X_i\}$ (14.11), we used subjective bandwidth choices.

Table 14.1.   Lower, median, and upper quartiles of mean integrated relative error (14.13), obtained from 200 samples, comparing longitudinally-predicted (14.9) and covariate-predicted (14.11) trajectories as estimates for actual trajectories.

Method	Lower	Median	Upper
Longitudinally-predicted	.026	.028	.032
Covariate-predicted	.052	.056	.062

The criterion used to measure the discrepancy between the actual observed individual trajectories $Y_i(t)$ and the estimated trajectories from either method is the mean integrated relative error (MIRE), defined as follows:

$$\text{MIRE} = \frac{1}{n} \sum_{i=1}^{n} \frac{\int \{Y_i(t) - \widehat{Y}_i(t)\}^2 dt}{\int Y_i^2(t) dt},$$

$$\text{or} \quad \text{MIRE} = \frac{1}{n} \sum_{i=1}^{n} \frac{\int \{Y_i(t) - \widehat{\mu}(X_i, t)\}^2 dt}{\int Y_i^2(t) dt}. \quad (14.13)$$

The lower, median and upper quartiles of the criterion MIRE (14.13) obtained from 200 simulation samples using the two estimation methods are reported in Table 14.1. Density estimates for the values of the MIRE (14.13) in the logarithmic scale for the two approaches are shown in Figure 14.1. These simulation comparisons clearly show that longitudinally-predicted trajectories are considerably more accurate than covariate-predicted trajectories, justifying the definition of residuals R_i as in (14.12).

14.4.2. *Application to longitudinal CD4 data*

These longitudinal data have been collected for 283 homosexual men who became HIV positive between 1984 and 1991. They consist of irregularly measured CD4 percentages with 1 to 14 measurements per subject after se-roconversion, with a median count of 7 measurements. These data were ana-lyzed previously with functional data analysis methodology in Yao, Müller, and Wang (2005b) and have also been analyzed by other authors. Details about the design of the study can be found in Kaslow *et al.*(1987).

In a first step, we applied the methods described in Yao, Müller, and Wang (2005b), estimating the model components of the longitudinal CD4 process, such as mean function, covariance surface and the eigenval-ues/eigenfunctions. In a second step, we then obtained the trajectory es-timates $\widehat{Y}_i(t)$ (14.8) by principal analysis through conditional expectation (PACE); these estimates correspond then to the estimated longitudinally predicted trajectories.

As covariate X_i we used Pre-CD4 percentage, which is a time-independent measurement that is available for each patient at the time of entry into the longitudinal study. We then investigated the influence of Pre-CD4 percentage on the shape of the longitudinal trajectories. As described above, two-dimensional smoothing is applied to obtain the esti-mates of covariate-predicted trajectories $\widehat{\mu}(x, t)$ (14.11) that target the true covariate-predicted trajectories $\tilde{\mu}(x, t)$ (14.10). Viewed for all (x, t) simul-taneously, the $\widehat{\mu}(x, t)$ form a surface that is shown in Figure 14.2. With few exceptions, the predicted trajectories are generally decreasing in t. The rate

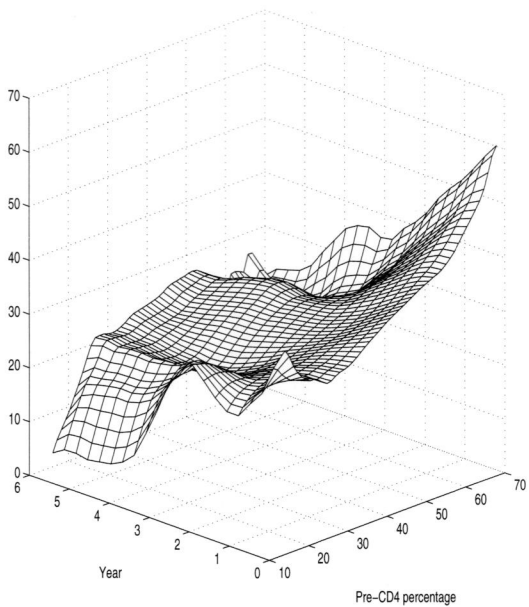

Figure 14.2.   Covariate-predicted trajectories (14.10), estimated via (14.11), as a function of time t in the study (in years) and Pre-CD4 level x.

of decrease appears to be related to the Pre-CD4 level, higher levels being associated with higher rates of decrease.

Estimated CD4 profiles comparing the estimates obtained for both longitudinally and covariate predicted trajectories (14.8) and (14.11) for four randomly selected patients are shown in Figure 14.3. As expected from the simulation results, the longitudinally predicted trajectories (solid lines) are seen to be considerably closer to the observed measurements. The residuals (14.12) are displayed in the left panel of Figure 14.4. These residual profiles can be subjected to a functional principal component analysis in order to study the behavior of these functional residuals.

The first three eigenfunctions resulting from such an eigen-analysis of residuals (the AIC criterion led to the selection of three components) are shown in the right panel of Figure 14.4. The variation of the residual profiles near the endpoints is of interest. If one wants to study lack of fit (bias) for any proposed fitting method, the usual goodness-of-fit plot known from

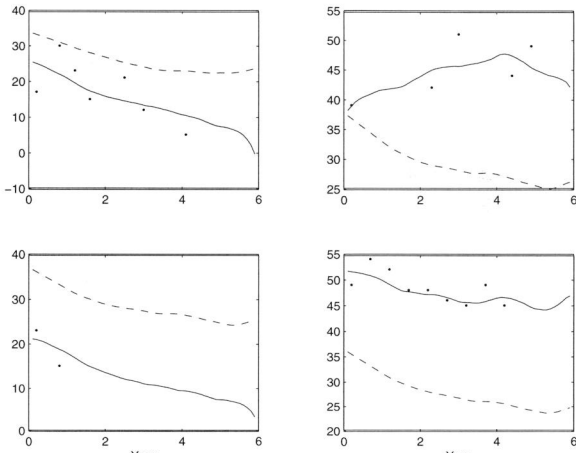

Figure 14.3. Observed measurements of CD4 percentage (dots) and estimated longitudinally predicted CD4 profiles (solid curves) as well as covariate-predicted profiles (dashed curves) for four randomly selected subjects.

multiple linear regression, plotting residuals versus the predicted values, is not easily feasible here since both residuals and predicted values are trajectories. Using functional principal component scores for the residual trajectories, one can for example plot the first two scores of the residuals against the covariate value (CD4 percentage) itself. This plot is shown in Figure 14.5 and does not provide evidence for systematic lack of fit for covariate-predicted trajectories.

Acknowledgments

This research was supported in part by NSF grants DMS03-54448 and DMS05-05537. We are grateful to a referee whose comments led to an improved version of the paper.

References

Cai, T., Hall, P. (2005). Prediction in functional linear regression. Preprint.

Cardot, H., Crambes, C., and Sarda, P. (2005). Quantile regression when the covariates are functions. Preprint.

Cardot, H., Ferraty, F., Mas, A., and Sarda, P. (2003). Testing hypotheses in the functional linear model. *Scand. J. Statist.* **30**, 241-255.

Cardot H., Sarda, P. (2005). Estimation in generalized linear models for functional data via penalized likelihood. *J. Multiv. Analysis* **92**, 24-41.

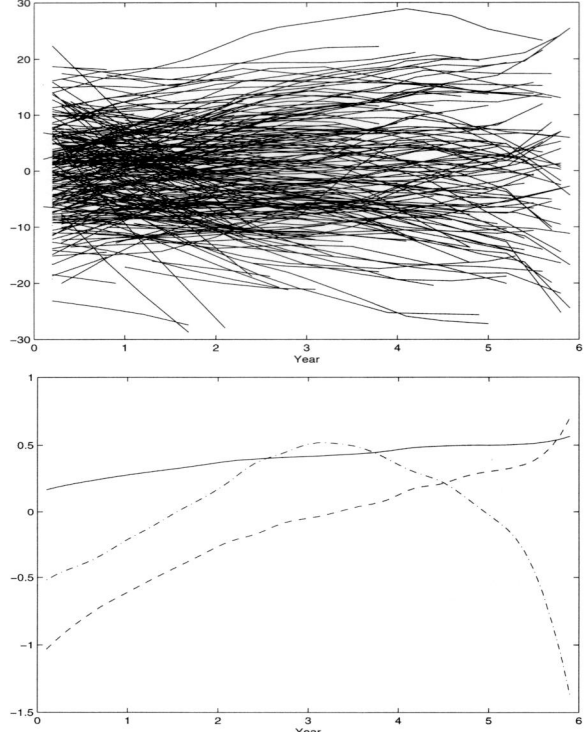

Figure 14.4. Left: Residual profiles obtained by subtracting covariate-predicted trajectories from longitudinally-predicted trajectories for the CD4 data. Right: First three eigenfunctions of the residual profiles (first eigenfunction, solid; second eigenfunction, dashed; third eigenfunction, dash-dotted).

Chiou, J. M., and Müller, H. G. (2004). Quasi-likelihood regression with multiple indices and smooth link and variance functions. *Scand. J. Statist.* **31**, 367-386.

Chiou, J. M., Müller, H. G., and Wang, J. L. (2003). Functional quasi-likelihood regression models with smooth random effects. *J. Roy. Statist. Soc. B* **65**, 405-423.

Chiou, J.M., Müller, H. G., and Wang, J. L. (2004). Functional response models. *Statistica Sinica* **14**, 675-693.

Chiou, J. M., Müller, H. G., Wang, J. L., Carey, J. R. (2003). A functional multiplicative effects model for longitudinal data, with application to reproductive histories of female medflies. *Statistica Sinica* **13**, 1119-1133.

Fan, J., Gijbels, I. (1996). *Local Polynomial Modelling and its Applications.* Chapman and Hall, London.

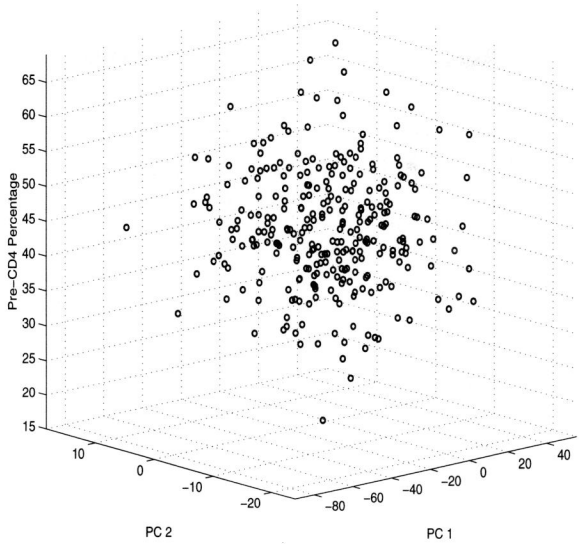

Figure 14.5.   Relationship between first (x-axis) and second (y-axis) functional principal component score of the residual profiles and the Pre-CD4 percentage (z-axis).

Fan, J., and Lin S. K. (1998). Test of significance when data are curves. *J. Amer. Statist. Assoc.* **93**, 1007-1021.

Faraway, J. J. (1997). Regression analysis for a functional response. *Technometrics* **39**, 254-262.

Friedman, J. H., Stuetzle, W. (1981). Projection pursuit regression. *J. Amer. Statist. Assoc.* **76**, 817-823.

Grenander, U. (1950). Stochastic processes and statistical inference. *Arkiv för Matematik*, 195-276.

Hall, P., Horowitz, J. L. (2005). Methodology and convergence rates for functional linear regression. Preprint.

Hall, P., Müller, H. G., Wang, J. L. (2005). Properties of principal component methods for functional and longitudinal data analysis. Preprint.

He, G., Müller, H. G., and Wang, J. L. (2000). Extending correlation and regression from multivariate to functional data. *Asymptotics in Statistics and Probability*, Ed. Puri, M.L., VSP International Science Publishers, pp. 301-315.

Hoffmann, M., Reiss, M. (2005). Nonlinear estimation for linear inverse problems with errors in the operator. Preprint.

James, G. (2002). Generalized linear models with functional predictors. *J. Roy. Statist. Soc. B* **64**, 411-432.

James, G., Hastie, T. G., and Sugar, C. A. (2000). Principal component models for sparse functional data. *Biometrika* **87**, 587-602.

James, G., and Silverman, B. W. (2005). Functional adaptive model estimation. *J. Amer. Statist. Assoc.* **100**, 565-576

Karhunen, K. (1946). Zur Spektraltheorie stochastischer Prozesse. *Ann. Acad. Sci. Fennicae* **A I 37**.

Lingjærde, O. C., Liestøl, K. (1998). Generalized projection pursuit regression. *SIAM J. Scientific Computing* **20**, 844-857.

Malfait N., and Ramsay J. O. (2003). The historical functional linear model. *Can. J. Statist.* **31**, 115-128.

Mardia, K. V., Kent, J. T., Bibby, J. M. (1979). *Multivariate Analysis.* London, Academic Press.

Morris, J. S., Vannucci, M., Brown, P. J., and Carroll, R. J. (2003). Wavelet-based nonparametric modeling of hierarchical functions in colon carcinogenesis (with discussion). *J. Amer. Statist. Assoc.* **98**, 573-597.

Müller, H. G. (2005). Functional modelling and classification of longitudinal data. *Scand. J. Statist.* **32**, 223-240.

Müller, H.G., and Stadtmüller, U. (2005). Generalized functional linear models. *Ann. Statist.* **33**, 774-805.

Müller, H.G., and Zhang, Y. (2005). Time-varying functional regression for predicting remaining lifetime distributions from longitudinal trajectories. *Biometrics* **59**, 676-685.

Ramsay, J. O., Dalzell, C. J. (1991). Some tools for functional data analysis. *J. Roy. Statist. Soc. B* **53**, 539-572.

Ramsay, J. O., Silverman, B. (2002). *Applied Functional Data Analysis.* New York, Springer.

Ramsay, J. O., Silverman, B. (2005). *Functional Data Analysis.* New York, Springer.

Rice, J. A. (2004). Functional and longitudinal data analysis: Perspectives on smoothing. *Statistica Sinica* **14**, 631-647.

Rice, J. A., and Silverman, B. (1991). Estimating the mean and covariance structure nonparametrically when the data are curves. *J. Roy. Statist. Soc. B*, **53**, 233-243.

Rice, J. A., Wu, C. (2000). Nonparametric mixed effects models for unequally sampled noisy curves. *Biometrics*, **57**, 253-259.

Shi, M., Weiss, R. E., Taylor, J. M. G. (1996). An analysis of paediatric CD4 counts for Acquired Immune Deficiency Syndrome using flexible random curves. *Appl. Statist.*, **45**, 151-163.

Yao, F., Lee, T. C. M. (2005). Penalized spline models for functional principal component analysis. *J. Roy. Statist. Soc. B*, to appear.

Yao, F., Müller, H.G., Wang, J.L. (2005a). Functional linear regression analysis for longitudinal data. *Ann. Statist.*, to appear.

Yao, F., Müller, H.G., Wang, J.L. (2005b). Functional data analysis for sparse longitudinal data. *J. Amer. Statist. Assoc.* **100**, 577-590.

Part V

Statistics in Science and Technology

CHAPTER 15

Statistical Physics and Statistical Computing: A Critical Link

James D. Servidea and Xiao-Li Meng

U.S. Department of Defense
Washington, DC

Department of Statistics
Harvard University
Cambridge, MA 02138
meng@stat.harvard.edu

The main purpose of this chapter is to demonstrate the fruitfulness of cross-fertilization between statistical physics and statistical computation, by focusing on the celebrated Swendsen-Wang algorithm for the Ising model and its recent perfect sampling implementation by Huber. In particular, by introducing Hellinger derivative as a measure of instantaneous changes of distributions, we provide probabilistic insight into the algorithm's critical slowing down at the phase transition point. We show that at or near the phase transition, an infinitesimal change in the temperature parameter of the Ising model causes an astronomical shift in the underlying state distribution. This finding suggests an interesting conjecture linking the critical slowing down in coupling time with the grave instability of the system as characterized by the Hellinger derivative (or equivalently, by Fisher information). It also suggests that we can approximate the critical point of the Ising model, a physics quantity, by monitoring the coupling time of Huber's bounding chain algorithm, an algorithmic quantity. This finding might provide an alternative way of approximating criticality of thermodynamic systems, which is typically intractable analytically. We also speculate that whether we can turn perfect sampling from a pet pony into a workhorse for general scientific computation may depend critically on how successful we can engage, in its development, researchers from statistical physics and related scientific fields.

15.1. MCMC Revolution and Cross-Fertilization

One of the spectacular successes in statistics, the Markov chain Monte Carlo (MCMC) revolution, began around 1990. The method, however, was in existence for more than four decades, beginning with the publication of Metropolis et. al. (1953) in the statistical physics literature. This well known fact is documented in Liu (2002), which also provides a detailed account of many recent advances made by physicists, statisticians and other scientists. Today, MCMC is used to study everything from single molecules to supernovas and from financial engineering to anti-spam software. This enormous success demonstrates the fruitfulness of cross-fertilization between statistics and an area of application, with the former serving as a "central hub" to enhance and transfer the method to essentially every field that requires quantitative investigations under uncertainty. Furthermore, the cross-fertilization also provides refinements and new methods for the application itself. This chapter documents one more example of this kind, along the way providing a brief review of Ising model, one of the most basic models in statistical physics, and of perfect sampling, a most recent development in the MCMC evolution.

15.2. The Ising Model

Introduced in the 1920s, the Ising model is widely used today in the areas of statistical mechanics and image analysis. It specifies a probability distribution (a Gibbs distribution, also called a Boltzmann distribution) on the magnetic spins of a lattice. In image analysis, the nodes of the lattice are viewed as image pixels and the spins on those nodes as colors. Let L denote a set of nodes (or more generally a graph). For two distinct nodes $i, j \in L$, we shall denote $i \sim j$ to mean that i and j are nearest neighbors in the lattice.

Associated with each node i of the lattice is a spin X_i that can be either down ($X_i = -1$) or up ($X_i = 1$). In image analysis, these correspond to white and black pixels, respectively. The probability mass function specified by the model is given by

$$p_\beta(x) = \frac{1}{Z(\beta)} \exp\left(J\beta \sum_{i \sim j} x_i x_j \right), \quad x \in \{-1, 1\}^{|L|}, \qquad (15.1)$$

where $|\cdot|$ denotes cardinality and the index set "$i \sim j$" is taken to mean $\{(i,j) \in L^2 : i \sim j\}$. Since $i \sim j$ if and only if $j \sim i$, we take $(i,j) \equiv$

(j, i), and therefore each pair of adjacent nodes is counted only once in the summation. In (15.1), J is known as the *coupling constant*, with $J = 1$ for a ferromagnetic model, where spins tend to align themselves with their neighbors, and with $J = -1$ for an anti-ferromagnetic model, where spins tend to be opposite to their neighbors (see Thompson, 1972). The case of $J = 0$ corresponds to independence among spins. The parameter $\beta = 1/(kT)$, where T is the absolute temperature of the system and k is Boltzmann's constant, controls the amount of attraction or repulsion. Here we focus on the more common case of $J = 1$. This is the usual case in image analysis, where pixels tend to be like their neighbors.

The normalizing constant $Z(\beta)$ is known to physicists as a partition function. For some simple lattices, the partition function can be calculated analytically, but, generally speaking, solutions that are closed-form or nearly so are elusive. Consequently, the evaluation of $Z(\beta)$ often requires Monte Carlo simulation.

15.3. The Swendsen-Wang Algorithm and Criticality

In order to generate Monte Carlo simulations, we need to sample from (15.1). The earliest attempt was christened the *heat bath algorithm*, which statisticians would characterize today as a Gibbs sampler. The idea was to update the spin of each node conditionally, given the spins of all other nodes in the lattice. Due to the spatial Markovian nature of (15.1), this is equivalent to updating each node's spin given the spins of its nearest neighbors. This rather simple scheme proves to be painfully slow when β is large, as its approach is "local" in nature.

Swendsen and Wang (1987) developed an MCMC algorithm to address this problem that can be viewed as an auxiliary variables technique (e.g., Higdon (1998)). For each pair of adjacent nodes (i, j), they define a binary random variable $U_{i,j}$ that takes values in $\{-1, 1\}$. Think of $U_{i,j}$ as an edge bond between i and j that might bind the two spins in the same direction $(U_{i,j} = 1)$ or leave them free $(U_{i,j} = -1)$. Each update of their algorithm consists of the following two steps.

STEP I:
  For each $(i, j) \in L^2$ such that $i \sim j$,
    Let $U_{i,j} := X_i X_j$.
    If $U_{i,j} = 1$, **Let** $U_{i,j} := -1$ with probability $\exp(-2\beta)$.
STEP II:

Let $\mathbb{C}$ be the set of all clusters $C \subset L$ such that the nodes of
C are connected by edges (i, j) with $U_{i,j} = 1$.
For each $C \in \mathbb{C}$,
  Let x be a random spin, -1 or 1, with probability $1/2$ each.
  For all $i \in C$, **Let** $X_i := x$.

By viewing $U \equiv \{U_{i,j} : i \sim j\}$ as auxiliary variables, the Swendsen-Wang algorithm can be seen as a Gibbs sampler where STEP I samples U given $X \equiv \{X_i : i \in L\}$ and STEP II samples X given U. Here the conditional distribution of U given X is constructed to be

$$f(u|x) \propto \left(1 - e^{-2\beta}\right)^{N_u} \left(e^{-2\beta}\right)^{N_x - N_u} \prod_{i \sim j} 1_{\{u_{i,j} \leq x_i x_j\}}, \tag{15.2}$$

where

$$N_u = \sum_{i \sim j} \frac{u_{i,j} + 1}{2} \quad \text{and} \quad N_x = \sum_{i \sim j} \frac{x_i x_j + 1}{2}. \tag{15.3}$$

This construction means that when $x_i \neq x_j$, we set $u_{i,j} = -1$, that is, the two nodes i and j are not bound. On the other hand, when $x_i = x_j$, it is only with probability $e^{-2\beta}$ that i and j are not bound (i.e., $u_{i,j} = -1$). The binomial part of (15.2) is easily seen when we note that there are exactly N_x edges with $x_i = x_j$ and N_u edges with $u_{i,j} = 1$.

Under the above joint specification $f(x, u) = f(u|x)f(x)$, noting that $f(x) \propto e^{2\beta N_x}$ from (15.1) and (15.3), we can easily see that $f(x|u) \propto \prod_{i \sim j} 1_{\{x_i x_j \geq u_{i,j}\}}$. This leads to

$$f(x|u) = \left(\frac{1}{2}\right)^{|\mathbb{C}_u|} \prod_{i \sim j} 1_{\{x_i x_j \geq u_{i,j}\}}, \tag{15.4}$$

where $\mathbb{C}_u$ is the set of all "clusters" defined by u and $|\mathbb{C}_u|$ is its cardinality. Therefore, conditioning on U, each cluster of lattice nodes is assigned a spin independently with equal probabilities for up and down. The product in (15.4) ensures $\{(i, j) : u_{i,j} = 1\} \subset \{(i, j) : x_i x_j = 1\}$, a compatibility requirement between corresponding edge bonds and nodes' spin directions.

It is not difficult to show that the Markov chain specified by the Swendsen-Wang algorithm is ergodic and has (15.1) as its stationary distribution. Moreover, this chain mixes substantially faster than the heat bath algorithm. But, both algorithms suffer from what physicists call *critical slowing down*, a phenomenon that we shall present in more probabilistic terms.

As mentioned earlier, the normalizing constant $Z(\beta)$ as a function of the system parameter β is an important tool in physics. By taking derivatives of simple functions of Z with respect to the parameters such as the temperature (hidden in β for our purposes), one obtains various important physical quantities. For example, the heat capacity C of a system can be computed by the formula (McCoy and Wu (1973))

$$C = -\frac{\partial}{\partial T}\frac{\partial}{\partial \beta} \log Z(\beta) = k\beta^2 \frac{\partial^2}{\partial \beta^2} \log Z(\beta), \qquad (15.5)$$

where the second identity is due to $\beta = 1/(kT)$. That is, the heat capacity is the rate of change of the expected energy of a system as a function of its temperature.

It turns out that, on many lattices, for a certain value of β, $Z(\beta)$ nearly fails to be analytic, thus leading to astronomically large values of physical quantities, the heat capacity in particular. Such a point is called *critical*, and, in this case, corresponds to a phase transition: the proportion of up spins as opposed to down goes from $1/2$ to virtually 0 or 1 with equal probability as β increases past its critical value. In the thermodynamic limit, for an infinitely large, square lattice, Z is not analytic at the critical point and the heat capacity (per node) at that point is infinite. This critical value of β is well known to be

$$\beta_c = \frac{1}{2}\operatorname{arcsinh}(1) = \frac{1}{2}\log(1+\sqrt{2}) \approx 0.44068679348. \qquad (15.6)$$

See Baxter (1982) for a derivation.

It is well known that when the parameters are near critical, the Swendsen-Wang algorithm, like many other Markov chain samplers, tends to mix very slowly. In the next section, we offer an intuitive explanation in probabilistic terms for why this happens.

15.4. Instantaneous Hellinger Distance and Heat Capacity

Given a measure space $(\Omega, \mathcal{F}, \mu)$, let $\mathcal{P}$ be the set of all probability densities with respect to μ. On $\mathcal{P}$, the *Hellinger distance* is the metric defined by

$$H(p,q) = \left[\int_\Omega (\sqrt{p} - \sqrt{q})^2 \, d\mu\right]^{\frac{1}{2}}, \quad p, q \in \mathcal{P}.$$

Now suppose $\Omega \subset \mathbb{R}^k$. The subfamily of $\mathcal{P}$ whose elements are of the form

$$p_\theta(\omega) = \frac{1}{c(\theta)} \exp(\theta \cdot \omega),$$

indexed by $\theta \in \mathbb{R}^k$ such that $\int \exp(\theta \cdot \omega)\, \mu(d\omega) < \infty$, is the *exponential family* with canonical parameter θ. The Hellinger distance between two distributions whose densities belong to the same exponential family reduces to

$$H(p_{\theta_1}, p_{\theta_2}) = \left[2 - 2\frac{c(\bar{\theta})}{\sqrt{c(\theta_1)c(\theta_2)}} \right]^{\frac{1}{2}}, \tag{15.7}$$

where $\bar{\theta} = (\theta_1 + \theta_2)/2$.

The Hellinger distance is a bounded metric that equals $\sqrt{2}$ if the two densities have no common support on Ω (almost everywhere with respect to μ). For our purposes, it will serve as a handy way of measuring overlap in the Gibbs distributions of the Ising model for different values of β. And since (15.1) has an exponential form, we can express the Hellinger distance between Ising model distributions as a function of the partition function Z using (15.7) above.

Consider the Hellinger "derivative" given by:

$$\partial H(\beta) = \lim_{\epsilon \downarrow 0} \frac{1}{2\epsilon} H(p_{\beta - \epsilon}, p_{\beta + \epsilon}). \tag{15.8}$$

For a finite lattice, $Z(\beta)$ is a finite sum of analytic functions and therefore is itself analytic. So, we can express $Z(\beta + \epsilon)$ and $Z(\beta - \epsilon)$ as Taylor series to obtain

$$Z(\beta + \epsilon)Z(\beta - \epsilon) = Z^2(\beta) + \epsilon^2 \left[Z(\beta)Z''(\beta) - [Z'(\beta)]^2 \right] + O(\epsilon^3).$$

Define $K = (Z(\beta)Z''(\beta) - [Z'(\beta)]^2)/Z^2(\beta)$ and apply (15.7) to see that

$$\partial H(\beta) = \lim_{\epsilon \downarrow 0} \frac{1}{\sqrt{2}} \left[\frac{1 - [1 + \epsilon^2 K + O(\epsilon^3)]^{-1/2}}{\epsilon^2} \right]^{\frac{1}{2}}.$$

Using L'Hôpital's rule, we finally obtain

$$\partial H(\beta) = \lim_{\epsilon \downarrow 0} \frac{1}{\sqrt{2}} \left[\frac{[1 + \epsilon^2 K + O(\epsilon^3)]^{-3/2}[\epsilon K + O(\epsilon^2)]}{2\epsilon} \right]^{\frac{1}{2}} = \frac{\sqrt{K}}{2}.$$

Note that K is simply the second derivative of $\log Z$, thus

$$\partial H(\beta) = \frac{1}{2} \left[\frac{\partial^2}{\partial \beta^2} \log Z(\beta) \right]^{\frac{1}{2}}. \tag{15.9}$$

Comparing (15.5) and (15.9), we see immediately that

$$C = 4k[\beta\, \partial H(\beta)]^2. \tag{15.10}$$

Therefore, the heat capacity of the Ising model is large wherever the Hellinger distance changes rapidly as a function of β. Conversely, the Hellinger distance changes very quickly wherever the heat capacity is large, in particular, whenever a phase transition occurs. Consequently, it is no surprise that it is so difficult to simulate the Ising model near the critical value. A tiny perturbation in the parameter β gives a new distribution that *effectively* shares almost no support with the old one. That is, the probability masses of the new and old distributions concentrate on different parts of the state space, even though technically both distributions have the same *mathematical* support.

For example, the above result suggests that, even for parallel tempering (Geyer (1991), Geyer and Thompson (1995), Tesi *et al.*, (1996)), one of today's most promising methods, we will need to use a very large number of intermediate temperatures in order to achieve a useful swapping probability at or near the critical value. Indeed, recent results by Madras and Zheng (2002) imposed the condition that the number of intermediate temperatures be proportional to the number of the nodes, in order for the parallel tempering to achieve polynomial mixing time, for a couple models simpler than (15.1). Our finding above makes us to believe that this condition is not only sufficient but also necessary. Moreover, because (15.10) is by no means limited to the Ising model — any thermodynamic model of this exponential type will have the same property — we conjecture that this necessary condition will hold quite generally. Putting it differently, it is very difficult to imagine that the parallel chains can swap with nontrivial probabilities without making the chains "infinitesimally closer" when their corresponding stationary distributions have "infinitesimally small" overlaps as measured by the Hellinger distance.

We remark that, in general, the Hellinger derivative, under mild regularity conditions, is half of the square root of the Fisher information. This can be seen by invoking a Taylor expansion of the log-likelihood function (when it is justified)

$$\ell(\theta + \delta|X) = \ell(\theta|X) + S(\theta|X)\delta - \frac{I(\theta|X)}{2}\delta^2 + O_p(\delta^3), \qquad (15.11)$$

where $S(\theta|X)$ is the score function and $I(\theta|X)$ is the *observed* Fisher information (where X is the observed data). Applying (15.11) to $p(x|\theta \pm \epsilon)$ in

(15.8) with $\delta = \pm\epsilon$ (and with θ in place of β) yields

$$\partial H(\theta) = \lim_{\epsilon \downarrow 0} \left[\int \frac{1 - \exp\{-I(\theta|X)\epsilon^2/2 + O_p(\epsilon^3)\}}{2\epsilon^2} p(X|\theta)\mu(dX) \right]^{\frac{1}{2}}.$$

(15.12)

When the limit operator in (15.12) can be interchanged with the integration operator, we can conclude that

$$\partial H(\theta) = \frac{1}{2} \left[\int I(\theta|X)p(X|\theta)\mu(dX) \right]^{\frac{1}{2}} = \frac{1}{2}\sqrt{I(\theta)}, \qquad (15.13)$$

where $I(\theta)$ is the *expected* Fisher information. This general result provides additional insight into the phase transition phenomenon, because Fisher information measures the ability to distinguish different distributions and a near infinite Fisher information indicates that the systems on "opposite" sides of a critical point are fundamentally different (and therefore trivial to distinguish). This result also implies that, for the Ising model (15.1), the Hellinger derivative is also the half the standard deviation of the "sufficient statistic" $\sum_{i \sim j} x_i x_j$ (this can also be obtained by observing that in an exponential family, the log of the normalizing constant is also the cumulant generating function). So, near the critical value, the sufficient statistic has essentially infinite variance. For a detailed treatment of the mixing properties of the Swendsen-Wang algorithm near critical points, see Borgs *et al.* (1999).

15.5. A Brief Overview of Perfect Sampling

A major contribution to Monte Carlo simulation was made by Propp and Wilson (1996), who established and popularized a general framework for converting Markov chain samplers into exact samplers. Whereas an MCMC algorithm generally produces authentic draws from the targeted stationary distribution only at the limit, Propp and Wilson showed how this can actually be achieved, for some problems, in a finite (but random) number of iterations with probability one. This seemingly impossible task was accomplished by coupling, *from the past* (i.e., with negative time), the same Markov chain with different starting points in a set $\mathcal{S}$ to detect a coalescence at time zero. As far as detecting limits goes, going from time $-\infty$ to time zero is the same as going from time zero to $+\infty$. The coupling is done by using the same random number sequence for all the chains, but by going backward in time from time zero, we do not need to go back infinitely many steps before detecting the coalescence.

One way to illustrate this is to consider a Markov chain $Y_t = \psi(Y_{t-1}, U_t)$, where $\{U_t, t = 1, 2, \ldots\}$ are i.i.d. random variables. This produces the so-called *time-forward* sequence

$$Y_t = \psi(\psi(\ldots \psi(Y_0, U_1) \ldots, U_{t-1}), U_t), \tag{15.14}$$

where $Y_0 \in \mathcal{S}$ is a starting point. In contrast, we can also build the *time-backward* sequence

$$\tilde{Y}_t = \psi(\psi(\ldots \psi(Y_0, U_t) \ldots, U_2), U_1). \tag{15.15}$$

Clearly, Y_t and $\tilde{Y}_t$ have identical distribution for any t, yet $\{\tilde{Y}_t, t \geq 1\}$ itself is not a Markov chain. As discussed in Craiu and Meng (2005), by giving up the Markovian property, we gain a better convergence property, because while $\{Y_t, t \geq 1\}$ converges in distribution, $\{\tilde{Y}_t, t \geq 1\}$ converges with probability 1 (under regularity conditions). This is because as t increases, $\tilde{Y}_t$ varies less and less as a function of Y_0, since U_t enters into deeper and deeper levels of the nested function compositions in (15.15). In other words, as t increases, $\tilde{Y}_t$ becomes increasingly restricted by the entire history of Us. Contrast this with (15.14), where U_t is always in the outside level of the composition. Indeed, it can be shown that (Foss and Tweedie (1998)) if ψ is such that Y_t is uniformly ergodic, then with probability one, $\tilde{Y}_t$, as a function of Y_0, will become a constant function with a finite (but random) t. That is, $\tilde{Y}_t$ can "hit and stay at" the limit in a *finite number* of iterations (Thorisson (2000)). This is really the essence of perfect or exact sampling — Propp and Wilson (1996) coupling from the past (CFTP) method is just a device for finding this finite stopping time by mapping t to $-t$ in (15.15) and thereby creating a forward-looking sequence going from time $-t$ to time zero.

While the idea of using negative time Markov chains was known previously in the theoretical context, Propp and Wilson (1996) appear to be the first to have brought this theoretical tool to the attention of the general Monte Carlo simulation community. One key observation made by Propp and Wilson (1996) is that, for many monotone, discrete-state space Markov chains, we do not need to choose the "starting set" $\mathcal{S}$ to be the whole state space. Here by "monotone" we mean $Y_t = \psi(Y_{t-1}, U_t)$ is monotone in Y_{t-1} with respect to a partial ordering. In such cases, if there is also a maximum state and minimum state with respect to the same partial ordering, we can choose $\mathcal{S}$ to contain only these two extreme states, that is, to trace only the maximum chain and minimum chain. The monotonicity of the updating function then ensures that if the two coupled chains from these two extreme states coalesce at time zero, all coupled chains starting from other states will automatically reach the same state at time zero because they will all be "squeezed" between the two extreme chains. For other clever tricks and strategies for implementing perfect sampling, consult the web site maintained by David Wilson, http://dimacs.rutgers.edu/~dbwilson/exact, which contains a comprehensive and up-to-date list of research papers on perfect sampling and related topics.

15.6. Huber's Bounding Chain Algorithm

When there is no such monotonicity that can be used, it is still possible to use the "squeezing" idea, as discussed in Häggström and Nelander (1998) and Huber (1998). The idea is to introduce an appropriate bounding chain such that the coalescence of the bounding chain implies the coalescence of the original chain, and the coalesced value from the original chain is a genuine draw from its stationary distribution. As an example, Huber (2002) provides such a bounding chain for the Ising model, using the Swendsen-Wang algorithm as its core. The key here is to design a Markov chain on the set of all nonempty subsets of the state space, such that when this chain arrives at a set of cardinality one, the one element of that set is a draw from the Ising model. See Huber (2002) for the derivation and proof. Here we give only the pseudocode. Note that in order to obtain exact samples from the Ising model, it should be implemented in the "backward" fashion as in Propp and Wilson (1996). However, to investigate empirically the coalescence time, we can run it forward as the forward and backward coalescence times have the same distribution (Propp and Wilson (1996)).

Specifically, for each $i \in L$, let Y_i be a set-valued random variable taking three possible values: $\{-1\}$, $\{1\}$, and $\{-1, 1\}$. Each update of Huber's bounding chain algorithm for the Ising model takes the following form:

> **For** each $(i, j) \in L^2$ such that $i \sim j$,
>> **If** $Y_i = Y_j$ and $|Y_i| = |Y_j| = 1$, **Let** $U_{i,j} := 1$.
>>> **Else If** $|Y_i \cap Y_j| \geq 1$, **Let** $U_{i,j} := 0$.
>>>> **Else Let** $U_{i,j} := -1$.
>> **If** $U_{i,j} \geq 0$, **Let** $U_{i,j} := -1$ with probability $e^{-2\beta}$.
>
> **Let** $\mathbb{C}$ be the set of all clusters $C \subset L$ such that the nodes of C are connected by edges (i, j) with $U_{i,j} = 1$.
>
> **Let** $\mathbb{C}'$ be the set of all clusters $C \subset L$ such that the nodes of C are connected by edges (i, j) with $U_{i,j} = 0$ or $U_{i,j} = 1$.

For each $C \in \mathbb{C}$,

   Let x be a random spin, -1 or 1, with probability $1/2$ each.
   For all $i \in C$, **Let** $X_i := x$.
Let π be a uniformly random ordering on $\mathbb{C}$.
For each $C \in \mathbb{C}$,
   Let C' be the unique set in $\mathbb{C}'$ such that $C \subset C'$.
   For each $i \in C$, **Let** X_D be the set (either $\{-1\}$ or $\{1\}$) that contains
      only the single common value of all x_i in the cluster D, and **Let**

$$Y_i := \bigcup_{\substack{D \in C \cap C': \\ \pi(D) > \pi(C)}} X_D.$$

This algorithm's roots in Swendsen-Wang shine through. The only complication is the classification of edges. In Swendsen-Wang, each edge either binds its end nodes ($U_{i,j} = 1$) or does not ($U_{i,j} = -1$). Here, there exists the third possibility ($U_{i,j} = 0$) that the current values of Y_i and Y_j do not conclusively determine whether nodes i and j should have the same spin.

In a private communication, Huber pointed out that the ordering π on $\mathbb{C}$ can be omitted. In such a case, the assignment of the new value of Y_i is the union over all $D \in \mathbb{C} \cap C'$. This modification also produces a valid bounding chain, but this chain will take longer to couple.

As mentioned, to actually obtain draws from (15.1), one must implement the updating algorithm above using the coupling from the past protocol of Propp and Wilson (1996). The easiest way to do this is to carefully keep track of the seeds used by the pseudorandom number generators for the previous pass through the algorithm (note each pass uses no more than $4|L|$ uniforms). Our implementation was essentially as follows.

$k := 1$. **Repeat**
   Generate new value for seed and
   Store it as the kth element in the list.
   Let $m := k$.   **Repeat**
      Let seed take the value of the mth
element in the list.
      Repeat 2^{m-1} times:
         Do one update according to the code above.
      Let $m := m - 1$.
   Until $m = 0$.
   Let $k := k + 1$.   **Until** $\|Y_i\| = 1$ for all

$i \in L$.

Return X such that $X_i \in Y_i$ for all $i \in L$.

End.

Implementation of this algorithm is not especially difficult when using a lattice with a regular structure. The only tricky parts are those operations performed on the clusters, that is, the subsets of L in $\mathbb{C}$ and $\mathbb{C}'$. We warn anyone interested in implementing this algorithm to be especially careful when programming. Telling the difference between Ising states with the right distribution and those with a wrong one (but not "too wrong") is nearly impossible when merely looking at the output.

Moreover, note that the implementation above is different from merely running the updating sequence until the bounding chains coalesce. Such an algorithm would be "forward" in nature and is not guaranteed to produce draws from (15.1). However, the draws from the forward implementation may not appear to be different from those of a backward implementation, at least on casual inspection. The following example illustrates how we can find lower dimensional summaries that can reveal the biases in a high-dimensional sampler, even when the biases arise in subtle ways.

Specifically, we made 1000 draws from each implementation with $\beta = 0.25$ for a 128×128 toroidal lattice, and recorded the sufficient statistic $\sum_{i \sim j} x_i x_j$ for each draw. Unsurprisingly, the sample statistics are approximately normally distributed, and therefore their first two moments are good summaries of their distributions. The forward draws had mean 2281 and sample variance 11,915. The backward draws had mean 2286 and sample variance 11,780. But, examination of a quantile-quantile plot (see Figure 15.1) reveals some differences in the tails. To see this more clearly, we divided our sample into 100 subsamples of 10 draws of each kind. For each subsample we computed the minimum and the maximum. The minima had mean 2113 and sample variance 3810 for the forward draws and mean 2122 and sample variance 5404 for the backward draws. Similarly for the maxima, we observed a mean of 2449 and sample variance 3666 for the forward draws and a mean of 2451 and a sample variance of 4595 for the backward draws. A common obstacle with most forward implementations is that the distribution of the resulting draw is condensed onto a subset of the draw's state space (i.e., it "gets stuck" in a subspace). We suspect that the diminished variances of the order statistics in the forward implementation reflects this problem.

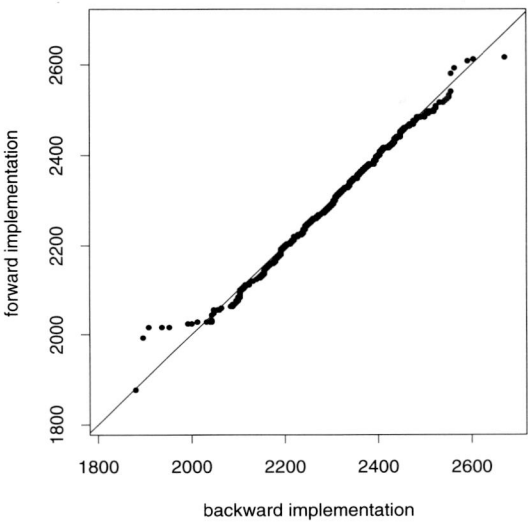

Figure 15.1.   Comparisons of forward and backward implementations.

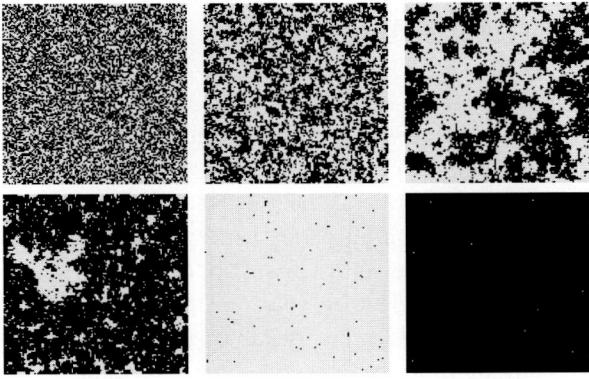

Figure 15.2.   Ising draws from Huber's algorithm, with $\beta = 0.05, 0.3, 0.42, 0.44, 0.7$ and 0.9, respectively.

15.7. Approximating Criticality via Coupling Time

Figure 15.2 shows six Ising draws on a 128×128 square lattice with toroidal ("wrap around") boundary conditions to reduce boundary effects. A white pixel indicates a down spin and a black pixel an up spin, though the distinction is unimportant given the symmetry of the distribution. The values of β corresponding to the draws are 0.05, 0.3, 0.42, 0.44, 0.7, and 0.9, respectively. Note, that for small β, there are approximately as many down spins as up spins. Also note that they are clumped together more than one would expect if each spin were assigned independently. In the $\beta = 0.42$ picture the clusters have grown larger, but we still have approximate parity of spin directions. This indicates that the phase transition has not occurred. When $\beta = 0.44$, we see that the parity of spin directions is long gone, indicating that the phase transition has occurred. As β increases, one spin direction dominates the lattice, with only a few brave spins in the opposite direction.

A natural question to ask is how many iterations are typically required for coalescence, particularly when β is near criticality. To answer this, for each β on a grid from 0.01 to 0.99 with increments of 0.01, and for each of four increasingly larger lattices, we made 100 draws using Huber's bounding chain *with a forward implementation*. This is justified by the fact that the distribution of the coupling time is the same for the forward or backward implementation. Moreover, a forward implementation counts the precise number of iterations, whereas a typical backward implementation counts on an exponential base 2 scale, the so-called "binary back-off" scale (Propp and Wilson, 1996). For each forward draw, we recorded the number of iterations.

The average number of iterations till coalescence are shown in Figure 15.3. The bottom curve was generated using a 16×16 square, toroidal lattice. The curves above it correspond to 32×32, 64×64, and 128×128 lattices, in that order. The scale of the plot is logarithmic, base 4. We chose base 4, since each lattice contains four times as many nodes as the next smaller one. The vertical line represents β_c of (15.6).

It is not difficult to conjecture that the peak corresponds to critical slowing down, although it does not occur precisely at β_c, which is for an infinite lattice, not for our finite lattices with toroidal boundary conditions. Given the well developed theory of "finite-size scaling" in statistical physics (e.g., Binder and Heermann, 1982; Landau and Binder, 2000), one way to empirically check this conjecture is to see if the peak as identified by the coupling method will approach the limiting value β_c as the lattice

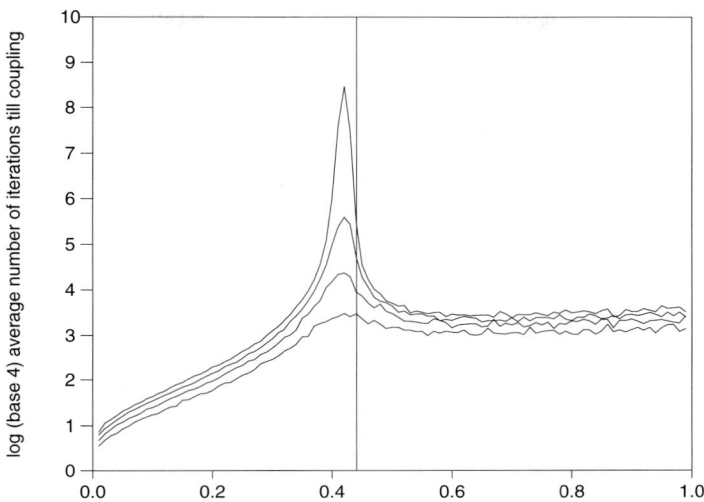

Figure 15.3. Average number of iterations till coupling of Huber's algorithm as a function of β for four lattice sizes: $2^k \times 2^k$, $k = 4, 5, 6, 7$ (from bottom up). The vertical line corresponds to the limiting critical value β_c with infinite lattice.

size increases, though we have neither the patience, nor the computational power, to verify this. The critical slowing is not surprising, since there is a strong relationship between the coupling time of a backward sampler and the mixing time of its underlying Markov chain, and the Swendsen-Wang algorithm is known to be subject to critical slowing, as discussed previously. Away from the peak, the graphs are more or less evenly spaced, suggesting a linear increase in the number of iterations till coupling as a function of lattice size. The bumpiness of the graphs on the high end indicates larger variation in the coupling time for big values of β. On the whole, we feel that the general shape of the graphs at that end is clear.

Using the information obtained when creating Figure 15.3 also enabled us to improve the quality of our backward implementation. Except when β is very small, starting with a negative t near zero for the backward coupling is usually a poor choice. In our implementation, we estimated the 99th percentile of the required number of steps till coupling, for each value of β. The initial number of steps was set to be this percentile depending on β. This allowed us to make only one pass through the outer loop for generating each picture of Figure 15.2. On occasion, of course, more than one pass may be necessary.

We emphasize that, although Huber's bounding chain algorithm does slow considerably near criticality, it is nevertheless a very useful algorithm. Our earlier discussion of criticality leaves little hope for the discovery of any algorithm that does not exhibit some kind of critical slowing. Moreover, by making plots similar to Figure 15.3, one can empirically determine critical values on any lattice for which the algorithm is implemented. This potentially can be useful, because the analytical computation of critical values quickly gets out of hand when the lattice is made even slightly more complex.

We conclude this section by noting that if our earlier conjecture is true, it should hold for MCMC algorithms other than Huber's algorithm, because our general results in Section 4 refer to a property of the target distribution, not any particular MCMC algorithm. That is, our general conjecture is that there is a close tie between the magnitude of the Hellinger derivative, with respect to a parameter indexing a family of distributions, and an MCMC algorithm whose stationary distribution is from the same family (and with the same parameter value). Of course, we are at a very early stage of exploring this relationship, for we yet need to formulate a measure to quantify this tie.

15.8. A Speculation

A perfect sampling algorithm such as Huber's is very useful even if one is unwilling to wait long to use it to obtain a large sample of independent draws, because it can be used to generate the initial state for a Markov chain sampler like Swendsen-Wang. Indeed, using a perfect sampling algorithm to generate an initial state is a generally recommended strategy because it eliminates potential biases due to insufficient "burn-in." For high-dimensional problems such as Ising models, this is particularly important because it is generally not possible to just choose a "reasonable looking" starting point (which is often feasible for univariate distributions), nor is it easy, if at all possible, to perform reliable diagnostic tests to determine an adequate burn-in period (for the entire state distribution, not just for some summary statistics).

Given this utility, we naturally wonder if we can construct perfect sampling algorithms for many, if not all, practical applications, such as those encountered in *routine* Bayesian computation (Gelman, Carlin, Stern, and

Rubin (2004); van Dyk and Meng (2001)). This turns out to be a very challenging problem in general, as demonstrated in Murdoch and Meng (2001). Indeed, the jury is still out on whether perfect sampling can ever be a tool as general as the general MCMC framework itself.

Historically, essentially all major Monte Carlo methods were originated, (re)discovered independently, or at least researched extensively by researchers in statistical physics or in related scientific areas. Perfect sampling seems to be the only exception, at least so far. While some of us might take pride in this observation, we wonder if this lack of involvement of these researchers, so far, is somewhat responsible for the fact that perfect sampling is still largely a pet pony, not a workhorse, for general scientific computation. Perhaps it can never be a workhorse as strong as Metropolis-Hastings or the Gibbs sampler. But nevertheless given the amazing success of statistical physicists and the like in the entire history of Monte Carlo development, it seems logical to speculate that the real breakthrough of perfect sampling as a general tool might depend on, possibly critically, the direct involvement of researchers from statistical physics and related scientific areas.

Acknowledgements

We thank J. Blanchet, G. Goswami, M. Huber, J.S. Liu, and S. Lalley for helpful exchanges, and a referee for a set of very insightful comments. This research was supported in part by a U.S. Department of Education grant (Servidea) and several National Science Foundation grants (Meng).

References

Baxter, R. J. (1982). *Exactly Solved Models in Statistical Mechanics*, New York: Academic Press, S6.1.

Borgs, C., Chayes, J. T., Frieze, A., Kim, J. H., Tetali, P., Vigoda, E., and Vu, V. H. (1999). Torpid mixing of some Monte Carlo Markov chain algorithms in statistical physics, *Proc. 40th IEEE Symposium on Foundations of Computer Science*, 218–229.

Craiu, R. and Meng, X.-L. (2005). Multiprocess parallel antithetic coupling for backward and forward Markov chain Monte Carlo, *Annals of Statistics* **33**, 661 - 697.

Foss, S. G. and Tweedie, R. L. (1998). "Perfect simulation and backward coupling," *Comm. Statist. Stochastic Models* **14**, 187-203.

Geyer, C. J. (1991). Markov chain Monte Carlo maximum likelihood, *Computing Science and Statistics: Proc. 23rd Symp. Interface,* 156-163.

Geyer, C. J. and Thompson, E. A. (1995). Annealing Markov chain Monte Carlo with applications to ancestral inference, *J Amer Statist Assoc* **90**, 909-920.

Gelman, A., Carlin, J, Stern, H. and Rubin, D. B. (2004). *Bayesian Data Analysis* (2nd Ed). Chapman and Hall.

Häggström, O. and Nelander, K., (1998). Exact sampling from anit-monotone systems, *Statistica Neerlandica* **52**, 360–380.

Higdon, D. M. (1998). Auxiliary variable methods for Markov chain Monte Carlo with applications, *J Amer Statist Assoc* **93**, 585–595.

Huber, M. L. (1998). Exact sampling and approximate counting techniques, *Proc 30th Symp Theory of Computing*, 31-40.

Huber, M. L. (2002). A bounding chain for Swendsen-Wang, *Random Struct Alg* **22**, 43-59.

Liu, J. S. (2002). *Monte Carlo Methods for Scientific Computation*, Springer Verlag, Berlin.

McCoy, B. M. and Wu, T. T. (1973). *The Two-Dimensional Ising Model*, Cambridge, Mass., Harvard University Press, §2.4-§2.5.

Metropolis, N., Rosenbluth, A. W., Rosenbluth, M. N., Teller, A. H. and Teller E. (1953). Equations of state calculations by fast computing machines, *J Chemical Physics* **21**(6), 1087–1091.

Madras, N. and Zheng, Z. (2002). On the swapping algorithm, *Random Struct Alg* **22**, 66-97.

Murdoch, D. and Meng, X.-L. (2001). Towards perfect sampling for Bayesian mixture prior, In *Bayesian Methods, with Applications to Science, Policy and Official Statistics* (E. George, ed.), 318-390. Eurostat.

Propp, J. G. and Wilson, D. B. (1996). Exact sampling with coupled Markov chains and applications to statistical mechanics, *Random Struct Alg* **9**, 223–252.

Swendsen, R. H. and Wang, J. S. (1987). Nonuniversal critical dynamics in Monte Carlo simulations, *Phys Rev Letters* **58**, 86–88.

Thompson. C. J. *Mathematical Statistical Mechanics.* Princeton University Press (1972).

Thorisson, H. (2000). *Coupling, Stationary, and Regeneration.* Springer, New York.

Tesi, M. C., van Rensburg, E. J., Orlandini, E. and Whittington, S. G. (1996). Monte Carlo study of the interacting self-avoiding walk model in three dimensions, *J. Stat. Phys.* **82**, 155-181.

van Dyk, D. and Meng, X. L. (2001). The art of data augmentation (with discussion), *J. Computational and Graphic Statistics* **10**, 1-111.

CHAPTER 16

Network Tomography: A Review and Recent Developments

Earl Lawrence

Statistical Sciences Group
Los Alamos National Laboratory
Los Alamos, NM 87545
earl@lanl.gov

George Michailidis, Vijayan N. Nair

Department of Statistics
The University of Michigan
Ann Arbor, MI 48109-1107
{gmichail,vnn}@umich.edu

Bowei Xi

Department of Statistics
Purdue University
West Lafayette, IN 47907
xbw@stat.purdue.edu

The modeling and analysis of computer communications networks give rise to a variety of interesting statistical problems. This chapter focuses on network tomography, a term used to characterize two classes of large-scale inverse problems. The first deals with passive tomography where aggregate data are collected at the individual router/node level and the goal is to recover path-level information. The main problem of interest here is the estimation of the origin-destination traffic matrix. The second, referred to as active tomography, deals with reconstructing link-level information from end-to-end path-level measurements obtained by actively probing the network. The primary application is estimation of quality-of-service parameters such as loss rates and delay distributions. The chapter provides a review of the statistical issues and developments in network tomography with an emphasis on active tomography. An application to internet telephony is used to illustrate the results.

16.1. Introduction

There has been a great deal of interest recently, in both the engineering and research communities, on the modeling and analysis of communications and computer networks. This paper provides a review of network tomography and describes some interesting statistical issues, challenges, and recent developments. The term network tomography, introduced by Vardi (1996), has been used in the literature to characterize two broad classes of inverse problems. The first is passive tomography where aggregate data are collected at the router level. The goal is to disaggregate these to obtain finer-level information. The most common application, which was the original problem studied in Vardi (1996), is estimation of the origin-destination traffic matrix of a network. The second is active tomography where the network is actively "probed" by sending packets from a source to several receiver nodes, all located on the periphery of the network. Here one can collect only end-to-end path-level information, and the goal is to use this to recover individual link-level information. We will provide a brief review of both of these areas but focus more on the latter, as this has been the subject of our own research. There are also several other interesting statistical problems, especially related to network data obtained from a single network link, that arise in the study of communications and computer networks. These will not be be discussed here. See, however, the collection of papers in Adler, Feldman, and Taqqu (1998), Park and Willinger (2000) and references therein.

The work in network tomography has been stimulated by the demand for sophisticated techniques and tools for monitoring network utilization and performance by network engineers and internet service providers (ISP). This need has increased further in recent years due to the complexity of new services (such as video-conferencing, internet telephony, and online games) that require high-level quality-of-service (QoS) guarantees. The tools and techniques are also important for network management tasks such as fault and congestion detection, ensuring service-level-agreement compliance, and dynamic replica management of web services, just to name a few: Coates, Castro, Gadhiok, King, Tsang, and Nowak (2002).

There are two categories of methods in network tomography: (i) node-oriented methods that collect packet and network flow information passively through monitoring agents located at local network devices such as routers, switches, and hosts; and (ii) path-oriented methods that collect information about connectivity and latency in a network by actively sending

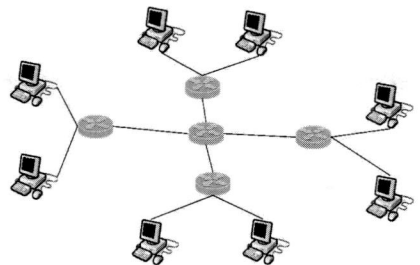

Figure 16.1.   Layout of a small network showing routers/nodes and links.

probe packets through the network from nodes located on its periphery. The first category of tools are geared towards network operators who use the information for capacity planning and management decisions. Their main shortcoming is that they require access to all the network elements (routers/switches) in order to deploy monitoring agents to collect the information. Furthermore, the amount of data generated can be substantial. The second category of tools collect data on network performance measures that are indirectly related to the parameters of interest and does not require cooperation from the internal nodes of the network. For both types, however, the collected data have to be appropriately processed (through the solution of different types of statistical inverse problems) to obtain the information of interest. (Castro, Coates, Liang, Nowak, and Yu (2004)).

We provide here a brief background on network traffic flow so that readers can follow the discussion. A more detailed and accessible introduction can be found in Marchette (2001). Throughout, we represent a network by a graph $\mathcal{G} = (\mathcal{V}, \mathcal{E})$ where $\mathcal{V}$ is the set of nodes and $\mathcal{E}$ the set of links. Figure 16.1 is an example of a small network with computers/workstations connected by routers and links. When a file is transferred from one location in the network to another (or one node to another), the file's content is first broken into pieces, called packets. Information about origin-destination, reassembly instructions (such as sequence numbers), and error-correcting features are also added to the packet. The origin-destination information is used by the network elements (routers and switches) to deliver the packets to the intended recipient. One can think of the routers (internal nodes in Figure 16.1) as the intersections in a road network. Packets are queued at routers, awaiting their transmission to the next router according to some protocol (first-in-first-out is common, but there are others). Physically, a queue consists of a block of computer memory that temporarily stores the packets. If the queue (memory) is full when a packet arrives, it is discarded and, depending on the transmission protocol, the sender may or may not be alerted. Otherwise, it waits until it reaches the front of the queue and is forwarded to the next router on the way to its destination. This queuing mechanism is responsible for observed packet losses and, to a large extent, for packet delays.

16.2. Passive Tomography

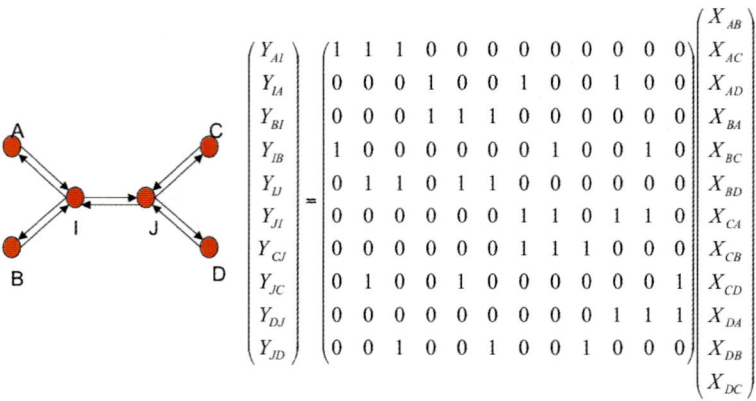

$$
\begin{pmatrix} Y_{AI} \\ Y_{IA} \\ Y_{BI} \\ Y_{IB} \\ Y_{IJ} \\ Y_{JI} \\ Y_{CJ} \\ Y_{JC} \\ Y_{DJ} \\ Y_{JD} \end{pmatrix} = \begin{pmatrix} 1 & 1 & 1 & 0 & 0 & 0 & 0 & 0 & 0 & 0 & 0 & 0 \\ 0 & 0 & 0 & 1 & 0 & 0 & 1 & 0 & 0 & 1 & 0 & 0 \\ 0 & 0 & 0 & 1 & 1 & 1 & 0 & 0 & 0 & 0 & 0 & 0 \\ 1 & 0 & 0 & 0 & 0 & 0 & 0 & 1 & 0 & 0 & 1 & 0 \\ 0 & 1 & 1 & 0 & 1 & 1 & 0 & 0 & 0 & 0 & 0 & 0 \\ 0 & 0 & 0 & 0 & 0 & 0 & 1 & 1 & 0 & 1 & 1 & 0 \\ 0 & 0 & 0 & 0 & 0 & 0 & 1 & 1 & 1 & 0 & 0 & 0 \\ 0 & 1 & 0 & 0 & 1 & 0 & 0 & 0 & 0 & 0 & 0 & 1 \\ 0 & 0 & 0 & 0 & 0 & 0 & 0 & 0 & 0 & 1 & 1 & 1 \\ 0 & 0 & 1 & 0 & 0 & 1 & 0 & 0 & 1 & 0 & 0 & 0 \end{pmatrix} \begin{pmatrix} X_{AB} \\ X_{AC} \\ X_{AD} \\ X_{BA} \\ X_{BC} \\ X_{BD} \\ X_{CA} \\ X_{CB} \\ X_{CD} \\ X_{DA} \\ X_{DB} \\ X_{DC} \end{pmatrix}
$$

Figure 16.2.   A small network and its associated origin-destination traffic matrix.

The goal in traffic (or origin-destination) matrix estimation is to obtain information about the distributions of traffic flowing from $\mathcal{V}_i$ to $\mathcal{V}_j$ for all pairs of nodes i and j in the network. Of course, one would also have to study various sources of variation such as time-of-day, day-of-the-week, and other effects to characterize the traffic pattern. This information is used by network engineers for capacity planning and network management operations.

In this application, monitoring agents are placed at the individual nodes, and data on total number of packet counts traversing the node are collected. The packets do contain origin-destination (OD) information, but due to volume of the data, it is impractical to access individual packets to collect this information. So only total packet count data are available and are obtained using the Simple Network Management Protocol (SNMP).

Let $|\mathcal{V}|$ and D denote the number of nodes and OD pairs in the network respectively. Let us restrict attention to a fixed time period where the traffic patterns are fairly homogeneous. Let Y_t be a $|\mathcal{V}|$ column vector containing the number of packets traversing all the nodes in period t for $t = 1, ..., T$ during the time of study, with $\bar{Y}$ being the average number of packets in the entire period. Finally, let R be a $|\mathcal{V}| \times D$ routing matrix (corresponding to the permissible routes through the network), and let X be a D column vector that represents the unknown OD flows traversing the network. The routing matrix can be deterministic (entires are 0-1) or random (entries are probabilities); the latter refers to the case with multiple paths in a network due to load balancing considerations. The routing matrix actually changes over time, usually on the order of a few hours (Paxson (1997)), and is typically estimated by computing shortest paths using the Interior Gate Protocol link weights that indicate congestion levels, together with known information about the network topology. An example of the traffic matrix problem is shown in Figure 16.2.

We can then write

$$\bar{Y} = RX.$$

The statistical inverse problem is to reconstruct the distribution of X from the aggregate level data $\bar{Y}$. In general, $D >> |\mathcal{V}|$, usually $D = \mathcal{O}(|\mathcal{V}|^2)$, so this is a highly ill-posed inverse problem and cannot be solved without additional assumptions or regularization. We provide below a review of several approaches in the literature for addressing this. See also Papagiannaki, Taft, and Lakhina (2004) for a discussion on why direct data collection and estimation of X is intractable using today's monitoring technologies.

Vardi (1996) modeled the traffic flows as Poisson, i.e., the X_js are independent Poisson random variables with means λ_js. The Poisson assumption provides additional estimating equations because the variance is equal to the mean, so the higher order information can be used for estimation. Vardi (1996) studied maximum likelihood estimation using the EM algorithm. However, as shown there, the EM algorithm may not converge to the MLE. More importantly, the algorithm becomes computationally intractable for large networks. Vardi (1996) studied several heuristic methods as alternatives, among which the following method-of-moments estimation was the most promising. Recall that $\bar{Y}_j$, $j = 1, ...|\mathcal{V}|$ provide $|\mathcal{V}|$ estimating equations. In addition, the sample variances and covariances $S_{ij} = \sum_{t=1}^{T} [Y_{it} - \bar{Y}_i][Y_{jt} - \bar{Y}_j]/(T - 1)$ provide another $\frac{|\mathcal{V}| \times (|\mathcal{V}|+1)}{2}$ equations. Note than $E(S_{ij})$ reduces to the expected value of the variance of the counts in the shared links, and so it is again a linear function of the λ_j's.

Hence, letting $\mathbf{S}$ denote a vector of the elements of the variance-covariance matrix and letting $\Lambda = (\lambda_1, ... \lambda_{|E|})^T$, where $|E|$ is the number of edges in the network, we have the linear model

$$\begin{bmatrix} E(\bar{Y}) \\ E(\mathbf{S}) \end{bmatrix} = \begin{bmatrix} R \\ B \end{bmatrix} \Lambda$$

for a suitable matrix B. Now, the data on the left-hand side are approximately normal when T is large, so we can use the resulting large-sample normal theory and weighted least-squares to estimate Λ and develop other inferential procedures. We note that Vardi proved identifiability for all practical networks. A Bayesian approach under the same framework was considered in Tebaldi and West (1998). The goal was slightly different, dealing with prediction of the actual OD traffic counts instead of the distribution of the counts. The Poisson assumption implies that the variance of X is proportional to the mean. Cao, Davis, Vander Wiel, and Yu (2000) relaxed this assumption by considering a general model of the form $Var(X) \propto E(X)^\alpha$ and developed estimation methods.

This first generation of models does not work well in estimating the distribution of X in high speed networks as they are very sensitive to the assumptions (Poisson, normal with a specific mean-variance relationship), which did not quite hold for real network traffic data (Medina *et al.* (2002)). This has led to a new generation of models that employ extra information or other assumptions. The two most prominent approaches are the *tomogravity* model (Zhang *et al.* (2003a and 2003b)) and the method of routing changes (Soule *et al.* (2004)). The tomogravity model is based on the premise that the OD flow $X(i, j)$ between nodes i and j is proportional to the total amount of traffic departing node j, $X(\cdot, j) = \sum_{i \in V} X(i, j)$, and the total amount of traffic entering node i, $X(i, \cdot)$; i.e. $X(i, j) \propto X(i, \cdot) \times X(\cdot, j)$ (Zhang, Roughan, Duffield, and Greenberg (2003)). This model assumes complete independence between the sources and destinations, which tends to be violated in backbone networks due to the so-called hot-potato routing policies adopted by their operators (operators offloading peering traffic at the nearest exit point). A modification of the simple tomogravity model capturing such issues was also proposed in Zhang, Roughan, Duffield, and Greenberg (2003). Another modification that embedded the tomogravity model in a regularization framework was proposed in Zhang *et al.* (2003b), where the problem was formulated as

$$\min ||Y - RX||_2^2 + \lambda^2 K(X|X'), \quad \text{subject to } X > 0,$$

where X' is the solution to the traffic estimation problem under the generalized gravity model, $K(X|X')$ the Kullback-Leibler divergence measure and $||\cdot||$ denotes the L_2 norm. In contrast, the route change method (Soule, Nucci, Leonardi, Cruz, and Taft (2004)) attempts to overcome the underconstrained nature of the problem by manipulating the link weights and thereby inducing additional routing matrices R.

A third generation of traffic matrix estimation methods incorporated temporal considerations, namely data are obtained over $t = 1, ..., T$ periods and the goal is to estimate the temporal evolution of the OD flows. Cao *et al.* (2000) developed two approaches for estimating the parameters of X as they evolved over time. The first was based on a a moving window with a locally time-homogeneous approach within each window while the second used a more formal temporal model. Soule *et al.* (2005) proposed the following state-space model:

$$Y_t = RX_t + u_t, \quad t = 1, \cdots, T \tag{16.1}$$
$$X_t = CX_{t-1} + e_t, \quad t = 1, \cdots, T, \tag{16.2}$$

where the first equation is the traditional traffic matrix model, with u_t representing measurement error, and the second equation posits both a temporal model for the underlying traffic state as well as spatial dependence between OD flows through the non-diagonal elements of the C matrix, and e_t captures the traffic system noise process. Liang *et al.* (2005) proposed a modification of the above model, and among other things, allowed for occasionally direct measurements for selected OD flows that aid in calibrating the parameters of the model.

The traffic matrix estimation problem has proved useful to network operators, especially for capacity planning purposes. In fact, many of the proposed techniques originated from telecommunication research groups such as AT&T and Sprint Labs and have been applied to networks involving up to 1000 nodes. The ill-posed nature of the problem requires the imposition of various modeling assumptions that in many cases have had a negative impact on the accuracy of the results (Medina *et al.* (2002), Roughan (2005)). The focus over the years has shifted to understanding both the temporal and spatial variability of OD flows, as attested to by the latest models. Some of the research challenges involve the quest for models that can capture more accurately the characteristics of the OD flows in today's high speed networks, fast and scalable estimation techniques and the simulation of realistic traffic matrices (Roughan (2005)).

16.3. Active Tomography

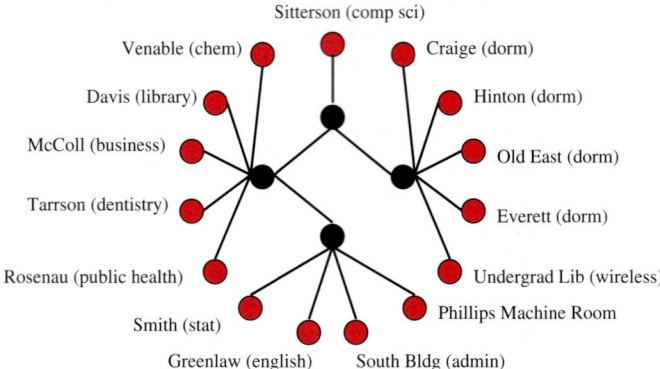

Figure 16.3.   The physical topology of the UNC campus network.

A second class of problems deals with estimation of the quality-of-service parameters of the network such as delays and loss rates at the individual link/router level. This information is used to characterize and monitor the performance of the network over time, detect congestions or other anomalies in the network, and ensure compliance with service-level agreements. The difficulty and the challenge arise from the fact that many service providers do not own the entire network and hence do not have access to the internal nodes. Active tomography provides an interesting and convenient alternative by "probing" the network from nodes located on the periphery and using this to recover information about the internal links.

16.3.1. *Background and probing experiments*

To describe the details, consider the network shown in Figure 16.3. It depicts part of the campus network at the University of North Carolina. We will come back to a real application dealing with this network, but in this section we will use it to describe the active tomography problem. We can study the performance of the internal links of this network by sending "probe" packets from a source (in this case Sitterson) to all the other nodes on the periphery of the network (receiver nodes). Special equipment placed on the source and receiver nodes is used to send the packets and collect end-to-end information on losses and delays. The packets can be sent from the source node to the receiver(s) using a unicast or multicast transmission protocol.

In a unicast protocol, the packets are sent to one receiver at a time; however, such schemes cannot estimate all the internal link-level parameters. In multicast protocols, packets are sent simultaneously to any specified set of receivers. The higher-order information in multicast schemes (performance of losses and delays in shared links) allows one to reconstruct internal link-level information. Some networks have turned off multicast transmission due to security reasons. In such situations, back-to-back unicast schemes, where packets are sent within nanoseconds of each other to two or more receivers, have been proposed in order to mimic multicast transmissions (Tsang, Coates, and Nowak (2003)).

The logical topology for the probing experiment associated with this UNC campus network is shown on the left panel of Figure 16.4. This corresponds to a tree topology with source node 0 (Sitterson) at the top and receiver nodes 4, 5, 6, 7, 8, 10, 11, 12, 13, 14, 15, 16, 17, and 18 as the leaves. Note that we can observe only end-to-end measurements (0-4, 0-10, 0-15, etc.) on losses and delays, and we have to use this information to reconstruct all the internal link-level information (0-1, 1-2, 2-9, etc.). This is the inverse problem. In practice, the number of nodes can vary from about a dozen for small area networks (such as a campus network) to several hundred in wide area networks. However, the investigators can reduce the size of the network by collapsing the links (combining links and nodes) if they are interested in just a coarse investigation of the network. A detailed study will require looking at all the nodes.

The traditional approach to probing experiments has been based on full multicast transmission where the probes are sent to all the receivers in the network (or back-to-back unicasts intended to mimic the multicast scheme). The difficultly is that this scheme is quite inflexible. One rarely wants to send the same number of probes to all the receiver nodes. Rather, we want to be able to investigate different regions of the network with different intensities and even possibly at different times. In Xi, Michailidis and Nair (2005) and Lawrence, Michailidis, and Nair (2005a), we have proposed the use of a flexible class of probing experiments (referred to as flexicast experiments) for active tomography. This consists of $\mathcal{C} = \{C_h, N_h\}$, a collection of sub-experiments C_h with probe size N_h that lead to identifiability of all the link-level parameters. The individual sub-experiment C_h covers only part of the network and by itself cannot estimate all the parameters in the subnetwork. However, by judiciously designing the subexperiments, we can estimate all the parameters in the entire network of interest. This class of experiments is particularly useful in network monitoring where we want to study different subregions of a network depending on where anomalies, such as congestion, occur.

We have developed necessary and sufficient conditions under which the flexicast experiments lead to identifiability (estimability) of all the link-level parameters. We first need the notion of a k-cast scheme and a splitting node. In a k-cast scheme, a probe packet is sent to a specified set of k receiver nodes. It is uniquely specified by the receiver nodes. For example, $\langle 15, 16, 17, 18 \rangle$ and $\langle 4, 5, 10, 11 \rangle$ are two four-cast schemes for the network in the left panel of Figure 16.4. A splitting node, as the name suggests, is an internal node at which a k-scheme splits. For example, the four-cast scheme $\langle 15, 16, 17, 18 \rangle$ splits at node 9 while $\langle 4, 510, 11 \rangle$ splits at nodes 1,2, and 3.

Proposition 16.1: *Let $\mathcal{C}$ be a collection of probing experiments $\{C_h, N_h\}$ and $\mathcal{T}$ be a general tree network topology. Then, all the internal link loss rates are identifiable if and only if (a) every internal node is a splitting node for some $C_h \in \mathcal{C}$ and (b) all receiver nodes are covered by $\mathcal{C}$. The same conditions are also necessary and sufficient for estimating link delay distributions provided they are discrete.*

Proofs can be found in Xi, Michailidis and Nair (2005) and Lawrence, Michailidis, and Nair (2005a). Additional conditions for identifiability of continuous delay distributions are given in Lawrence (2005).

To get some insight into the Proposition, consider the logical topology (tree) on the left panel of Figure 16.4. Suppose that $\mathcal{C}$ consists of the following three subexperiments: $C_1 = \langle 4, 5, 6, 7, 8 \rangle$, $C_2 = \langle 10, 11, 12, 13, 14, 15 \rangle$, and $C_3 = \langle 15, 16, 17, 18 \rangle$. All the receiver nodes are covered, and 2, 3, and 9 are splitting nodes for C_1, C_2 and C_3 respectively. However, the internal node 1 is not a splitting node, so this experiment will not be able to recover all the link-level parameters. A modified experiment with C_1 as before and $C_2' = \langle 10, 15 \rangle, C_3' = \langle 11, 12, 13, 14, 15 \rangle$, and $C_4' = \langle 16, 17, 18 \rangle$ will allow for estimation of all the parameters. Of course, there are many other ways of modifying the original experiment to get identifiability.

This raises the question of whether there are "optimal" approaches to constructing the flexicast experiments. This is a difficult question in general, as there are many ways to define optimality. From the point of view of statistical efficiency, a multicast scheme that sends probes to all the receivers in the network is the most optimal as it provides the highest order of dependence among the shared links and hence is most informative. However, as we have already noted, it is not very flexible. Moreover, it generates a lot of

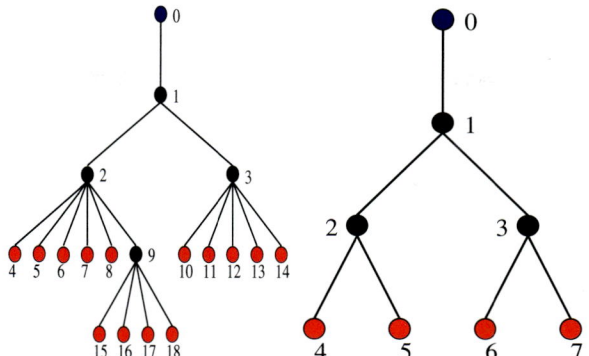

Figure 16.4.   The logical topology for the UNC study (left); A 3-layer symmetric binary tree (right).

probing traffic. Among the flexicast experiments, a collection of bicast (or two-cast) and unicast subexperiments has the least data complexity since the highest dimension is that of a bicast scheme which is multinomial with dimension four. For such a collection, one can find *minimal* experiments (smallest collections) that lead to identifiability of all the internal link parameters as follows: (a) For each internal node s, use exactly *one* bicast pair b whose splitting node is s; (b) Choose these bicast pairs to maximize the number of receiver nodes that are covered; and (c) Choose unicast schemes to cover the remaining receiver nodes $r \in \mathbb{R}$ that are not covered by the bicast pairs.

To understand the details, consider the three-layer symmetric binary tree in the right panel of Figure 16.4. The full multicast experiment sends packets to all the seven receiver nodes and hence is a seven-dimensional multinomial experiment with 2^7 outcomes. A bicast experiment can be based on all possible pairs (21 pairs). However, a minimal experiment that can estimate all the internal links requires only three bicast pairs, for example, $C_1 = \langle 4, 5 \rangle$, $C_2 = \langle 6, 7 \rangle$ and $C_3 = \langle 5, 6 \rangle$. This is not unique as we can replace C_3 by $C_3' = \langle 4, 7 \rangle$ or several others.

The active tomography problem has been studied by several authors in the literature. The problem for loss rates was formulated in Caceres *et al.* (1999), where the multicast transmission scheme was also introduced and an algorithm that computes estimators that are asymptotically equivalent to the MLES was derived. The problem for delays was discussed in Lo Presti, Duffield, Horowitz, and Towsley (2002), where a heuristic algorithm was proposed for calculating a nonparametric estimate of the link delay distributions. Liang and Yu (2003) developed a pseudo-likelihood approach for the delay problem by considering all possible pairwise probes from the full-multicast experiment. Shih and Hero (2003) presented an estimator for the back-to-back probing scheme that models link delay using a point mass at zero and a finite mixture of Gaussian distributions.

16.3.2. *Inference for loss rates*

Inference for loss rates has been studied in the literature under the following stochastic framework. Let $Z_r(m) = 1$ if the m-th probe packet sent from the source node reached receiver node $r \in \mathbb{R}$, the set of all receiver nodes, and 0 otherwise. Define hypothetical random variables as follows: $X_i(m) = 1$ if packet m traverses successfully link $i \in \mathcal{E}$, the set of all links, and 0 otherwise. The collected data are analyzed under the following model (Caceres *et al.*, (1999), Lo Presti, Duffield, Horowitz, and Towsley (2002), Castro *et al.* (2003), and others). For the loss rate problem, let $\alpha_i(m) = P(X_i(m) = 1)$, i.e., the probability that the probe packet traverses successfully the link between nodes $f(i)$ and i, and reaches node $i \in \mathcal{T} - \{0\}$. It is assumed that the $X_i(m)$'s are independent across i and m. Further, $\alpha_i(m) \equiv \alpha_i$ for all probes m (temporal homogeneity). Then, $P(Z_r(m) = 1) = \Pi_{s \in \mathcal{P}(0,r)} \alpha_s$. Further, $P(X_j(m) = 1,$ for all $j \in \mathcal{D}(i)) = \Pi_{s \in \mathcal{P}(0,i)} \alpha_s \times \Pi_{j \in \mathcal{D}(i)} \alpha_j$.

Some comments about these assumptions are in order. The temporal homogeneity assumption is not critical as the time frame for the probing experiment is in the order of minutes, but the effect of spatial dependence merits further study as examples using the network simulator tool and real data sets indicate.

Recall that the flexicast experiment $\mathcal{C}$ is made up of a collection of independent subexperiments C_h. Each subexperiment is a k-cast experiment (for some k), so it can be viewed as a k-dimensional multinomial experiment. More specifically, each outcome is of the form $\{Z_{r_1}, ..., Z_{r_k}\}$ where $Z_{r_j} = 1$ or 0 depending on whether the probe reached receiver node r_j or not. Let $N_{(r_1,...,r_k)}$ denote the number of outcomes corresponding to this event, and let $\gamma_{(r_1,...,r_k)}$ be the probability of this event. Then the log-likelihood for the experiment C_h is proportional to $\gamma_{(r_1,...,r_k)} \log(N_{(r_1,...,r_k)}$ and that for $\mathcal{C}$ is the sum of the log-likelihoods for the individual experiments. However, the $\gamma_{(r_1,...,r_k)}$'s are complicated functions of the αs, the link-level loss rates, so one has to use numerical methods to obtain the MLEs.

This belongs to the class of missing data problems, so the EM-algorithm is a natural approach to computing the MLEs (Coates and Nowak (2000);

Coates, Castro, Gadhiok, King, Tsang, and Nowak (2002); Castro, Coates, Liang, Nowak, and Yu (2004); Xi (2004)). Xi, Michailidis and Nair (2005) developed the structure of the EM-algorithm explicitly for flexicast experiments described above. In our experience, the EM algorithm works well when coupled with a collection of bicast and unicast experiments for small to moderate networks. For large networks, however, they are computationally not practical.

A class of fast estimation methods based on least-squares has been developed in Michailidis, Nair and Xi (2005). This is done by transforming the loss-estimation problem to a linear inverse problem as follows. Consider the the 3-layer symmetric binary tree in the right panel of Figure 16.4, and suppose we use a three-cast experiment to the receivers $\langle 4, 5, 6 \rangle$. There are eight possible outcomes $(1, 1, 1), (1, 1, 0), (1, 0, 0), ... (0, 0, 0)$; denote the corresponding number of the outcomes by $N_{(1,1,1)}, N_{(1,1,0)}$ and so on. We can ignore the last one as there are only seven linearly independent observations. Consider the one-to-one transformation of these seven events to the following: $(1, 1, 1), (1, 1, +), (1, +, +), ...$ where a '+' indicates either a '1' or a '0'. The new outcomes are obtained by replacing all the '0's with '+'s. Let $M_{(i,j,k)}$ denote the number of these outcomes. Now, if N_h denotes the total number of probes for the subexperiment h, we can write $E(M_{i,j,k})$ as N_h times the product of appropriate link-level α's. For instance, $E(M_{(1,1,+)}) = N_h \alpha_1 \alpha_2 \alpha_4 \alpha_5$. Similarly, $E(M_{(1,+,1)}) = N_h \alpha_1 \alpha_2 \alpha_3 \alpha_4 \alpha_6$. This expression where the expectations are products of the appropriate link-level loss rates holds in general for any $k-$cast experiment. This suggests fitting a log-linear model to the estimated probabilities $Y^h_{(i_1,...i_k)} = M^h_{(i_1,...i_k)}/N_h$. In other words, if Y^h denotes the vector of estimated probabilities for subexperiment h, then

$$Y^h = R^h \beta^h + \epsilon^h,$$

where $\beta_j = \log(\alpha_j)$, R^h is a matrix of ones and zeros that depend on the logical topology of the subexperiment and ϵ^h is a vector of errors. The expected value of the errors tends to zero as the probe size $N_h \to \infty$. Also, the errors are correlated in general, although the variance-covariance structure can be obtained easily due to its block-diagonal form.

Now, by stacking up the vectors of Y^hs for all the subexperiments, we get a linear system of equations that can be used to estimate all the internal link-level parameters α_js. The ordinary least-squares algorithm provides a non-iterative and very fast estimation scheme. Since the ϵ^hs are correlated, a more efficient estimation scheme is based on iteratively-reweighted LS.

These and other schemes and their properties are studied in Michailidis, Nair and Xi (2005). It was found that the IRWLS estimators are very close to the MLEs even in reasonably small samples. One can also compute the asymptotic variance-covariance matrix of the estimators based on the LS schemes easily, leading to explicit construction of standard errors and hypothesis tests. This is another advantage of these LS schemes over the MLEs obtained using the EM algorithm.

16.3.3. *Inference for delay distributions*

Let X_k denote the (unobservable) delay on link k, and let the cumulative delay accumulated from the root node to the receiver node r be $Y_r = \sum_{k \in \mathcal{P}_{0,r}} X_k$, where $\mathcal{P}_{0,r}$ denotes the path from node 0 to node r. The observed data are end-to-end delays consisting of Y_r for all the receiver nodes. A common approach for inference that can accommodate the heavy-tailed nature of internet measurements is based on discretizing the continuous delays using a common bin size q. Let $X_k \in \{0, q, 2q, \ldots, bq\}$ be the discretized delay accumulated on link k where bq is the maximum delay. Let $\alpha_k(i) = P\{X_k = iq\}$. Our objective is to estimate the delay distributions or the $\alpha_k(i)$'s for $k \in \mathcal{E}$ and i in $\{0, 1, \ldots, b\}$ using the end-to-end data Y_rs.

Let $\vec{\alpha}_k = [\alpha_k(0), \alpha_k(1), \ldots \alpha_k(b)]'$ and let $\vec{\alpha} = [\vec{\alpha}_0', \vec{\alpha}_1', \ldots, \vec{\alpha}_{|E|}']'$. The observed end-to-end measurements consist of the number of times each possible outcome $\vec{y}$ was observed from the set of outcomes $\mathcal{Y}^h$ for a given scheme h. Let $N_{\vec{y}}^h$ denote these counts. These are distributed as multinomial random variables with corresponding path-level probability $\gamma_c(\vec{y}; \vec{\alpha})$. So the log-likelihood is given by

$$l(\vec{\alpha}; \mathbf{Y}) = \sum_{c \in \mathcal{C}} \sum_{\vec{y} \in \mathcal{Y}^c} N_{\vec{y}}^c \log[\gamma_c(\vec{y}; \vec{\alpha})].$$

In principle, one can maximize it numerically to get the MLEs. As in the loss case, this can be viewed as an instance of a missing data problem, and the EM algorithm provides a convenient approach for computing the MLEs. This has been studied in the literature (See Lawrence, Michailidis, and Nair (2003) for full multicast experiments and Lawrence, Michailidis, and Nair (2005a) for flexicast experiments). However, the computations in the E-step are quite involved, so it can work only with very small networks for full multicast experiments. The use of the EM algorithm is more manageable when coupled with the flexicast experiments, but even then it is practical only in moderate-sized networks. There have been heuristic estimation methods that have been proposed in the literature for the full multicast situation. The first, by Lo Presti *et al.* (2002), tries to mimic the clever algorithm for the loss case in Caceres *et al.* (1999) and relies on solving higher-order polynomials. However, this algorithm does not use all the data and can be very inefficient (Lawrence, Michailidis, and Nair (2005a)). Liang and Yu (2003) proposed a pseudo-likelihood method where one considers only data from all pairs of probes and ignores the third and higher order information. The all-pairs-bicasts by Liang and Yu (2003) is similar in spirit to a flexicast experiment with all pairs of bicasts, although they will all be independent in the flexicast set up. Also, as we showed earlier, one can use many fewer independent bicasts than all possible pairs to estimate the link delay parameters. Even then, the complexity of the EM algorithm grows exponentially with the number of layers in the tree.

To handle larger networks, Lawrence *et al.* (2005a) developed a *grafting* method which fits the EM to subtrees and uses a heuristic method based on a fixed point algorithm to combine the results across the subtrees. This is a very fast algorithm, and extensive numerical work has shown that its small sample performance is favorable compared to the estimator in Lo Presti, Duffield, Horowitz, and Towsley (2002) and the pseudo-likelihood estimator in Liang and Yu (2003). Further, the efficiency loss is relatively small compared to the full MLE.

Lawrence *et al.* (2005a) also studied inference for continuous delay distributions and developed moment-based estimation methods for the means and variances of the delay distribution assuming a mixture model for the individual link-delay distributions of the form

$$F_j(x) = p_j \delta_{\{0\}} + (1 - p_j)G_j(x).$$

Here $\delta_{\{0\}}$ denotes point mass at 0 (i.e., no delay with probability p_j) and the continuous part has a mean-variance relationship of the form $V_j = \phi \mu_j^\theta$ where μ_j and V_j are, respectively the mean and variance of $G_j(\cdot)$.

16.4. An Application

We illustrate the results using real data collected from the campus network at the University of North Carolina, Chapel Hill. The loss rates in this network were negligible, so we will focus on delays. We have collected extensive amounts of data but report here only selected results. See also Xi, Michailidis and Nair (2005) for an application of the results to network monitoring.

Voice over IP (VoIP) or internet telephony is a technology that turns analog voice signals into digital packets and then uses the internet to transmit them to the intended receivers. The main difference with classical telephony is that the call does not use a dedicated connection with reserved bandwidth, but instead packets carrying the voice data are multiplexed in the network with other traffic. The quality of service (QoS) requirements in terms of packet losses and delays for this application are significantly more stringent than other non-real time applications, such as email. Hence, assessing network links to ensure that they are capable of supporting VoIP telephony is an important part of the technology. The University of North Carolina (UNC) is currently in the planning phase of deploying VoIP telephony. As part of this effort, monitoring equipment and software capable of placing such phone calls were installed throughout the campus network. Specifically, the software allowed the emulation of VoIP calls between the monitoring devices. It can then synchronize their clocks and obtain very accurate packet loss and delay measurements along the network paths.

Fifteen monitoring devices had been deployed in a variety of buildings and on a range of different capacity links through the UNC network. The locations included dorms, libraries, and various academic buildings. The links included large capacity gigabit links, smaller 100 megabit links, and one wireless link. Monitoring VoIP transmissions between these buildings allowed one to examine traffic influenced by the physical conditions of the link and the demands of various groups of users. Figure 16.3 gives the logical connectivity of the UNC network. Each of the nodes on the circle have a basic machine that can place a VoIP phone call to any of the other endpoints. The three nodes in the middle are part of the core (main routers) of the network. One of these internal nodes, the upper router linked to Sitterson Hall, also connects to the gateway that exchanges traffic with the rest of the internet.

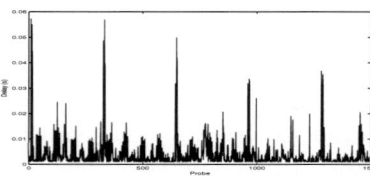

Figure 16.5.   Traffic trace of packet delays generated by a single phone call across the UNC network.

The data were collected using a tool designed by Avaya Labs for testing a network's readiness for VoIP. There are two parts to this tool. First, there are the monitoring devices that are computers deployed throughout the network with the capability of exchanging VoIP-style traffic. These devices run an operating system that allows them to accurately measure the time at which packets are sent and received. The machines collect these time stamps and report them back to the second part of the system: the collection software. This software remotely controls the devices and determines all the features of each call, such as source-destination devices, start time, duration, and protocol which includes the inter-packet time intervals. The software collects the time stamps when the calls are finished and processes them. The processing consists of adjusting the time stamps to account for the difference among the machines' clocks, and then calculating the one-way end-to-end delays.

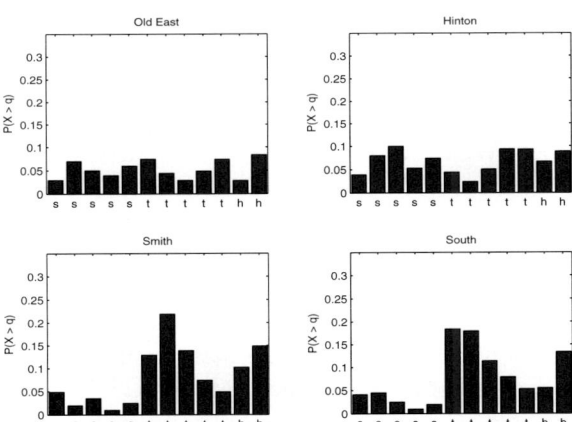

Figure 16.6.   Probability of large delay throughout 2/26/2005 (s), 3/1/2005 (t), and 3/3/2005 (h) at two dorms, Old East and Hinton and two university buildings, South and Smith. s, t and h, denote Saturday, Tuesday and Thursday, respectively. On the y-axis, the probability of a delay larger than 1 ms is depicted.

Figure 16.5 shows the delays for the packets of one phone call (flow) between two devices. The data contain information about the entire path between a pair of end-points, which spans several links. For example, a phone call placed between the dorm and the library follows a path that goes through three main routers. Many of the features found in other types of network data can be seen here: heavy-tailed marginal distributions and significant autocorrelation between consecutive observations. We will see that, by using the techniques developed in this paper, we are able to reconstruct link-level information about delays from the end-to-end path-level data.

For the data collection, Sitterson served as the root, and we used seven bicast pairs to cover the 14 receiver nodes:

$$\mathcal{C} = \{\langle 4, 5\rangle, \langle 6, 7\rangle, \langle 8, 10\rangle, \langle 11, 12\rangle, \langle 13, 14\rangle, \langle 15, 16\rangle, \langle 17, 18\rangle\}.$$

The network allowed only unicast transmission protocol, so back-to-back probing was used to simulate multicast transmissions. The span of time between the two packets comprising the back-to-back probe was on the order of a few nanoseconds while the time between successive probes was one tenth of a second. Prior experimentation using the call synthesis tool and this probing method leads us to believe that the correlation between the two packets on the shared links is close to one. Most of the probing sessions resulted in 200 packets to each pair (one session ran considerably longer and produced more probes due to operator error).

In this paper, we consider data collected on 2/26/2005, 3/1/2005, and 3/3/2005 (corresponding to a Saturday, Tuesday, and Thursday respectively) during the Winter semester at UNC. In addition to methodology confirmation, this data will allow us to contrast the weekend/weekday behavior of the network. The data collection is somewhat irregular, but there are five collections throughout the day on Saturday and Tuesday, and a morning and noon collection on Thursday. Analysis was conducted using the discrete delay MLE approach. The unit size was chosen as approximately 1 ms. This results in most of the mass occurring in the 'zero' bin, which gives us a useful statistic to track over time.

Figure 16.6 gives some selected results from this analysis. Each bar represents the probability of a delay of one unit or larger in each of four locations for each time period. Some interesting results can be noted. The university buildings South and Smith both show very few delays due to limited traffic on the Saturday collections; there are very small probabilities of delays, half of a percent or less, throughout the day. The weekdays show a typically diurnal pattern with a peak midday that tapers off. In many respects, the dorms, Old East and Hinton, show opposite patterns. The activity on the weekend is not much less than during the week. During the week, the traffic, particularly at Hinton the large freshman dorm, actually dips during the day when the students are busy with classroom activities and rises at night.

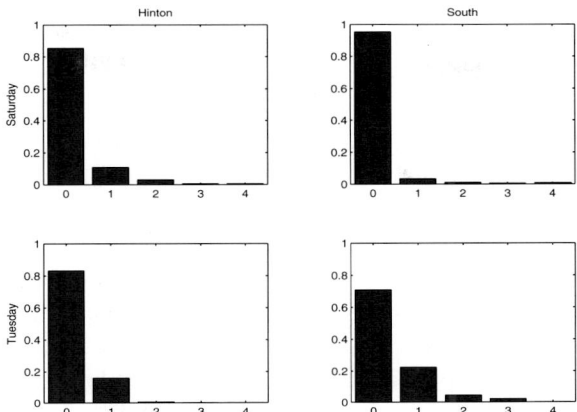

Figure 16.7. Detailed distributions with unit sizes of .5 ms at Hinton and South on Saturday, 2/26, and Tuesday, 3/1, during the mid-afternoon.

Figure 16.7 shows partial results from a more detailed analysis of the data. Here the unit size is about half that of the previous analysis. The plots give the first five bins on the Hinton and South links on Saturday and Tuesday. We see that the whole distribution is relatively stable on the dorm link despite the weekend/weekday difference. In the office building, we see the mass shift outward from Saturday to Tuesday, but most of the mass still falls within the first five bins indicating fairly good link performance. Additional analyses of the UNC network data can be found in Lawrence, Michailidis, and Nair (2005a) and Lawrence, Michailidis, and Nair (2005b).

16.5. Concluding Remarks

This paper has provided an overview of network tomography and some of the interesting statistical issues that arise from the inverse problems. The discussion so far assumed that the routing matrix (logical topology) is known or can be determined easily. This is sometimes possible as the traceroute tool, based on the internet control message protocol (ICMP), reports all the network devices (routers and hosts) along a node. Unfortunately, many routers have disabled the protocol and do not respond to the traceroute requests. As a result, there has been a lot of interest in developing tools for topology identification. Various statistical techniques such as clustering, maximum likelihood, and Bayesian methods have been used, based on measurements from active tomography experiments (see Coates *et al.* (2002b), Castro, Coates, Liang, Nowak, and Yu (2004), Rabbat, Nowak, and Coates (2004), Shih and Hero (2004) and references therein). However, the ill-posed nature of the problem and its exponential complexity make topology identification very challenging.

The tools and techniques that have been developed thus far have proved useful for characterizing network performance, detecting anomalies such as congestion, and capacity planning. Nevertheless, some of the underlying assumptions, especially those dealing with the spatio-temporal behavior, are somewhat restrictive and merit further study. Another ongoing challenge (largely unaddressed in the literature) is the lack of distributed algorithms. All the proposed methods require the existence of a central data repository, which limits the applicability of tomography techniques to online network monitoring. Moreover, fast algorithms are critical for implementing the techniques in real time. Therefore, useful methodology needs to strike a balance between computational complexity and statistical efficiency.

Acknowledgments

The research was supported in part by NSF grants DMS-0204247, CCR-0325571, and DMS-0505535. The authors would like to thank the following for their help on the collection, modeling, and analysis of data from the UNC network: Jim Landwehr, Lorraine Denby and Jean Meloche from the Avaya Labs, Yinghan Yang, Jim Gogan and his team from the Information Technology Division at UNC, Don Smith from CS Department at UNC, and Steve Marron from the Statistics Department at UNC.

References

Adler, R. J., Feldman, R. and Taqqu, M. S. (1998), *A practical guide to heavy tails: statistical techniques and applications*, Birkhauser, Boston

Cao, J., Davis, D., Vander Wiel, S. and Yu, B. (2000), Time varying network tomography: router link data, *Journal of the American Statistical Association*, 95, 1063-1072.

Castro, R., Coates, M. J., Liang, G., Nowak, R. D. and Yu, B. (2004), Network tomography: recent developments, *Statistical Science*, 19, 499-517.

Castro, R., Coates, M. J., and Nowak, R.D (2004), Likelihood based hierarchical clustering, *IEEE Transactions on Signal Processing*, 52, 2308-2321.

Coates, M. J., Hero, A. O., Nowak, R. D. and Yu, B. (2002), Internet tomography, *Signal Processing Magazine*, 19, 47-65.

Coates, M. J., Castro, R., Gadhiok, M., King, R., Tsang Y., and Nowak, R. (2002), Maximum likelihood network topology identification from edge-based unicast measurements, *ACM Sigmetrics Proceedings*, Marina Del Rey, CA.

Duffield, N. G., Horowitz, J. L., Lo Presti, F., and Towsley, D. (2002), Multicast topology inference from measured end-to-end loss, *IEEE Transactions in Information Theory*, 48, 26-45.

Lawrence E. (2005), *Flexicast Network Delay Tomography*, Unpublished doctoral dissertation, Department of Statistics, The University of Michigan.

Lawrence, E., Michailidis, G. and Nair, V. N. (2003), Maximum likelihood estimation of internal network link delay distributions using multicast measurements, *Proceedings of the Conference on Information Systems and Sciences*, Johns Hopkins University, March 12-14, 2003.

Lawrence, E., Michailidis, G. and Nair, V.N. (2005a), Flexicast delay tomography (submitted).

Lawrence, E., Michailidis, G. and Nair, V.N. (2005b), Local area network analysis using end-to-end delay tomography, *Proceedings Large Scale Network Inference Workshop,* Banff, Canada and also to appear in *Performance Evaluation Review*

Liang, G. and Yu, B. (2003), Maximum pseudo-likelihood estimation in network tomography, *IEEE Transactions on Signal Processing*, 51, 2043-2053.

Liang, G., Taft, N. and Yu, B. (2005), A fast lightweight approach to origin destination IP traffic estimation using partial measurements, (submitted).

Lo Presti, F., Duffield, N. G., Horowitz, J. L. and Towsley, D. (2002), Multicast based inference of network internal delay distributions, *ACM/IEEE Transactions on Networking*, 10, 761-775.

Marchette, D. J. (2001), *Computer intrusion detection and network monitoring: a statistical viewpoint*, Springer, New York.

Michailidis, G., Nair, V. N. and Xi, B. (2005), Fast least squares algorithms for estimating and monitoring network link losses based on active probing schemes, (preprint).

Papagiannaki, K., Taft, N. and Lakhina, A. (2004), A distributed approach to measure traffic matrices, *ACM Internet Measurement Conference Proceedings*, Taormina, Italy.

Park, K. and Willinger W. (2000), *Self-similar network traffic and performance evaluation*, Wiley Interscience, New York.

Paxson, V. (1997), End-to-end routing behavior in the Internet, *IEEE/ACM Transactions on Networking*, 5, 601-615.

Rabbat, M., Nowak, R., and Coates, M. J. (2004), Multiple source, multiple destination network tomography, Proceedings IEEE Infocom, Hong Kong.

Roughan, M. (2005), First order characterization of Internet traffic matrices, Proceedings ISI, Sydney, Australia.

Shih, M. F. and Hero, A. O. (2003), Unicast based inference of network delay distributions with finite mixture models, *IEEE Transactions on Signal Processing*, 51, 2219-2228.

Shih, M.F., and Hero, A.O. (2004), Network Topology Discovery Using Finite Mixture Models, Proceedings of IEEE Int. Conf. on Acoust. Speech and Sig. Proc., Montreal.

Soule, A., Nucci, A., Leonardi, E. Cruz, R. and Taft, N. (2004), How to identify and estimate the largest traffic matrix elements in a dynamic environment, *ACM Sigmetrics Proceedings*, New York, NY.

Soule, A., Lakhina, A., Taft, N., Papagiannaki, K., Salamatian, K., Nucci, A., Crovella, M. and Diot, C. (2005), Traffic matrices; balancing measurements, inference and modeling, *ACM Sigmetrics Proceedings*, Banff, Canada.

Tebaldi, C. and West, M. (1998), Bayesian inference of network traffic using link count data (with discussion), *Journal of the American Statistical Association*, 93, 557-576.

Tsang, Y., Coates, M. J. and Nowak, R.D. (2003), Network delay tomography, *IEEE Transactions on Signal Processing*, 51, 2125-2135.

Vardi, Y. (1996), Estimating source-destination traffic intensities from link data, *Journal of the American Statistical Association*, 91, 365-377.

Xi, B. (2004), *Estimating internal link loss rates using active network tomography*, Unpublished doctoral dissertation, Department of Statistics, The University of Michigan.

Xi, B., Michailidis, G. and Nair, V.N. (2005), Estimating network loss rates using active tomography, (submitted).

Zhang, Y., Roughan, M., Duffield, N. and Greenberg, A. (2003), Fast accurate computation of large scale IP traffic matrices from link loads, *ACM Sigmetrics Proceedings*, San Diego, CA.

Zhang, Y., Roughan, M., Lund, C. and Donoho, D. (2003) An information theoretic approach to traffic matrix estimation, *ACM SIGCOMM Proceedings*, Karlsruhe, Germany.

Part VI

Financial Econometrics

CHAPTER 17

Likelihood Inference for Diffusions: A Survey

Yacine Aït-Sahalia

Department of Economics and Bendheim Center for Finance
Princeton University and NBER
Princeton, NJ 08544
yacine@princeton.edu

This chapter reviews closed form expansions for discretely sampled diffusions, and their use for likelihood inference and testing. The method is applicable to a large class of models, covering both univariate and multivariate processes, and with a state vector that is either fully or partially observed. Examples are included.

17.1. Introduction

This chapter surveys recent results on closed form likelihood expansions for discretely sampled diffusions. The basic model is written in the form of a stochastic differential equation for the state vector X

$$dX_t = \mu(X_t; \theta)dt + \sigma(X_t; \theta)dW_t \qquad (17.1)$$

where W_t is an $m-$dimensional standard Brownian motion. In the parametric case, the functions μ and σ are known, but not the parameter vector θ which is the object of interest. Available data are discrete observations on the process sampled at dates $\Delta, 2\Delta, ..., N\Delta$, with Δ fixed. The case where Δ is either deterministic and time-varying or random (as long as independent from X) introduces no further difficulties.

One major impediment to both theoretical modeling and empirical work with continuous-time models is the fact that in most cases little can be said about the implications of the instantaneous dynamics (17.1) for X_t for longer time intervals Δ. One cannot in general characterize in closed form an object as simple, yet fundamental for everything from prediction to estimation and derivative pricing, as the conditional density of $X_{t+\Delta}$

given the current value X_t, also known as the transition function of the process. For a list of the rare exceptions, see Wong (1964). In finance, the well-known models of Black and Scholes (1973) (the geometric Brownian motion $dX_t = \beta X_t dt + \sigma X_t dW_t$), Vasicek (1977) (the Ornstein-Uhlenbeck process $dX_t = \beta (\alpha - X_t) dt + \sigma dW_t$) and Cox, Ingersoll, and Ross (1985) (Feller's square root process $dX_t = \beta (\alpha - X_t) dt + \sigma X_t^{1/2} dW_t$) rely on these existing closed-form expressions.

In many cases that are relevant in finance, however, the transition function is unknown: see for example the models used in Courtadon (1982) ($dX_t = \beta (\alpha - X_t) dt + \sigma X_t dW_t$), Marsh and Rosenfeld (1982) ($dX_t = (\alpha X_t^{-(1-\delta)} + \beta) dt + \sigma X_t^{\delta/2} dW_t$), Cox (1975) and the more general version of Chan, Karolyi, Longstaff, and Sanders (1992) ($dX_t = \beta (\alpha - X_t) dt + \sigma X_t^\gamma dW_t$), Constantinides (1992) ($dX_t = (\alpha_0 + \alpha_1 X_t + \alpha_2 X_t^2) dt + (\sigma_0 + \sigma_1 X_t) dW_t$), the affine class of models in Duffie and Kan (1996) and Dai and Singleton (2000) ($dX_t = \beta (\alpha - X_t) dt + \sqrt{\sigma_0 + \sigma_1 X_t} dW_t$), the nonlinear mean reversion model in Aït-Sahalia (1996) ($dX_t = (\alpha_0 + \alpha_1 X_t + \alpha_2 X_t^2 + \alpha_{-1}/X_t) dt + (\beta_0 + \beta_1 X_t + \beta_2 X_t^{\beta_3}) dW_t$). While it is possible to write down the continuous-time likelihood function for the full sample sample path, ignoring the difference between sampling at a fixed time interval Δ and seeing the full sample path can lead to inconsistent estimators of the parameter vector θ.

In Aït-Sahalia (1999) (examples and application to interest rate data), Aït-Sahalia (2002) (univariate theory) and Aït-Sahalia (2001) (multivariate theory), I developed a method which produces accurate approximations *in closed form* to the unknown transition function $p_X(x|x_0, \Delta; \theta)$, that is, the conditional density that $X_{n\Delta} = x$ given $X_{(n-1)\Delta} = x_0$ in an amount of time Δ implied by the model in equation (17.1).

Bayes' rule combined with the Markovian nature of the process, which the discrete data inherit, imply that the log-likelihood function has the simple form

$$\ell_N(\theta) \equiv \sum_{i=1}^N l_X \left(X_{i\Delta} | X_{(i-1)\Delta}, \Delta; \theta \right) \tag{17.2}$$

where $l_X \equiv \ln p_X$, and the asymptotically irrelevant density of the initial observation, X_0, has been left out. As is clear from (17.2), the availability of tractable formulae for p_X is what makes likelihood inference feasible under these conditions.

The rest of this paper is devoted to reviewing these methods and their applications. I start with the univariate case in Section 17.2, then move on to the multivariate case in Section 17.3. Section 17.4 shows a connection between this method and saddlepoint approximations. I then provide two examples, one of a nonlinear univariate model, and one of a multivariate model, in Sections 17.5 and 17.6 respectively. Section 17.7 discusses inference using this method when the state vector is only partially observed, as in stochastic volatility or term structure models. Section 17.8 outlines the use of this method in specification testing while Section 17.9 sketches derivative pricing applications. Finally, Section 17.10 discusses likelihood inference for continuous time models when the underlying process is nonstationary.

17.2. The Univariate Case

Existing methods to derive MLE for discretely sampled diffusions required solving numerically the Fokker-Planck-Kolmogorov partial differential equation satisfied by p_X (see e.g., Lo (1988)), or simulating a large number of sample paths along which the process is sampled very finely (Pedersen (1995), Brandt and Santa-Clara (2002)). Neither methods produce a closed-form expression, so they both result in a large computational effort since the likelihood must be recomputed for each observed realization of the state vector, and each value of the parameter vector θ along the maximization. Both methods deliver a sequence of approximations to $\ell_N(\theta)$ which become increasingly accurate as some control parameter J tends to infinity.

By contrast, the closed form likelihood expressions that I will describe here make MLE a feasible choice for estimating θ in practical applications. The method involves no simulations and no PDE to solve numerically. Like these two methods, I also construct a sequence $\ell_N^{(J)}$ for $J = 1, 2, ...$ of approximations to ℓ_N, but the essential difference is that $\ell_N^{(J)}$ will be obtained in closed-form. It converges to ℓ_N as $J \to \infty$ and maximizing $\ell_N^{(J)}$ in lieu of the true but incomputable ℓ_N results in an estimator which converges to the true MLE. Since $\ell_N^{(J)}$ is explicit, the effort involved is minimal.

17.2.1. The density approximation sequence

To understand the construction of the sequence of approximations to p_X, the following analogy may be helpful. Consider a standardized sum of random variables to which the Central Limit Theorem (CLT) applies. Often, one is willing to approximate the actual sample size n by infinity and use the $N(0, 1)$ limiting distribution for the properly standardized transformation of the data. If not, higher order terms of the limiting distribution (for example the classical Edgeworth expansion based on Hermite polynomials) can be calculated to improve the small sample performance of the approximation. Consider now approximating the transition density of a diffusion, and think of the sampling interval Δ as playing the role of the sample size N in the CLT. If we properly standardize the data, then we can find out the limiting distribution of the standardized data as Δ tends to 0 (by analogy with what happens in the CLT when N tends to ∞). Properly standardizing the data in the CLT means summing them and dividing by $N^{1/2}$; here it involves transforming the original diffusion X into another one, called Z below. In both cases, the appropriate standardization makes $N(0,1)$ the leading term of the approximation. This $N(0,1)$ approximation is then refined by "correcting" for the fact that Δ is not 0 (just like in practical applications of the CLT N is not infinity). As in the CLT case, it is natural to consider higher order terms based on Hermite polynomials, which are orthogonal with respect to the leading $N(0,1)$ term.

This is not a standard Edgeworth expansion, however: we want convergence as $J \to \infty$, not $N \to \infty$. Further, in general, p_X cannot be approximated for fixed Δ around a Normal density by standard series because the distribution of X is too far from that of a Normal: for instance, if X follows a geometric Brownian motion, the right tail of p_X is too thick, and the Edgeworth expansion diverges as $J \to \infty$. Therefore there is a need for a transformation of X that standardizes the tails of its distribution.

Since Z is a known transformation of X, one can then revert the transformation from X to Z and obtain an expansion for the density of X. As a result of transforming Z back into X, which in general is a nonlinear transformation (unless $\sigma(x;\theta)$ is independent of the state variable x), the leading term of the expansion for the transition function of X will be a *deformed*, or stretched, normal density rather than the $N(0,1)$ leading term of the expansion for p_Z.

The first step towards constructing the sequence of approximations to p_X consists in standardizing the diffusion function of X, i.e., transforming X into Y defined as

$$Y_t \equiv \gamma(X_t;\theta) = \int^{X_t} \frac{du}{\sigma(u;\theta)} \tag{17.3}$$

so that

$$dY_t = \mu_Y(Y_t;\theta)\,dt + dW_t \tag{17.4}$$

where

$$\mu_Y(y;\theta) = \frac{\mu\left(\gamma^{-1}(y;\theta);\theta\right)}{\sigma\left(\gamma^{-1}(y;\theta);\theta\right)} - \frac{1}{2}\frac{\partial\sigma}{\partial x}\left(\gamma^{-1}(y;\theta);\theta\right). \qquad (17.5)$$

Let p_Y denote the transition function of Y. The tails of p_Y have a Gaussian-like upper bound, so Y is "closer" to a Normal variable than X is. But it is still not practical to expand p_Y. This is due to the fact that p_Y gets peaked around the conditioning value y_0 when Δ gets small. And a Dirac mass is not a particularly appealing leading term for an expansion. For that reason, a further transformation is performed, defining the "pseudo-normalized" increment of Y as

$$Z_t \equiv \Delta^{-1/2}\left(Y_t - y_0\right). \qquad (17.6)$$

Given the density of Y, we can work back to the density of X by applying the Jacobian formula:

$$p_X(x|x_0,\Delta;\theta) = \frac{p_Y\left(\gamma(x;\theta)\mid\gamma(x_0;\theta),\Delta;\theta\right)}{\sigma\left(\gamma(x;\theta);\theta\right)} \qquad (17.7)$$

where p_Y can itself be deduced from the density p_Z of Z :

$$p_Y(y|y_0,\Delta;\theta) = \Delta^{-1/2}p_Z\left(\Delta^{-1/2}(y - y_0)\Big|y_0,\Delta;\theta\right). \qquad (17.8)$$

So this leaves us with the need to approximate the density function p_Z. Consider a Hermite series expansion for the conditional density of the variable Z_t, which has been constructed precisely so that it be close enough to a $N(0,1)$ variable for an expansion around a $N(0,1)$ density to converge. The classical Hermite polynomials are

$$H_j(z) \equiv e^{z^2/2}\frac{d^j}{dz^j}\left[e^{-z^2/2}\right], \quad j \geq 0, \qquad (17.9)$$

and let $\phi(z) \equiv e^{-z^2/2}/\sqrt{2\pi}$ denote the $N(0,1)$ density function. Also, define

$$p_Z^{(J)}(z|y_0,\Delta;\theta) \equiv \phi(z)\sum_{j=0}^{J}\eta_j(\Delta,y_0;\theta)\,H_j(z) \qquad (17.10)$$

as the Hermite expansion of the density function $z \mapsto p_Z(z|y_0,\Delta;\theta)$ (for fixed Δ, y_0 and θ). The coefficients $\eta_Z^{(j)}$ are given by:

$$\eta_Z^{(j)}(\Delta, y_0; \theta) = (1/j!) \int_{-\infty}^{+\infty} H_j(z) \, p_Z(z|y_0, \Delta; \theta) \, dz$$

$$= (1/j!) \int_{-\infty}^{+\infty} H_j(z) \, \Delta^{1/2} p_Y\left(\Delta^{1/2} z + y_0 \middle| y_0, \Delta; \theta\right) dz$$

$$= (1/j!) \int_{-\infty}^{+\infty} H_j\left(\Delta^{-1/2}(y - y_0)\right) p_Y(y|y_0, \Delta; \theta) \, dy$$

$$= (1/j!) E\left[H_j\left(\Delta^{-1/2}(Y_{t+\Delta} - y_0)\right) \middle| Y_t = y_0; \theta \right] \quad (17.11)$$

so that the coefficients $\eta_Z^{(j)}$ are specific conditional moments of the process Y. As such, they can be computed in a number of ways, including for instance Monte Carlo integration.

A particularly attractive alternative, however, is to calculate explicitly a Taylor series expansion in Δ for the coefficients $\eta_Z^{(j)}$. Let $f(y, y_0)$ be a polynomial. Polynomials and their iterates obtained by repeated application of the generator $\mathcal{A}$ are in $D(\mathcal{A})$ under regularity assumptions on the boundary behavior of the process. $\mathcal{A}$ is the operator which under regularity conditions returns

$$\mathcal{A} \cdot f = \frac{\partial f}{\partial \delta} + \mu(y)\frac{\partial f}{\partial y} + \frac{1}{2}\sigma^2(y)\frac{\partial^2 f}{\partial y^2} \quad (17.12)$$

when applied to functions $f(\delta, y, y_0)$ that are sufficiently differentiable and display an appropriate growth behavior (this includes the Hermite polynomials under mild restrictions on (μ, σ)). For such an f, we have

$$E[f(\Delta, Y_{t+\Delta}, y_0) \,|\, Y_t = y_0] = \sum_{k=0}^{K} \mathcal{A}^k(\theta) \bullet f(0, y_0, y_0) \frac{\Delta^k}{k!}$$

$$+ O(\Delta^{K+1}), \quad (17.13)$$

which is then applied to the Taylor-expand (17.11) in powers of Δ. This can be viewed as an expansion in small time, although one that is fully explicit since it merely requires the ability to differentiate repeatedly (μ, σ).

17.2.2. *Explicit expressions for the transition function expansion*

I now apply the method just described. Let $p_Z^{(J,K)}$ denote the Taylor series up to order K in Δ of $p_Z^{(J)}$, formed by using the Taylor series in Δ, up to order K, of the coefficients $\eta_Z^{(j)}$. The series $\eta_Z^{(j,K)}$ of the first seven Hermite coefficients $(j = 0, \ldots, 6)$ are given by $\eta_Z^{(0)} = 1$, and, to order $K = 3$, by:

$$\eta_Z^{(1,3)} = -\mu_Y \Delta^{1/2} - \left(2\mu_Y \mu_Y^{[1]} + \mu_Y^{[2]}\right) \Delta^{3/2}/4$$
$$\times \left(4\mu_Y \mu_Y^{[1]2} + 4\mu_Y^2 \mu_Y^{[2]} + 6\mu_Y^{[1]}\mu_Y^{[2]} + 4\mu_Y \mu_Y^{[3]} + \mu_Y^{[4]}\right) \Delta^{5/2}/24$$

$$\eta_Z^{(2,3)} = \left(\mu_Y^2 + \mu_Y^{[1]}\right) \Delta/2 + \left(6\mu_Y^2\mu_Y^{[1]} + 4\mu_Y^{[1]2} + 7\mu_Y\mu_Y^{[2]} + 2\mu_Y^{[3]}\right) \Delta^2/12$$
$$+ \left(28\mu_Y^2\mu_Y^{[1]2} + 28\mu_Y^2\mu_Y^{[3]} + 16\mu_Y^{[1]3} + 16\mu_Y^3\mu_Y^{[2]}\right.$$
$$\left. + 88\mu_Y\mu_Y^{[1]}\mu_Y^{[2]} + 21\mu_Y^{[2]2} + 32\mu_Y^{[1]}\mu_Y^{[3]} + 16\mu_Y\mu_Y^{[4]} + 3\mu_Y^{[5]}\right) \Delta^3/96$$

$$\eta_Z^{(3,3)} = -\left(\mu_Y^3 + 3\mu_Y\mu_Y^{[1]} + \mu_Y^{[2]}\right) \Delta^{3/2}/6 - \left(12\mu_Y^3\mu_Y^{[1]} + 28\mu_Y\mu_Y^{[1]2} + 22\mu_Y^2\mu_Y^{[2]}\right.$$
$$\left. + 24\mu_Y^{[1]}\mu_Y^{[2]} + 14\mu_Y\mu_Y^{[3]} + 3\mu_Y^{[4]}\right) \Delta^{5/2}/48$$

$$\eta_Z^{(4,3)} = \left(\mu_Y^4 + 6\mu_Y^2\mu_Y^{[1]} + 3\mu_Y^{[1]2} + 4\mu_Y\mu_Y^{[2]} + \mu_Y^{[3]}\right) \Delta^2/24$$
$$+ \left(20\mu_Y^4\mu_Y^{[1]} + 50\mu_Y^3\mu_Y^{[2]} + 100\mu_Y^2\mu_Y^{[1]2} + 50\mu_Y^2\mu_Y^{[3]} + 23\mu_Y\mu_Y^{[4]}\right.$$
$$\left. + 180\mu_Y\mu_Y^{[1]}\mu_Y^{[2]} + 40\mu_Y^{[1]3} + 34\mu_Y^{[2]2} + 52\mu_Y^{[1]}\mu_Y^{[3]} + 4\mu_Y^{[5]}\right) \Delta^3/240$$

$$\eta_Z^{(5,3)} = -\left(\mu_Y^5 + 10\mu_Y^3\mu_Y^{[1]} + 15\mu_Y\mu_Y^{[1]2} + 10\mu_Y^2\mu_Y^{[2]}\right.$$
$$\left. + 10\mu_Y^{[1]}\mu_Y^{[2]} + 5\mu_Y\mu_Y^{[3]} + \mu_Y^{[4]}\right) \Delta^{5/2}/120$$

$$\eta_Z^{(6,3)} = \left(\mu_Y^6 + 15\mu_Y^4\mu_Y^{[1]} + 15\mu_Y^{[1]3} + 20\mu_Y^3\mu_Y^{[2]} + 15\mu_Y^{[1]}\mu_Y^{[3]} + 45\mu_Y^2\mu_Y^{[1]2}\right.$$
$$\left. + 10\mu_Y^{[2]2} + 15\mu_Y^2\mu_Y^{[3]} + 60\mu_Y\mu_Y^{[1]}\mu_Y^{[2]} + 6\mu_Y\mu_Y^{[4]} + \mu_Y^{[5]}\right) \Delta^3/720$$

where $\mu_Y^{[k]m} \equiv \left(\partial^k \mu_Y(y_0; \theta)/\partial y_0^k\right)^m$.

Different ways of gathering the terms are available (as in the CLT, where for example both the Edgeworth and Gram-Charlier expansions are based on a Hermite expansion). Here, if we gather all the terms according to increasing powers of Δ instead of increasing order of the Hermite polynomials, and let $\tilde{p}_Z^{(K)} \equiv p_Z^{(\infty,K)}$ –and similarly for Y– we obtain an explicit representation of $\tilde{p}_Y^{(K)}$, given by:

$$\tilde{p}_Y^{(K)}(y|y_0, \Delta; \theta) = \Delta^{-1/2}\phi\left(\frac{y - y_0}{\Delta^{1/2}}\right) \exp\left(\int_{y_0}^{y} \mu_Y(w; \theta)\,dw\right)$$
$$\times \sum_{k=0}^{K} c_k(y|y_0; \theta) \frac{\Delta^k}{k!} \tag{17.14}$$

where $c_0 = 1$ and for all $j \geqslant 1$

$$
c_k(y|y_0;\theta) = k(y-y_0)^{-k} \int_{y_0}^{y} (w-y_0)^{k-1} \Big\{ \lambda_Y(w;\theta)c_{k-1}(w|y_0;\theta)
$$
$$
+ \left(\partial^2 c_{k-1}(w|y_0;\theta)/\partial w^2\right)/2 \Big\} dw \tag{17.15}
$$

with

$$
\lambda_Y(y;\theta) \equiv -\left(\mu_Y^2(y;\theta) + \partial\mu_Y(y;\theta)/\partial y\right)/2. \tag{17.16}
$$

This equation allows the determination of the coefficients c_k recursively starting from c_0. These calculations are easily amenable to an implementation using software such as *Mathematica*. This implementation is typically the most convenient and accurate in empirical applications. Of course, that calculation needs only be done once for a particular model; once the formulae are obtained, they can be used in a standard MLE routine.

The first two coefficients are given by

$$
c_1(y|y_0;\theta) = \frac{\int_{y_0}^{y} \lambda_Y(u;\theta)du}{y-y_0} \tag{17.17}
$$

$$
c_2(y|y_0;\theta) = \frac{1}{(y-y_0)^2} \int_{y_0}^{y} \frac{dw}{(y_0-w)^2} \Big\{ 2\left(\int_{y_0}^{w} \lambda_Y(u;\theta)du\right)
$$
$$
\times \left(\lambda_Y(w;\theta)(y_0-w)^2 + 1\right)
$$
$$
+ 2\lambda_Y(w;\theta)(y_0-w) + (y_0-w)^2\lambda_Y'(w;\theta) \Big\}. \tag{17.18}
$$

These formulae solve the FPK equations up to order Δ^K, both forward and backward:

$$
\frac{\partial \tilde{p}_Y^{(K)}}{\partial \Delta} + \frac{\partial}{\partial y}\left\{\mu_Y(y;\theta)\tilde{p}_Y^{(K)}\right\} - \frac{1}{2}\frac{\partial^2 \tilde{p}_Y^{(K)}}{\partial y^2} = O\left(\Delta^K\right) \tag{17.19}
$$

$$
\frac{\partial \tilde{p}_Y^{(K)}}{\partial \Delta} - \mu_Y(y_0;\theta)\frac{\partial \tilde{p}_Y^{(K)}}{\partial y_0} - \frac{1}{2}\frac{\partial^2 \tilde{p}_Y^{(K)}}{\partial y_0^2} = O\left(\Delta^K\right). \tag{17.20}
$$

The boundary behavior of $\tilde{p}_Y^{(K)}$ is similar to that of p_Y: $\lim_{y \to \underline{y}\ or\ \bar{y}} p_Y = 0$. The expansion is designed to deliver an approximation of the density function $y \mapsto p_Y(\Delta, y|y_0;\theta)$ for a fixed value of conditioning variable y_0. It is not designed to reproduce the limiting behavior of p_Y in the limit where y_0 tends to the boundaries.

By applying (17.7), we obtain the corresponding expression for $\tilde{p}_X^{(K)}$. For instance, at order $K = 1$ we get

$$
\tilde{p}_X^{(1)}(x|x_0, \Delta; \theta) = \sigma\left(\gamma(x; \theta); \theta\right)^{-1} \Delta^{-1/2} \phi\left(\Delta^{-1/2} \int_{x_0}^{x} \frac{du}{\sigma(u; \theta)}\right)
$$

$$
\times \exp\left(\int_{x_0}^{x} \mu_Y(\gamma(u; \theta); \theta) du / \sigma(u; \theta)\right)
$$

$$
\times \left(1 + c_1\left(\gamma(x; \theta)\,|\,\gamma(x_0; \theta); \theta\right)\Delta\right)
$$

$$
= \left(\frac{\sigma(x_0; \theta)}{2\pi\Delta\sigma^3(x; \theta)}\right)^{1/2} \exp\left\{-\frac{1}{2\Delta}\left(\int_{x_0}^{x} \frac{du}{\sigma(u; \theta)}\right)^2\right.
$$

$$
+ \left.\int_{x_0}^{x} \frac{\mu(u; \theta)}{\sigma^2(u; \theta)} du\right\}\left(1 + c_1\left(\gamma(x; \theta)\,|\,\gamma(x_0; \theta); \theta\right)\Delta\right) \quad (17.21)
$$

where

$$
c_1\left(\gamma(x; \theta)\,|\,\gamma(x_0; \theta); \theta\right) = \frac{\int_{x_0}^{x} \lambda_Y(\gamma(u; \theta); \theta) du / \sigma(u; \theta)}{\int_{x_0}^{x} du / \sigma(u; \theta)}. \quad (17.22)
$$

17.2.3. *Convergence of the density sequence*

Aït-Sahalia (2002) shows that the resulting expansion converges as more correction terms are added. Under regularity conditions, there exists $\bar{\Delta} > 0$ such that for every $\Delta \in (0, \bar{\Delta})$, $\theta \in \Theta$ and $(x, x_0) \in D_X^2$:

$$
p_X^{(J)}(\Delta, x|x_0; \theta) \to p_X(\Delta, x|x_0; \theta) \quad \text{as } J \to \infty.
$$

In addition, the convergence is uniform in θ over Θ, in x over D_X, and in x_0 over compact subsets of D_X.

Finally, maximizing $\ell_N^{(J)}(\theta)$ instead of the true $\ell_N(\theta)$ results in an estimator $\widehat{\theta}_N^{(J)}$ which converges to the true (but incomputable) MLE $\widehat{\theta}_N$ as $J \to \infty$ and inherits all its asymptotic properties. In general, the expansion $\tilde{p}_X^{(K)}$ will converge to p_X as $\Delta \to 0$.

17.2.4. *Extensions and comparison with other methods*

Jensen and Poulsen (2002), Stramer and Yan (2005) and Hurn, Jeisman, and Lindsay (2005) conducted extensive comparisons of different techniques for approximating the transition function and demonstrated that the method described is both the most accurate and the fastest to implement for the types of problems and sampling frequencies one encounters in finance. The method has been extended to time inhomogeneous processes by Egorov, Li, and Xu (2003) and to jump-diffusions by Schaumburg (2001) and Yu (2003). DiPietro (2001) has extended the methodology to make it applicable in a Bayesian setting. Bakshi and Yu (2002) propose an alternative centering to (17.6) in the univariate case. Li (2005) considers the case of "damped diffusion" processes.

17.3. Multivariate Likelihood Expansions

Of course, many models of interest in finance are inherently multivariate. The main difficulty in the multivariate case is that the transformation from X to Y that played a crucial role in the construction of the expansions in the univariate case above is, in general, not possible.

17.3.1. Reducibilty

As defined in Aït-Sahalia (2001), a diffusion X is *reducible* if and if only if there exists a one-to-one transformation of the diffusion X into a diffusion Y whose diffusion matrix σ_Y is the identity matrix. That is, there exists an invertible function $\gamma(x;\theta)$ such that $Y_t \equiv \gamma(X_t;\theta)$ satisfies the stochastic differential equation

$$dY_t = \mu_Y(Y_t;\theta)\,dt + dW_t. \tag{17.23}$$

Every univariate diffusion is reducible, through the transformation (17.3). Whether or not a given multivariate diffusion is reducible depends on the specification of its σ matrix. Specifically, Proposition 1 of Aït-Sahalia (2001) provides a necessary and sufficient condition for reducibility: the diffusion X is reducible if and only if the inverse diffusion matrix $\sigma^{-1} = \left[\sigma_{i,j}^{-1}\right]_{i,j=1,\ldots,m}$ satisfies on $\mathcal{S}_X \times \Theta$ the condition that

$$\frac{\partial \sigma_{ij}^{-1}(x;\theta)}{\partial x_k} = \frac{\partial \sigma_{ik}^{-1}(x;\theta)}{\partial x_j} \tag{17.24}$$

for each triplet $(i,j,k) = 1,\ldots,m$ such that $k > j$, or equivalently

$$\sum_{l=1}^{m} \frac{\partial \sigma_{ik}(x;\theta)}{\partial x_l}\sigma_{lj}(x;\theta) = \sum_{l=1}^{m} \frac{\partial \sigma_{ij}(x;\theta)}{\partial x_l}\sigma_{lk}(x;\theta). \tag{17.25}$$

Whenever a diffusion is reducible, an expansion can be computed for the transition density p_X of X by first computing it for the density p_Y of Y and then transforming Y back into X (see Section 17.3.2). When a diffusion is not reducible, the situation is going to be more involved (see Section 17.3.3), although it still leads to a closed form expression.

17.3.2. Determination of the coefficients in the reducible case

The expansion for l_Y is of the form

$$l_Y^{(K)}(\Delta, y|y_0; \theta) = -\frac{m}{2}\ln(2\pi\Delta) + \frac{C_Y^{(-1)}(y|y_0; \theta)}{\Delta}$$
$$+ \sum_{k=0}^{K} C_Y^{(k)}(y|y_0; \theta)\frac{\Delta^k}{k!}. \qquad (17.26)$$

The coefficients of the expansion are given explicitly by:

$$C_Y^{(-1)}(y|y_0; \theta) = -\frac{1}{2}\sum_{i=1}^{m}(y_i - y_{0i})^2 \qquad (17.27)$$

$$C_Y^{(0)}(y|y_0; \theta) = \sum_{i=1}^{m}(y_i - y_{0i})\int_0^1 \mu_{Yi}(y_0 + u(y - y_0); \theta)\,du \qquad (17.28)$$

and, for $k \geq 1$,

$$C_Y^{(k)}(y|y_0; \theta) = k\int_0^1 G_Y^{(k)}(y_0 + u(y - y_0)|y_0; \theta)u^{k-1}du \qquad (17.29)$$

where

$$G_Y^{(1)}(y|y_0; \theta) = -\sum_{i=1}^{m}\frac{\partial\mu_{Yi}(y; \theta)}{\partial y_i} - \sum_{i=1}^{m}\mu_{Yi}(y; \theta)\frac{\partial C_Y^{(0)}(y|y_0; \theta)}{\partial y_i}$$
$$+ \frac{1}{2}\sum_{i=1}^{m}\left\{\frac{\partial^2 C_Y^{(0)}(y|y_0; \theta)}{\partial y_i^2} + \left[\frac{\partial C_Y^{(0)}(y|y_0; \theta)}{\partial y_i}\right]^2\right\} \qquad (17.30)$$

and for $k \geq 2$

$$G_Y^{(k)}(y|y_0; \theta) = -\sum_{i=1}^{m}\mu_{Yi}(y; \theta)\frac{\partial C_Y^{(k-1)}(y|y_0; \theta)}{\partial y_i} + \frac{1}{2}\sum_{i=1}^{m}\frac{\partial^2 C_Y^{(k-1)}(y|y_0; \theta)}{\partial y_i^2}$$
$$+ \frac{1}{2}\sum_{i=1}^{m}\sum_{h=0}^{k-1}\binom{k-1}{h}\frac{\partial C_Y^{(h)}(y|y_0; \theta)}{\partial y_i}\frac{\partial C_Y^{(k-1-h)}(y|y_0; \theta)}{\partial y_i}. \qquad (17.31)$$

Given an expansion for the density p_Y of Y, an expansion for the density p_X of X can be obtained by a direct application of the Jacobian formula:

$$l_X^{(K)}(\Delta, x|x_0; \theta) = -\frac{m}{2}\ln(2\pi\Delta) - D_v(x; \theta) + \frac{C_Y^{(-1)}(\gamma(x; \theta)|\gamma(x_0; \theta); \theta)}{\Delta}$$
$$+ \sum_{k=0}^{K} C_Y^{(k)}(\gamma(x; \theta)|\gamma(x_0; \theta); \theta)\frac{\Delta^k}{k!} \qquad (17.32)$$

from $l_Y^{(K)}$ given in (17.26), using the coefficients $C_Y^{(k)}$, $k = -1, 0, ..., K$ given above, and where

$$v(x; \theta) \equiv \sigma(x; \theta)\sigma(x; \theta)^T \tag{17.33}$$

$$D_v(x; \theta) \equiv \frac{1}{2} \ln\left(Det[v(x; \theta)]\right). \tag{17.34}$$

17.3.3. Determination of the coefficients in the irreducible case

In the irreducible case, the expansion of the log likelihood is taken in the form

$$l_X^{(K)}(\Delta, x|x_0; \theta) = -\frac{m}{2}\ln(2\pi\Delta) - D_v(x; \theta)$$
$$+ \frac{C_X^{(-1)}(x|x_0; \theta)}{\Delta} + \sum_{k=0}^{K} C_X^{(k)}(x|x_0; \theta)\frac{\Delta^k}{k!}. \tag{17.35}$$

The approach is now to calculate a Taylor series in $(x-x_0)$ of each coefficient $C_X^{(k)}$, at order j_k in $(x - x_0)$. Such an expansion will be denoted by $C_X^{(j_k, k)}$ at order $j_k = 2(K - k)$, for $k = -1, 0, ..., K$.

The resulting expansion will then be

$$\tilde{l}_X^{(K)}(\Delta, x|x_0; \theta) = -\frac{m}{2}\ln(2\pi\Delta) - D_v(x; \theta)$$
$$+ \frac{C_X^{(j-1, -1)}(x|x_0; \theta)}{\Delta} + \sum_{k=0}^{K} C_X^{(j_k, k)}(x|x_0; \theta)\frac{\Delta^k}{k!}. \tag{17.36}$$

Such a Taylor expansion was unnecessary in the reducible case: the expressions given in Section 17.3.2 provide the explicit expressions of the coefficients $C_Y^{(k)}$ and then in (17.32) we have the corresponding ones for $C_X^{(k)}$. However, even for an irreducible diffusion, it is still possible to compute the coefficients $C_X^{(j_k, k)}$ explicitly.

With $v(x; \theta) \equiv \sigma(x; \theta)\sigma^T(x; \theta)$, define the following functions of the coefficients and their derivatives:

$$
G_X^{(0)}(x|x_0;\theta) = \frac{m}{2} - \sum_{i=1}^{m} \mu_i(x;\theta) \frac{\partial C_X^{(-1)}(x|x_0;\theta)}{\partial x_i}
$$
$$
+ \sum_{i=1}^{m} \sum_{j=1}^{m} \frac{\partial v_{ij}(x;\theta)}{\partial x_i} \frac{\partial C_X^{(-1)}(x|x_0;\theta)}{\partial x_j}
$$
$$
+ \frac{1}{2} \sum_{i=1}^{m} \sum_{j=1}^{m} v_{ij}(x;\theta) \frac{\partial^2 C_X^{(-1)}(x|x_0;\theta)}{\partial x_i \partial x_j}
$$
$$
- \sum_{i=1}^{m} \sum_{j=1}^{m} v_{ij}(x;\theta) \frac{\partial C_X^{(-1)}(x|x_0;\theta)}{\partial x_i} \frac{\partial D_v(x;\theta)}{\partial x_j}, \quad (17.37)
$$

$$
G_X^{(1)}(x|x_0;\theta) = - \sum_{i=1}^{m} \frac{\partial \mu_i(x;\theta)}{\partial x_i} + \frac{1}{2} \sum_{i=1}^{m} \sum_{j=1}^{m} \frac{\partial^2 v_{ij}(x;\theta)}{\partial x_i \partial x_j}
$$
$$
- \sum_{i=1}^{m} \mu_i(x;\theta) \left(\frac{\partial C_X^{(0)}(x|x_0;\theta)}{\partial x_i} - \frac{\partial D_v(x;\theta)}{\partial x_i} \right)
$$
$$
+ \sum_{i=1}^{m} \sum_{j=1}^{m} \frac{\partial v_{ij}(x;\theta)}{\partial x_i} \left(\frac{\partial C_X^{(0)}(x|x_0;\theta)}{\partial x_j} - \frac{\partial D_v(x;\theta)}{\partial x_j} \right)
$$
$$
+ \frac{1}{2} \sum_{i=1}^{m} \sum_{j=1}^{m} v_{ij}(x;\theta) \left\{ \frac{\partial^2 C_X^{(0)}(x|x_0;\theta)}{\partial x_i \partial x_j} - \frac{\partial^2 D_v(x;\theta)}{\partial x_i \partial x_j} \right.
$$
$$
+ \left(\frac{\partial C_X^{(0)}(x|x_0;\theta)}{\partial x_i} - \frac{\partial D_v(x;\theta)}{\partial x_i} \right)
$$
$$
\left. \times \left(\frac{\partial C_X^{(0)}(x|x_0;\theta)}{\partial x_j} - \frac{\partial D_v(x;\theta)}{\partial x_j} \right) \right\} \quad (17.38)
$$

and for $k \geq 2$:

$$G_X^{(k)}(x|x_0;\theta) = -\sum_{i=1}^{m} \mu_i(x;\theta) \frac{\partial C_X^{(k-1)}(x|x_0;\theta)}{\partial x_i} \qquad (17.39)$$

$$+ \sum_{i=1}^{m} \sum_{j=1}^{m} \frac{\partial v_{ij}(x;\theta)}{\partial x_i} \frac{\partial C_X^{(k-1)}(x|x_0;\theta)}{\partial x_j}$$

$$+ \frac{1}{2} \sum_{i=1}^{m} \sum_{j=1}^{m} v_{ij}(x;\theta) \frac{\partial^2 C_X^{(k-1)}(x|x_0;\theta)}{\partial x_i \partial x_j} + \frac{1}{2} \sum_{i=1}^{m} \sum_{j=1}^{m} v_{ij}(x;\theta)$$

$$\times \left\{ 2 \left(\frac{\partial C_X^{(0)}(x|x_0;\theta)}{\partial x_i} - \frac{\partial D_v(x;\theta)}{\partial x_i} \right) \frac{\partial C_X^{(k-1)}(x|x_0;\theta)}{\partial x_j} \right.$$

$$\left. + \sum_{h=1}^{k-2} \binom{k-2}{h} \frac{\partial C_X^{(h)}(x|x_0;\theta)}{\partial x_i} \frac{\partial C_X^{(k-1-h)}(x|x_0;\theta)}{\partial x_j} \right\}.$$

For each $k = -1, 0, ..., K$, the coefficient $C_X^{(k)}(x|x_0;\theta)$ in (17.35) solves the equation

$$f_X^{(k-1)}(x|x_0;\theta) = 0 \qquad (17.40)$$

where

$$f_X^{(-2)}(x|x_0;\theta) = -2C_X^{(-1)}(x|x_0;\theta)$$

$$-\sum_{i=1}^{m} \sum_{j=1}^{m} v_{ij}(x;\theta) \frac{\partial C_X^{(-1)}(x|x_0;\theta)}{\partial x_i} \frac{\partial C_X^{(-1)}(x|x_0;\theta)}{\partial x_j}$$

$$f_X^{(-1)}(x|x_0;\theta) = -\sum_{i=1}^{m} \sum_{j=1}^{m} v_{ij}(x;\theta) \frac{\partial C_X^{(-1)}(x|x_0;\theta)}{\partial x_i} \frac{\partial C_X^{(0)}(x|x_0;\theta)}{\partial x_j}$$

$$- G_X^{(0)}(x|x_0;\theta)$$

and for $k \geq 1$

$$f_X^{(k-1)}(x|x_0;\theta) = C_X^{(k)}(x|x_0;\theta) - \sum_{i=1}^{m} \sum_{j=1}^{m} v_{ij}(x;\theta) \frac{\partial C_X^{(-1)}(x|x_0;\theta)}{\partial x_i}$$

$$\times \frac{\partial C_X^{(k)}(x|x_0;\theta)}{\partial x_j} - G_X^{(k)}(x|x_0;\theta),$$

where the functions $G_X^{(k)}$, $k = 0, 1, ..., K$ are given above. $G_X^{(k)}$ involves only the coefficients $C_X^{(h)}$ for $h = -1, ..., k-1$, so this system of equation can be utilized to solve recursively for each coefficient at a time. Specifically, the equation $f_X^{(-2)} = 0$ determines $C_X^{(-1)}$; given $C_X^{(-1)}$, $G_X^{(0)}$ becomes known and the equation $f_X^{(-1)} = 0$ determines $C_X^{(0)}$; given $C_X^{(-1)}$ and $C_X^{(0)}$, $G_X^{(1)}$ becomes known and the equation $f_X^{(0)} = 0$ then determines $C_X^{(1)}$, etc. It turns out that this results in a system of linear equations in the coefficients of the polynomials $C_X^{(j_k,k)}$, so each one of these equations can be solved explicitly in the form of the Taylor expansion $C_X^{(j_k,k)}$ of the coefficient $C_X^{(k)}$, at order j_k in $(x - x_0)$. Convergence results for the expansion are proved in Aït-Sahalia (2001).

As in the univariate case, these calculations are straightforward to implement using software. For actual implementation of this method to practical problems in various contexts and with various datasets, see Aït-Sahalia and Kimmel (2002), Aït-Sahalia and Kimmel (2004), Thompson (2004), Cheridito, Filipović, and Kimmel (2005), Mosburger and Schneider (2005), Takamizawa (2005) and Schneider (2006).

17.4. Connection to Saddlepoint Approximations

Aït-Sahalia and Yu (2005) developed an alternative strategy for constructing closed form approximations to the transition density of a continuous time Markov process. Instead of expanding the transition function in orthogonal polynomials around a leading term, we rely on the saddlepoint method, which originates in the work of Daniels (1954). We show that, in the case of diffusions, it is possible by expanding the cumulant generating function of the process to obtain an alternative closed form expansion of its transition density. We also show there that this approach provides an alternative gathering of the correction terms beyond the leading term that is equivalent at order Δ to the irreducible expansion of the transition density just described.

To understand the connection to the saddlepoint approach, it is useful to contrast it with the Hermite-based method described in Section 17.2. That expansion can be viewed as analogous to a small sample correction to the CLT. As in the CLT case, it is natural to consider higher order terms based on Hermite polynomials, which are orthogonal with respect to the leading $N(0,1)$ term. This is an Edgeworth (or Gram–Charlier, depending upon how the terms are gathered) type of expansion. By contrast, saddlepoint expansions rely on first tilting the original density — *transforming it into another one* — and then applying an Edgeworth-like expansion to the tilted density. If the tilted density is chosen wisely, the resulting approximation can be quite accurate in the tails, and applicable fairly generally. In order to be able to calculate a saddlepoint approximation, one needs to be able to calculate the Laplace transform or characteristic function of the process of interest. This requirement is a restriction on the applicability of the method, but as we will see, one that is possible to satisfy in many cases in our context of Markov processes. But even when such a computation is not possible explicitly, we go one step further by showing how useful approximations can be obtained by replacing the characteristic function by an expansion in small time. Expansions in small time, which involve the infinitesimal generator of the Markov process, are a key element shared with the Hermite-based expansions described above.

The key to our approach is to approximate the Laplace transform of the process, and the resulting saddlepoint, as a Taylor series in Δ around their continuous-time limit. This will result in an approximation (in Δ) to the saddlepoint (which itself is an approximation to the true but unknown transition density of the process). By applying (17.13) to the function $f(\delta, x, x_0) = \exp(ux)$, u treated as a fixed parameter, we can compute the expansion of the Laplace transform $\varphi(\Delta, u|x_0)$ in Δ. At order $n_2 = 1$, the result is

$$\varphi^{(1)}(u|x_0, \Delta; \theta) = e^{ux_0}\left(1 + \left(\mu(x_0; \theta)u + \frac{1}{2}\sigma^2(x_0; \theta)u^2\right)\Delta\right).$$

Then, by taking its log, we see that the expansion at order Δ of the cumulant transform K is simply

$$K^{(1)}(u|x_0, \Delta; \theta) = ux_0 + \left(\mu(x_0; \theta)u + \frac{1}{2}\sigma^2(x_0; \theta)u^2\right)\Delta.$$

The first order saddlepoint $\widehat{u}^{(1)}$ solves $\partial K^{(1)}(u|x_0, \Delta; \theta)/\partial u = x$, that is

$$\widehat{u}^{(1)}(x|x_0, \Delta; \theta) = \frac{x - (x_0 + \mu(x_0; \theta)\Delta)}{\sigma^2(x_0; \theta)\Delta}$$

and, when evaluated at $x = x_0 + z\Delta^{1/2}$, we have

$$\widehat{u}^{(1)}\left(x_0 + z\Delta^{1/2}|x_0, \Delta; \theta\right) = \frac{z}{\sigma^2(x_0; \theta)\Delta^{1/2}} + O(1) \qquad (17.41)$$

and

$$K^{(1)}\left(\widehat{u}^{(1)}\left(x_0 + z\Delta^{1/2}|x_0, \Delta; \theta\right)|x_0, \Delta; \theta\right)$$
$$- \widehat{u}^{(1)}\left(x_0 + z\Delta^{1/2}|x_0, \Delta; \theta\right) \cdot \left(x_0 + z\Delta^{1/2}\right) = -\frac{z^2}{2\sigma^2(x_0; \theta)} + O(\Delta^{1/2}).$$

Similarly, a second order expansion in Δ of $K^{(2)}$ is obtained as

$$K^{(2)}\left(u|x_0, \Delta; \theta\right) = ux_0 + \Delta\left(\mu(x_0; \theta)u + \frac{1}{2}\sigma^2(x_0; \theta)u^2\right)$$

$$+ \frac{\Delta^2 u}{8}\left\{4\mu(x_0; \theta)\mu'(x_0; \theta) + 2\sigma^2(x_0; \theta)\mu''(x_0; \theta)\right.$$
$$+ u\left(4\sigma^2(x_0; \theta)\mu'(x_0; \theta) + 2\mu(x_0; \theta)(\sigma^2)'(x_0; \theta) + \sigma^2(x_0; \theta)(\sigma^2)''(x_0; \theta)\right)$$
$$\left. + 2u^2\sigma^2(x_0; \theta)(\sigma^2)'(x_0; \theta)\right\} + O(\Delta^3).$$

The second order saddlepoint $\widehat{u}^{(2)}$ solves $\partial K^{(2)}\left(u|x_0, \Delta; \theta\right)/\partial u = x$, which is a quadratic equation explicitly solvable in u, and we see after some calculations that

$$\widehat{u}^{(2)}\left(x_0 + z\Delta^{1/2}|x_0, \Delta; \theta\right)$$

$$= \frac{z}{\sigma^2(x_0; \theta)\Delta^{1/2}} - \left\{\frac{\mu(x_0; \theta)}{\sigma^2(x_0; \theta)} + \frac{3(\sigma^2)'(x_0; \theta)}{4\sigma^4(x_0; \theta)}z^2\right\} + O(\Delta^{1/2}) \quad (17.42)$$

and

$$K^{(2)}\left(\widehat{u}^{(2)}\left(x_0 + z\Delta^{1/2}|x_0, \Delta; \theta\right)|x_0, \Delta; \theta\right)$$

$$- \widehat{u}^{(2)}\left(x_0 + z\Delta^{1/2}|x_0, \Delta; \theta\right) \cdot \left(x_0 + z\Delta^{1/2}\right)$$

$$= -\frac{z^2}{2\sigma^2(x_0; \theta)} + \left\{\frac{\mu(x_0; \theta)}{\sigma^2(x_0; \theta)}z + \frac{(\sigma^2)'(x_0; \theta)}{4\sigma^4(x_0; \theta)}z^3\right\}\Delta^{1/2} + O(\Delta).$$

The way the correction terms in $\varphi^{(2)}\left(u|x_0, \Delta; \theta\right)$ are grouped is similar to that of an Edgeworth expansion. Higher order approximate Laplace transforms can be constructed (see Aït-Sahalia and Yu (2005)). Write $p^{(n_1, n_2)}$ to indicate a saddlepoint approximation of order n_1 using a Taylor expansion in Δ of the Laplace transform φ, that is correct at order n_2 in Δ. When the expansions in Δ are analytic at zero, then $p^{(n_1, \infty)} = p^{(n_1)}$. First, the leading term of the saddlepoint approximation at the first order in Δ and with a Gaussian base coincides with the classical Euler approximation of the transition density,

$$p_X^{(0,1)}\left(x|x_0, \Delta; \theta\right) = \left(2\pi\Delta\sigma^2(x_0; \theta)\right)^{-1/2}$$

$$\times \exp\left(-\frac{(x - x_0 - \mu(x_0; \theta)\Delta)^2}{\sigma^2(x_0; \theta)\Delta}\right). \quad (17.43)$$

The first order saddlepoint approximation at the first order in Δ and with a Gaussian base is

$$
p_X^{(1,1)} \left(x_0 + z\Delta^{1/2} | x_0, \Delta; \theta \right)
$$

$$
= \frac{\exp\left(-\frac{z^2}{2\sigma^2(x_0;\theta)} + e_{1/2}(z|x_0;\theta)\Delta^{1/2} + e_1(z|x_0;\theta)\Delta\right)}{\sqrt{2\pi}\sigma(x_0;\theta)\Delta^{1/2}\left\{1 + d_{1/2}(z|x_0;\theta)\Delta^{1/2} + d_1(z|x_0;\theta)\Delta\right\}}
$$
$$
\times \left\{1 + c_1(z|x_0;\theta)\Delta\right\} \tag{17.44}
$$

where

$$
e_{1/2}(z|x_0;\theta) = \frac{z\mu(x_0;\theta)}{\sigma^2(x_0;\theta)} + \frac{z^3(\sigma^2)'(x_0;\theta)}{4\sigma^4(x_0;\theta)}
$$

$$
e_1(z|x_0;\theta) = -\frac{\mu(x_0;\theta)^2}{2\sigma^2(x_0;\theta)}
$$

$$
+ \frac{z^2\left(12\sigma^2(x_0;\theta)\left(4\mu'(x_0;\theta) + (\sigma^2)''(x_0;\theta)\right) - 48\mu(x_0;\theta)(\sigma^2)'(x_0;\theta)\right)}{96\sigma^4(x_0;\theta)}
$$

$$
+ \frac{z^4\left(8\sigma^2(x_0;\theta)(\sigma^2)''(x_0;\theta) - 15(\sigma^2)'(x_0;\theta)^2\right)}{96\sigma^6(x_0;\theta)} \tag{17.45}
$$

$$
d_{1/2}(z|x_0;\theta) = \frac{3z\sigma'(x_0;\theta)}{2\sigma(x_0;\theta)}
$$

$$
d_1(z|x_0;\theta) = \frac{\mu'(x_0;\theta)}{2} - \frac{\mu(x_0;\theta)\sigma'(x_0;\theta)}{\sigma(x_0;\theta)} + \frac{\sigma'(x_0;\theta)^2}{4} + \frac{\sigma(x_0;\theta)\sigma''(x_0;\theta)}{4}
$$

$$
+ z^2\left(\frac{5\sigma'(x_0;\theta)^2}{8\sigma(x_0;\theta)^2} + \frac{\sigma''(x_0;\theta)}{\sigma(x_0;\theta)}\right) \tag{17.46}
$$

$$
c_1(z|x_0;\theta) = \frac{1}{4}(\sigma^2)''(x_0;\theta) - \frac{3}{32}\frac{(\sigma^2)'(x_0;\theta)^2}{\sigma^2(x_0;\theta)}.
$$

The expression (17.44) provides an alternative gathering of the correction terms beyond the leading term that is equivalent at order Δ to the irreducible expansion of the transition density resulting from the irreducible method described in Section 17.3.3.

17.5. An Example with Nonlinear Drift and Diffusion Specifications

The likelihood expansions are given for many specific models in Aït-Sahalia (1999), including the Ornstein-Uhlenbeck specification of Vasicek (1977), the Feller square root model of Cox, Ingersoll, and Ross (1985), the linear drift with CEV diffusion model of Cox (1975) and the more general version of Chan, Karolyi, Longstaff, and Sanders (1992),

$$dX_t = \kappa \left(\alpha - X_t \right) dt + \sigma X_t^\rho dW_t$$

a double well model

$$dX_t = \left(\alpha_1 X_t - \alpha_3 X_t^3 \right) dt + dW_t$$

and a simpler version of the nonlinear model of Aït-Sahalia (1996),

$$dX_t = \left(\alpha_{-1} X_t^{-1} + \alpha_0 + \alpha_1 X_t + \alpha_2 X_t^2 \right) dt + \sigma X_t^\rho dW_t.$$

One example that was not included in full generality in Aït-Sahalia (1999), however, is the general model proposed for the short term interest rate in Aït-Sahalia (1996)

$$dX_t = \left(\alpha_{-1} X_t^{-1} + \alpha_0 + \alpha_1 X_t + \alpha_2 X_t^2 \right) dt$$
$$+ (\beta_0 + \beta_1 X_t + \beta_2 X_t^{\beta_3}) dW_t \qquad (17.47)$$

because even in the univariate case the transformation $X \mapsto Y$ does not lead to an explicit integration in (17.3). But, as discussed in Aït-Sahalia (2001), one can use the irreducible method in that case, thereby bypassing that transformation. For instance, at order $K = 1$ in Δ, the irreducible expansion for the generic model $dX_t = \mu(X_t)dt + \sigma(X_t)dW_t$ is given by (17.36) with $m = 1$, namely:

$$\tilde{l}_X^{(1)} \left(\Delta, x | x_0; \theta \right) = -\frac{1}{2} \ln \left(2\pi\Delta \right) - D_v \left(x; \theta \right) + \frac{C_X^{(4,-1)} \left(x | x_0; \theta \right)}{\Delta}$$
$$+ C_X^{(2,0)} \left(x | x_0; \theta \right) + C_X^{(0,1)} \left(x | x_0; \theta \right) \Delta$$

with $D_v \left(x; \theta \right) = \ln(\sigma \left(x; \theta \right))$. The coefficients $C_X^{(j_k, k)}$, $k = -1, 0, 1$ are given by

$$C_X^{(4,-1)} \left(x | x_0; \theta \right) = -\frac{1}{2\sigma(x_0; \theta)^2} (x - x_0)^2 + \frac{\sigma'(x_0; \theta)}{2\sigma(x_0; \theta)^3} (x - x_0)^3$$
$$+ \frac{\left(4\sigma(x_0; \theta)\sigma''(x_0; \theta) - 11\sigma'(x_0; \theta)^2 \right)}{24\sigma(x_0; \theta)^4} (x - x_0)^4$$

$$C_X^{(2,0)} \left(x | x_0; \theta \right) = \frac{(2\mu(x_0; \theta) - \sigma(x_0; \theta)\sigma'(x_0; \theta))}{2\sigma(x_0; \theta)^2} (x - x_0)$$
$$+ \frac{1}{4\sigma(x_0; \theta)^3} \Big\{ \left(\sigma'(x_0; \theta)^2 + 2\mu'(x_0; \theta) \right) \sigma(x_0; \theta)$$
$$- 4\mu(x_0; \theta)\sigma'(x_0; \theta) - \sigma''(x_0; \theta)\sigma(x_0; \theta)^2 \Big\} (x - x_0)^2$$

$$C_X^{(0,1)}(x|x_0;\theta) = \frac{1}{8}\left(2\sigma(x_0;\theta)\sigma''(x_0;\theta) - \frac{4\mu(x_0;\theta)^2}{\sigma(x_0;\theta)^2}\right.$$
$$\left. + \frac{8\sigma'(x_0;\theta)\mu(x_0;\theta)}{\sigma(x_0;\theta)} - \sigma'(x_0;\theta)^2 - 4\mu'(x_0;\theta)\right) \quad (17.48)$$

In the case of model (17.47), this specializes to the following expressions:

$$C_X^{(4,-1)}(x|x_0;\theta) = -\frac{1}{2\left(\beta_2 x_0^{\beta_3} + \beta_1 x_0 + \beta_0\right)^2}(x-x_0)^2$$

$$+ \frac{\left(\beta_2\beta_3 x_0^{\beta_3-1} + \beta_1\right)}{2\left(\beta_2 x_0^{\beta_3} + \beta_1 x_0 + \beta_0\right)^3}(x-x_0)^3$$

$$+ \frac{1}{24\left(\beta_2 x_0^{\beta_3} + \beta_1 x_0 + \beta_0\right)^4}\left\{4\beta_2(\beta_3-1)\beta_3 x_0^{\beta_3-2}\right.$$

$$\left. \times \left(\beta_2 x_0^{\beta_3} + \beta_1 x_0 + \beta_0\right) - 11\left(\beta_2\beta_3 x_0^{\beta_3-1} + \beta_1\right)^2\right\}(x-x_0)^4$$

$$C_X^{(2,0)}(x|x_0;\theta) = \frac{1}{2\left(\beta_2 x_0^{\beta_3} + \beta_1 x_0 + \beta_0\right)^2}\left\{\left(-\beta_2\beta_3 x_0^{\beta_3-1} - \beta_1\right)\right.$$

$$\left. \times \left(\beta_2 x_0^{\beta_3} + \beta_1 x_0 + \beta_0\right) + 2\left(\alpha_0 + x_0(\alpha_1 + \alpha_2 x_0) + \frac{\alpha_{-1}}{x_0}\right)\right\}$$

$$\times (x-x_0) + \left\{\frac{1}{4\left(\beta_2 x_0^{\beta_3} + \beta_1 x_0 + \beta_0\right)^3}\left\{-\beta_2(\beta_3-1)\beta_3\right.\right.$$

$$\times \left(\beta_2 x_0^{\beta_3} + \beta_1 x_0 + \beta_0\right)^2 x_0^{\beta_3-2}$$

$$\left. -4\left(\beta_2\beta_3 x_0^{\beta_3-1} + \beta_1\right)\left(\alpha_0 + x_0(\alpha_1 + \alpha_2 x_0) + \frac{\alpha_{-1}}{x_0}\right)\right\}$$

$$+ \frac{\left(\beta_2\beta_3 x_0^{\beta_3-1} + \beta_1\right)^2 + 2\left(\alpha_1 + 2\alpha_2 x_0 - \frac{\alpha_{-1}}{x_0^2}\right)}{4\left(\beta_2 x_0^{\beta_3} + \beta_1 x_0 + \beta_0\right)^2}\left.\right\}(x-x_0)^2$$

$$C_X^{(0,1)}(x|x_0;\theta) = \frac{1}{8}\Big\{2\beta_2(\beta_3-1)\beta_3\big(\beta_2{x_0}^{\beta_3}+\beta_1 x_0+\beta_0\big){x_0}^{\beta_3-2}$$

$$-\big(\beta_2\beta_3{x_0}^{\beta_3-1}+\beta_1\big)^2-4\big(\alpha_1+2\alpha_2 x_0-\frac{\alpha_{-1}}{{x_0}^2}\big)\Big\}$$

$$+\frac{1}{8\big(\beta_2{x_0}^{\beta_3}+\beta_1 x_0+\beta_0\big)^2}\Big\{8\big(\beta_2\beta_3{x_0}^{\beta_3-1}+\beta_1\big)\big(\alpha_0+x_0(\alpha_1+\alpha_2 x_0)+\frac{\alpha_{-1}}{x_0}\big)$$

$$\times\big(\beta_2{x_0}^{\beta_3}+\beta_1 x_0+\beta_0\big)-4\big(\alpha_0+x_0(\alpha_1+\alpha_2 x_0)+\frac{\alpha_{-1}}{x_0}\big)^2\Big\}.$$

Bakshi, Ju and Qu-Yang (2006) provide an application to equity volatility dynamics for a variety of models.

17.6. An Example with Stochastic Volatility

Consider as a second example a typical stochastic volatility model

$$\begin{pmatrix} dX_{1t} \\ dX_{2t} \end{pmatrix} = \begin{pmatrix} \mu \\ \kappa\,(\alpha-X_{2t}) \end{pmatrix} dt + \begin{pmatrix} \gamma_{11}\exp(X_{2t}) & 0 \\ 0 & \gamma_{22} \end{pmatrix} \begin{pmatrix} dW_{1t} \\ dW_{2t} \end{pmatrix} \quad (17.49)$$

where X_{1t} plays the role of the log of an asset price and $\exp(X_{2t})$ is the stochastic volatility variable. While the term $\exp(X_{2t})$ violates the linear growth condition, it does not cause explosions due to the mean reverting nature of the stochastic volatility. This model has no closed-form solution.

The diffusion (17.49) is in general not reducible, so I will apply the irreducible method described above to derive the expansion. The expansion at order $K = 3$ is of the form (17.35), with the coefficients $C_X^{(j_k,k)}$, $k = -1, 0, ..., 3$ given explicitly by:

$$C_X^{(8,-1)}(x|x_0;\theta) = -\frac{1}{2}\frac{(x_1-x_{01})^2}{e^{2x_{02}}\gamma_{11}^2} - \frac{1}{2}\frac{(x_2-x_{02})^2}{\gamma_{22}^2} + \frac{(x_1-x_{01})^2(x_2-x_{02})}{2e^{2x_{02}}\gamma_{11}^2}$$

$$-\frac{(x_1-x_{01})^2(x_2-x_{02})^2}{6e^{2x_{02}}\gamma_{11}^2} + \frac{(x_1-x_{01})^4\gamma_{22}^2}{24e^{4x_{02}}\gamma_{11}^4} - \frac{(x_1-x_{01})^4(x_2-x_{02})\gamma_{22}^2}{12e^{4x_{02}}\gamma_{11}^4}$$

$$+\frac{(x_1-x_{01})^2(x_2-x_{02})^4}{90e^{2x_{02}}\gamma_{11}^2} + \frac{(x_1-x_{01})^4(x_2-x_{02})^2\gamma_{22}^2}{15e^{4x_{02}}\gamma_{11}^4} - \frac{(x_1-x_{01})^6\gamma_{22}^4}{180e^{6x_{02}}\gamma_{11}^6}$$

$$-\frac{(x_1-x_{01})^4(x_2-x_{02})^3\gamma_{22}^2}{45e^{4x_{02}}\gamma_{11}^4} + \frac{(x_1-x_{01})^6(x_2-x_{02})\gamma_{22}^4}{60e^{6x_{02}}\gamma_{11}^6}$$

$$-\frac{(x_1-x_{01})^2(x_2-x_{02})^6}{945e^{2x_{02}}\gamma_{11}^2} - \frac{(x_1-x_{01})^4(x_2-x_{02})^4\gamma_{22}^2}{630e^{4x_{02}}\gamma_{11}^4}$$

$$-\frac{3(x_1-x_{01})^6(x_2-x_{02})^2\gamma_{22}^4}{140e^{6x_{02}}\gamma_{11}^6} + \frac{(x_1-x_{01})^8\gamma_{22}^6}{1120e^{8x_{02}}\gamma_{11}^8}$$

$$C_X^{(6,0)}(x|x_0;\theta) = \frac{\mu(x_1 - x_{01})}{e^{2x_{02}}\gamma_{11}^2} + (x_2 - x_{02})\left(\frac{1}{2} + \frac{\kappa(\alpha - x_{02})}{\gamma_{22}^2}\right)$$

$$- \frac{\mu(x_1 - x_{01})(x_2 - x_{02})}{e^{2x_{02}}\gamma_{11}^2} - \frac{(x_1 - x_{01})^2\gamma_{22}^2}{12e^{2x_{02}}\gamma_{11}^2} - \frac{(x_2 - x_{02})^2\left(6\kappa + \gamma_{22}^2\right)}{12\gamma_{22}^2}$$

$$+ \frac{\mu(x_1 - x_{01})(x_2 - x_{02})^2}{3e^{2x_{02}}\gamma_{11}^2} - \frac{\mu(x_1 - x_{01})^3\gamma_{22}^2}{6e^{4x_{02}}\gamma_{11}^4} + \frac{(x_1 - x_{01})^2(x_2 - x_{02})\gamma_{22}^2}{12e^{2x_{02}}\gamma_{11}^2}$$

$$+ \frac{(x_2 - x_{02})^4}{360} + \frac{\mu(x_1 - x_{01})^3(x_2 - x_{02})\gamma_{22}^2}{3e^{4x_{02}}\gamma_{11}^4} - \frac{(x_1 - x_{01})^2(x_2 - x_{02})^2\gamma_{22}^2}{45e^{2x_{02}}\gamma_{11}^2}$$

$$+ \frac{7(x_1 - x_{01})^4\gamma_{22}^4}{720e^{4x_{02}}\gamma_{11}^4} - \frac{\mu(x_1 - x_{01})(x_2 - x_{02})^4}{45e^{2x_{02}}\gamma_{11}^2} - \frac{4\mu(x_1 - x_{01})^3(x_2 - x_{02})^2\gamma_{22}^2}{15e^{4x_{02}}\gamma_{11}^{\leq}}$$

$$- \frac{(x_1 - x_{01})^2(x_2 - x_{02})^3\gamma_{22}^2}{180e^{2x_{02}}\gamma_{11}^2} + \frac{\mu(x_1 - x_{01})^5\gamma_{22}^4}{30e^{6x_{02}}\gamma_{11}^6} - \frac{7(x_1 - x_{01})^4(x_2 - x_{02})\gamma_{22}^4}{360e^{4x_{02}}\gamma_{11}^4}$$

$$- \frac{(x_2 - x_{02})^6}{5670} + \frac{4\mu(x_1 - x_{01})^3(x_2 - x_{02})^3\gamma_{22}^2}{45e^{4x_{02}}\gamma_{11}^4} + \frac{(x_1 - x_{01})^2(x_2 - x_{02})^4\gamma_{22}^2}{315e^{2x_{02}}\gamma_{11}^2}$$

$$- \frac{\mu(x_1 - x_{01})^5(x_2 - x_{02})\gamma_{22}^4}{10e^{6x_{02}}\gamma_{11}^6} + \frac{223(x_1 - x_{01})^4(x_2 - x_{02})^2\gamma_{22}^4}{15120e^{4x_{02}}\gamma_{11}^4}$$

$$- \frac{71(x_1 - x_{01})^6\gamma_{22}^6}{45360e^{6x_{02}}\gamma_{11}^6}$$

$$C_X^{(2,2)}(x|x_0;\theta) = \frac{1}{180e^{2x_{02}}\gamma_{11}^2}\left\{-30e^{2x_{02}}\kappa^2\gamma_{11}^2 - 30e^{2x_{02}}\alpha\kappa^2\gamma_{11}^2\right.$$

$$+ 30e^{2x_{02}}\kappa^2 x_{02}\gamma_{11}^2 - 30\mu^2\gamma_{22}^2 + e^{2x_{02}}\gamma_{11}^2\gamma_{22}^4\Big\} + \frac{(x_2 - x_{02})\left(e^{2x_{02}}\kappa^2\gamma_{11}^2 + 2\mu^2\gamma_{22}^2\right)}{12e^{2x_{02}}\gamma_{11}^2}$$

$$- \frac{\mu(x_1 - x_{01})}{90e^{4x_{02}}\gamma_{11}^4}\left\{30e^{2x_{02}}\alpha\kappa^2\gamma_{11}^2 - 30e^{2x_{02}}\kappa^2 x_{02}\gamma_{11}^2 + 30\mu^2\gamma_{22}^2 + e^{2x_{02}}\gamma_{11}^2\gamma_{22}^4\right\}$$

$$+ \frac{\mu(x_1 - x_{01})(x_2 - x_{02})}{90e^{4x_{02}}\gamma_{11}^4}\left\{15e^{2x_{02}}\kappa^2\gamma_{11}^2 + 30e^{2x_{02}}\alpha\kappa^2\gamma_{11}^2 - 30e^{2x_{02}}\kappa^2 x_{02}\gamma_{11}^2\right.$$

$$+ 60\mu^2\gamma_{22}^2 + e^{2x_{02}}\gamma_{11}^2\gamma_{22}^4\Big\} - \frac{(x_1 - x_{01})^2\gamma_{22}^2}{3780e^{4x_{02}}\gamma_{11}^4}\left\{-105e^{2x_{02}}\kappa^2\gamma_{11}^2 - 21e^{2x_{02}}\alpha\kappa^2\gamma_{11}^2\right.$$

$$+ 21e^{2x_{02}}\kappa^2 x_{02}\gamma_{11}^2 - 441\mu^2\gamma_{22}^2 + 4e^{2x_{02}}\gamma_{11}^2\gamma_{22}^4\Big\} - \frac{(x_2 - x_{02})^2}{3780e^{2x_{02}}\gamma_{11}^2}\left\{-21e^{2x_{02}}\kappa^2\gamma_{11}^2\right.$$

$$- 42e^{2x_{02}}\alpha\kappa^2\gamma_{11}^2 + 42e^{2x_{02}}\kappa^2 x_{02}\gamma_{11}^2 + 168\mu^2\gamma_{22}^2 + 4e^{2x_{02}}\gamma_{11}^2\gamma_{22}^4\Big\}$$

$$C_X^{(4,1)}(x|x_0;\theta) = -\frac{1}{24e^{2x_{02}}\gamma_{11}^2\gamma_{22}^2}\left\{12e^{2x_{02}}\alpha^2\kappa^2\gamma_{11}^2 - 24e^{2x_{02}}\alpha\kappa^2 x_{02}\gamma_{11}^2\right.$$

$$+ 12e^{2x_{02}}\kappa^2 x_{02}\gamma_{11}^2 + 12\mu^2\gamma_{22}^2 12e^{2x_{02}}\kappa\gamma_{11}^2\gamma_{22}^2 + e^{2x_{02}}\gamma_{11}^2\gamma_{22}^4\right\} + \frac{\mu(x_1-x_{01})\gamma_{22}^2}{6e^{2x_{02}}\gamma_{11}^2}$$

$$- \frac{(x_2-x_{02})}{2e^{2x_{02}}\gamma_{11}^2\gamma_{22}^2}\left\{-e^{2x_{02}}\alpha\kappa^2\gamma_{11}^2 + e^{2x_{02}}\kappa^2 x_{02}\gamma_{11}^2 - \mu^2\gamma_{22}^2\right\}$$

$$- \frac{\mu(x_1-x_{01})(x_2-x_{02})\gamma_{22}^2}{6e^{2x_{02}}\gamma_{11}^2} - \frac{(x_1-x_{01})^2}{360e^{4x_{02}}\gamma_{11}^4}\left\{-30e^{2x_{02}}\alpha\kappa^2\gamma_{11}^2 + 30e^{2x_{02}}\kappa^2 x_{02}\gamma_{11}^2\right.$$

$$- 90\mu^2\gamma_{22}^2 - e^{2x_{02}}\gamma_{11}^2\gamma_{22}^4\right\} + \frac{(x_2-x_{02})^2}{360e^{2x_{02}}\gamma_{11}^2\gamma_{22}^2}\left\{-60e^{2x_{02}}\kappa^2\gamma_{11}^2 - 60\mu^2\gamma_{22}^2\right.$$

$$+ e^{2x_{02}}\gamma_{11}^2\gamma_{22}^4\right\} + \frac{2\mu(x_1-x_{01})(x_2-x_{02})^2\gamma_{22}^2}{45e^{2x_{02}}\gamma_{11}^2} - \frac{7\mu(x_1-x_{01})^3\gamma_{22}^4}{180e^{4x_{02}}\gamma_{11}^4}$$

$$- \frac{(x_1-x_{01})^2(x_2-x_{02})}{360e^{4x_{02}}\gamma_{11}^4}\left\{15e^{2x_{02}}\kappa^2\gamma_{11}^2 + 30e^{2x_{02}}\alpha\kappa^2\gamma_{11}^2 - 30e^{2x_{02}}\kappa^2 x_{02}\gamma_{11}^2\right.$$

$$+ 180\mu^2\gamma_{22}^2 + e^{2x_{02}}\gamma_{11}^2\gamma_{22}^4\right\} + \frac{\mu(x_1-x_{01})(x_2-x_{02})^3\gamma_{22}^2}{90e^{2x_{02}}\gamma_{11}^2}$$

$$+ \frac{7\mu(x_1-x_{01})^3(x_2-x_{02})\gamma_{22}^4}{90e^{4x_{02}}\gamma_{11}^4} - \frac{(x_2-x_{02})^4(-42\mu^2 + e^{2x_{02}}\gamma_{11}^2\gamma_{22}^2)}{3780e^{2x_{02}}\gamma_{11}^2}$$

$$+ \frac{(x_1-x_{01})^2(x_2-x_{02})^2}{2520e^{4x_{02}}\gamma_{11}^4}\left\{98e^{2x_{02}}\kappa^2\gamma_{11}^2 + 56e^{2x_{02}}\alpha\kappa^2\gamma_{11}^2 - 56e^{2x_{02}}\kappa^2 x_{02}\gamma_{11}^2\right.$$

$$+ 1008\mu^2\gamma_{22}^2 + e^{2x_{02}}\gamma_{11}^2\gamma_{22}^4\right\} - \frac{(x_1-x_{01})^4\gamma_{22}^2}{10080e^{6x_{02}}\gamma_{11}^6}\left\{42e^{2x_{02}}\kappa^2\gamma_{11}^2 + 112e^{2x_{02}}\alpha\kappa^2\gamma_{11}^2\right.$$

$$- 112e^{2x_{02}}\kappa^2 x_{02}\gamma_{11}^2 + 840\mu^2\gamma_{22}^2 + 5e^{2x_{02}}\gamma_{11}^2\gamma_{22}^4\right\}$$

and

$$C_X^{(0,3)}(x|x_0;\theta) = \frac{1}{7560e^{4x_{02}}\gamma_{11}^4\gamma_{22}^2}\left\{1890\mu^4\gamma_{22}^4\right.$$

$$+ 126e^{2x_{02}}\mu^2\gamma_{11}^2\gamma_{22}^2\left(30\kappa^2(\alpha-x_{02})\gamma_{22}^4\right) + e^{4x_{02}}\gamma_{11}^4\left(1890\kappa^4(x_{02}-\alpha)^2\right.$$

$$\left. - 63\kappa^2(1-2\alpha+2x_{02})\gamma_{22}^4 - 16\gamma_{22}^8\right)\right\}.$$

17.7. Inference When the State is Partially Observed

In many cases, the state vector is of the form $X_t = [S_t; V_t]'$, where the $(m-q)$–dimensional vector S_t is observed but the q–dimensional V_t is not. Two typical examples in finance consist of stochastic volatility models, such as the example just discussed, where V_t is the volatility state variable(s), and term structure models, where V_t is a vector of factors or yields. One can conduct likelihood inference in this setting, without resorting to the statistically sound but computationally infeasible integration of the latent variables from the likelihood function. The idea is simple: write down in closed form an expansion for the log-likelihood of the state vector X, including its unobservable components. Then enlarge the observation state by adding variables that are observed and functions of X. For example, in the stochastic volatility case, an option price or an option-implied volatility; in term structure models, as many bonds as there are factors. Then, using the Jacobian formula, write down the likelihood function of the pair consisting of the observed components of X and the additional observed variables, and maximize it.

Identification of the parameter vector must be ensured. In fact, identifying a multivariate continuous-time Markov process from discrete-time data can be problematic when the process is not reversible, as an aliasing problem can be present in the multivariate case (see Philips (1973) and Hansen and Sargent (1983)). As for the distributional properties of the resulting estimator, a fixed interval sample of a time-homogenous continuous-time Markov process is a Markov process in discrete time. Given that the Markov state vector is observed and the unknown parameters are identified, properties of the MLE follow from what is known about ML estimation of discrete-time Markov processes (see Billingsley (1961)).

17.7.1. *Likelihood inference for stochastic volatility models*

In a stochastic volatility model, the asset price process S_t follows

$$dS_t = (r - \delta) S_t dt + \sigma_1 (X_t; \theta) dW_t^Q \qquad (17.50)$$

where r is the riskfree rate, δ is the dividend yield paid by the asset (both taken to be constant for simplicity only), σ_1 denotes the first row of the matrix σ and Q denotes the equivalent martingale measure (see e.g., Harrison and Kreps (1979)). The volatility state variables V_t then follow a SDE on their own. For example, in the Heston (1993) model, $m = 2$ and $q = 1$:

$$dX_t = d \begin{bmatrix} S_t \\ V_t \end{bmatrix} = \begin{bmatrix} (r - \delta) S_t \\ \kappa (\gamma - V_t) \end{bmatrix} dt$$

$$+ \begin{bmatrix} \sqrt{(1 - \rho^2) V_t} S_t & \rho \sqrt{V_t} S_t \\ 0 & \sigma \sqrt{V_t} \end{bmatrix} d \begin{bmatrix} W_1^Q (t) \\ W_2^Q (t) \end{bmatrix}. \qquad (17.51)$$

The model is completed by the specification of a vector of market prices of risk for the different sources of risk (W_1 and W_2 here), such as

$$\Lambda\left(X_t; \theta\right) = \left[\lambda_1 \sqrt{(1 - \rho^2)\, V_t},\ \lambda_2 \sqrt{V_t}\right]', \tag{17.52}$$

which characterizes the change of measure from Q back to the physical probability measure P.

Likelihood inference for this and other stochastic volatility models is discussed in Aït-Sahalia and Kimmel (2004). Given a time series of observations of both the asset price, S_t, and a vector of option prices (which, for simplicity, we take to be call options) C_t, the time series of V_t can then be inferred from the observed C_t. If V_t is multidimensional, sufficiently many options are required with varying strike prices and maturities to allow extraction of the current value of V_t from the observed stock and call prices. Otherwise, only a single option is needed. For reasons of statistical efficiency, we seek to determine the joint likelihood function of the observed data, as opposed to, for example, conditional or unconditional moments. We employ the closed-form approximation technique described above, which yields in closed form the joint likelihood function of $[S_t; V_t]'$. From there, the joint likelihood function of the observations on $G_t = [S_t; C_t]' = f\left(X_t; \theta\right)$ is obtained simply by multiplying the likelihood of $X_t = [S_t; V_t]'$ by the Jacobian term J_t:

$$\begin{aligned}
\ln p_G\left(g|g_0, \Delta; \theta\right) = {} & -\ln J_t\left(g|g_0, \Delta; \theta\right) \\
& + l_X\!\left(f^{-1}\left(g; \theta\right) | f^{-1}\left(g_0; \theta\right); \Delta, \theta\right)
\end{aligned} \tag{17.53}$$

with l_X obtained as described above.

If a proxy for V_t is used directly, this last step is not necessary. Indeed, we can avoid the computation of the function f by first transforming C_t into a proxy for V_t. The simplest one consists in using the Black-Scholes implied volatility of a short-maturity at-the-money option in place of the true instantaneous volatility state variable. The use of this proxy is justified in theory by the fact that the implied volatility of such an option converges to the instantaneous volatility of the logarithmic stock price as the maturity of the option goes to zero. An alternate proxy (which we call the integrated volatility proxy) corrects for the effect of mean reversion in volatility during the life of an option. If V_t is the instantaneous variance of the logarithmic stock price, we can express the integral of variance from time t to T as

$$V\left(t, T\right) = \int_t^T V_u\, du. \tag{17.54}$$

If the volatility process is instantaneously uncorrelated with the logarithmic stock price process, then we can calculate option prices by taking the expected value of the Black-Scholes option price (with $V(t,T)$ as implied variance) over the probability distribution of $V(t,T)$ (see Hull and White (1987)). If the two processes are correlated, then the price of the option is a weighted average of Black-Scholes prices evaluated at different stock prices and volatilities (see Romano and Touzi (1997)).

The proxy we examine is determined by calculating the expected value of $V(t,T)$ first, and substituting this value into the Black-Scholes formula as implied variance. This proxy is model-free, in that it can be calculated whether or not an exact volatility can be computed and results in a straightforward estimation procedure. On the other hand, this procedure is in general approximate, first because the volatility process is unlikely to be instantaneously uncorrelated with the logarithmic stock price process, and second, because the expectation is taken before substituting $V(t,T)$ into the Black-Scholes formula rather than after and we examine in Monte Carlo simulations the respective impact of these approximations, with the objective of determining whether the trade-off involved between simplicity and exactitude is worthwhile.

The idea is to adjust the Black-Scholes implied volatility for the effect of mean reversion in volatility, essentially undoing the averaging that takes place in equation (17.54). Specifically, if the Q-measure drift of Y_t is of the form $a + bY_t$ (as it is in many of the stochastic volatility models in use), then the expected value of $V(t,T)$ is given by:

$$E_t\left[V\left(t,T\right)\right] = \left(\frac{e^{b(T-t)}-1}{b}\right)\left(V_t + \frac{a}{b}\right) - \frac{a}{b}\left(T-t\right). \qquad (17.55)$$

A similar expression can be derived in the special case where $b = 0$. By taking the expected value on the left-hand side to be the observed implied variance $V_{\text{imp}}(t,T)$ of a short maturity T at-the-money option, our adjusted proxy is then given by:

$$V_t \approx \frac{bV_{\text{imp}}(t,T) + a\left(T-t\right)}{e^{b(T-t)}-1} - \frac{a}{b}. \qquad (17.56)$$

Then we can simply take $[S_t; V_{\text{imp}}(t,T)]'$ as the state vector, write its likelihood from that of $[S_t; V_t]'$ using a Jacobian term for the change of variable (17.56).

It is possible to refine the implied volatility proxy by expressing it in the form of a Taylor series in the "volatility of volatility" parameter σ in the case of the CEV model, where the Q-measure drift of Y_t is of the form

$a+bY_t$, and the Q-measure diffusion of Y_t is of the form σY_t^β (Lewis (2000)). However, unlike (17.56), the relationship between the observed $V_{\text{imp}}(t, T)$ and the latent Y_t is not invertible without numerical computation of the parameter-dependent integral.

17.7.2. *Likelihood inference for term structure models*

Another example of a class of models where the state can be only partially observed consist of term structure models. A multivariate term structure model specifies that the instantaneous riskless rate r_t is a deterministic function of an $m-$dimensional vector of state variables, X_t

$$r_t = r\left(X_t; \theta\right) \tag{17.57}$$

which will typically not be fully observable. Under the equivalent martingale measure Q, the state vector X follows the dynamics given in (17.1). In order to avoid arbitrage opportunities, the price at t of a zero-coupon bond maturing at T is given by the Feynman-Kac representation:

$$P\left(x, t, T; \theta\right) = E^Q\left[\exp\left(-\int_t^T r_u du\right)\bigg| X_t = x\right]. \tag{17.58}$$

An affine yield model is any model where the short rate (17.57) is an affine function of the state vector and the risk-neutral dynamics (17.1) are affine:

$$dX_t = \left(\tilde{A} + \widetilde{B}X_t\right) dt + \Sigma\sqrt{S\left(X_t; \alpha, \beta\right)} dW_t^Q \tag{17.59}$$

where $\tilde{A}$ is an $m-$element column vector, $\widetilde{B}$ and Σ are $m \times m$ matrices, and $S\left(X_t; \alpha, \beta\right)$ is the diagonal matrix with elements $S_{ii} = \alpha_i + X_t'\beta_i$, with each α_i a scalar and each β_i an $m \times 1$ vector, $1 \leq i \leq m$ (see Dai and Singleton (2000)).

It can then be shown that, in affine models, bond prices have the exponential affine form

$$P\left(x, t, T; \theta\right) = \exp\left(-\gamma_0\left(\tau; \theta\right) - \gamma\left(\tau; \theta\right)' x\right) \tag{17.60}$$

where $\tau = T - t$ is the bond's time to maturity. That is, bond yields (non-annualized, and denoted by $g\left(x, t, T; \theta\right) = -\ln\left(P\left(x, t, T; \theta\right)\right)$) are affine functions of the state vector:

$$g(x, t, T; \theta) = \gamma_0\left(\tau; \theta\right) + \gamma\left(\tau; \theta\right)' x. \tag{17.61}$$

Alternatively, one can start with the requirement that the yields be affine, and show that the dynamics of the state vector must be affine (see Duffie and Kan (1996)).

The final condition for the bond price implies that $\gamma_0(0;\theta) = \gamma(0;\theta) = 0$, while

$$r_t = \delta_0 + \delta' x. \tag{17.62}$$

Affine yield models owe much of their popularity to the fact that bond prices can be calculated quickly as solutions to a system of ordinary differential equations. Under non-linear term structure models, bond prices will normally be solutions to a partial differential equation that is far more difficult to solve.

Aït-Sahalia and Kimmel (2002) consider likelihood inference for affine term structure models. They derive the likelihood expansions for the nine canonical models of Dai and Singleton (2000) in dimensions $m = 1, 2$ and 3. For instance, in dimension $m = 3$, the four canonical models are respectively

$$\begin{pmatrix} dX_{1t} \\ dX_{2t} \\ dX_{3t} \end{pmatrix} = \begin{pmatrix} \kappa_{11} & 0 & 0 \\ \kappa_{21} & \kappa_{22} & 0 \\ \kappa_{31} & \kappa_{32} & \kappa_{33} \end{pmatrix} \begin{pmatrix} -X_{1t} \\ -X_{2t} \\ -X_{3t} \end{pmatrix} dt + \begin{pmatrix} dW_{1t} \\ dW_{2t} \\ dW_{3t} \end{pmatrix},$$

$$\begin{pmatrix} dX_{1t} \\ dX_{2t} \\ dX_{3t} \end{pmatrix} = \begin{pmatrix} \kappa_{11} & 0 & 0 \\ \kappa_{21} & \kappa_{22} & \kappa_{23} \\ \kappa_{31} & \kappa_{32} & \kappa_{33} \end{pmatrix} \begin{pmatrix} \theta_1 - X_{1t} \\ -X_{2t} \\ -X_{3t} \end{pmatrix} dt$$
$$+ \begin{pmatrix} X_{1t}^{1/2} & 0 & 0 \\ 0 & (1+\beta_{21}X_{1t})^{\frac{1}{2}} & 0 \\ 0 & 0 & (1+\beta_{21}X_{1t})^{\frac{1}{2}} \end{pmatrix} \begin{pmatrix} dW_{1t} \\ dW_{2t} \\ dW_{3t} \end{pmatrix},$$

$$\begin{pmatrix} dX_{1t} \\ dX_{2t} \\ dX_{3t} \end{pmatrix} = \begin{pmatrix} \kappa_{11} & \kappa_{12} & 0 \\ \kappa_{21} & \kappa_{22} & 0 \\ \kappa_{31} & \kappa_{32} & \kappa_{33} \end{pmatrix} \begin{pmatrix} \theta_1 - X_{1t} \\ \theta_2 - X_{2t} \\ -X_{3t} \end{pmatrix} dt$$
$$+ \begin{pmatrix} X_{1t}^{1/2} & 0 & 0 \\ 0 & X_{2t}^{1/2} & 0 \\ 0 & 0 & (1+\beta_{31}X_{1t}+\beta_{32}X_{2t})^{\frac{1}{2}} \end{pmatrix} \begin{pmatrix} dW_{1t} \\ dW_{2t} \\ dW_{3t} \end{pmatrix},$$

$$\begin{pmatrix} dX_{1t} \\ dX_{2t} \\ dX_{3t} \end{pmatrix} = \begin{pmatrix} \kappa_{11} & \kappa_{12} & \kappa_{13} \\ \kappa_{21} & \kappa_{22} & \kappa_{23} \\ \kappa_{31} & \kappa_{32} & \kappa_{33} \end{pmatrix} \begin{pmatrix} \theta_1 - X_{1t} \\ \theta_2 - X_{2t} \\ \theta_3 - X_{3t} \end{pmatrix} dt + \begin{pmatrix} X_{1t}^{1/2} & 0 & 0 \\ 0 & X_{2t}^{1/2} & 0 \\ 0 & 0 & X_{3t}^{1/2} \end{pmatrix} \begin{pmatrix} dW_{1t} \\ dW_{2t} \\ dW_{3t} \end{pmatrix}.$$

MLE in this case requires evaluation of the likelihood of an observed panel of yield data for each parameter vector considered during a search procedure. The procedure for evaluating the likelihood of the observed yields at a particular value of the parameter vector consists of four steps. First, we extract the value of the state vector X_t (which is not directly observed) from those yields that are treated as observed without error. Second, we evaluate the joint likelihood of the series of implied observations of the state vector X_t, using the closed-form approximations to the likelihood function described above. Third, we multiply this joint likelihood by a Jacobian term, to find the likelihood of the panel of observations of the yields observed without error. Finally, we calculate the likelihood of the observation errors for those yields observed with error, and multiply this likelihood by the likelihood found in the previous step, to find the joint likelihood of the panel of all yields.

The first task is therefore to infer the state vector X_t at date t from the cross-section of bond yields at date t with different maturities. Affine yield models, as their name implies, make yields of zero coupon bonds affine functions of the state vector. Given this simple relationship between yields and the state vector, the likelihood function of bond yields is a simple transformation of the likelihood function of the state vector.

If the number of observed yields at that point in time is smaller than the number N of state variables in the model, then the state is not completely observed, and the vector of observed yields does not follow a Markov process, even if the (unobserved) state vector does, enormously complicating maximum likelihood estimation. On the other hand, if the number of observed yields is larger than the number of state variables, then some of the yields can be expressed as deterministic functions of other observed yields, without error. Even tiny deviations from the predicted values have a likelihood of zero. This problem can be avoided by using a number of yields exactly equal to the number of state variables in the underlying model, but, in general, the market price of risk parameters will not all be identified. Specifically, there are affine yield models that generate identical dynamics for yields with a given set of maturities, but different dynamics for yields with other maturities. A common practice (see, for example, Duffee (2002)) is to use more yields than state variables, and to assume that certain benchmark yields are observed precisely, whereas the other yields are observed with measurement error. The measurement errors are generally held to be i.i.d., and also independent of the state variable processes.

We take this latter approach, and use $N + H$ observed yields, $H \geq 0$, in the postulated model, and include observation errors for H of those yields. At each date t, the state vector X_t is then exactly identified by the yields observed without error, and these N yields jointly follow a Markov process. Denoting the times to maturity of the yields observed without error as $\tau_1, ..., \tau_N$, the observed values of these yields, on the left-hand side, are equated with the predicted values (from (17.61)) given the model parameters and the current values of the state variables, X_t:

$$g_t = \Gamma_0\left(\theta\right) + \Gamma\left(\theta\right)' X_t. \tag{17.63}$$

The current value of the state vector X_t is obtained by inverting this equation:

$$X_t = \left[\Gamma\left(\theta\right)'\right]^{-1}\left[g_t - \Gamma_0\left(\theta\right)\right]. \tag{17.64}$$

While the only parameters entering the transformation from observed yields to the state variables are the parameters of the risk-neutral (or Q-measure) dynamics of the state variables, once we have constructed our time series of values of X_t sampled at dates $\tau_0, \tau_1, ..., \tau_n$, the dynamics of the state variable that we will be able to infer from this time series are the dynamics under the physical measure (denoted by P). The first step in the estimation procedure is the only place where we rely on the tractability of the affine bond pricing model. In particular, we can now specify freely (that is, without regard for considerations of analytical tractability) the market prices of risk of the different Brownian motions

$$\begin{aligned} dX_t &= \mu^P\left(X_t; \theta\right) dt + \sigma\left(X_t; \theta\right) dW_t^P \\ &= \left\{\mu^Q\left(X_t; \theta\right) + \sigma\left(X_t; \theta\right)\Lambda\left(X_t; \theta\right)\right\} dt + \sigma\left(X_t; \theta\right) dW_t^P. \end{aligned} \tag{17.65}$$

We adopt the simple specification for the market price of risk

$$\Lambda\left(X_t; \theta\right) = \sigma\left(X_t; \theta\right)' \lambda \tag{17.66}$$

with λ an $m \times 1$ vector of constant parameters, so that under P, the instantaneous drift of each state variables is its drift under the risk-neutral measure, plus a constant times its volatility squared. Under this specification, the drift of the state vector is then affine under both the physical and risk-neutral measures, since

$$\mu^P\left(X_t; \theta\right) = \left(\tilde{A} + \tilde{B}X_t\right) + \Sigma S\left(X_t; \beta\right)' \Sigma' \lambda \equiv A + BX_t. \tag{17.67}$$

An affine μ^P is not required for our likelihood expansions. Since we can derive likelihood expansions for arbitrary diffusions, μ^P may contain terms that are non-affine, such as the square root of linear functions of the state vector, as in Duarte (2004) for instance. Duffee (2002) and Cheridito, Filipović, and Kimmel (2005) also allow for a more general market price of risk specifications than Dai and Singleton (2000), but retain the affinity of μ^Q and μ^P (and also of the diffusion matrix). However, we do rely on the affine character of the dynamics under Q because those allow us to go from state to yields in the tractable manner given by (17.64).

These closed form likelihood expansions are used in various contexts by Thompson (2004), Takamizawa (2005) and Schneider (2006) for interest rate and term structure models of affine or more general type.

17.8. Application to Specification Testing

Aït-Sahalia, Fan, and Peng (2005) develop a specification test for the transition density of the process, based on a direct comparison of the nonparametric estimate of the transition function to the parametric transition function $p_X(x|x_0, \Delta; \theta)$ implied by the model in order to test

$$H_0 : p_X(x|x_0, \Delta) = p_X(x|x_0, \Delta; \theta)$$
$$\text{vs.} \quad H_1 : p_X(x|x_0, \Delta) \neq p_X(x|x_0, \Delta; \theta). \quad (17.68)$$

As in the parametric situation of (17.2), note that the logarithm of the likelihood function of the observed data $\{X_1, \cdots, X_{n+\Delta}\}$ is

$$\ell(p_X) = \sum_{i=1}^{N} \ln p_X(X_{i\Delta}|X_{(i-1)\Delta}, \Delta),$$

after ignoring the stationary density of X_0. A natural test statistic is then to compare the likelihood ratio under the null and alternative hypotheses. This leads to the test statistic

$$T_0 = \sum_{i=1}^{N} \ln \left(\widehat{p}_X(X_{i\Delta}|X_{(i-1)\Delta}, \Delta)/p_X(X_{i\Delta}|X_{(i-1)\Delta}, \Delta; \widehat{\theta}) \right)$$
$$\times w(X_{(i-1)\Delta}, X_{i\Delta}) \quad (17.69)$$

where w is a weight function, $\widehat{p}_X$ a nonparametric estimator of the transition function based on locally linear polynomials (see Fan,Yao and Tong (1996)) and $p_X(\cdot, \widehat{\theta})$ a parametric estimator based on the closed form expressions described above. Aït-Sahalia, Fan, and Peng (2005) consider other distance measures and tests, and derive their asymptotic properties.

A complementary approach to this is the one proposed by Hong and Li (2005), who use the fact that under the null hypothesis, the random variables $\{P(X_{i\Delta}|X_{(i-1)\Delta}, \Delta, \theta)\}$ are a sequence of i.i.d. uniform random variables; see also Thompson (2004), Chen and Gao (2004) and Corradi and Swanson (2005). That approach will only work in the univariate case, however, unlike one based on (17.69).

17.9. Derivative Pricing Applications

Consider a generic derivative security with payoff function $\Psi(X_\Delta)$ at time Δ. If the derivative is written on a traded underlying asset, with price process X and risk-neutral dynamics

$$dX_t/X_t = \{r - q\}\,dt + \sigma(X_t; \theta)\,dW_t \tag{17.70}$$

where r is the risk-free rate, q the dividend yield paid by that asset, both viewed as constant, then with complete markets, absence of arbitrage opportunities implies that the price at time 0 of the derivative is

$$
\begin{aligned}
P_0 &= e^{-r\Delta} E\left[\Psi(X_\Delta)\middle|\, X_0 = x_0\right] \\
&= e^{-r\Delta} \int_0^{+\infty} \Psi(x)\, p_X(\Delta, x|x_0; \theta)\, dx.
\end{aligned}
\tag{17.71}
$$

In general, the transition function p_X corresponding to the dynamics (17.70) is unknown and one will either solve numerically the PDE solved by P or perform Monte Carlo integration of (17.71).

But we can instead, as long as Δ is not too large, use the $p_X^{(J)}$ corresponding to the SDE (17.70), and get a closed form approximation of the derivative price in the form

$$P_0^{(J)} = e^{-r\Delta} \int_0^{+\infty} \Psi(x)\, p_X^{(J)}(\Delta, x|x_0; \theta)\, dx \tag{17.72}$$

which is of a different nature than the ad hoc "corrections" to the Black-Scholes Merton formula (as in for example Jarrow and Rudd (1982)), which break the link between the derivative price and the dynamic model for the underlying asset price by assuming directly a functional form for p_X. By contrast, (17.72) is the option pricing formula (of order J in Δ) which matches the dynamics of the underlying asset. Being in closed form, comparative statics, etc. are possible. Being an expansion in small time, accuracy will be limited to relatively small values of Δ (of the order of 3 months in practical applications).

17.10. Likelihood Inference for Diffusions under Nonstationarity

There is an extensive literature applicable to discrete-time stationary Markov processes starting with the work of Billingsley (1961). The asymptotic covariance matrix for the ML estimator is the inverse of the score covariance or information matrix where the score at date t is $\partial \ln p(X_{t+\Delta}|X_t, \Delta, \theta)/\partial\theta$ where $\ln p(\cdot|x, \Delta, \theta)$ is the logarithm of the conditional density over an interval of time Δ and a parameter value θ. In the stationary case, the MLE will under standard regularity conditions converge at speed $n^{1/2}$ to a normal distribution whose variance is given by the inverse of Fisher's information matrix.

When the underlying Markov process is nonstationary, the score process inherits this nonstationarity. The rate of convergence and the limiting distribution of the maximum likelihood estimator depends upon growth properties of the score process (e.g. see Hall and Heyde (1980) Chapter 6.2). A nondegenerate limiting distribution can be obtained when the score process behaves in a sufficiently regular fashion. The limiting distribution can be deduced by showing that general results pertaining to time series asymptotics (see e.g., Jeganathan (1995)) can be applied to the present context. One first establishes that the likelihood ratio has the locally asymptotically quadratic (LAQ) structure, then within that class separates between the locally asymptotically mixed Normal (LAMN), locally asymptotically Normal (LAN) and locally asymptotically Brownian functional (LABF) structures. As we have seen, when the data generating process is stationary and ergodic, the estimation is typically in the LAN class. The LAMN class can be used to justify many of the standard inference methods given the ability to estimate the covariance matrix pertinent for the conditional normal approximating distribution. Rules for inference are special for the LABF case. These structures are familiar from the linear time series literature on unit roots and co-integration. Details for the case of a nonlinear Markov process can be found in Aït-Sahalia (2002).

As an example of the types of results that can be derived, consider the Ornstein-Uhlenbeck specification, $dX_t = -\kappa X_t dt + \sigma dW_t$, where $\theta = (\kappa, \sigma^2)$. The sampled process is a first-order scalar autoregression, which has received extensive attention in the literature on time series. The discrete-time process obtained by sampling at a fixed interval Δ is a Gaussian first-order autoregressive process with autoregressive parameter $\exp(-\kappa\Delta)$ and innovation variance $\sigma^2 \left(1 - e^{-2\kappa\Delta}\right)/(2\kappa)$. White (1958) and Anderson

(1959) originally characterized the limiting distribution for the discrete-time autoregressive parameter when the Markov process is not stationary. Alternatively, by specializing the general theory of the limiting behavior of the ML estimation to this model, one obtains the following asymptotic distribution for the MLE of the continuous-time parameterization (see Corollary 2 in Aït-Sahalia (2002)):

- If $\kappa > 0$ (LAN, stationary case):

$$\sqrt{N}\left(\begin{pmatrix}\widehat{\kappa}_N\\\widehat{\sigma}_N^2\end{pmatrix}-\begin{pmatrix}\kappa\\\sigma^2\end{pmatrix}\right)\longrightarrow N\left(\begin{pmatrix}0\\0\end{pmatrix}\right),$$

$$\left(\begin{matrix}\dfrac{e^{2\kappa\Delta}-1}{\Delta^2} & \dfrac{\sigma^2\left(e^{2\kappa\Delta}-1-2\kappa\Delta\right)}{\kappa\Delta^2}\\\dfrac{\sigma^2\left(e^{2\kappa\Delta}-1-2\kappa\Delta\right)}{\kappa\Delta^2} & \dfrac{\sigma^4\left(\left(e^{2\kappa\Delta}-1\right)^2+2\kappa^2\Delta^2\left(e^{2\kappa\Delta}+1\right)+4\kappa\Delta\left(e^{2\kappa\Delta}-1\right)\right)}{\kappa^2\Delta^2\left(e^{2\kappa\Delta}-1\right)}\end{matrix}\right)\right) \tag{17.73}$$

- If $\kappa < 0$ (LAMN, explosive case), assume $X_0 = 0$, then:

$$\frac{e^{-(N+1)\kappa\Delta}\Delta}{e^{-2\kappa\Delta}-1}\left(\widehat{\kappa}_N-\kappa\right)\to G^{-1/2}\times N\left(0,1\right)$$

$$\sqrt{N}\left(\widehat{\sigma}_N^2-\sigma^2\right)\to N\left(0,2\sigma^4\right) \tag{17.74}$$

where G has a $\chi^2[1]$ distribution independent of the $N(0,1)$. $G^{-1/2}\times N(0,1)$ is a Cauchy distribution.

- If $\kappa = 0$ (LABF, unit root case), assume $X_0 = 0$, then:

$$N\widehat{\kappa}_N\to\left(1-W_1^2\right)\left(2\Delta\int_0^1 W_t^2 dt\right)^{-1}$$

$$\sqrt{N}\left(\widehat{\sigma}_N^2-\sigma^2\right)\to N\left(0,2\sigma^4\right) \tag{17.75}$$

where N is the sample size and $\{W_t : t \geq 0\}$ is a standard Brownian motion.

Acknowledgments

Financial support from the NSF under grants SES-0350772 and DMS-0532370 is gratefully acknowledged.

References

Aït-Sahalia, Y. (1996). Testing continuous-time models of the spot interest rate. *Review of Financial Studies* **9**, 385–426.

Aït-Sahalia, Y. (1999). Transition densities for interest rate and other nonlinear diffusions. *Journal of Finance* **54**, 1361–1395.

Aït-Sahalia, Y. (2001). Closed-form likelihood expansions for multivariate diffusions. Tech. rep., Princeton University.

Aït-Sahalia, Y. (2002). Maximum-likelihood estimation of discretely-sampled diffusions: A closed-form approximation approach. *Econometrica* **70**, 223–262.

Aït-Sahalia, Y., Fan, J., Peng, H. (2005). Nonparametric transition-based tests for diffusions. Tech. rep., Princeton University.

Aït-Sahalia, Y., Kimmel, R. (2002). Estimating affine multifactor term structure models using closed-form likelihood expansions. Tech. rep., Princeton University.

Aït-Sahalia, Y., Kimmel, R. (2004). Maximum likelihood estimation of stochastic volatility models. *Journal of Financial Economics* forthcoming.

Aït-Sahalia, Y., Yu, J. (2005). Saddlepoint approximations for continuous-time markov processes. *Journal of Econometrics* forthcoming.

Anderson, T. W. (1959). On asymptotic distributions of estimates of parameters of stochastic difference equations. *Annals of Mathematical Statistics*, **30** 676–687.

Bakshi, G. S., Ju, N. and Qu-Yang, H. (2006). Estimation of continuous-time models with an application to equity volatility dynamics, *Journal of Financial Economics*, forthcoming.

Bakshi, G. S., Yu, N. (2005). A refinement to Aït-Sahalia's (2002) "Maximum likelihood estimation of discretely sampled diffusions: A closed-form approximation approach". *Journal of Business* **78**, 2037–2052.

Billingsley, P. (1961). Statistical Inference for Markov Processes. The University of Chicago Press, Chicago.

Black, F., Scholes, M. (1973). The pricing of options and corporate liabilities. *Journal of Political Economy* **81**, 637–654.

Brandt, M., Santa-Clara, P. (2002). Simulated likelihood estimation of diffusions with an application to exchange rate dynamics in incomplete markets. *Journal of Financial Economics* **63**, 161–210.

Chan, K. C., Karolyi, G. A., Longstaff, F. A., Sanders, A. B. (1992). An empirical comparison of alternative models of the short-term interest rate. *Journal of Finance* **48**, 1209–1227.

Chen, S. X., Gao, J. (2004). An adaptive empirical likelihood test for time series models. Tech. rep., Iowa State University.

Cheridito, P., Filipović, D., Kimmel, R. (2005). Market price of risk specifications for affine models: Theory and evidence. *Journal of Financial Economics* forthcoming.

Constantinides, G. M. (1992). A theory of the nominal term structure of interest rates. *Review of Financial Studies* **5**, 531–552.

Corradi, V., Swanson, N. R. (2005). A bootstrap specification test for diffusion processes. *Journal of Econometrics*, forthcoming.

Courtadon, G. (1982). The pricing of options on default free bonds. *Journal of Financial and Quantitative Analysis* **17**, 75–100.

Cox, J. C. (1975). The constant elasticity of variance option pricing model. The Journal of Portfolio Management, Special Issue published in 1996, 15–

17.

Cox, J. C., Ingersoll, J. E., Ross, S. A. (1985). A theory of the term structure of interest rates. *Econometrica* **53**, 385–408.

Dai, Q., Singleton, K. J. (2000). Specification analysis of affine term structure models. *Journal of Finance* **55**, 1943–1978.

Daniels, H. (1954). Saddlepoint approximations in statistics. *Annals of Mathematical Statistics* **25**, 631–650.

DiPietro, M. (2001). Bayesian inference for discretely sampled diffusion processes with financial applications. Ph.D. thesis, Department of Statistics, Carnegie-Mellon University.

Duarte, J. (2004). Evaluating an alternative risk preference in affine term structure models. *Review of Financial Studies* **17**, 379–404.

Duffee, G. R. (2002). Term premia and interest rate forecasts in affine models. *Journal of Finance* **57**, 405–443.

Duffie, D., Kan, R. (1996). A yield-factor model of interest rates. *Mathematical Finance* **6**, 379–406.

Egorov, A. V., Li, H., Xu, Y. (2003). Maximum likelihood estimation of time inhomogeneous diffusions. *Journal of Econometrics* **114**, 107–139.

Fan, J., Yao, Q., Tong, H. (1996). Estimation of conditional densities and sensitivity measures in nonlinear dynamical systems. *Biometrika* **83**, 189–206.

Hall, P., Heyde, C. C. (1980). Martingale Limit Theory and Its Application. Academic Press, Boston.

Hansen, L. P., Sargent, T. J. (1983). The dimensionality of the aliasing problem in models with rational spectral densities. *Econometrica* **51**, 377–387.

Harrison, M., Kreps, D., 1979. Martingales and arbitrage in multiperiod securities markets. *Journal of Economic Theory* **20**, 381–408.

Heston, S. (1993). A closed-form solution for options with stochastic volatility with applications to bonds and currency options. *Review of Financial Studies* **6**, 327–343.

Hong, Y., Li, H. (2005). Nonparametric specification testing for continuous-time models with applications to term structure of interest rates. *Review of Financial Studies* **18**, 37–84.

Hull, J., White, A. (1987). The pricing of options on assets with stochastic volatilities. *Journal of Finance* **42**, 281–300.

Hurn, A. S., Jeisman, J., Lindsay, K. (2005). Seeing the wood for the trees: A critical evaluation of methods to estimate the parameters of stochastic differential equations. Tech. rep., School of Economics and Finance, Queensland University of Technology.

Jarrow, R., Rudd, A. (1982). Approximate option valuation for arbitrary stochastic processes. *Journal of Financial Economics* **10**, 347–369.

Jeganathan, P. (1995). Some aspects of asymptotic theory with applications to time series models. *Econometric Theory* **11**, 818–887.

Jensen, B., Poulsen, R. (2002). Transition densities of diffusion processes: Numerical comparison of approximation techniques. *Journal of Derivatives* **9**, 1–15.

Lewis, A. L. (2000). Option Valuation under Stochastic Volatility. Finance Press,

Newport Beach, CA.

Li, M. (2005). A damped diffusion framework for financial modeling and closed-form maximum likelihood estimation. Tech. rep. Georgia Institute of Technology.

Lo, A. W. (1988). Maximum likelihood estimation of generalized Itô processes with discretely sampled data. *Econometric Theory* **4**, 231–247.

Marsh, T., Rosenfeld, E. (1982). Stochastic processes for interest rates and equilibrium bond prices. *Journal of Finance* **38**, 635–646.

Mosburger, G., Schneider, P., 2005. Modelling international bond markets with affine term structure models. Tech. rep., Vienna University of Economics and Business Administration.

Pedersen, A. R. (1995). A new approach to maximum-likelihood estimation for stochastic differential equations based on discrete observations. *Scandinavian journal of Statistics* **22**, 55–71.

Philips, P. C. B. (1973). The problem of identification in finite parameter continuous time models. *Journal of Econometrics* **1**, 351–362.

Romano, M., Touzi, N. (1997). Contingent claims and market completeness in a stochastic volatility model. *Mathematical Finance* **7**, 399–412.

Schaumburg, E. (2001). Maximum likelihood estimation of jump processes with applications to finance. Ph.D. thesis, Princeton University.

Schneider, P. (2006). Approximations of transition densities for nonlinear multivariate diffusions with an application to dynamic term structure models. Tech. rep. Vienna University of Economics and Business Administration.

Stramer, O., Yan, J. (2005). On simulated likelihood of discretely observed diffusion processes and comparison to closed-from approximation. Tech. rep., University of Iowa.

Takamizawa, H. (2005). Is nonlinear drift implied by the short-end of the term structure? Tech. rep., Hitotsubashi Uninversity.

Thompson, S. (2004). Identifying term structure volatility from the LIBOR-swap curve. Tech. rep., Harvard University.

Vasicek, O. (1977). An equilibrium characterization of the term structure. *Journal of Financial Economics* **5**, 177–188.

White, J. S. (1958). The limiting distribution of the serial correlation coefficient in the explosive case. *Annals of Mathematical Statistics* **29**, 1188–1197.

Wong, E. (1964). The construction of a class of stationary Markoff processes. In: Bellman, R. (Ed.), Sixteenth Symposium in Applied Mathematics - Stochastic Processes in Mathematical Physics and Engineering. American Mathematical Society, Providence, RI, 264 – 276.

Yu, J. (2003). Closed-form likelihood estimation of jump-diffusions with an application to the realignment risk premium of the Chinese yuan. Ph.D. thesis, Princeton University.

CHAPTER 18

Nonparametric Estimation of Production Efficiency

Byeong U. Park, Seok-Oh Jeong and Young Kyung Lee

Department of Statistics
Seoul National University
Seoul 151-747, Korea
bupark@stats.snu.ac.kr

Department of Statistics
Hankuk University of Foreign Studies
Gyeong-Gi 449-791, Korea
seokohj@hufs.ac.kr

Department of Statistics
Seoul National University
Seoul 151-747, Korea
itsgirl@hanmail.net

This paper gives an overview of the recently developed theory for several promising nonparametric estimators of frontiers or boundaries in productivity analysis. The estimators considered here include two popular envelop estimators that are based on data envelopment analysis and free disposability. Also discussed are the recently proposed nonparametric estimation of order-m frontiers, and an extension of the methods and theory to the case where there exist exogenous variables that affect the production process.

18.1. The Frontier Model

Efficiency analysis stems from the works of Koopmans (1951) and Debreu (1951) on activity analysis. Farrell (1957) is considered to be the first empirical work for this, and Shephard (1970) provides a modern economic formulation of this problem.

Consider a production technology where the producer uses inputs $x \in \mathbb{R}^p_+$ to produce outputs $y \in \mathbb{R}^q_+$. The production set Ψ of all technically

407

feasible input-output pairs $(\boldsymbol{x}, \boldsymbol{y})$ is defined by

$$\Psi = \{(\boldsymbol{x}, \boldsymbol{y}) \in \mathbb{R}_+^p \times \mathbb{R}_+^q \,|\, \boldsymbol{x} \text{ can produce } \boldsymbol{y}\}.$$

When $q = 1$, the set Ψ can also be characterized by a function g, called a frontier function, i.e., $\Psi = \{(\boldsymbol{x}, \boldsymbol{y}) \in \mathbb{R}_+^{p+1} \,|\, y \le g(\boldsymbol{x})\}$. The set Ψ is usually assumed to be free disposable, *i.e.*, if $(\boldsymbol{x}, \boldsymbol{y}) \in \Psi$, then $(\boldsymbol{x}', \boldsymbol{y}') \in \Psi$ as long as $\boldsymbol{x}' \ge \boldsymbol{x}$ and $\boldsymbol{y}' \le \boldsymbol{y}$. This means that the production scenario with more input and less output is always technically feasible.

For $\boldsymbol{y} \in \mathbb{R}_+^q$, the input set $X(\boldsymbol{y})$ is the set of all input vectors which are used to yield at least $\boldsymbol{y}$, i.e.,

$$X(\boldsymbol{y}) = \{\boldsymbol{x} \in \mathbb{R}_+^p \,|\, (\boldsymbol{x}, \boldsymbol{y}) \in \Psi\}.$$

Similarly, for $\boldsymbol{x} \in \mathbb{R}_+^p$, the output set $Y(\boldsymbol{x})$ is defined by the set of all output vectors obtainable from the input $\boldsymbol{x}$:

$$Y(\boldsymbol{x}) = \{\boldsymbol{y} \in \mathbb{R}_+^q \,|\, (\boldsymbol{x}, \boldsymbol{y}) \in \Psi\}.$$

As far as efficiency is concerned, the boundaries of $X(\boldsymbol{y})$ and $Y(\boldsymbol{x})$ are of our interest. The efficient boundaries(or frontiers) of these sets are defined in a radial way as follows:

$$\partial X(\boldsymbol{y}) = \{\boldsymbol{x} \in \mathbb{R}_+^p \,|\, \boldsymbol{x} \in X(\boldsymbol{y}), \theta\boldsymbol{x} \notin X(\boldsymbol{y}) \text{ for all } \theta \in (0,1)\},$$
$$\partial Y(\boldsymbol{x}) = \{\boldsymbol{y} \in \mathbb{R}_+^q \,|\, \boldsymbol{y} \in Y(\boldsymbol{x}), \lambda\boldsymbol{y} \notin Y(\boldsymbol{x}) \text{ for all } \lambda \in (1,\infty)\}.$$

For a given level of input and output $(\boldsymbol{x}, \boldsymbol{y}) \in \Psi$, the corresponding (radial) measures of efficiency are defined by

$$\theta(\boldsymbol{x}, \boldsymbol{y}) = \inf\{\theta \in (0,1) \,|\, (\theta\boldsymbol{x}, \boldsymbol{y}) \in X(\boldsymbol{y})\};$$
$$\lambda(\boldsymbol{x}, \boldsymbol{y}) = \sup\{\lambda \in (1,\infty) \,|\, (\boldsymbol{x}, \lambda\boldsymbol{y}) \in Y(\boldsymbol{x})\}.$$

Thus, $\theta(\boldsymbol{x}, \boldsymbol{y}) = 1$ means that the producer performs efficiently in terms of input efficiency, and $\theta(\boldsymbol{x}, \boldsymbol{y}) < 1$ suggests that the producer may reduce $\boldsymbol{x}$ to produce $\boldsymbol{y}$. Similarly, $\lambda(\boldsymbol{x}, \boldsymbol{y}) = 1$ means that the producer performs efficiently in terms of output efficiency, and $\lambda(\boldsymbol{x}, \boldsymbol{y}) > 1$ suggests that the producer may increase $\boldsymbol{y}$ using $\boldsymbol{x}$. Once these efficiency measures are given, one may define the most efficient levels of inputs and outputs by

$$x^\partial(\boldsymbol{x}, \boldsymbol{y}) = \theta(\boldsymbol{x}, \boldsymbol{y})\boldsymbol{x}, \quad y^\partial(\boldsymbol{x}, \boldsymbol{y}) = \lambda(\boldsymbol{x}, \boldsymbol{y})\boldsymbol{y}.$$

Note that, given $(\boldsymbol{x}, \boldsymbol{y}) \in \Psi$, $x^\partial(\boldsymbol{x}, \boldsymbol{y})$ is the point where the ray $\{\theta\boldsymbol{x} \,|\, \theta \ge 0\}$ penetrates the efficient boundary $\partial X(\boldsymbol{y})$, and likewise $y^\partial(\boldsymbol{x}, \boldsymbol{y})$ is the point where the ray $\{\lambda\boldsymbol{y} \,|\, \lambda \ge 0\}$ intersects with the efficient boundary $\partial Y(\boldsymbol{x})$.

In practice the production set is not known. Neither are the input, output sets and the efficiency measures. The next two sections are devoted to discussion on the problem of estimating frontiers and efficiency measures in nonparametric ways.

18.2. Envelope Estimators

Let $\mathcal{X}_n = \{(\mathbf{X}_i, \mathbf{Y}_i) \mid i = 1, 2, \ldots, n\}$ be a random sample from a production set. Below we limit our discussion to the case of input orientation. The case of output orientation can be treated in a similar way.

Efficiency of a production $(\boldsymbol{x}, \boldsymbol{y})$ is measured by its distance to the frontier. It is estimated by the distance to the estimated frontier. There are two nonparametric methods to do this, among others: the data envelopment analysis(DEA) and the free disposal hull(FDH). The FDH estimator is based only on the free disposability assumption on the production set Ψ, and DEA estimator relies upon the additional assumption of convexity of Ψ. We assume a deterministic frontier model for the data generating process, i.e. no noise in each sample point $(\mathbf{X}_i, \mathbf{Y}_i)$, $i = 1, 2, \ldots, n$ is assumed so that $P(\mathcal{X}_n \subset \Psi) = 1$.

When one allows errors in sample points, there may be some points $(\mathbf{X}_i, \mathbf{Y}_i) \notin \Psi$. This is the stochastic frontier model studied by Aigner, Lovell and Schmidt (1977) and Meeusen and van den Broeck (1977). We do not consider this case in the present paper. For a detailed account of the stochastic frontier model, see Kumbhakar and Lovell (2000). For semi-parametric treatments of the stochastic frontier model, see Park and Simar (1994), Park, Sickles and Simar (1998), Park, Sickles and Simar (2003), Park, Sickles and Simar (2006), and Kumbhakar, Park, Simar and Tsionas (2006), among others.

18.2.1. *DEA Estimator*

The data envelopment analysis approach was proposed by Farrell (1957), and popularized in terms of linear programming by Charnes, Cooper and Rhodes (1978). When the free disposability and convexity are assumed on the production set Ψ, one would naturally think of the smallest free disposable and convex set enveloping the whole sample points for the estimation of Ψ. The DEA estimator of Ψ is defined as

$$\widehat{\Psi}_{\text{DEA}} = \{(\boldsymbol{x}, \boldsymbol{y}) \in \mathbb{R}_+^{p+q} \mid \boldsymbol{x} \geq \mathbb{X}\boldsymbol{\gamma}, \ \boldsymbol{y} \leq \mathbb{Y}\boldsymbol{\gamma},$$
$$\text{for some } \boldsymbol{\gamma} \in \mathbb{R}_+^n \text{ such that } \mathbf{1}_n'\boldsymbol{\gamma} = 1\}$$

where $\mathbb{X} = \left[\mathbf{X}_1, \mathbf{X}_2, \ldots, \mathbf{X}_n \right]$, $\mathbb{Y} = \left[\mathbf{Y}_1, \mathbf{Y}_2, \ldots, \mathbf{Y}_n \right]$ and $\mathbf{1}_n$ is the n-vector with all elements equal 1. The DEA efficiency, $\widehat{\theta}_{\mathrm{DEA}}(\boldsymbol{x}, \boldsymbol{y})$, of a given level of input-output $(\boldsymbol{x}, \boldsymbol{y})$ is then defined as

$$\widehat{\theta}_{\mathrm{DEA}}(\boldsymbol{x}, \boldsymbol{y}) = \inf\{\theta \in (0,1) \,|\, (\theta \boldsymbol{x}, \boldsymbol{y}) \in \widehat{\Psi}_{\mathrm{DEA}}\}$$

It can be computed by solving the following linear programming problem:

$$\widehat{\theta}_{\mathrm{DEA}}(\boldsymbol{x}, \boldsymbol{y}) = \min\{\theta \geq 0 \,|\, \theta \boldsymbol{x} \geq \mathbb{X}\boldsymbol{\gamma}, \ \boldsymbol{y} \leq \mathbb{Y}\boldsymbol{\gamma},$$
$$\text{for some } \boldsymbol{\gamma} \in \mathbb{R}^n_+ \text{ such that } \mathbf{1}'_n \boldsymbol{\gamma} = 1\}.$$

18.2.2. FDH estimator

As is often the case, convexity of the production set Ψ is denied. Then, one cannot adhere to the DEA approach since it does not give a consistent estimator in that case. By dropping the convexity assumption on the production set, Deprins, Simar, and Tulkens (1984) proposed a more flexible estimator, the free disposable hull estimator. It is defined by the smallest free disposable set containing $\mathcal{X}_n$:

$$\widehat{\Psi}_{\mathrm{FDH}} = \{(\boldsymbol{x}, \boldsymbol{y}) \in \mathbb{R}^{p+q}_+ \,|\, \boldsymbol{x} \geq \mathbf{X}_i, \ \boldsymbol{y} \leq \mathbf{Y}_i, \ i = 1, 2 \ldots, n\}.$$

As in the case of the DEA estimator, the FDH estimator of the efficiency at an input-output level $(\boldsymbol{x}, \boldsymbol{y})$ is defined by

$$\widehat{\theta}_{\mathrm{FDH}}(\boldsymbol{x}, \boldsymbol{y}) = \inf\{\theta \in (0,1) \,|\, (\theta \boldsymbol{x}, \boldsymbol{y}) \in \widehat{\Psi}_{\mathrm{FDH}}\},$$

and, by a straight forward calculation, we can see that

$$\widehat{\theta}_{\mathrm{FDH}}(\boldsymbol{x}, \boldsymbol{y}) = \min_{i: \boldsymbol{Y}_i \geq \boldsymbol{y}} \max_{1 \leq k \leq p} \left(\frac{\mathbf{e}^T_{p,k} \mathbf{X}_i}{\mathbf{e}^T_{p,k} \boldsymbol{x}} \right)$$

where $\mathbf{e}_{p,k}$ is the unit vector in $\mathbb{R}^p$ with its k-th element equal 1.

By construction we have $\widehat{\Psi}_{\mathrm{FDH}} \subset \widehat{\Psi}_{\mathrm{DEA}} \subset \Psi$, and so, for any $(\boldsymbol{x}, \boldsymbol{y}) \in \widehat{\Psi}_{\mathrm{FDH}}$, $0 \leq \theta(\boldsymbol{x}, \boldsymbol{y}) \leq \widehat{\theta}_{\mathrm{DEA}}(\boldsymbol{x}, \boldsymbol{y}) \leq \widehat{\theta}_{\mathrm{FDH}}(\boldsymbol{x}, \boldsymbol{y}) \leq 1$.

18.2.3. Asymptotic properties of DEA and FDH estimators

Below we collect the assumptions that are needed for the asymptotic properties of the DEA and FDH estimators. For the DEA estimator, one needs

(A1) $\mathcal{X}_n$ is a set of n independent and identically distributed copies of a random variable $(\mathbf{X}, \mathbf{Y})$ on Ψ;

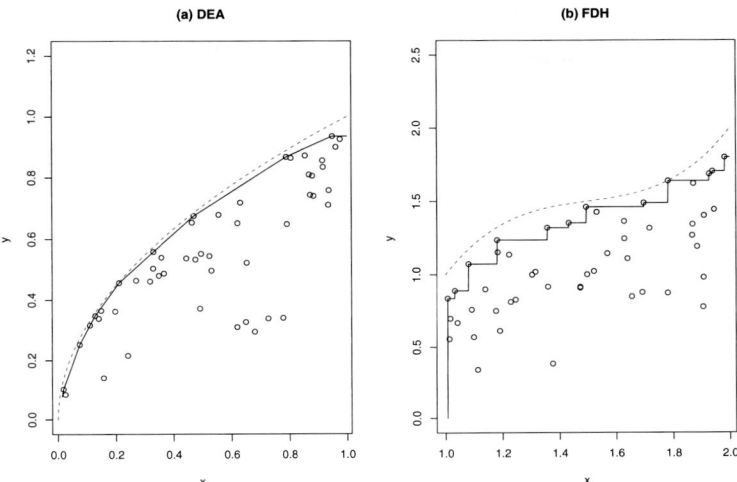

Figure 18.1.   Illustrations of (a) the DEA and (b) the FDH estimates in the case where $p = q = 1$. The circles are the simulated data points, the solid curves depict the estimates of the frontier of Ψ, and the dotted represent the true frontiers.

(A2) $(\mathbf{X}, \mathbf{Y})$ possesses a density f with support $\mathcal{D} \subset \Psi$ such that f is continuous on $\mathcal{D}$ and that $f(\theta(\mathbf{x}, \mathbf{y})\mathbf{x}, \mathbf{y}) > 0$ for all $(\mathbf{x}, \mathbf{y}) \in \mathcal{D}$;
(A3) Ψ is free disposable, closed and convex;
(A4) For $(\mathbf{x}, \mathbf{y})$ in the interior of $\mathcal{D}$, $\theta(\cdot, \cdot)$ is twice continuously differentiable in a neighborhood of $(\mathbf{x}, \mathbf{y})$, and $\theta''(\mathbf{x}, \mathbf{y})$, the Hessian matrix of θ at $(\mathbf{x}, \mathbf{y})$, is positive definite.

For the FDH estimator the assumptions (A3) and (A4) are modified as

(A3$'$) Ψ is free disposable and closed;
(A4$'$) For $(\mathbf{x}, \mathbf{y})$ in the interior of $\mathcal{D}$, $\theta(\cdot, \cdot)$ is continuously differentiable in a neighborhood of $(\mathbf{x}, \mathbf{y})$, and every element of $\theta'(\mathbf{x}, \mathbf{y})$, the vector of the partial derivatives of θ at $(\mathbf{x}, \mathbf{y})$, is nonzero.

Assumption (A2) means that the density f is bounded away from zero along the frontier. The case where the values of f decrease to zero or increase to infinity at a speed of a power of the distance from the frontier has been treated by Härdle, Park and Tsybakov (1995), and Hwang, Park and Ryu (2002).

DEA estimator

The asymptotic analysis of the DEA estimator is now fully available. Kneip, Park and Simar (1998) showed that, under Assumptions (A1)–(A4), the DEA efficiency estimator is a consistent estimator with the convergence rate of $n^{-2/(p+q+1)}$. Gijbels, Mammen, Park and Simar (1999) derived an explicit formula for the asymptotic distribution of the DEA estimator $\hat{g}_{\mathrm{DEA}}$ for the boundary function g when $p = q = 1$. As an extension of this result, Jeong and Park (2006) derived a large sample approximation of the asymptotic distribution when $p \geq 1$ and $q = 1$. For analyzing the statistical properties of the DEA efficiency estimator in the fully multivariate setup where $p \geq 1$ and $q \geq 1$, one can translate the original problem to that in the setup of $p \geq 1$ but $q = 1$. Once this translation is made, one can directly apply the results of Jeong and Park (2006) to obtain the distributional properties of the DEA estimator in the fully general setup. The idea is suggested and used by Jeong (2004) and Jeong and Simar (2005), which we introduce below first.

Fix $(\boldsymbol{x}_0, \boldsymbol{y}_0) \in \Psi$. Let $P \equiv P(\boldsymbol{x}_0)$ be a $p \times (p-1)$ matrix whose columns constitute an orthonormal basis for $\boldsymbol{x}_0^{\perp}$. Consider the transformation which maps $\boldsymbol{x} \in \mathbb{R}^p$ to $(\mathbf{z}^T, v)^T \in \mathbb{R}^{p-1} \times \mathbb{R}$, where $\mathbf{z} = P^T \boldsymbol{x}$ and $v = \boldsymbol{x}_0^T \boldsymbol{x}/|\boldsymbol{x}_0|$, where $|\cdot|$ denotes the Euclidean norm of vectors. Note that v is the distance between $\boldsymbol{x}$ and $\boldsymbol{x}_0^{\perp}$, and $(\mathbf{z}^T, v)^T$ is simply the vector, which corresponds to $\boldsymbol{x}$, in the new coordinate system with axes being $\boldsymbol{x}_0$ and the columns of P. The transformation is one-to-one and its inverse is given by $= P\mathbf{z} + v(\boldsymbol{x}_0/|\boldsymbol{x}_0|)$. Define a function $g(\cdot) \equiv g(\cdot; \boldsymbol{x}_0)$ by

$$g(\mathbf{z}, \boldsymbol{w}) = \inf \left\{ v > 0 \,\bigg|\, \left(P\mathbf{z} + v \frac{\boldsymbol{x}_0}{|\boldsymbol{x}_0|}, \boldsymbol{w} + \boldsymbol{y}_0 \right) \in \Psi \right\}. \qquad (18.1)$$

It can be seen from the definitions of θ and g that

$$\theta(\boldsymbol{x}_0, \boldsymbol{y}_0) = g(\mathbf{0}_{p-1+q})/|\boldsymbol{x}_0|. \qquad (18.2)$$

Thus, estimation of the function g at $(\mathbf{0}_{p-1+q})$ directly leads to an estimator of $\theta(\boldsymbol{x}_0, \boldsymbol{y}_0)$.

The function g is convex. To see this, consider two fixed points $(\mathbf{z}_1, \boldsymbol{w}_1)$ and $(\mathbf{z}_2, \boldsymbol{w}_2)$ and let v_1 and v_2 be arbitrary positive numbers such that

$$v_1 \in \left\{ v > 0 \,\bigg|\, \left(P\mathbf{z}_1 + v \frac{\boldsymbol{x}_0}{|\boldsymbol{x}_0|}, \boldsymbol{w}_1 + \boldsymbol{y}_0 \right) \in \Psi \right\},$$

$$v_2 \in \left\{ v > 0 \,\bigg|\, \left(P\mathbf{z}_2 + v \frac{\boldsymbol{x}_0}{|\boldsymbol{x}_0|}, \boldsymbol{w}_2 + \boldsymbol{y}_0 \right) \in \Psi \right\}.$$

Then, by the convexity of Ψ

$$\lambda v_1 + (1 - \lambda) v_2 \in \left\{ v > 0 \,\middle|\, \left(P\left[\lambda \mathbf{z}_1 + (1 - \lambda) \mathbf{z}_2\right] + v \frac{\boldsymbol{x}_0}{|\boldsymbol{x}_0|}, \right.\right.$$
$$\left.\left. \lambda \boldsymbol{w}_1 + (1 - \lambda) \boldsymbol{w}_2 + \boldsymbol{y}_0 \right) \in \Psi \right\}$$

for any $0 \leq \lambda \leq 1$. This implies

$$\lambda v_1 + (1 - \lambda) v_2 \geq g\left(\lambda \mathbf{z}_1 + (1 - \lambda) \mathbf{z}_2, \lambda \boldsymbol{w}_1 + (1 - \lambda) \boldsymbol{w}_2\right)$$

by the definition of the function g. By taking infimums of v_1 and v_2 in the corresponding set of positive numbers, we get

$$\lambda g(\mathbf{z}_1, \boldsymbol{w}_1) + (1 - \lambda) g(\mathbf{z}_2, \boldsymbol{w}_2) \geq g\left(\lambda \mathbf{z}_1 + (1 - \lambda) \mathbf{z}_2, \lambda \boldsymbol{w}_1 + (1 - \lambda) \boldsymbol{w}_2\right).$$

Now consider the collection of all transformed sample points:

$$\tilde{\mathcal{X}}_n = \left\{ \begin{pmatrix} \mathbf{U}_i \\ V_i \end{pmatrix} \,\middle|\, \mathbf{U}_i = \begin{pmatrix} \mathbf{Z}_i \\ \boldsymbol{w}_i \end{pmatrix}, \ \mathbf{Z}_i = P^T \mathbf{X}_i, \ \mathbf{W}_i = \boldsymbol{Y}_i - \boldsymbol{y}_0, \right.$$
$$\left. V_i = \frac{\boldsymbol{x}_0^T \mathbf{X}_i}{|\boldsymbol{x}_0|}, \ (\mathbf{X}_i, \boldsymbol{Y}_i) \in \mathcal{X}_n, \ i = 1, 2, \ldots, n \right\}.$$

From the definition of the function g, all the points in $\tilde{\mathcal{X}}_n$ lie in the region

$$\{(\mathbf{z}^T, \boldsymbol{w}^T, v)^T \mid v \geq g(\mathbf{z}, \boldsymbol{w})\}.$$

Hence, the original problem of estimating $\theta(\boldsymbol{x}_0, \boldsymbol{y}_0)$ from $\mathcal{X}_n$ is reduced to that of estimating the convex frontier function g from the transformed data $\tilde{\mathcal{X}}_n$ where $\mathbf{U}_i$ are treated as inputs and V_i as outputs. Let $\hat{g}(\mathbf{z}, \boldsymbol{y})$ be the version of $g(\mathbf{z}, \boldsymbol{y})$ obtained by replacing Ψ with $\widehat{\Psi}_{\mathrm{DEA}}$ in (18.1). Then it can be shown that

$$\widehat{\theta}_{\mathrm{DEA}}(\boldsymbol{x}_0, \boldsymbol{y}_0) = \hat{g}(\mathbf{0}_{p-1+q}) / |\boldsymbol{x}_0| \tag{18.3}$$

with probability tending to one. The relations (18.2) and (18.3) imply that we can directly apply the approach of Jeong and Park (2006) to derive a large sample approximation of the distribution of $\widehat{\theta}_{\mathrm{DEA}}(\boldsymbol{x}_0, \boldsymbol{y}_0)$ as is described below.

Note that Assumptions (A1) and (A4) jointly imply that $g(\cdot, \cdot)$ is convex and twice continuously differentiable in a neighborhood of $(\mathbf{z}, \boldsymbol{w}) = \mathbf{0}_{p-1+q}$. Consider the following transformation on $\tilde{\mathcal{X}}_n$:

$$\mathbf{U}_i^* = n^{1/(p+q+1)} \left(G_2/2\right)^{1/2} \mathbf{U}_i, \quad V_i^* = n^{2/(p+q+1)} \left(V_i - g_0 - \mathbf{g}_1^T \mathbf{U}_i\right),$$

where $g_0 = g(\mathbf{0}_{p-1+q})$, $\mathbf{g}_1 = \nabla g(\mathbf{0}_{p-1+q})$ and $G_2 = \nabla^2 g(\mathbf{0}_{p-1+q})$ with ∇ denoting the partial differential operator in $(\mathbf{z}, \boldsymbol{w})$. Denote the transformed sample by $\mathcal{X}_n^*$.

Let $Z_{\text{conv}}^*(\cdot)$ be the lower boundary to the convex hull of $\{(\mathbf{U}_i^*, V_i^*) \mid i = 1, 2, \ldots, n\}$, *i.e.*, for $\mathbf{u}^* \in \mathbb{R}^{p-1+q}$

$$Z_{\text{conv}}^*(\mathbf{u}^*) = \min \left\{ \sum_{i=1}^n \gamma_i V_i^* \,\middle|\, \mathbf{u}^* = \sum_{i=1}^n \gamma_i \mathbf{U}_i^* \right.$$
$$\left. \text{for some } (\gamma_1, \gamma_2, \ldots, \gamma_n) \in \mathbb{R}_+^n \text{ such that } \sum_{i=1}^n \gamma_i = 1 \right\}.$$

Then it can be shown that with probability tending to one

$$Z_{\text{conv}}^*(\mathbf{0}_{p-1+q}) = n^{2/(p+q+1)} \left[\widehat{g}(\mathbf{0}_{p-1+q}) - g_0 \right],$$

see Jeong and Park (2006). The transformed data $\mathcal{X}_n^*$ has the new boundary $v^* = g_n(\mathbf{u}^*)$ in the new coordinate system $(\mathbf{u}^*, v^*)$ such that

$$g_n(\mathbf{u}^*) = \mathbf{u}^{*T}\mathbf{u}^* + o(1) \tag{18.4}$$

as $n \to \infty$ uniformly for $\mathbf{u}^*$ in any compact set. Also, writing $f_n(\cdot, \cdot)$ for the density of the transformed data in the new coordinate system we have by (A3)

$$\sup{}' \left| n \|G_2/2\|^{1/2} f_n(\mathbf{u}^*, v^*) - f_0 \right| \to 0 \tag{18.5}$$

for any sequence of positive real numbers ε_n that goes down to 0, where $\sup'$ means the supremum over $(\mathbf{u}^*, v^*)$ such that $|\mathbf{u}^*| \le \varepsilon_n n^{1/(p+q+1)}$ and $\mathbf{u}^{*T}\mathbf{u}^* \le v^* \le \varepsilon_n n^{2/(p+q+1)}$, f_0 is the density of $\widetilde{\mathcal{X}}_n$ at the boundary point $(\mathbf{0}_{p-1+q}, g_0)$, and $\| \cdot \|$ denotes the determinant of a matrix.

Define $\kappa = (\|G_2/2\|/f_0^2)^{1/(p+q+1)}$ and

$$\mathcal{B}_\kappa = \{(\mathbf{u}^*, v^*) \mid \mathbf{u}^* \in [-(\sqrt{\kappa}/2)n^{1/(p+q+1)}, (\sqrt{\kappa}/2)n^{1/(p+q+1)}]^{p-1+q}$$
$$\text{and } \mathbf{u}^{*T}\mathbf{u}^* \le y^* \le \mathbf{u}^{*T}\mathbf{u}^* + \kappa n^{2/(p+q+1)}\}.$$

Consider a new sample $\{(\mathbf{U}_i^{\text{u}}, V_i^{\text{u}}), i = 1, 2, \ldots, n\}$ which is generated from the uniform distribution on $\mathcal{B}_\kappa$. Note that the uniform density on $\mathcal{B}_\kappa$ is given by $n^{-1}\kappa^{-(p+q+1)/2}$ which equals $n^{-1}\|G_2/2\|^{-1/2}f_0$, the approximation of f_n near the boundary point $(\mathbf{0}_{p-1+q}, g_0)$ as is given at (18.5).

Let $Z_{\text{conv}}^{\text{u}}(\cdot)$ be the version of $Z_{\text{conv}}^*(\cdot)$ constructed by the new sample $\{(\mathbf{U}_i^{\text{u}}, V_i^{\text{u}}), i = 1, 2, \ldots, n\}$. By (18.4) and the fact that $Z_{\text{conv}}^*(\mathbf{0}_{p-1+q})$ and $Z_{\text{conv}}^{\text{u}}(\mathbf{0}_{p-1+q})$ are completely determined by those sample points near $(\mathbf{0}_{p-1+q}, 0)$, it can be shown that $Z_{\text{conv}}^*(\mathbf{0}_{p-1+q})$ and $Z_{\text{conv}}^{\text{u}}(\mathbf{0}_{p-1+q})$ have the same limit distribution, see Jeong and Park (2006). Therefore, when n

is large enough, the limit distribution of the DEA estimator can be well approximated by the distribution of $Z_{\text{conv}}^{\text{u}}(\mathbf{0}_{p-1+q})$.

Theorem 18.1: Suppose Assumptions (A1)–(A4) hold. For $z > 0$,

$$P\left[n^{2/(p+q+1)}\left\{\widehat{\theta}_{\text{DEA}}(\boldsymbol{x}_0, \boldsymbol{y}_0) - \theta(\boldsymbol{x}_0, \boldsymbol{y}_0)\right\} \leq z\right] \longrightarrow F(z)$$

as $n \to \infty$, where

$$F(z) = \lim_{n\to\infty} P\left[|\boldsymbol{x}_0|^{-1} Z_{\text{conv}}^{\text{u}}(\mathbf{0}_{p-1+q}) \leq z\right].$$

We note that, once κ is determined(or estimated), one may simulate the limit distribution F by a Monte Carlo method. We close this subsection by mentioning that Kneip, Simar and Wilson (2003) also derived a large sample approximation of the distribution of DEA estimator in a fully multidimensional setup. Their results are not directly applicable in practice, however. They proposed to use a subsampling bootstrap scheme to approximate the sampling distribution of the DEA estimator.

FDH estimator

Contrary to the case of the DEA estimator, the asymptotic distribution of the FDH estimator can be obtained explicitly. The work has been done by Park, Simar and Weiner (2000) for the output efficiency measure. Below, we give a result for the input oriented case. There are no essential differences between the treatments for the input and the output oriented cases.

Write $\theta_0 = \theta(\boldsymbol{x}_0, \boldsymbol{y}_0)$ and $\widehat{\theta}_0 = \widehat{\theta}_{\text{FDH}}(\boldsymbol{x}_0, \boldsymbol{y}_0)$ for brevity. For $z > 0$,

$$P\left[n^{1/(p+q)}(\widehat{\theta}_0 - \theta_0) > z\right]$$
$$= P\left[(\mathbf{X}_i, \mathbf{Y}_i) \notin \text{NW}((\theta_0 + z')\boldsymbol{x}_0, \boldsymbol{y}_0) \text{ for any } i = 1, 2, \cdots, n\right]$$
$$= \left[1 - P\left((\mathbf{X}, \mathbf{Y}) \in \text{NW}((\theta_0 + z')\boldsymbol{x}_0, \boldsymbol{y}_0)\right)\right]^n,$$

where $z' = n^{-1/(p+q)}z$ and $\text{NW}(\boldsymbol{x}, \boldsymbol{y}) = \{(\boldsymbol{x}', \boldsymbol{y}') \,|\, \boldsymbol{x}' \leq \boldsymbol{x}, \ \boldsymbol{y}' \geq \boldsymbol{y}\} \cap \Psi$. By Assumption (A2),

$$P\left[(\mathbf{X}, \mathbf{Y}) \in \text{NW}((\theta_0 + z')\boldsymbol{x}_0, \boldsymbol{y}_0)\right]$$
$$= \int\int f(\boldsymbol{x}, \boldsymbol{y}) I\left[(\boldsymbol{x}, \boldsymbol{y}) \in \text{NW}((\theta_0 + z')\boldsymbol{x}_0, \boldsymbol{y}_0)\right] d\boldsymbol{x} d\boldsymbol{y}$$
$$= f(\theta_0\boldsymbol{x}_0, \boldsymbol{y}_0) \cdot \mu_{\text{nw}}(z') \cdot \left[1 + o(1)\right]$$

as $n \to \infty$, where $\mu_{\text{nw}}(z') = \text{mes}\left[\text{NW}((\theta_0 + z')\boldsymbol{x}_0, \boldsymbol{y}_0)\right]$ and mes(A) denotes the Lebesgue measure of a set A.

Now we compute $\mu_{\text{nw}}(z')$. For a p-vector $\boldsymbol{x}$, let $\boldsymbol{x}_{-p}$ be the $(p-1)$-vector with elements being the first $p-1$ entries of $\boldsymbol{x}$ and let $\boldsymbol{x}_{-p}(\eta) = (\boldsymbol{x}_{-p}^T, \eta)^T$, the p-vector obtained by adding η to $\boldsymbol{x}_{-p}$ at the pth position. Consider the function g on $\mathbb{R}^{p-1+q}$ defined by

$$g(\boldsymbol{x}_{-p}, \boldsymbol{y})) = \min\left\{\eta \mid \left(\boldsymbol{x}_{-p}(\eta), \boldsymbol{y}\right) \in \Psi\right\}. \tag{18.6}$$

By the free disposability of Ψ, g is decreasing in $\boldsymbol{x}_{-p}$ and increasing in $\boldsymbol{y}$. It follows that

$$\left\{\left(\boldsymbol{x}_{-p}(\eta), \boldsymbol{y}\right) \mid \eta \geq g(\boldsymbol{x}_{-p}, \boldsymbol{y})\right\} \equiv \Psi, \tag{18.7}$$

see Lemma A1 in Park, Simar and Weienr (2000). From (18.7) one can write

$$\mu_{\text{nw}}(z') = \int I\left(x_p \geq g(\boldsymbol{x}_{-p}, \boldsymbol{y}), \ \boldsymbol{x} \leq (\theta_0 + z')\boldsymbol{x}_0, \ \boldsymbol{y} \geq \boldsymbol{y}_0\right) d\boldsymbol{x} d\boldsymbol{y}.$$

By changing the variables $(\boldsymbol{x}, \boldsymbol{y})$ in the integral to $(\mathbf{u}, v)$ where $\mathbf{u} = \left((\boldsymbol{x}_{-p} - \theta_0 \boldsymbol{x}_{0,-p})^T, (\boldsymbol{y} - \boldsymbol{y}_0)^T\right)^T$ and $v = x_p - \theta_0 x_{0,p}$, we obtain

$$\mu_{\text{nw}}(z') = \int I\Big[z'x_{0,p} \geq v \geq \left\{g\left((\theta_0 \boldsymbol{x}_{0,-p}, \boldsymbol{y}_0) + \mathbf{u}\right) - \theta_0 x_{0,p}\right\},$$
$$\mathbf{u} \in (\mathbf{0}_{p-1}, z'\boldsymbol{x}_{0,-p}] \times [\mathbf{0}_q, \infty)\Big] d\mathbf{u} dv.$$

Note that by the definition of g at (18.6), $g\left(\theta_0 \boldsymbol{x}_{0,-p}, \boldsymbol{y}_0\right) = \theta_0 x_{0,p}$, and that the effective integration region of $\mathbf{u}$ is contained in a set of $\mathbf{u}$ such that $|\mathbf{u}| \leq Cn^{-1/(p+q)}$ for some constant $C > 0$. The latter enable us to approximate $g\left((\theta_0 \boldsymbol{x}_{0,-p}, \boldsymbol{y}_0) + \mathbf{u}\right)$ as n tends to infinity. Let G_1 be the $(p-1+q)$-diagonal matrix whose diagonal elements are the partial derivatives of g at $(\theta_0 \boldsymbol{x}_{0,-p}, \boldsymbol{y}_0)$. Then, uniformly for $\mathbf{u}$ in the integration region

$$g\left((\theta_0 \boldsymbol{x}_{0,-p}, \boldsymbol{y}_0) + \mathbf{u}\right) = \theta_0 x_{0,p} + \mathbf{1}_{p-1+q}^T G_1 \mathbf{u} + o(n^{-1/(p+q)}),$$

so that $\mu_{\text{nw}}(z')$ can be expressed as

$$\mu_{\text{nw}}(z') = \int I\Big[\mathbf{1}_{p-1+q}^T G_1 \mathbf{u} \leq v \leq z'x_{0,p},$$
$$\mathbf{u} \in (\mathbf{0}_{p-1}, z'\boldsymbol{x}_{0,-p}] \times [\mathbf{0}_q, \infty)\Big] d\mathbf{u} \, dv \, (1 + o(1)). \tag{18.8}$$

Let G_{11} and G_{12} denote, respectively, the top-left $(p-1) \times (p-1)$, and the bottom-right $q \times q$ diagonal matrices in G_1. Write $\mathbf{u} = (\mathbf{u}_1^T, \mathbf{u}_2^T)^T$ where $\mathbf{u}_1$ is a $(p-1)$-vector and $\mathbf{u}_2$ is a q-vector. Note that all diagonal elements of G_{11} is negative while those of G_{12} is positive. Make the following transformation in the integral (18.8): $\mathbf{t}_1 = n^{1/(p+q)} G_{11}\left(\mathbf{u}_1 - z'\boldsymbol{x}_{0,-p}\right)$, $\mathbf{t}_2 = n^{1/(p+q)} G_{12}\mathbf{u}_2$, $w = n^{1/(p+q)}\left(z'\boldsymbol{x}_{0,p} - v\right)$. Then, we obtain

$$\mu_{\text{nw}}(z') = n^{-1} \int_{[0,\infty)^{p+q}} I\left[w + \mathbf{1}_{p-1+q}^T \mathbf{t} \le z\zeta(\boldsymbol{x}_0)\right] dt \, dw \tag{18.9}$$
$$\times (-1)^{p-1}\|G_1\|^{-1}\left(1 + o(1)\right),$$

where $\zeta(\boldsymbol{x}_0) = x_{0,p} + \mathbf{1}_{p-1}^T(-G_{11})\boldsymbol{x}_{0,-p}$. Note that $(-1)^{p-1}\|G_1\| > 0$ since all the $(p-1)$ entries of the diagonal matrix G_{11} are negative. The integral at (18.9) is the volume of a $(p+q)$-dimensional simplex with edges of length $\zeta(\boldsymbol{x}_0)z$, thus it equals $[\zeta(\boldsymbol{x}_0)z]^{p+q}/(p+q)!$. Writing $f_0 = f(\theta_0\boldsymbol{x}_0, \boldsymbol{y}_0)$, we have the following theorem.

Theorem 18.2: Suppose Assumptions (A1), (A2), (A3') and (A4') hold. For $z > 0$, as $n \to \infty$

$$P\left[n^{1/(p+q)}\left\{\widehat{\theta}_{\text{FDH}}(\boldsymbol{x}_0, \boldsymbol{y}_0) - \theta(\boldsymbol{x}_0, \boldsymbol{y}_0)\right\} \le z\right]$$
$$\longrightarrow 1 - \exp\left(-\frac{f_0\zeta(\boldsymbol{x}_0)^{p+q}z^{p+q}}{(-1)^{p-1}\|G_1\|(p+q)!}\right).$$

18.3. Order-*m* Estimators

Nonparametric envelope estimators are appealing because they rely on very few assumptions on the frontier. However, by construction, they are very sensitive to outliers. In this section, we discuss robust estimation of the frontier, proposed by Cazals, Florens and Simar (2002) and generalized by Daraio and Simar (2005a), which provides another benchmarking frontier for production units to be compared.

18.3.1. *Alternative formulation of efficiencies*

Let $(\mathbf{X}, \mathbf{Y})$ be a pair of random vectors in $\mathbb{R}_+^p \times \mathbb{R}_+^q$, and define $H(\cdot, \cdot)$ by

$$H(\boldsymbol{x}, \boldsymbol{y}) = P(\mathbf{X} \le \boldsymbol{x}, \mathbf{Y} \ge \boldsymbol{y}).$$

Under Assumptions (A2) and (A3') in Section 2.3, the support of H is the same as that of the joint density f of $(\mathbf{X}, \mathbf{Y})$. Note that this joint probability can be decomposed as follows:

$$H(\boldsymbol{x}, \boldsymbol{y}) = P(\mathbf{X} \le \boldsymbol{x} \mid \mathbf{Y} \ge \boldsymbol{y})P(\mathbf{Y} \ge \boldsymbol{y}),$$

provided that the conditional probability $P(\mathbf{X} \leq \boldsymbol{x} \,|\, \mathbf{Y} \geq \boldsymbol{y})$, denoted by $F(\boldsymbol{x} \,|\, \boldsymbol{y})$, exists. One can define equivalently the efficiency measure $\theta(\boldsymbol{x}, \boldsymbol{y})$ for $(\boldsymbol{x}, \boldsymbol{y}) \in \Psi$ by

$$\theta(\boldsymbol{x}, \boldsymbol{y}) = \inf\{\theta > 0 \,|\, F(\theta\boldsymbol{x} \,|\, \boldsymbol{y}) > 0\}. \tag{18.10}$$

A nonparametric estimator of $\theta(\boldsymbol{x}, \boldsymbol{y})$ is then obtained by simply replacing $F(\cdot \,|\, \cdot)$ by its empirical version $\widehat{F}_n(\cdot \,|\, \cdot)$, where

$$\widehat{F}_n(\boldsymbol{x} \,|\, \boldsymbol{y}) = \frac{\sum_{i=1}^{n} I(\mathbf{X}_i \leq \boldsymbol{x}, \mathbf{Y}_i \geq \boldsymbol{y})}{\sum_{i=1}^{n} I(\mathbf{Y}_i \geq \boldsymbol{y})} \tag{18.11}$$

and $I(\cdot)$ denotes the indicator function. Interestingly, as shown in Cazals, Florens and Simar (2002), the resulting estimator of $\theta(\boldsymbol{x}, \boldsymbol{y})$ coincides with the FDH estimator $\widehat{\theta}_{\mathrm{FDH}}(\boldsymbol{x}, \boldsymbol{y})$, i.e.,

$$\widehat{\theta}_{\mathrm{FDH}}(\boldsymbol{x}, \boldsymbol{y}) = \inf\{\theta > 0 \,|\, \widehat{F}_n(\theta\boldsymbol{x} \,|\, \boldsymbol{y}) > 0\}. \tag{18.12}$$

18.3.2. *Order-m frontiers*

For $\boldsymbol{y}_0$ in the interior of the support of the marginal distribution of $\mathbf{Y}$, consider m independent and identically distributed random vectors $\mathbf{X}_i^c$, $i = 1, 2, \ldots, m$, which are generated by the conditional distribution function $F(\cdot \,|\, \boldsymbol{y}_0)$. Let $\Psi_m(\boldsymbol{y}_0)$ be the set defined by

$$\Psi_m(\boldsymbol{y}_0) = \bigcup_{i=1}^{m}\{(\boldsymbol{x}, \boldsymbol{y}) \in \mathbb{R}_+^{p+q} \,|\, \boldsymbol{x} \geq \mathbf{X}_i^c, \ \boldsymbol{y} \geq \boldsymbol{y}_0\}.$$

For any $\boldsymbol{x}_0$, define

$$\tilde{\theta}_m(\boldsymbol{x}_0, \boldsymbol{y}_0) = \inf\{\theta > 0 \,|\, (\theta\boldsymbol{x}_0, \boldsymbol{y}_0) \in \Psi_m(\boldsymbol{y}_0)\}.$$

It can be seen that

$$\tilde{\theta}_m(\boldsymbol{x}_0, \boldsymbol{y}_0) = \min_{1 \leq i \leq m} \max_{1 \leq k \leq p} \frac{\mathbf{e}_{p,k}^T \mathbf{X}_i^c}{\mathbf{e}_{p,k}^T \boldsymbol{x}_0}. \tag{18.13}$$

Define $\theta_m(\boldsymbol{x}_0, \boldsymbol{y}_0)$, called the (expected) order-m efficiency measure at $(\boldsymbol{x}_0, \boldsymbol{y}_0)$, by

$$\theta_m(\boldsymbol{x}_0, \boldsymbol{y}_0) = E\big[\tilde{\theta}_m(\boldsymbol{x}_0, \boldsymbol{y}_0) \,|\, \mathbf{Y} \geq \boldsymbol{y}_0\big],$$

provided that the conditional expectation exists. In the case where $p = 1$, it equals $E\big(\min\{X_1^c, \ldots, X_m^c\} \,|\, \mathbf{Y} \geq \boldsymbol{y}_0\big)/x_0$. Roughly speaking, $\boldsymbol{x}_0\theta_m(\boldsymbol{x}_0, \boldsymbol{y}_0)$ is the expected minimal input level among a set of m production units that produce an output higher than the level $\boldsymbol{y}_0$. By definition, it follows that $\theta_m(\boldsymbol{x}_0, \boldsymbol{y}_0) \geq \theta(\boldsymbol{x}_0, \boldsymbol{y}_0)$ for all $m \geq 1$. Thus, it is a more conservative measure of efficiency than $\theta(\boldsymbol{x}_0, \boldsymbol{y}_0)$.

From the representation (18.13), it can be shown that $P\big[\tilde{\theta}_m(\boldsymbol{x}_0, \boldsymbol{y}_0) \leq \theta \,\big|\, \boldsymbol{Y} \geq \boldsymbol{y}_0\big] = 1 - \big[1 - F(\theta \boldsymbol{x}_0 \,|\, \boldsymbol{y}_0)\big]^m$. This yields

$$\theta_m(\boldsymbol{x}_0, \boldsymbol{y}_0) = \int_0^\infty \big[1 - F(\theta \boldsymbol{x}_0 | \boldsymbol{y}_0)\big]^m \, d\theta. \tag{18.14}$$

By (18.10) one has for any $\epsilon > 0$

$$\big|\theta_m(\boldsymbol{x}_0, \boldsymbol{y}_0) - \theta(\boldsymbol{x}_0, \boldsymbol{y}_0)\big| \leq 3\epsilon + \int_{\theta(\boldsymbol{x}_0, \boldsymbol{y}_0)+\epsilon}^\infty \big[1 - F(\theta \boldsymbol{x}_0 | \boldsymbol{y}_0)\big]^m \, d\theta. \tag{18.15}$$

The integral on the right hand side of (18.15) tends to zero as m goes to infinity, again by the representation (18.10). Thus, $\theta_m(\boldsymbol{x}_0, \boldsymbol{y}_0)$ converges to $\theta(\boldsymbol{x}_0, \boldsymbol{y}_0)$ as m goes to infinity.

A nonparametric estimator of $\theta_m(\boldsymbol{x}_0, \boldsymbol{y}_0)$ may be constructed by replacing $F(\cdot \,|\, \cdot)$ in (18.14) by its empirical version $\widehat{F}_n(\cdot \,|\, \cdot)$ at (18.11):

$$\widehat{\theta}_{m,n}(\boldsymbol{x}_0, \boldsymbol{y}_0) = \int_0^\infty \big[1 - \widehat{F}_n(\theta \boldsymbol{x}_0 \,|\, \boldsymbol{y}_0)\big]^m \, d\theta.$$

If one applies the same arguments leading to (18.15), one can see from (18.12) that for each fixed n, $\widehat{\theta}_{m,n}(\boldsymbol{x}_0, \boldsymbol{y}_0)$ converges to $\widehat{\theta}_{\mathrm{FDH}}(\boldsymbol{x}_0, \boldsymbol{y}_0)$ as m goes to infinity.

Theorem 18.3: Suppose that $\boldsymbol{x}_0 \in \mathbb{R}_+^p$ and $\boldsymbol{y}_0$ is an interior point in the support of the marginal distribution of $\boldsymbol{Y}$. Assume that $\theta_m(\boldsymbol{x}_0, \boldsymbol{y}_0)$ exists. Then, as $m \to \infty$, $\theta_m(\boldsymbol{x}_0, \boldsymbol{y}_0) \to \theta(\boldsymbol{x}_0, \boldsymbol{y}_0)$, and for each fixed n, $\widehat{\theta}_{m,n}(\boldsymbol{x}_0, \boldsymbol{y}_0) \to \widehat{\theta}_{\mathrm{FDH}}(\boldsymbol{x}_0, \boldsymbol{y}_0)$ almost surely.

The asymptotic properties of $\widehat{\theta}_{m,n}(\boldsymbol{x}_0, \boldsymbol{y}_0)$ as n tends to infinity for fixed m has been derived by Cazals, Florens and Simar (2002) for the case where $p = 1$. A direct extension of their results to general p and q is contained in the following theorem. To state the theorem, define $F_{\boldsymbol{Y}}(\boldsymbol{y})$ be the marginal distribution function of $\boldsymbol{Y}$, and for $m \geq 1$

$$\sigma_m^2(\boldsymbol{x}_0, \boldsymbol{y}_0) = E\bigg[\frac{m}{(1 - F_{\boldsymbol{Y}}(\boldsymbol{y}_0))^m} \int_0^\infty P\Big(\boldsymbol{X} > \theta \boldsymbol{x}_0, \boldsymbol{Y} > \boldsymbol{y}_0\Big)^{m-1}$$
$$\times I(\boldsymbol{X} > \theta \boldsymbol{x}_0, \boldsymbol{Y} > \boldsymbol{y}_0) \, d\theta - \frac{m \theta_m(\boldsymbol{x}_0, \boldsymbol{y}_0)}{1 - F_{\boldsymbol{Y}}(\boldsymbol{y}_0)} I(\boldsymbol{Y} > \boldsymbol{y}_0)\bigg]^2.$$

Theorem 18.4: Assume that the support of the distribution of $(\boldsymbol{X}, \boldsymbol{Y})$ is compact. Let $\boldsymbol{y}_0$ be an interior point in the support of the marginal distribution of $\mathbf{Y}$. Then for all fixed $m \geq 1$,

$$\sqrt{n}\Big(\widehat{\theta}_{m,n}(\boldsymbol{x}_0, \boldsymbol{y}_0) - \theta_m(\boldsymbol{x}_0, \boldsymbol{y}_0)\Big) \xrightarrow[d]{} N\Big(0, \sigma_m^2(\boldsymbol{x}_0, \boldsymbol{y}_0)\Big).$$

From Theorem 3.2, we see that $\widehat{\theta}_{m,n}(\boldsymbol{x}_0, \boldsymbol{y}_0)$ is a $\sqrt{n}$-consistent estimator of $\theta_m(\boldsymbol{x}_0, \boldsymbol{y}_0)$, regardless of the dimensions of the input and output. A final remark for the estimator of the order-m efficiency is that it is less affected by outlying observations than envelop estimators such as the DEA and FDH. It is robust to extreme values, noise and outliers, as indicated by Figure 18.2.

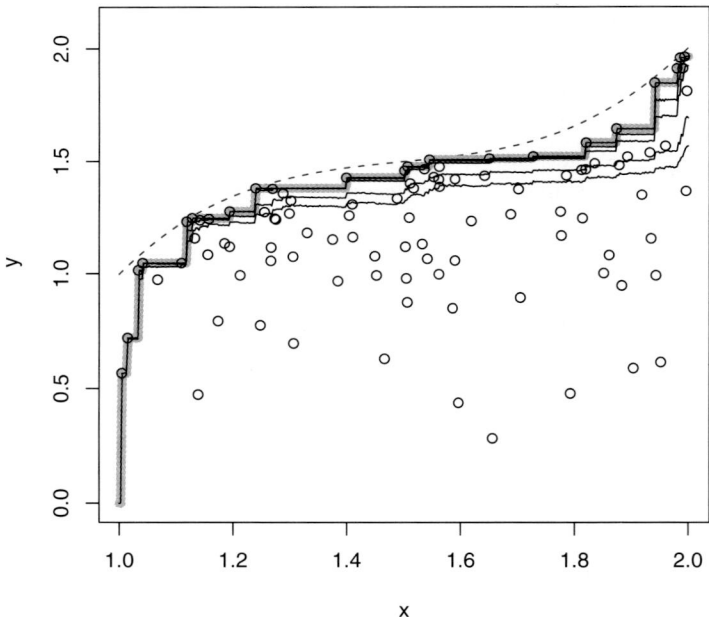

Figure 18.2. Illustration of the order-m estimates in the case where $p = q = 1$. The circles are the simulated data points, the thick gray curve depicts the FDH estimate, the solid represent the order-m frontier estimates ($m = 5, 10, 50, 100, 400$ in the order of bottom to top), and the dotted is the true frontier.

18.4. Conditional Frontier Models

The probabilistic formulation in Section 3.1 can be extended to a more general model involving additional variable $\mathbf{Z} \in \mathbb{R}^r$ which is exogenous to the production process in the sense that one cannot technically control its value in the production process. Although this exogenous variable $\mathbf{Z}$ is neither an input nor an output in the production process, it may affect the efficiency of the production process. Hence it is important to analyze the effect of this environmental variable on the performance of production units to identify economic conditions that cause inefficiency. See Daraio and Simar (2005a) and Daraio and Simar (2005b) for detailed discussion and some examples.

The basic idea is conditioning the production process on a given value of $\mathbf{Z} = \mathbf{z}$, which was proposed by Cazals, Florens and Simar (2002) and Daraio and Simar (2005a). The conditional distribution of $(\mathbf{X}, \mathbf{Y})$ given $\mathbf{Z} = \mathbf{z}$ defines the production process when $\mathbf{Z} = \mathbf{z}$. Conditioning on $\mathbf{Z} = \mathbf{z}$, the efficiency measure for a production unit $(\boldsymbol{x}, \boldsymbol{y})$ is now defined as

$$\theta(\boldsymbol{x}, \boldsymbol{y} \,|\, \mathbf{z}) = \inf\{\theta \,|\, F(\theta \boldsymbol{x} \,|\, \boldsymbol{y}, \mathbf{z}) > 0\}, \qquad (18.16)$$

where $F(\boldsymbol{x} \,|\, \boldsymbol{y}, \mathbf{z}) = P(\mathbf{X} \leq \boldsymbol{x} \,|\, \mathbf{Y} \geq \boldsymbol{y}, \mathbf{Z} = \mathbf{z})$.

18.4.1. *Conditional FDH estimator*

Suppose we have a set of independent and identically distributed copies, $\{(\mathbf{X}_i, \mathbf{Y}_i, \mathbf{Z}_i) \,|\, i = 1, 2, \ldots, n\}$, of $(\mathbf{X}, \mathbf{Y}, \mathbf{Z})$. An FDH analogue for the conditional efficiency measure $\theta(\boldsymbol{x}, \boldsymbol{y} \,|\, \mathbf{z})$ can be obtained by plugging the following empirical version of $F(\cdot \,|\, \cdot, \cdot)$ into the definition (18.16):

$$\widehat{F}_{n,h}(\boldsymbol{x} \,|\, \boldsymbol{y}, \mathbf{z}) = \frac{\sum_{i=1}^{n} I(\mathbf{X}_i \leq \boldsymbol{x}, \mathbf{Y}_i \geq \boldsymbol{y}, |\mathbf{Z}_i - \mathbf{z}| \leq h)}{\sum_{i=1}^{n} I(\mathbf{Y}_i \geq \boldsymbol{y}, |\mathbf{Z}_i - \mathbf{z}| \leq h)}, \qquad (18.17)$$

where $h > 0$ is the bandwidth of appropriate size. Thus,

$$\widehat{\theta}_{\mathrm{FDH}}(\boldsymbol{x}, \boldsymbol{y} \,|\, \mathbf{z}) = \inf\{\theta \,|\, \widehat{F}_{n,h}(\theta \boldsymbol{x} \,|\, \boldsymbol{y}, \mathbf{z}) > 0\}.$$

It can be checked that

$$\widehat{\theta}_{\mathrm{FDH}}(\boldsymbol{x}, \boldsymbol{y} \,|\, \mathbf{z}) = \min\left\{ \max_{1 \leq k \leq p} \frac{\mathbf{e}_{p,k}^T \mathbf{X}_i}{\mathbf{e}_{p,k}^T \boldsymbol{x}} \,\middle|\, \mathbf{Y}_i \geq \boldsymbol{y}, \, |\mathbf{Z}_i - \mathbf{z}| \leq h \right\}.$$

See Figure 18.3 for a graphical illustration of a conditional FDH frontier built by $\widehat{\theta}_{\mathrm{FDH}}(\boldsymbol{x}, \boldsymbol{y} \,|\, \mathbf{z})$. Jeong, Park and Simar (2005) showed that $\widehat{\theta}_{\mathrm{FDH}}(\boldsymbol{x}, \boldsymbol{y} \,|\, \mathbf{z})$ is a consistent estimator and that asymptotically it has a Weibull-distribution. Daraio and Simar (2005b) gave a unified approach to cover the conditional DEA approach as well.

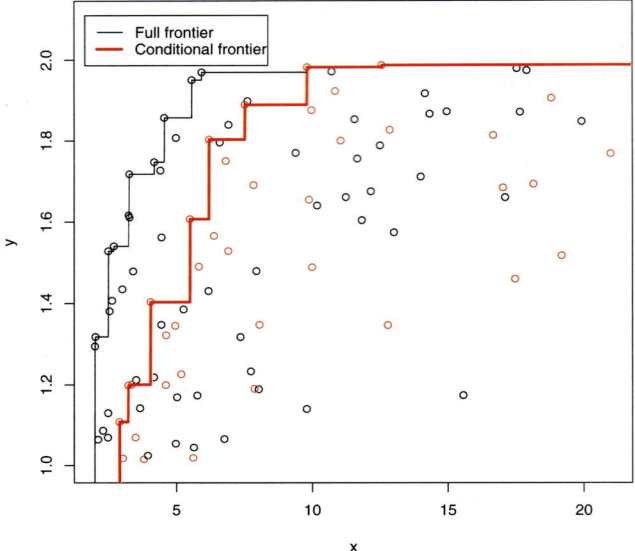

Figure 18.3.   Illustration of the conditional FDH estimate in the case where $p = q = 1$ for a simulated data set. The model is: $X_i = Y_i^2 Z_i^2 / U_i$ where $Y_i \sim \text{Unif}[1,2]$, $Z_i \sim \text{Unif}[1,2]$ and $U_i^3 \sim \text{Unif}[0,1]$, $i = 1, 2, \cdots, 100$. The circles are the simulated data points, the red circles are the points such that $|Z_i - z| \le h$ with $z = 1.5$ and $h = 0.2$, the thin solid line is the FDH estimate, and the thick red line is the conditional FDH frontier estimate.

18.4.2. *Conditional order-m estimator*

One may extend the idea of the conditional efficiency to the conditional order-m frontiers. Here, we fix $(\boldsymbol{x}_0, \boldsymbol{y}_0, \mathbf{z}_0)$. Consider a set of m independent and identically distributed random variables $\mathbf{X}_i^c$, $i = 1, 2, \ldots, m$, which are generated by the conditional distribution function $F(\cdot \,|\, \boldsymbol{y}_0, \mathbf{z}_0)$. Define the set

$$\Psi_m(\boldsymbol{y}_0 \,|\, \mathbf{z}_0) = \bigcup_{i=1}^{m} \Big\{ (\boldsymbol{x}, \boldsymbol{y}) \in \mathbb{R}_+^{p+q} \,\Big|\, x \ge \mathbf{X}_i^c, \boldsymbol{y} \ge \boldsymbol{y}_0 \Big\}.$$

Also define

$$\tilde{\theta}_m(\boldsymbol{x}_0, \boldsymbol{y}_0 \,|\, \mathbf{z}_0) = \inf \Big\{ \theta > 0 \,\Big|\, (\theta \boldsymbol{x}_0, \boldsymbol{y}_0) \in \Psi_m(\boldsymbol{y}_0 \,|\, \mathbf{z}_0) \Big\},$$

$$\theta_m(\boldsymbol{x}_0, \boldsymbol{y}_0 \,|\, \mathbf{z}_0) = E \Big(\tilde{\theta}_m(\boldsymbol{x}_0, \boldsymbol{y}_0 \,|\, \mathbf{z}_0) \,\Big|\, \boldsymbol{Y} \ge \boldsymbol{y}_0, \boldsymbol{Z} = \mathbf{z}_0 \Big),$$

whenever the latter exists. An analogue of Theorem 3.1 is given by the following theorem.

Theorem 18.5: Suppose that $x_0 \in \mathbb{R}_+^p$ and (y_0, z_0) is an interior point in the support of the joint distribution of (Y, Z). Assume that $\theta_m(x_0, y_0 \mid z_0)$ exists. Then, as $m \to \infty$, $\theta_m(x_0, y_0 \mid z_0) \to \theta(x_0, y_0 \mid z_0)$, and for each fixed n, $\widehat{\theta}_{m,n}(x_0, y_0 \mid z_0) \to \widehat{\theta}_{\text{FDH}}(x_0, y_0 \mid z_0)$ almost surely.

Furthermore, an analogue of $\widehat{\theta}_{m,n}(x_0, y_0)$ in the unconditional case can be obtained by using $\widehat{F}_{n,h}(\cdot \mid \cdot, \cdot)$ at (18.17):

$$\widehat{\theta}_{m,n}(x_0, y_0 \mid z_0) = \int_0^\infty \left[1 - \widehat{F}_{n,h}(\theta x_0 \mid y_0, z_0)\right]^m d\theta.$$

The asymptotic properties of $\widehat{\theta}_{m,n}(x_0, y_0 \mid z_0)$ as n tends to infinity for fixed m has also been derived by Cazals, Florens and Simar (2002) for the case where $p = 1$. A straightforward extension to general p and q is demonstrated in the following theorem. For this let r be the dimension of the exogenous variable Z. Define $F_{Y,Z}(y, z)$ be the marginal distribution function of (Y, Z). Also, let $f_Z(z)$ be the marginal density of Z, and $\nabla_{z_0} g = \left[\partial^r g(z)/(\partial z_1 \cdots \partial z_r)\right]_{z=z_0}$. Finally, for $m \geq 1$ define

$$\sigma_m^2(x_0, y_0 \mid z_0) = 2 f_Z(z_0) \, Var\Bigg[\frac{m}{\left(f_Z(z_0) - \nabla_{z_0} F_{Y,Z}(y_0, \cdot)\right)^m} $$
$$\times \int_0^\infty \left\{\nabla_{z_0} P\Big(X > \theta x_0, Y > y_0, Z \leq \cdot\Big)\right\}^{m-1} $$
$$\times I(X > \theta x_0, Y > y_0)\, d\theta - \frac{m\theta_m(x_0, y_0 \mid z_0)}{f_Z(z_0) - \nabla_{z_0} F_{Y,Z}(y_0, \cdot)} $$
$$\times I(Y > y_0)\,\Big| Z = z_0\Bigg].$$

Theorem 18.6: Assume the support of the conditional distribution of (X, Y) given $Z = z_0$ is compact. Let y_0 be an interior point in the support of the conditional distribution of Y given $Z = z_0$. Suppose the bandwidth h satisfies $nh^r \to \infty$ and $nh^{r+4} \to 0$ as $n \to \infty$. Then for all fixed $m \geq 1$,

$$\sqrt{nh^r}\left(\widehat{\theta}_{m,n}(x_0, y_0 \mid z_0) - \theta_m(x_0, y_0 \mid z_0)\right) \xrightarrow{d} N\left(0, \sigma_m^2(x_0, y_0 \mid z_0)\right).$$

18.5. Outlook

The asymptotic properties of the DEA and FDH in the unconditional frontier models is now fully available. The conditional frontier models, however, needs further study for the statistical properties of the conditional DEA and FDH estimators. For the order-m frontiers, the choice of m should be considered as an important issue. There have been no knowledge on how to give any additional restriction on the shape of the frontier such as convexity.

References

Aigner, D. C., Lovell, C. A. K., and Schmidt, P. (1977). Formulation and estimation of stochastic frontier production function models. *Journal of Econometrics* **6**, 21-37.

Cazals, C., Florens, J. P. and Simar, L. (2002). Nonparametric frontier estimation: a robust approach. *Journal of Econometrics* **106**, 1-25.

Charnes, A., Cooper, W. W. and Rhodes, E. (1978). *European Journal of Operational Research* **2**, 429.

Daraio, C. and Simar, L. (2005a). *Journal of Productivity Analysis* **24**, 93.

Daraio, C. and Simar, L. (2005b). Discussion Paper no. 0502, Institut de Statistique, UCL, Belgium.

Debreu, G. (1951). The Coefficient of Resource Utilization. *Econometrica* **19**, 273-292.

Deprins, D., Simar, L., and Tulkens, H. (1984). *The Performance of Public Enterprises: Concepts and measurements*, ed. by M. Marchand, P. Pestieau and H. Tulkens. Amsterdam: North-Holland.

Farrell, M. J., (1957). The Measurement of Productive Efficiency. *Journal of the Royal Statistical Society, Series A* **120**, 253-290.

Gijbels, I, Mammen, E., Park, B. U., and Simar, L. (1999). On Estimation of Monotone and Concave Frontier Functions. *Journal of the American Statistical Association* **94**, 220-228.

Härdle, W., Park, B. U. and Tsybakov, A. B. (1995). Estimation of non-sharp support boundaries. *Journal of Multivariate Analysis* **55**, 205-218.

Hwang, J. H.,Park, B. U. and Ryu, W. (2002). Limit Theorems for Boundary Function Estimators. *Statistics and Probability Letters* **59**, 353-360.

Jeong, S.-O. (2004). *Journal of the Korean Statistical Society* **33**, 449.

Jeong, S.-O. and Park, B. U. (2006). *Scandinavian Journal of Statistics*, to appear.

Jeong, S.-O. and Simar, L. (2005). Discussion Paper no. 0501, Institut de Statistique, UCL, Belgium.

Jeong, S.-O., Park, B. U. and Simar, L. (2005). Unpublished manuscript.

Kneip, A., Park, B. U. and Simar, L. (1998). A note on the convergence of nonparametric DEA estimators for production efficiency scores *Econometric Theory* **14**, 783-793.

Kneip, A., Simar, L. and Wilson, P. (2003). Discussion Paper no. 0317, Institut de Statistique, UCL, Belgium.

Koopmans, T. C. (1951). *Activity Analysis of Production and Allocation*, ed. by T.C. Koopmans, Cowles Commission for Research in Economics, Monograph 13. New York: John-Wiley and Sons, Inc.

Kumbhakar, S. K. and Lovell, C. A. K. (2000). *Stochastic Frontier Analysis*.

Cambridge University Press (2000).

Kumbhakar, S. K., Park, B. U., Simar, L. and Tsionas, E. G. (2006). *Journal of Econometrics*, to appear.

Meeusen, W. and van den Broeck, J. (1977). *International Economic Review* **18**, 435.

Park, B. U. and Simar, L. (1994). Efficient semiparametric estimation in stochastic frontier model. *Journal of the American Statistical Association* **89**, 929-936.

Park, B. U., Sickles, R. C. and Simar, L. (1998). Stochastic panel frontiers : a semiparametric approach. *Journal of Econometrics* **84**, 273-301.

Park, B. U., Sickles, R. C. and Simar, L. (2003). Semiparametric efficient estimation of AR(1) panel data models. *Journal of Econometrics* **117**, 279-309.

Park, B. U., Sickles, R. C. and Simar, L. (2006). *Journal of Econometrics*, to appear.

Park, B. U., Simar, L., and Weiner, Ch. (2000). The FDH estimator for productivity efficiency scores: Asymptotic properties. *Econometric Theory* **16**, 855-877.

Shephard, R. W. (1970). *Theory of Cost and Production Function*. Princeton University Press, Princeton.

Part VII

Parametric Technique and Inferences

Chromatic Technique and Influences

CHAPTER 19

Convergence and Consistency of Newton's Algorithm for Estimating Mixing Distribution

Jayanta K. Ghosh[a,b] and Surya T. Tokdar[a]

a. Department of Statistics
Purdue University
150 N University St, W Lafayette, IN 47907
ghosh@stat.purdue.edu

b. Indian Statistical Institute
203 B T Road, Kolkata, India 700108.

We provide a new convergence and consistency proof of Newton's algorithm for estimating a mixing distribution under some rather strong conditions. An auxiliary result used in the proof shows that the Kullback Leibler divergence between the estimate and the true mixing distribution converges as the number of observations tends to infinity. This holds under much weaker conditions. It is pointed out that Newton's proof of convergence, based on a representation of the algorithm as a nonhomogeneous weakly ergodic Markov chain, is incomplete. Our proof is along quite different lines. We also study various other aspects of the estimate, including its claimed superiority to the Bayes estimate based on a Dirichlet mixture.

19.1. Introduction

Newton (2002) introduces an interesting recursive algorithm to estimate an unknown mixing density $f(\theta)$ when observations are available from the mixture density $\int p(x|\theta)f(\theta)d\mu(\theta)$. The sampling density $p(x|\theta)$ is assumed known and would generally belong to a known parametric family dominated by either the counting measure or the Lebesgue measure on the space of observations x_is. This algorithm is simple to implement numerically and its computation time is much smaller compared to any state of the art nonparametric Bayesian or classical estimation procedure. As observed by Newton (2002), the estimate produced by his recursive algorithm is not a Bayes estimate but is based on a predictive updating procedure that is closely related to the first step in the calculation of posterior for a Dirichlet process prior (Newton, Quintana and Zhang (1998), Newton and Zhang (1999)). Newton (2002) observes that his estimate of mixing distribution for the thumbtack tossing data (Beckett and Diaconis (1994)) compares well with the estimate obtained in Liu (1996) who uses a Dirichlet process prior on the mixing distribution and employs a sequential imputation technique to evaluate the posterior. This seems to be supported by our simulations in the continuous case; see Figure 19.6. A connection with Lindsay (1995)'s nonparametric MLE is noted in Newton (2002).

Notwithstanding these attractive properties, very little is known either about the convergence of the algorithm or the consistency of the estimate as the number of observations tends to infinity. Newton (2002) shows that his algorithm can be viewed as a nonhomogeneous Markov chain and presents a proof for convergence, but not necessarily consistency, when the true mixing distribution is known to sit on a finite set Θ. Unfortunately, as indicated in our Section 19.3, it appears that the proof is incomplete and not easily rectified.

In the present study we provide an alternative proof of both convergence and consistency of Newton's estimate when the mixing distribution is identifiable and has a finite support Θ. This result, namely Theorem 19.3, is based on an auxiliary result, Theorem 19.2, which shows that the Kullback-Leibler divergence between the true mixing distribution and its estimate obtained from Newton's algorithm converges to a finite limit almost surely. This result does not require Θ to be finite. The above results are proved in Section 19.4. Other properties of the algorithm are examined in Section 19.5. New simulations for a number of different mixing distributions as well as a comparison with the Bayes estimate is presented in Section 19.6. It is clear that much further work is needed to see if Newton's estimate can compete with existing Bayesian or classical methods.

We end the introduction with a few related comments. Ghosh and Ramamoorthi (2003), p. 161-162) contains a discussion of posterior consistency for the mixing distribution of a normal location mixture. This is essentially a simpler version of the deconvolution problem. A good reference to results on rates of convergence for deconvolution in classical statistics is Fan (1991). For discussion of the nonparametric MLE for normal location mixtures see Lindsay (1995). Posterior consistency for the density of x_is is discussed in Ghosal, Ghosh and Ramamoorthi (1999).

19.2. Newton's Estimate of Mixing Distributions

We start with a formal description of Newton's recursive algorithm and the estimate obtained from it. Suppose a sample of observations $x_1, \cdots, x_n$ are obtained from the mixture density $\int p(x|\theta) f(\theta) d\mu(\theta)$, where the sampling density $p(x|\theta)$ is known but the mixing density $f(\theta)$ is unknown. Newton (2002) suggests the following recursive algorithm to obtain an estimate of $f(\theta)$.

1 Start with an initial estimate $f_0(\theta)$.
2 Fix a set of weights $w_1, \cdots, w_n \in [0, 1]$.
3 For $i = 1, \cdots, n$ compute

$$f_i(\theta) = (1 - w_i) f_{i-1}(\theta) + w_i \frac{p(x_i|\theta) f_{i-1}(\theta)}{c(x_i, f_{i-1})} \tag{19.1}$$

where $c(x, f) = \int p(x|\theta) f(\theta) d\mu(\theta)$.
4 Report $f_n(\theta)$ as the final estimate.

Newton (2002) notes that, if after observing $x_1, \cdots, x_{i-1}$, a Bayesian models the unknown mixing density f as $f \sim Dir(\frac{1-w_i}{w_i} f_{i-1})$ then the posterior expectation of f given a new observation x_i is precisely given by f_i defined in (19.1). This type of predictive updating was first introduced in Newton, Quintana and Zhang (1998) and further studied in Newton and Zhang (1999).

The choice of the weights w_i plays an important role in determining the effectiveness of the updating step given in (19.1). For example, one would typically like the updates to be less drastic towards a later stage of the recursion, i.e., one would prefer $f_i - f_{i-1}$ to tend to zero in some appropriate sense. For this to happen a necessary condition is that $w_i \to 0$. Furthermore, an alternative representation of the recursive estimate f_n can be given as,

$$f_n(\theta) = v_{n,0} f_0(\theta) + \sum_{i=1}^{n} v_{n,i} \frac{p(x_i|\theta) f_{i-1}(\theta)}{c(x_i, f_{i-1})},$$

where

$$v_{n,i} = \begin{cases} \Pi_{j=1}^{n}(1 - w_j) & \text{if } i = 0 \\ w_i \Pi_{j=i+1}^{n}(1 - w_j) & \text{if } 1 \leq i < n \\ w_n & \text{if } i = n. \end{cases}$$

Therefore for the estimate f_n to be asymptotically independent of the initial guess f_0, it is necessary that $v_{n,0} = \Pi_{j=1}^{n}(1 - w_j) \to 0$ as $n \to \infty$. In other words, one would like to choose the weights w_ns such that $\sum_n w_n = \infty$. The sequence $w_i = (1+i)^{-1}$ satisfies these conditions and the corresponding f_n has been studied a lot in the simulations presented in Section 19.6. Also see Proposition 19.1 which shows that if $w_i = 1$ for all i, then f_n converges but not to the true mixing density f^*.

19.3. Review of Newton's Result on Convergence

While Newton's estimate bears some resemblance with Bayes estimate obtained using a Dirichlet process prior, the two quantities are fundamentally different. For example, unlike the Bayes estimate, Newton's estimate f_n depends on the order of $x_1, \cdots, x_n$. Therefore standard results on consistency of non-parametric estimates do not apply to the estimate obtained by Newton's algorithm and different techniques are needed to assess convergence. Lemma 1 of Newton (2002) states an *everywhere* (and not almost everywhere) convergence result.

Theorem 19.1: Assume that Θ is finite and that there exist $0 < \underline{p} < \bar{p} < \infty$ such that $\underline{p} < p(x|\theta) < \bar{p}$,   for all $x \in \mathcal{X}, \theta \in \Theta$. Also assume that $\sum_n w_n = \infty$. Then for every sequence $x_1, x_2, \cdots$ there exists a density f_∞ on Θ such that $f \to f_\infty$.

Remark 19.1: It may be noted that Newton (2002) was unable to identify f_∞ as f^*.

Newton (2002) presents a proof of the above result based on the theory of non-homogeneous Markov chains. However, the arguments seem to be incomplete. Newton (2002) notes that given a sequence $x_1, x_2, \cdots$, one can represent f_n as the n-step marginal distribution of the Markov chain $\{z_n\}$ given by,

$$z_0 \sim f_0$$
$$z_n = \begin{cases} z_{n-1} & \text{with probability } 1 - w_n \\ y_n & \text{with probability } w_n, \end{cases}$$

where $y_n \sim p(x_n|\theta) f_{n-1}(\theta)/c(x_n, f_{n-1})$ and are mutually independent. From the assumption $\sum_n w_n = \infty$, one can argue that the Markov chain $\{z_n\}$ is weakly ergodic (i.e., for all $m \geq 0, \theta_1, \theta_2 \in \Theta, \sum_\theta |\Pr(z_n = \theta|z_m = \theta_1) - \Pr(z_n = \theta|z_m = \theta_2)| \to 0$ as $n \to \infty$). But from this it does not follow that the chain is strongly ergodic (i.e., $\exists p(\theta)$ such that for all $m \geq 0, \theta_1 \in \Theta, P(z_n = \theta|z_m = \theta_1) \to p(\theta)$ as $n \to \infty$) even when Θ is finite. A simple counter-example is the Markov chain $\tilde{z}_n$ on $\Theta = \{-1, 1\}$

given by, $\tilde{z}_n = (-1)^n$. More examples are available in Isaacson and Madsen (1976, Ch. V).

19.4. Convergence Results

In this section we provide an alternative proof of convergence of f_n and in fact identify the limit as the true mixing density f^* under certain conditions. More precisely we show that f_n converges to f^* through the Kullback-Leibler divergence $K(f,g) = -\int f(\theta) \log(g(\theta)/f(\theta))d\theta$. The motivation for using this divergence comes from a simple Taylor expansion given in (19.2) that seems to fit quite well with Newton's algorithm. We first give an intermediate result which proves convergence of $K(f^*, f_n)$ under mild conditions.

Theorem 19.2: Assume that,

I. $\exists \underline{p}, \, \bar{p} \in (0, \infty)$ such that $\underline{p} < p(x|\theta) < \bar{p}$, for all $x \in \mathcal{X}, \theta \in \Theta$.
II: $\sum_n w_n^2 < \infty$.

Then, $\exists K$ with $E(K) < \infty$ such that $K(f^*, f_n) \to K$ a.s.

Proof We introduce the notation

$$R(x) = \frac{\log(1+x) - x}{x^2}, \quad x > -1,$$

the error term in the first order Taylor expansion of $\log(1+x)$. Note that, $|R(x)| \le (1/2) \max\{1, (1+x)^{-2}\}$. Use (19.1) to write,

$$
\begin{aligned}
&K(f^*, f_i) - K(f^*, f_{i-1}) \\
&= -\int f^*(\theta) \log \frac{f_i(\theta)}{f_{i-1}(\theta)} d\mu(\theta) \\
&= -\int f^*(\theta) \log \left\{ 1 - w_i \left(1 - \frac{p(x_i|\theta)}{c(x_i, f_{i-1})} \right) \right\} d\mu(\theta) \\
&= -\int f^*(\theta) \left\{ -w_i \left(1 - \frac{p(x_i|\theta)}{c(x_i, f_{i-1})} \right) + w_i^2 R_i(\theta) \right\} d\mu(\theta) \\
&= w_i \left(1 - \frac{c(x_i, f^*)}{c(x_i, f_{i-1})} \right) + w_i^2 \int f^*(\theta) R_i(\theta) d\mu(\theta), \quad\quad (19.2)
\end{aligned}
$$

where

$$R_i(\theta) = \left(1 - \frac{p(x_i|\theta)}{c(x_i, f_{i-1})} \right)^2 R \left(-w_i \left\{ 1 - \frac{p(x_i|\theta)}{c(x_i, f_{i-1})} \right\} \right).$$

Define,

$$M_i = E\left(1 - \frac{c(x_i, f^*)}{c(x_i, f_{i-1})} \mid x_1, \cdots, x_{i-1}\right) = 1 - \int \frac{c(x, f^*)}{c(x, f_{i-1})} c(x, f^*) dx,$$

$$V_i = \left(1 - \frac{c(x, f^*)}{c(x, f_{i-1})}\right) - M_i, \qquad E_i = \int f^*(\theta) R_i(\theta) d\mu(\theta).$$

Then, (19.2) can be rewritten as,

$$K(f^*, f_i) - K(f^*, f_{i-1}) = w_i V_i + w_i M_i + w_i^2 E_i.$$

Application of a telescoping sum to above gives

$$K(f^*, f_n) = K(f^*, f_0) + \sum_{i=1}^{n} w_i V_i + \sum_{i=1}^{n} w_i M_i + \sum_{i=1}^{n} w_i^2 E_i. \qquad (19.3)$$

We will show that each of the three sums in (19.3) converges surely or almost surely. We start with the third term. Note that, by Assumption I, for all $i \geq 1$ and for all $\theta \in \Theta$,

$$1 - w_i\left\{1 - \frac{p(x|\theta)}{c(x, f)}\right\} \geq 1 - w_i(1 - (\underline{p}/\bar{p})) \geq \underline{p}/\bar{p}$$

and hence, for all $i \geq 1$, $\theta \in \Theta$,

$$\left|R\left(-w_i\left\{1 - \frac{p(x_i|\theta)}{c(x_i, f_{i-1})}\right\}\right)\right| \leq \frac{1}{2}\max\{1, (\underline{p}/\bar{p})^{-2}\} = \frac{1}{2}(\bar{p}/\underline{p})^2,$$

and so, by Assumption I, for all $i \geq 1$, $\theta \in \Theta$,

$$|R_i(\theta)| = \left(1 - \frac{p(x_i|\theta)}{c(x_i, f_{i-1})}\right)^2 \left|R\left(-w_i\left\{1 - \frac{p(x_i|\theta)}{c(x_i, f_{i-1})}\right\}\right)\right|$$

$$\leq (1/2)(1 + \bar{p}/\underline{p})^2 (\bar{p}/\underline{p})^2.$$

Hence, one can conclude that for every sequence $x_1, x_2, \cdots$ and every $i \geq 1$,

$$|E_i| \leq \sup_{\theta \in \Theta} |R_i(\theta)| \leq (1/2)(1 + \bar{p}/\underline{p})^2 (\bar{p}/\underline{p})^2.$$

This fact, together with the Assumption II, ensures the existance of a random variable E_∞ with $|E_\infty| < (1/2)(1 + \bar{p}/\underline{p})^2 (\bar{p}/\underline{p})^2 \sum_n w_n^2 < \infty$ such that for every sequence $x_1, x_2, \cdots$, $\sum_{i=1}^{n} w_i^2 E_i \to E_\infty$.

Next, note that, $S_n = \sum_{i=1}^{n} w_i V_i$ is a Martingale sequence w.r.t. $x_1, x_2, \cdots$. Also, since

$$|V_i| = \left|\frac{c(x_i, f^*)}{c(x_i, f_{i-1})} - \int \frac{c(x, f^*)}{c(x, f_{i-1})} c(x, f^*) dx\right| \leq 2(\bar{p}/\underline{p}),$$

one has $\sup_n E(S_n^2) \le [2(\bar{p}/\underline{p})]^2 \sum_n w_n^2 < \infty$. Therefore using Martingale Convergence Theorem (Durrett (2005), p. 250) we can find a random variable S_∞ with $E|S_\infty| < \infty$ such that $S_n \to S_\infty$ a.s.

Now, by applying Jensen's inequality to the function $\phi(x) = x^{-1}$, $x > 0$, one obtains

$$M_i = 1 - \int \frac{c(x, f^*)}{c(x, f_{i-1})} c(x, f^*) dx \le 1 - \left(\int \frac{c(x, f_{i-1})}{c(x, f^*)} c(x, f^*) dx \right)^{-1} = 0.$$

Therefore, $\sum_{i=1}^n w_i M_i$ is a decreasing sequence in n. Also, rewriting the equality (19.3) and using $K(f^*, f_n) \ge 0$ one obtains

$$0 \ge \liminf_{n \to \infty} \sum_{i=1}^n w_i M_i \ge \liminf_{n \to \infty} \left\{ -K(f^*, f_0) - \sum_{i=1}^n w_i V_i - \sum_{i=1}^n w_i^2 E_i \right\}$$

$$= -K(f^*, f_0) - \lim_{n \to \infty} \sum_{i=1}^n w_i V_i - \lim_{n \to \infty} \sum_{i=1}^n w_i^2 E_i$$

$$> -\infty \text{ a.s.}$$

and hence $\sum_{i=1}^n w_i M_i$ converges to a finite limit $\tilde{M}_\infty$. Also, by above $E|\tilde{M}_\infty| \le K(f^*, f_0) + E(|S_\infty|) + E(|E_\infty|) < \infty$. From this, the statement of Theorem 19.2 follows with K defined as

$$K = K(f^*, f_0) + S_\infty + \tilde{M}_\infty + E_\infty.$$

Remark 19.2: Note that this result guarantees that the (Kullback-Leibler) distance between the true density f^* and its estimate f_n converges to some limit, but does not say anything about whether this limit is 0 almost surely. Also note that this theorem does not need $\sum_n w_n = \infty$ which has been argued before as a valid requirement for the f_n's to behave in a good way.

In light of the above remark it would be natural to ask "what can be gained from the above result if one adds the extra condition $\sum_n w_n = \infty$?" Two immediate consequences are given below.

Corollary 19.1: *If in addition to assumptions I and II one also assumes,*

III. $\qquad \sum_n w_n = \infty,$

then almost surely there is a subsequence n_k (which depends on $x_1, x_2, \cdots$) such that $M_{n_k} \to 0$.

Proof   It follows from the arguments given in the proof of Theorem 19.2, that for almost every sequence $x_1, x_2, \cdots$,

$$0 \geq \sum_n w_n M_n > -\infty. \tag{19.4}$$

Now, suppose that there exists a $\delta > 0$ and an $N \geq 1$ such that $|M_n| > \delta$ for all $n > N$. But since $M_n \leq 0$ for all $n \geq 1$, this is equivalent to saying that for all $n > N$, $M_n < -\delta$ and hence for all $n > N$, $\sum_{i=1}^n w_i M_i \leq \sum_{i=N+1}^n w_i M_i < -\delta \sum_{i=N+1}^n w_i$. But this gives a contradiction to (19.4) since $\sum_n w_n = \infty$ by Assumption III. Therefore for all $\delta > 0$ and for all $N \geq 1$ there exists an $n > N$ such that $|M_n| < \delta$. This proves the result. For the next corollary, let

$$K(c(\cdot, f^*), c(\cdot, f_n)) = \int c(x, f^*) \log \left(\frac{c(x, f^*)}{c(x, f_n)} \right) dx,$$

stand for the Kullback-Leibler divergence between $c(\cdot, f^*)$ and $c(\cdot, f_n)$.

Corollary 19.2: *Under assumptions I, II and III, almost surely there is a subsequence n_k (which depends on $x_1, x_2, \cdots$) such that $K(c(\cdot, f^*), c(\cdot, f_{n_k})) \to 0$.*

Proof   Use the inequality $\log(x) \leq x - 1$ to see that,

$$K(c(\cdot, f^*), c(\cdot, f_n)) = \int c(x, f^*) \log \frac{c(x, f^*)}{c(x, f_n)} dx$$

$$\leq \int c(x, f^*) \left\{ \frac{c(x, f^*)}{c(x, f_n)} - 1 \right\} dx = -M_n$$

Hence the result follows directly from Corollary 19.1.

From this and the above theorem, we could readily conclude that $K(f^*, f_n) \to 0$ if only we could assert, $K(c(\cdot, f^*), c(\cdot, f_n)) \to 0 \Rightarrow K(f^*, f_n) \to 0$. However such a result is not readily available for the Kullback-Leibler divergence. But one can get around this problem by assuming that Θ is finite.

Lemma 19.1: *Suppose Θ is finite and μ is the counting measure on Θ. In addition, assume that the map $f \mapsto c(\cdot, f)$ is identifiable, i.e.,*

IV. $c(\cdot, f_1) = c(\cdot, f_2)$ a.e. $\Rightarrow f_1 = f_2$.

Then,

$$K(c(\cdot, f^*), c(\cdot, f_n)) \to 0 \Rightarrow K(f^*, f_n) \to 0. \tag{19.5}$$

Proof   Since Θ is finite, it suffices to prove that $f_n \to f^*$ pointwise. We would prove the latter assertion by the method of contradiction. Suppose $f_n \not\to f^*$ pointwise. Then, there exists a $\theta^* \in \Theta$ such that $f_n(\theta^*) \not\to f^*(\theta^*)$. Hence we can find an $\epsilon > 0$ and a subsequence $\{n_k\}$ such that for all $k \geq 1$, $|f_{n_k}(\theta^*) - f^*(\theta^*)| > \epsilon$. Now, since Θ is finite, f_{n_k} is a bounded sequence. Hence, there exists a probability density $\tilde{f}$ on Θ and a further subsequence $\{n_{k_l}\}$ of $\{n_k\}$ such that, $f_{n_{k_l}} \to \tilde{f}$ pointwise. So, trivially,

$$|\tilde{f}(\theta^*) - f^*(\theta^*)| \geq \epsilon. \tag{19.6}$$

Now observe that,

$$f_{n_{k_l}} \to \tilde{f} \text{ pointwise} \Rightarrow c(\cdot, f_{n_k}) \to c(\cdot, \tilde{f}) \text{ weakly,}$$

and,

$$K(c(\cdot, f^*), c(\cdot, f_n)) \to 0 \Rightarrow K(c(\cdot, f^*), c(\cdot, f_{n_{k_l}})) \to 0$$
$$\Rightarrow c(\cdot, f_{n_{k_l}}) \to c(\cdot, f^*), \quad \text{weakly,}$$

and hence, $c(\cdot, \tilde{f}) = c(\cdot, f^*)$. a.e. Therefore $\tilde{f} = f^*$ by Assumption IV. But this gives a contradiction to (19.6)! So we must have $f_n \to f^*$ pointwise.

Next, we summarize Theorem 19.2, Corollary 19.2 and Lemma 19.1 into the following result.

Theorem 19.3: If Θ is finite and assumptions I, II, III and IV hold, then $K(f^*, f_n) \to 0$ a.s.

Proof   From Theorem 19.2 and Corollary 19.2, for almost every sequence $x_1, x_2, \cdots$, we have $K(f^*, f_n) \to K$ and $K(c(\cdot, f^*), c(\cdot, f_{n_k})) \to 0$ for some subsequence n_k. Combining with Lemma 19.1 this statement can be rewritten as: for almost every sequence $x_1, x_2, \cdots$, we have $K(f^*, f_n) \to K$ and $K(f^*, f_{n_k}) \to 0$ for some subsequence n_k. But this implies $K = 0$ a.s. and hence the result follows.

Remark 19.3: Theorem 19.3 would have an immediate extension to the case of compact Θ if one could extend Lemma 19.1 to this case. It is true that under identifiability, the weaker assertion $c(\cdot, f_n) \to c(\cdot, f)$ *weakly*, holds in the compact case but unless one can prove equicontinuity of $\{f_n\}$ (or at least f_n are bounded away from zero) the limiting result (19.5) does not follow.

Remark 19.4: Since we compare Newton's estimate with the estimate based on Dirichlet location mixture of normals, it is worth pointing out that the latter estimates the mixing distribution consistently under the weak topology. Consider the set up of p. 161 of Ghosh and Ramamoorthi (2003), i.e. P is compactly supported and $f_P(x) = \int p(x|\theta)dP(\theta)$, where $p(x|\theta)$ is the density of $N(\theta, h_0^2)$ for some fixed $h_0 > 0$. Consider a fixed compact set $K \in \mathbb{R}$ such that support of all P under consideration is contained in K. The class M of all such P is compact in the weak topology. Let $Dir(cG_0)$ denote the Dirichlet process prior with base measure G_0 whose support is K. In this case, all Ps with support contained in K are in the weak support of $Dir(cG_0)$. Then Theorem 1.3.4 of Ghosh and Ramamoorthi (2003) is applicable to $\theta = P$ and $\Theta = M$, and hence we have the posterior consistency at any $P_0 \in M$. This does not contradict the remarks in Ghosh and Ramamoorthi (2003), p. 162, that the Kullback-Leibler support of a Dirichlet process prior is empty. The empty Kullback-Leibler support may be the reason why the Bayes estimate for the mixing density seems to be inferior to Newton's estimate; see Figure 19.6 below.

19.5. Other Results

In this section, we present a few results to illustrate the importance of the weights w_i. These results are presented in the context of specific examples but can be easily generalized.

Proposition 19.1: *Consider the case when $w_i = 1$ for all $i \geq 1$, $p(x|\theta) = N(\theta, 1)$ and $f_0 = N(0, 1)$. Suppose the true mixing density f^* is symmetric density and $c(\cdot, f^*)$ admits a first moment. Then the estimate f_n obtained from Newton's algorithm converges weakly to δ_0 (the point mass at 0).*

Proof   For this case, it can be readily seen that

$$f_n(\theta) = \frac{\prod_j p(x_j|\theta)f_0(\theta)}{\int \prod_j p(x_j|\theta)f_0(\theta)d\theta}.$$

This is the same as the posterior of the parameter θ when the data are modeled as $x_i \overset{iid}{\sim} N(\theta, 1)$ and θ is given the prior $f_0(\theta)$. Therefore, $f_n = N((n/(n+1))\bar{x}, 1/(n+1))$, where $\bar{x} = \sum_j x_j/n$. Now, under the true model, $x_1, \cdots, x_n$ is a random sample from $c(\cdot, f^*)$ which has mean 0 since f^* is symmetric. Therefore by the strong law of large numbers, $\bar{x} \to 0$ a.s. as $n \to \infty$. From this the result follows directly.

More generally if $w_i = 1$ then Newton's estimate would converge weakly to the point mass at θ^* provided θ^* is the unique minimizer of $K(c(\cdot, f^*), p(\cdot, \theta))$, see Bunke and Milhaud (1998). These results point out that it is indeed important to have $w_i \to 0$.

For any element $d = (\delta_1, \cdots, \delta_n) \in \{0,1\}^n$ let $d' = (\delta_1, \cdots, \delta_{n-1})$. For any $n \geq 1$ and $d \in \{0,1\}^n$ define

$$r_{n,d} = \frac{\int \prod_{j=1}^{n} p(x_j|\theta)^{\delta_j} f_0(\theta) d\mu(\theta)}{\int \prod_{j=1}^{n-1} p(x_j|\theta)^{\delta_j} f_0(\theta) d\mu(\theta)}.$$

Consider the triangular system of weights $c_{n,d}$ defined as

$$c_{1,(0)} = 1 - w_1, \; c_{1,(1)} = w_1,$$

$$c_{n,d} = w_n^{\delta_n} (1 - w_n)^{1-\delta_n} \frac{c_{n-1,d'} r_{n,d}}{\sum_{\tilde{d} \in \{0,1\}^n, \, \tilde{\delta}_n = \delta_n} c_{n-1,\tilde{d}'} r_{n,\tilde{d}}}.$$

The following result gives an alternative representation of f_n.

Proposition 19.2: *Newton's estimate can be represented as the mixture*

$$f_n(\theta) = \sum_{d \in \{0,1\}^n} c_{n,d} \frac{\prod_{j=1}^{n} p(x_j|\theta)^{\delta_j} f_0(\theta)}{\int \prod_{j=1}^{n} p(x_j|\theta)^{\delta_j} f_0(\theta) d\mu(\theta)}. \qquad (19.7)$$

Proof   Follows by induction on n using the recursion given in (19.1).

From Proposition 19.2 Newton's estimate can be interpreted as a mixture whose components (indexed by $d \in \{0,1\}^n$) are formed by all possible subsets of $\{x_1, \cdots, x_n\}$. The weights of these components are precisely given by $c_{n,d}$. A little calculation shows that

$$c_{n,d} = \left(\prod_{j=1}^{n} w_j^{\delta_j} (1 - w_j)^{1-\delta_j} \right) \frac{\int \prod_{j=1}^{n} p(x_j|\theta)^{\delta_j} f_0(\theta) d\mu(\theta)}{\prod_{j=1}^{n} \sum_{\tilde{d} \in \{0,1\}^d, \tilde{\delta}_j = \delta_j} c_{j-1,\tilde{d}'} r_{j,\tilde{d}}}.$$

The factor $\prod_j w_j^{\delta_j} (1 - w_j)^{1-\delta_j}$ acts as a prefixed penalizing function on the component sizes. To see this, assume that δ_j are independent $Bernoulli(w_j)$ random variables. Then the weight of the component corresponding to the index $d = (\delta_1, \cdots, \delta_n)$ would be precisely $\prod_j w_j^{\delta_j} (1 - w_j)^{1-\delta_j}$. Also note that the number of x_i's used by such a component is simply $\sum_j \delta_j$. The next proposition gives the distribution of this quantity.

Proposition 19.3: *Suppose δ_j, $1 \leq j \leq n$, are independent $Bernoulli(w_j)$ random variables. If $\sum_n w_n = \infty$ and $\sum_n w_n^2 < \infty$, then $\sum_{j=1}^{n} \delta_j \sim N(\sum_{j=1}^{n} w_j, \sum_{j=1}^{n} w_j)$ asymptotically.*

Proof   Since δ_js are uniformly bounded and $s_n^2 = V(\delta_1) + \cdots + V(\delta_n) = \sum_{j=1}^{n} w_j(1 - w_j) = \sum_{j=1}^{n} w_j - \sum_{j=1}^{n} w_j^2 \to \infty$, it is easy to check that the Lindeberg's condition $\lim_{n\to\infty} s_n^{-2} \sum_{j=1}^n E[\delta_j^2 I(|\delta_j|) > \epsilon s_n] = 0,$ for all $\epsilon > 0$ holds. Therefore,

$$\frac{\sum_{j=1}^n \delta_j - \sum_{j=1}^n w_j}{\sqrt{\sum_{j=1}^n w_j - \sum_{j=1}^n w_j^2}} \xrightarrow{\mathcal{L}} N(0,1).$$

But since, $\sum_{j=1}^n w_j / (\sum_{j=1}^n w_j - \sum_{j=1}^n w_j^2) \to 1$, one can rewrite (19.8) as

$$\frac{\sum_{j=1}^n \delta_j - \sum_{j=1}^n w_j}{\sqrt{\sum_{j=1}^n w_j}} \xrightarrow{\mathcal{L}} N(0,1).$$

The result follows trivially from this.

For $w_i = 1/(i+1)$, Proposition 19.3 would seem to imply that most components in (19.7) use only up to $\log n + O(\sqrt{\log n})$ of the observations. This seems to be wasteful of the later observations in that the components with those observations will have relatively small weights. This suggests a potential improvement of the algorithm. Divide up the n observations into different clusters according to, say, the k-means algorithm. Choose about $\log n$ x_i's which represent the clusters best. Permute x_i's to bring these representative x_i's among the first $\log n$' observations. The idea of re-permuting x_i's to improve f_n has been considered in unpublished work of Newton and Lavine (personal communications).

19.6. Simulation

In this section we present simulation studies of Newton's algorithm. Our main interest is in comparing Newton's estimate with the Bayes estimate obtained by using a Dirichlet process prior on the mixing distribution $f(\theta)$. Specifically, for a sample of observations $x_1, \cdots, x_n$ from $\int p(x|\theta) f^*(\theta) d\mu(\theta)$, we would obtain Newton's estimate with an initial guess f_0 and weights $w_i = 1/(i+1)$ and also compute the Bayes estimate using a $Dir(f_0)$ prior.

Example 19.1: Finite Support

We begin with an example that satisfies assumptions I through IV of Theorem 19.3. Namely, we take:

- $\Theta = \{-6, -5, -4, -3, -2, -1, 0, 1, 2, 3, 4, 5, 6\}$;
- $p(x|\theta) = TN_{[-30,30]}(\theta, 1)$, the normal density with mean θ and variance 1, truncated at $[-30, 30]$;
- $w_i = 1/(i+1)$; and

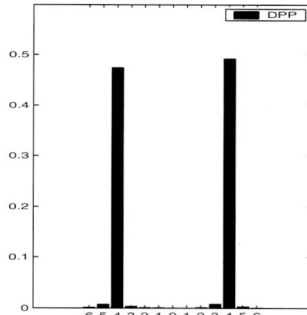 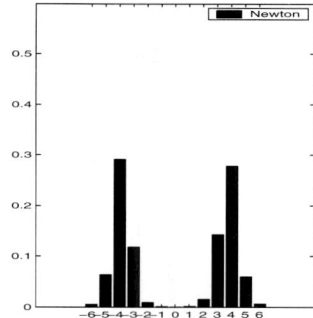

Figure 19.1. Estimates of mixing distribution when Θ is finite. Left: Bayes estimate using Dirichlet Process Prior. Right: Estimate obtained from Newton's method.

- $f_0(\theta) = 1/13$, for all $\theta \in \Theta$, the uniform distribution on the parameter space.

A sample of size 100 is generated from the density $0.5TN_{[-30,30]}(-4,1) + 0.5TN_{[-30,30]}(4,1)$, i.e., the true mixing density (w.r.t counting measure) is $f^* = 0.5\delta_{-4} + 0.5\delta_4$. The plots in Figure 19.6 show the estimated densities. Apparently the Dirichlet mixture estimate outperforms the estimate obtained from Newton's algorithm.

Example 19.2: Continuous Support

This example illustrates the case when Θ is continuous. Specifically we consider:

- $\Theta = (-\infty, \infty)$;
- $p(x|\theta) = N(\theta, 1)$;
- $w_i = 1/(i+1)$; and
- $f_0 = N(0, 4)$.

A sample of size 100 is generated with the true mixing density $f^* = 0.5N(-4,1) + 0.5N(4,1)$, i.e., f^* itself is a mixture. Figure 2 gives the plots of the densities estimated from these two methods. It seems that while Dirichlet does a bad job estimating the mixing density, the estimate obtained from Newton's algorithm works really well. One reason for the poor estimation of the mixing density by Dirichlet process prior could be that it sits only on the discrete distributions. However, the Dirichlet does a good job in estimating the density of x_is, nearly as well as Newton's estimate.

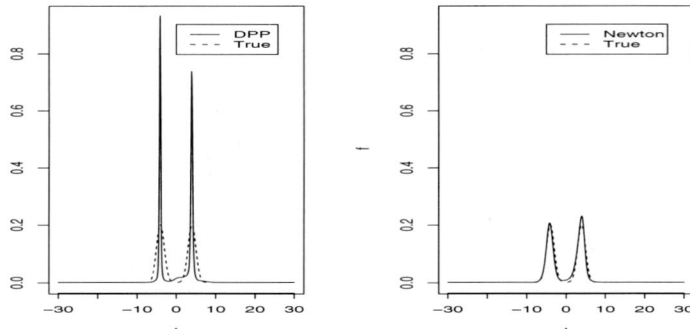

Figure 19.2.   Estimates of mixing distribution when Θ is continuous. Left: Bayes estimate using Dirichlet Process Prior. Right: Estimate obtained from Newton's method.

Remark 19.5: It may be noted that the time required to compute Newton's estimate is much smaller (of the order 10^{-3}) compared to the time needed to compute the Bayes estimate from a Dirichlet process prior. For our Example 19.1, with 100 observations and finite Θ, Newton's estimate is obtained in 0.0160 seconds, whereas, it takes 0.0297 seconds to obtain a single MCMC sample for the Dirichlet process prior. For Example 19.2, with 100 observations and $\Theta = (-\infty, \infty)$, it takes 0.0310 seconds to compute Newton's estimate, whereas computation of each Monte Carlo sample for the Dirichlet process prior takes 0.0360 seconds. For the MCMC to converge in any of these examples, one needs a few thousands of the MCMC samples.

Remark 19.6: While we have not calculated the NPMLE, the algorithm given in Lindsay (1995) seems fairly computation intensive.

References

Beckett, L. and Diaconis, P. (1994). Spectral analysis for discrete longitudinal data. *Adv. Math.*, **103**, 107-128.

Bunke, O. and Milhaud, X. (1998). Asymptotic behavior of Bayes estimates under possibly incorrect models. *Ann. Statist.*, **26(2)**, 617-644.

Durrett, R. (2005). Probability: Theory and examples. Thomson, Brooks/Cole, Belmont, CA.

Fan, J. (1991). On the optimal rates of convergence for nonparametric deconvolution problems. *Ann. Statist.* **19(3)**, 1257-1272.

Ghosal, S., Ghosh, J. K. and Ramamoorthi, R. V. (1999). Posterior consistency of Dirichlet Mixtures in density estimation. *Ann. Statist.* **27**, 143-158.

Ghosh, J. K. and Ramamoorthi, R .V. (2003). Bayesian nonparametrics. Springer-Verlag, New York.

Isaacson, D. L. and Madsen, R. W. (1976). Markov chains: Theory and applications. Wiley, New York.

Lindsay, B. (1995). Mixture models: Theory, geometry and applications. *Institute of Mathematical Statistics*, Hayward, CA.

Liu, J. S. (1996).Nonparametric hierarchical Bayes via sequential imputations. *Ann. Statist.*, **24**, 911-930.

Newton, M. A., Quintana, F. A. and Zhang, Y. (1998). Nonparametric Bayes methods using predictive updating. In *Practical Nonparametric and Semiparametric Bayesian Statistics*, Eds. D. Dey, P. Muller and D. Sinha, 45-61. Springer, New York.

Newton, M. A. and Zhang, Y. (1999). A recursive algorithm for nonparametric analysis with missing data. *Biometrika*, **86**, 15-26.

Newton, M.A. (2002) On a nonparametric recursive estimator of the mixing distribution. *Sankhyā, Series A*, **64**, 306-322.

CHAPTER 20

Mixed Models: An Overview

Jiming Jiang and Zhiyu Ge

University of California, Davis and HSBC Securities (USA) Inc.

We give an overview of linear, generalized linear and nonlinear mixed effects models with emphases on recent developments and challenges.

20.1. Introduction

The term "mixed model" is sometimes confused with "mixture model", although the two are not unrelated. The main difference is that, while a mixture model is often defined through conditional distributions, a mixed model almost always involves random effects.

There is no general consensus among mixed-model users on the roles that the random effects play. Some believe that the random effects represent unobserved variables of practical interest, which for good reasons should be considered random. This is what we call the first-type usage. Others use the random effects as ways of modelling the correlations among the observations, but are not interested in the random effects themselves. Such a usage is called the second-type. Robinson (1991) gives a wide-ranging account of the first-type usage of random effects. As for the second-type, one of the areas is the analysis of longitudinal data (e.g., Laird and Ware (1982), Diggle, Liang and Zeger (1996)). Several books have been published on mixed models in general or with emphases in special applications. See, Rao and Kleffe (1988), Burdick and Graybill (1992), Searle, Casella and McCulloch (1992), Davidian and Giltinan (1995), Vonesh and Chinchilli (1997), Khuri, Mathew and Sinha (1998), McCulloch and Searle (2000), McCulloch (2003), and Rao (2003), among others. In addition, several authors have provided reviews on the application of mixed models in specific areas such as genetics, small area estimation, longitudinal data analysis and pharmacokinetics. See, for example, Shaw (1987), Ghosh and Rao (1994) and Yuh, Beal, Davidian, Harrison, Hester, Kowalski, Vonesh and Wolfinger (1994).

For the most part, there are three classes of mixed models that are widely used in practice. These are linear mixed models, generalized linear mixed models and nonlinear mixed models. In the subsequent sections, we shall focus on recent developments and challenges, including the authors' own research over the past decade, in these fields.

20.2. Linear Mixed Models

The best way to understand a linear mixed model, or mixed linear model in some earlier literature, is to first recall a linear regression model. The latter can be expressed as $y = X\beta + \epsilon$, where y is a vector of observations, X is a matrix of known covariates, β is a vector of unknown regression coefficients and ϵ is a vector of (unobservable random) errors. In this model, the regression coefficients are considered fixed. However, there are cases in which it makes sense to assume that some of these coefficients are random. These cases typically occur when the observations are correlated. For example, in medical studies, repeated measures are often collected from the same individuals over times. It may be reasonable to assume that correlations exist among the observations from the same individual, especially if the times at which the observations are collected are relatively close. Such correlations may be taken into account in a linear mixed model.

A general linear mixed model may be expressed as

$$y = X\beta + Z\alpha + \epsilon, \qquad (20.1)$$

where y is a vector of observations, X is a matrix of known covariates, β is a vector of unknown regression coefficients, which are often called the fixed effects, Z is a known matrix, α is a vector of random effects and ϵ is a vector of errors. Note that both α and ϵ are unobservable. Compared with the linear regression model, it is clear that the difference is $Z\alpha$, which may take many different forms, thus creates a rich class of models. The basic assumptions for (20.1) are that the random effects and errors have mean zero and finite variances. Typically, the covariance matrices $G = \text{Var}(\alpha)$ and $R = \text{Var}(\epsilon)$ depend on some unknown dispersion parameters, or variance components. It is also assumed that α and ϵ are uncorrelated.

If, in addition, the normality assumption is made, the model is called a Gaussian linear mixed model, or simply Gaussian mixed model. A Gaussian mixed model may be defined with or without random effects. In the latter case, it is assumed that the (marginal) distribution of y is multivariate normal with mean $X\beta$ and covariance matrix V, which is specified up to a vector θ of variance components, i.e., $y \sim N(X\beta, V)$, where $V = V(\theta)$. Such a model is also called a *marginal model*.

20.2.1. *Estimation of variance components*

A main problem in the analysis of linear mixed models is the estimation of the variance components. In many cases (e.g., quantitative genetics), the variance components are of main interest (e. g., Shaw (1987)). In some other cases (e.g., longitudinal data analysis), the variance components themselves are not of main interest, but they need to be estimated in order to assess the variability of estimators of other quantities of main interest, such as the fixed effects. Some of the earlier methods in mixed model analysis did not require the normality assumption. These include the analysis of variance (ANOVA) method, or Henderson's methods (Henderson 1953), and minimum norm quadratic unbiased estimation (MINQUE) method, proposed by C. R. Rao (e.g., Rao (1972)). However, the ANOVA method is known to produce inefficient estimators of the variance components when the data is unbalanced. The MINQUE method, on the other hand, depends on some initial values of the variance components. Also, both ANOVA and MINQUE can result in estimators that fall outside the parameter space.

If normality is assumed, the efficient estimators of the variance components are the maximum likelihood estimators (MLE). However, the latter had not been in serious use in linear mixed models, until Hartley and Rao (1967). The main reason was that, unlike the ANOVA estimator, the MLE under a linear mixed model was not easy to handle computationally in the earlier days. There was also an issue regarding the asymptotic behavior of the MLE, because, unlike the traditional i.i.d. case, the observations are correlated under a linear mixed model. Both issues were addressed by the Hartley-Rao paper. Asymptotic properties of the MLE were further studied by Miller (1977) for a wider class of models.

The MLE of the variance components are, in general, biased. Such a bias will not vanish as the sample size increases, if the number of the fixed effects is proportional to the sample size. In fact, in the latter case the MLE will be inconsistent, as the Neyman-Scott example showed (Neyman and Scott(1948)). Furthermore, in some cases such as animal genetics the parameters of main interest are the variance components, while the fixed effects are considered as nuisance parameters. It would be nice to have a method that can focus on the variance components without having to simultaneously estimate the nuisance parameters. Thompson (1962) proposed a method, which was later put on a broader basis by Patterson and Thompson (1971), known as restricted or residual maximum likelihood, or REML in either case. The idea is to consider a transformation of the data that is orthogonal to the design matrix of the fixed effects, that is, X. In other words, each component of such a transformation is an *error contrast*. To formulate the REML, let the dimensions of y and β be n and p, respectively. Without loss of generality, assume that rank$(X) = p$. Let A be a $n \times (n - p)$ matrix of full rank such that $A'X = 0$. The REML estimators of the variance components are simply the MLE based on $z = A'y$. It is easy to show that the REML estimators do not depend on the choice of A. Furthermore, several authors have argued that there is no loss of information in REML for estimating the variance components (e.g., Patterson and Thompson (1971), Harville (1977), Jiang (1996)). Different derivations of REML have been given by Harville (1974), Barndorff-Nielsen (1983), Verbyla (1990), Heyde (1994), and Jiang (1996). In addition, several authors have written review articles on REML, see Harville (1977), Khuri and Sahai (1985), Robinson (1987) and Speed (1997).

The ML and REML methods were developed under the normality assumption. However, the latter is likely to be violated in real life. For example, Lange and Ryan (1989) gave several examples showing that nonnormality of the random effects is, indeed, encountered in practice. Due to such concerns, some researchers have taken a quasi-likelihood approach. The idea is to use the Gaussian ML or REML estimators in nonnormal situations. See Richardson and Welsh (1994), Jiang (1996), Jiang (1997a), Heyde (1994), Heyde (1997), among others. Throughout this review, these estimators will also be called ML and REML estimators, even if normality does not hold. In his Ph.D dissertation under the supervision of Professor Bickel, Jiang (1996) and Jiang (1997a) established consistency and asymptotic normality of REML estimators in nonnormal situations. In exactly the same situation, the author also gave necessary and sufficient conditions for the consistency and asymptotic normality of the MLE. These conditions are in terms of the rate at which p goes to infinity with n. In particular, Jiang (1996) derived the asymptotic covariance matrix (ACM) of the REML estimator of the variance components. See Jiang (1998b) for the ML analogue. The ACM is important to various inferences about the model, including interval estimation and hypothesis testing. Unfortunately, the ACM under nonnormality involves parameters other than the variance components, that is, the third and fourth moments of the random effects and errors. Note that standard procedures such as ML and REML do not produce estimators of these additional parameters. Jiang (2005a) proposed a method known as partially observed information for estimating the ACM.

20.2.2. *Prediction of random effects*

The prediction of random effects, or mixed effects in a more general context, has a long history dating back to C. R. Henderson in his early work in animal breedings (e.g., Henderson (1948)). The best-known method is the best linear unbiased prediction, or BLUP. Robinson (1991) gives a wide-ranging account of BLUP with examples and applications.

A mixed effect may be expressed as $\eta = b'\beta + a'\alpha$, where a, b are known vectors. If the fixed effects and variance components are both known, the best predictor for η, under the normality assumption, is given by $\tilde{\eta}$, where $\tilde{\eta}$ is η with α replace by $E(\alpha|y)$, i.e., $\tilde{\alpha} = GZ'V^{-1}(y - X\beta)$. Without assuming normality, $\tilde{\eta}$ is the best linear predictor of η in the sense that it minimizes the mean squared error (MSE) of a predictor that is linear in y (e.g. Searle, Casella and McCulloch (1992), Section 7.3). Of course, β is unknown in practice. It is then customary to replace β by $\tilde{\beta} = (X'V^{-1}X)^{-1}X'V^{-1}y$, which is the MLE of β under normality, provided that θ is known. The result is BLUP, in other words, the BLUP of η is given by $\tilde{\eta}$ with β replaced by $\tilde{\beta}$. The original derivation of BLUP was given by Henderson (1950), in which he proposed to find the "maximum likelihood estimates" of the random effects. Of course, these are not the MLE in the usual sense, because the random effects are different from fixed parameters. Later, in Henderson (1973), he showed that BLUP is, indeed, what is meant by the name, that is, (i) it is linear in y; (ii) its expected value is equal to that of η; and (iii) it minimizes the MSE among all linear unbiased predictors of η. Different derivations of BLUP were also given by Harville (1990) and Jiang (1997b), among others.

The expression of BLUP involves θ, the vector of variance components, which is typically unknown in practice. It is customary to replace θ by a consistent estimator, $\widehat{\theta}$. The resulting predictor is often called empirical BLUP, or EBLUP, denoted by $\widehat{\eta}$. Kackar and Harville (1981) showed that, if $\widehat{\theta}$ is an even and translation invariant estimator and the data is normal, EBLUP remains unbiased. Some of the well-known estimators of θ, including ANOVA, ML and REML estimators, are even and translation invariant. In their arguments, however, Kackar and Harville had assumed existence of the expected value of EBLUP, which is not obvious. The existence of the expected value was later proved by Jiang (2000a).

One of the areas in which EBLUP has been extensively used, and studied, is small area estimation (SAE). See, for example, Ghosh and Rao (1994) and Rao (2003). While EBLUP is fairly easy to calculate, its MSE is complicated. On the other hand, the MSE of EBLUP is of significant practical interest. Kackar and Harville (1984) gave an approximation to the MSE of EBLUP under the linear mixed model (20.1), taking into account the variability in $\hat{\theta}$, and proposed an estimator of $\text{MSE}(\hat{\eta})$ based on this approximation. But the approximation is somewhat heuristic, and the accuracy of the approximation and the associated MSE estimator was not studied. Prasad and Rao (1990) studied the accuracy of a second order approximation to $\text{MSE}(\hat{\eta})$ for two important special cases of linear mixed models that are frequently used in SAE, that is, the Fay-Herriot model (Fay and Herriot (1979)) and nested error regression model (e.g., Battese, Harter, and Fuller (1988)). The results were extended by Das, Jiang and Rao (2004) to general linear mixed models. Alternatively, Jiang, Lahiri and Wan (2002) proposed a jackknife method which led to second order approximation and estimation of the MSE of EBLUP in the case of longitudinal linear mixed models.

20.2.3. *Other types of inference*

Most of the literature on analysis of linear mixed models has been focusing on estimation and prediction problems. However, virtually every other types of analysis were also considered. These include: *(i) Bayesian inference.* See, for example, Hill (1965), Tiao and Tan (1965), Tiao and Tan (1966), Gianola and Fernando (1986), and Berger and Deely (1988). *(ii) Tests in linear mixed models.* See, for example, Khuri, Mathew and Sinha (1998) for tests under normality; and Jiang (2003a), Jiang (2005a) for tests under non-Gaussian linear mixed models. *(iii) Confidence intervals.* See Burdick and Graybill (1992). *(iv) Prediction intervals.* See, for example, Jeske and Harville (1988), Jiang and Zhang (2002). *(v) Mixed model diagnostics.* See, for example, Dempster and Ryan (1985), Lange and Ryan (1989)), Calvin and Sedransk (1991), Jiang, Lahiri and Wu (2001) and Jiang (2001a). *(vi) Mixed model selection.* See Jiang and Rao (2003). Further research is needed in the last two areas.

20.3. Generalized Linear Mixed Models

For the most part, linear mixed models have been used in situations where the observations are continuous. However, in many cases the responses are correlated as well as discrete or categorical, so that neither linear mixed models nor generalized linear models (GLM; McCullagh and Nelder (1989)) would apply. Note that a basic assumption in GLM is that the responses are independent. It is clear that one needs a new class of models that combines linear mixed models and GLM.

20.3.1. *From linear mixed model to GLMM*

To motivate the extension, let us first consider an alternative expression of the Gaussian mixed model. Suppose that, given a vector of random effects, α, the observations $y_1, \ldots, y_n$ are (conditionally) independent such that $y_i \sim N(x_i'\beta + z_i'\alpha, \tau^2)$, where x_i and z_i are known vectors, β is an unknown vector of regression coefficients, and τ^2 is an unknown variance. Furthermore, suppose that α is multivariate normal with mean 0 and covariance matrix G, which depends on a vector θ of unknown variance components. Let X and Z be the matrices whose ith rows are x_i' and z_i', respectively. This leads to the linear mixed model (20.1) with normality and $R = \tau^2 I$.

The key elements in the above that define a Gaussian mixed model are (i) conditional independence given the random effects with a conditional distribution; (ii) the distribution of the random effects. These are also the essential parts of a generalized linear mixed model, or GLMM. Suppose that given a vector of random effects, α, the responses, $y_1, \ldots, y_n$ are conditionally independent such that the conditional distribution of y_i given α is a member of the exponential family with pdf $f_i(y_i|\alpha) = \exp[a_i^{-1}(\phi)\{y_i\xi_i - b(\xi_i)\} + c_i(y_i, \phi)]$, where $b(\cdot)$, $a_i(\cdot)$, $c_i(\cdot, \cdot)$ are known functions, and ϕ is a dispersion parameter which may or may not be known. Then, ξ_i is associated with $\mu_i = \mathrm{E}(y_i|\alpha)$, which, in term, is associated with a linear predictor $\eta_i = x_i'\beta + z_i'\alpha$ through a known link function $g(\cdot)$ such that $g(\mu_i) = \eta_i$. Here x_i and z_i are known vectors, and β is a vector of unknown parameters (the fixed effects). Furthermore, it is assumed that $\alpha \sim N(0, G)$, where G depends on a vector θ of unknown variance components. Note that, according to the properties of the exponential family, one has $b'(\xi_i) = \mu_i$. In particular, under the so-called *canonical* link, one has $\xi_i = \eta_i$, that is, $g = h^{-1}$, where $h(\cdot) = b'(\cdot)$.

One of the earlier examples of GLMM was given by McCullagh and Nelder (1989), Section 14.5, involving some salamander mating experiments. The models have since received considerable attention because of their usefulness in various fields, including biology, medical research and surveys. See, for example, Breslow and Clayton (1993), Lee and Nelder

(1996), Malec *et al.* (1997), Ghosh *et al.* (1998) and McCulloch and Searle
(2000).

Despite the usefulness of these models, inference about GLMM has encountered some difficulties. This is because, unlike linear mixed models, the
likelihood function under a GLMM typically does not have a closed-form
(with, of course, the exception of the normal case). In fact, such a likelihood may involve high dimensional integrals which cannot be evaluated
analytically. Because of the computational difficulties, there have been two
main approaches in inference about GLMM. The first focuses on developing computational method for the maximum likelihood. The second tries
to avoid the computational difficulties of the likelihood-based inference by
considering approximate or other alternative methods.

20.3.2. *Likelihood-based inference*

For relatively simple models, the likelihood function may be evaluated by
numerical integration techniques. See, for example, Hinde (1982), Crouch
and Spiegelman (1990). However, numerical integration is generally intractable in GLMM if the dimension of integrals involved is greater than
two. Alternatively, the integrals may be evaluated by Monte Carlo methods. It should be pointed out that, for problems involving irreducibly high-
dimensional integrals, naive Monte Carlo usually does not work. McCulloch (1994) proposed a Monte Carlo EM algorithm, in which the E-step
was implemented by a Gibbs Markov chain. Later, in McCulloch (1997),
the author improved his earlier algorithm by replacing the Gibbs sampler
with a Metropolis-Hastings algorithm to fit more general models. Subsequently, Booth and Hobert (1999) used importance sampling and rejection
sampling to generate i.i.d. samples for the approximation at the E-step.
Furthermore, the latter authors proposed a rule for automatically increasing the Monte-Carlo sample size whenever necessary as the algorithm proceeds. They showed that their new algorithms have several advantages over
the Markov chain methods, including easier assessment of the Monte Carlo
errors and faster speed.

In a related development, Zeger and Karim (1991) used Gibbs sampling to approximate Bayesian posterior means of the parameters under a
GLMM. Also see Malec *et al.* (1997), Ghosh *et al.* (1998).

Recently, Song, Fan and Kalbfleisch (2005) proposed a maximization by
parts method, which is potentially applicable to the computation of MLE
in some GLMMs.

20.3.3. *Non-likelihood-based inference*

1. Approximate inference. When the exact likelihihood function is difficult to compute, approximation becomes one of the natural alternatives. A well-known method is Laplace approximation (LA) to integrals. Several authors have used this method to approximate the likelihood function, and then treat the approximate likelihood as the true likelihood for inference. The method may also be viewed as estimation of both fixed and random effects via maximization of the joint density functions of the observations and random effects, penalized quasi-likelihood (PQL), or maximum hierarchical likelihood. See, Schall (1991), Breslow and Clayton (1993), Wolfinger and O'Connell (1993), McGilchrist (1994), Lee and Nelder (1996), and Lin and Breslow (1996), among others. Unfortunately, the LA-based methods are known to have some unsatisfactory properties. In particular, the resulting estimators are inconsistent under standard asymptotic assumptions (e.g. Kuk (1995), Jiang (1998a), Jiang (1999)).

2. Estimating equations. On the other hand, generalized estimating equations (GEE) have been used in the analysis of longitudinal data (Liang and Zeger (1986), Prentice (1988)). The GEE applies to a special class of GLMM, in which the observations are independently clustered, hence the covariance matrix is block-diagonal. Jiang (1998a) proposed estimating equations that apply to GLMMs not necessarily having a block-diagonal covariance structure, such as the one in the salamander mating problem. He showed that the resulting estimators are consistent but inefficient. Later, Jiang and Zhang (2001) proposed a two-step procedure to obtain more efficient estimators. They considered a broader class of models, which does not require full specification of the conditional distribution of the responses given the random effects. Therefore, their method applies to a broader class of models than the GLMM.

20.3.4. *Prediction of random effects*

The method of joint estimation of the fixed and random effects discussed earlier provides estimators, or predictors, of the random effects. Jiang, Jia and Chen (2001) took another look at the method as maximization of a posterior of the random effects (under a non-informative prior).

The joint estimates of the fixed and random effects are typically obtained by solving a system of nonlinear equations $\partial l_J / \partial \beta = 0$, $\partial l_J / \partial \alpha = 0$, where l_J is the logarithm of the joint density. However, in many cases the number of random effects involved is quite large. This means that one has to solve a large system of nonlinear equations. It is well known that standard methods of solving nonlinear systems such as Newton-Raphson is inefficient in such problems. Jiang (2000b) proposed a nonlinear Gauss-Seidel algorithm for effectively solving the nonlinear system. The author proved global convergence of the algorithm. Alternatively, Breslow and Clayton (1993) proposed an an iterative procedure for solving the nonlinear system by modifying the Fisher scoring algorithm. An attractive feature of the latter procedure is that it exploits a close correspondence with the well-known mixed model equations (Henderson *et al.* (1959)).

One of the main areas in which prediction of random effects, or mixed effects, is of main interest is SAE (e.g. Rao (2003)). In this context, a method known as empirical best prediction (EBP) has been developed. It is a two-step procedure. In the first-step, one derives an expression for the best predictor which is the conditional expectation of the mixed effect given the data. The expression is likely dependent on a vector of unknown parameters. Therefore, in the second-step, one replace the unknown parameters by consistent estimators. See Jiang and Lahiri (2001), Jiang (2003b). A feature of EBP is that it is model-based. If the assumed model fails, the predictor may perform poorly. Jiang and Lahiri (2005) developed a model-assisted EBP method that has the property of design-consistency, that is, even under model failure, the predictor of a domain mean is approximately equal to a design-based estimator as long as the domain sample size is large. The design-based estimator (e.g. Hajek (1971)), on the other hand, is known to approach the true domain mean when the domain sample size is large regardless of the assumed model. In both the EBP and model-assisted EBP, the authors have obtained estimators of the MSE of the EBP whose bias is corrected to the second-order.

20.3.5. *Future research and open problems*

There is a lack of studies on theoretical properties, such as asymptotic behavior, of the MLE under a GLMM, despite considerable effort made in developing computing algorithms for these estimators. The problem is relatively straightforward for longitudinal GLMMs, in which the responses may be divided into independent clusters (e.g., Jiang (2001b)). What is really challenging is the case with crossed random effects, such that there is no independent clusters or groups. Note that, unlike linear mixed models, the likelihood function can only be expressed as (multi-dimensional) integrals under a typical GLMM. General asymptotic results such as those in linear mixed models (e.g., Miller (1977), Jiang (1996)) do not exist for GLMMs. For example, the salamander mating data has been analyzed by many authors; some others use the same model and data structure for simulations (e.g., McCullagh and Nelder (1989), Karim and Zeger (1992), Drum and McCullagh (1993), Lin and Breslow (1996), Lin (1997), Jiang (1998a), and Jiang and Zhang (2001)). Furthermore, Monte-Carlo EM algorithms have been developed to obtain the MLE with the salamander data (e. g., Booth and Hobert (1999)). However, one fundamental question has yet to be answered: Is the MLE consistent? The answer is not obvious at all.

Unlike estimation problems, the literature on testing in GLMM is not extensive. For example, Lin (1997) used PQL method for testing the hypothesis that all the variance components are zero. While PQL is known to yield inconsistent point estimators, its usage in testing such a hypothesis is entirely appropriate. This is because the random effects become zero under the null hypothesis, so a GLMM reduces to a GLM, for which the likelihood-based methods are well justified. Also see Lin and Carroll (1999). However, it is difficult to extend the method to testing complex hypotheses, under which the random effects may not vanish. Note that the lack of asymptotic theory on MLE also contributed to the difficulties in testing. On the other hand, once again, testing for longitudinal GLMMs is relatively easier (e.g., Song and Jiang (2000)).

Other underdeveloped areas include GLMM diagnostics and section as well as some prediction problems.

20.4. Nonlinear Mixed Effects Models

Nonlinear mixed effects (NLME) modeling methodology and applications are reviewed extensively in two monographs, one by Davidian and Giltinan (1995) and the other by Vonesh and Chinchilli (1997), in addition to the collection of numerous research articles.

A subclass of the general NLME models has received the most attention in the literature. These models have the form

$$\underset{m_i \times 1}{\boldsymbol{y}_i} = \boldsymbol{f}_i(\underset{l_i \times 1}{\boldsymbol{x}_i}, \underset{r \times 1}{\boldsymbol{\beta}}, \underset{d \times 1}{\boldsymbol{b}_i}) + \underset{s_i \times 1}{\boldsymbol{\epsilon}_i} \qquad i = 1, 2, \cdots, n, \qquad (20.2)$$

where $\boldsymbol{y}_i$ is the response or observation vector for stratum i, $\boldsymbol{f}_i$ is a known function, $\boldsymbol{x}_i$ is a vector of covariates, $\boldsymbol{\beta}$ is the population parameter vector, $\boldsymbol{b}_i$ is a vector of unobserved latent variables which are random across strata, and $\boldsymbol{\epsilon}_i$ represents a random perturbation such as measurement error. Observations from different strata are usually assumed independent.

For simplicity of our discussion below, we consider constant dimensions across i for $\boldsymbol{y}_i$, $\boldsymbol{x}_i$ and $\boldsymbol{\epsilon}_i$, respectively. We also drop the subscript i in $\boldsymbol{f}_i$, thereby assume the $\boldsymbol{f}_i$'s are the same across strata.

With this model formulation, if $\boldsymbol{f}_i$ is nonlinear in $\boldsymbol{b}_i$, then tracking the marginal likelihood based on the observed $\boldsymbol{y}_i$'s may be a daunting task. The marginal likelihood will involve integration of the joint density of $(\boldsymbol{y}_i, \boldsymbol{b}_i)$ over the random effect space of $\boldsymbol{b}_i$, and often there is no closed form for such integration. This is the primary source of theoretical and numerical difficulties associated with NLME models.

20.4.1. *Parametric modeling*

In the parametric setting, we usually assume that $\boldsymbol{y}_i|\boldsymbol{b}_i$ has density $p_i(\cdot|\boldsymbol{b}_i, \boldsymbol{\beta})$ and that $\boldsymbol{b}_i$ has density $g_i(\cdot|\boldsymbol{\beta})$, where $p_i(\cdot|\cdot, \cdot)$ and $g_i(\cdot|\cdot)$ are known functions. Like the GLMMs discussed previously, there are also likelihood-based and approximate-likelihood(AL)-based inferences available for parametric NLME models.

1. Likelihood-based inference. This approach includes the traditional EM algorithm, Monte Carlo based methods, and hybrid Monte Carlo EM algorithms.

The traditional EM algorithm essentially has the same difficulty that the exact maximum likelihood procedure has in handling NLME models. It faces computational challenges in both the E-step and the M-step, therefore practical usefulness is limited.

Monte Carlo methods are used to handle the numerical integration in the marginal likelihood. Standard techniques such as Markov chain Monte Carlo and importance sampling can be applied to address simulation and computation complexity issues, especially for high dimensional random effects. Gelman, Bois and Jiang (1996) provided an example of Monte Carlo application in which the effective dimension of the random effects is 13. Monte Carlo methods in general are computationally intensive.

Hybrid methods combining elements of the traditional EM algorithms and Monte Carlo methods may be able to handle some specific models more efficiently. Walker (1996) proposed a Monte Carlo EM algorithm for a class of NLME models. The algorithm performs Monte Carlo integration to handle conditional expectations in the E-step.

2. AL-based inference. Linearization algorithms for NLME models are by far the most popular parametric approaches. They are natural extensions of the linearization algorithm for classical nonlinear regression. These include the first order algorithm and the conditional first order algorithm. In model (20.2), if we assume that $\boldsymbol{\epsilon}_i = R_i^{1/2}(\boldsymbol{\beta}, \boldsymbol{b}_i)\boldsymbol{e}_i$, where $\boldsymbol{e}_i$ has a known distribution, then the first order algorithm approximates the original model by $\boldsymbol{y}_i = \boldsymbol{f}(\boldsymbol{x}_i, \boldsymbol{\beta}, \boldsymbol{0}) + (\partial \boldsymbol{f}/\partial \boldsymbol{b})(\boldsymbol{x}_i, \boldsymbol{\beta}, \boldsymbol{0})\boldsymbol{b}_i + R_i^{1/2}(\boldsymbol{\beta}, \boldsymbol{0})\boldsymbol{e}_i$.

In comparison, the conditional first order algorithm approximates the original model by $\boldsymbol{y}_i = \boldsymbol{f}(\boldsymbol{x}_i, \boldsymbol{\beta}, \boldsymbol{b}_i^*) + (\partial \boldsymbol{f}/\partial \boldsymbol{b})(\boldsymbol{x}_i, \boldsymbol{\beta}, \boldsymbol{b}_i^*)(\boldsymbol{b}_i - \boldsymbol{b}_i^*) + R_i^{1/2}(\boldsymbol{\beta}, \boldsymbol{b}_i^*)\boldsymbol{e}_i$, where $\boldsymbol{b}_i^*$ is the current predictor of $\boldsymbol{b}_i$.

Different types of linearization algorithms proposed in the literature are variations of the ones presented above. Linearization methods are numerically simple, because they avoid complicated numerical integrations. They are implemented in popular software packages such as NONMEM (Beal and Sheiner (1992)) and *nlme*() in S-PLUS. Despite their popularity, it should be noted that the asymptotic properties of linearization algorithms are not well established in the presence of random effect variability. In the most common modeling situation with limited number of observations per stratum, these linearization methods give consistent estimates of fixed parameters ($\boldsymbol{\beta}$) only as the number of strata n tends to infinity and the variability of random effects tends to zero at the same time. With large random effects variability, estimates given by linearization approximations may carry non-ignorable systematic bias, and bias correction methods based on higher order Taylor expansions rely inevitably on the assumption of small variability of random effects and are therefore not satisfactory.

Similar to the GLMM case, Laplace-approximation (LA) has been used, which has different variations. Wolfinger (1993)'s LA treats the marginal likelihood as an integral with respect to both the random effects and the fixed effects with a tacitly present flat prior for the fixed effects. This approach includes the Lindstrom and Bates (1990)' linearization algorithm as a special case. Another type of LA approaches treat the marginal likelihood as an integral with respect to the random effects only. Like that of the linearization algorithms, consistency of the LA-based estimators is questionable when the variability in random effects is large.

The Gaussian quadrature approach, which is based on the well-known Gauss-Hermite quadrature rule, can be viewed as a deterministic version of the Monte Carlo integration. It uses a fixed set of abscissae and associated weights, and approximates the marginal likelihood integral by a linear combination of values of the quadrature function evaluated at the abscissae. The quadrature function must satisfy certain conditions for the Gauss-Hermite rule to apply, otherwise the numerical behavior may not be predictable. Despite this potential problem, this approach seems to work fine in many applications.

Pinhero and Bates (1995) compared existing popular AL algorithms, including several discussed above, by means of data analyses and numerical simulations. They described these algorithms, discussed theoretical connections among them, and provided a good summary of the potential relative strengths and weaknesses of these approaches in terms of consistency and numerical efficiency. Some of their findings, however, may be specific to their simulation settings.

An AL approach to NLME models via spline approximation was proposed recently by Ge, Bickel and Rice (2004). In this approach, the conditional likelihood $\mathcal{L}(\boldsymbol{y}_i|\boldsymbol{b}_i)$ is approximated by a piecewise polynomial in the random effects $\boldsymbol{b}$, with the coefficients taken as functions of $(\boldsymbol{y}_i, \boldsymbol{\beta})$. For commonly adopted parametric assumptions on random effect distributions such as normality, evaluation of the resulting AL does not require complicated numerical integration, and therefore the numerical difficulties associated with the exact ML approach can be greatly reduced. In the mean time, the asymptotic differences between the AL estimates and the MLE can be controlled by choosing an appropriate level of approximation. The conditional likelihood carries more information than the conditional first and second moments combined, and therefore a direct approximation to the conditional likelihood is expected to do better than the linearization algorithms. The numerical efficiency of the spline approximation approach lies between those of the linearization algorithms and the exact ML.

20.4.2. *Semi-parametric modeling*

In a semi-parametric setting, the distribution of the random effects is assumed to be non-parametric, while the parametric assumptions on the conditional distribution of the observed data given the random effects are maintained. Methods for semi-parametric modeling include an extension of the classic nonparametric ML estimation (NPMLE), a semiparametric modeling approach in combination with Gauss-Hermite quadrature proposed by Davidian and Gallant (1993), and an extension of the spline approximation approach proposed by Ge, Bickel and Rice (2004).

The NPMLE based approach assumes a simplex distribution for the random effects concentrated on a set of no more than n points in the random effect space, where n is the number of strata. This set of points together with the corresponding probability weights are part of the model parameters to be estimated numerically. Therefore this approach often involves high dimensional optimization, which can be computationally intensive and unstable. Nonetheless, the robust simplistic nonparametric modeling of the random effect distribution is attractive for the purpose of exploring and revealing information about the random effects.

Davidian and Gallant (1993) proposed a semiparametric modeling approach in which the density of the random effects comes from a rich family. The functional form of this family can easily accommodate the Gauss-Hermite quadrature rule in the numerical integrations for the marginal likelihood. This approach is appealing for exploring the distribution of the random effects and its numerical simplicity. The theoretical framework was provided by Gallant and Nychka (1987).

The spline approximation approach to the parametric modeling proposed by Ge, Bickel and Rice (2004) also has a natural extention to the semiparametric case. The density of the random effects can be represented as a histogram or a spline, and this representation can be combined with the spline approximation to the conditional likelihood to result in an AL function. With respect to the approximation to the random effect distribution, this method can be viewed as a method of sieves. It attempts to obtain consistent estimators of fixed effects and the density of the random effects and, at the mean time, reduce the number of parameters in the numerical optimization associated with NPMLE. The asymptotic properties of this method are yet to be established.

Another direction of the semiparametric modeling is to specify the random effect distribution parametrically, leaving the conditional distribution not fully parametrically specified. If only the first and second moments of the conditional distribution are specified, the GEE approach is applicable. In this case, Ge, Bickel and Rice (2004) used spline approximations to the conditional first and second moments to address integration difficulties associated with the unconditional first and second moments. The resulting approximate GEE estimator converges to the original GEE estimator asymptotically.

20.4.3. *Other topics*

Similar to the GLMM case, there is limitted literature on the asymptotic behavior of the MLE for NLME models. Researchers are generally content with the assumption that the MLEs are consistent and efficient, and their attention is primarily focused on getting good approximations of MLEs given the numerical difficulties associated with the integrations.

Despite extensive literature on computing algorithms, asymptotic behaviors of these algorithms have been established with mathematical rigor only in some restricted cases. General asymptotic quantifications of the bias of the AL-based methods are not yet available. Furthermore, validity of the classical testing procedures applied to NLME has not been adequately addressed, and testing under the framework of AL-based inference adds another level of complexity. Regarding practical issues such as model selection, diagnostics and prediction, although methods have been proposed as extensions of the standard procedures for linear models, their validity still needs to be scrutinized.

Acknowledgements

The authors wish to thank, once again, Professor Peter J. Bickel for his guidance of the authors' Ph.D dissertations and the many helps he has provided during the authors' professional and academic careers.

References

Barndorff-Nielsen, O. (1983), On a formula for the distribution of the maximum likelihood estimator, *Biometrika* 70, 343-365.

Battese, G. E., Harter, R. M., and Fuller, W. A. (1988), An error-components model for prediction of county crop areas using survey and satellite data, *J. Amer. Statist. Assoc.* 80, 28-36.

Berger, J. O. and Deely, J. (1988), A Bayesian approach to ranking and selection of related means with alternatives to analysis-of-variance methodology, *J. Amer. Statist. Assoc.* 83, 364-373.

Booth, J. G. and Hobert, J. P. (1999), Maximum generalized linear mixed model likelihood with an automated Monte Carlo EM algorithm, *J. Roy. Statist. Soc. B* 61, 265-285.

Breslow, N. E. and Clayton, D. G. (1993), Approximate inference in generalized linear mixed models, *J. Amer. Statist. Assoc.* 88, 9-25.

Burdick, R. K. and Graybill, F. A. (1992), *Confidence Intervals on Variance Components*, Dekker, New York.

Calvin, J. A., and Sedransk, J. (1991), Bayesian and frequentist predictive inference for the patterns of care studies, *J. Amer. Statist. Assoc.* 86, 36-48.

Crouch, E. A. C. and Spiegelman, D. (1990), The evaluation of integrals of the form $\int f(t) \exp(-t^2)dt$: Application to logistic normal models, *J. Amer. Statist. Assoc.* 85, 464-469.

Das, K., Jiang, J. and Rao, J. N. K. (2004), Mean squared error of empirical predictor, *Ann. Statist.* 32, 818-840.

Davidian, M. and Gallant, A. R. (1993), The nonlinear mixed effects model with a smooth random effects density, *Biometrika* 80, 475-488.

Davidian, M. and Giltinan, D. M. (1995), *Nonlinear Models for Repeated Measurement Data*, Chapman & Hall.

Dempster, A. P., and Ryan, L. M. (1985), Weighted normal plots, *J. Amer. Ststist. Assoc.* 80, 845-850.

Diggle, P. J., Liang, K. Y. and Zeger, S. L. (1996). *Analysis of Longitudinal Data*. Oxford Univ. Press.

Drum, M. L. and McCullagh, P. (1993), REML estimation with exact covariance in the logistic mixed model, *Biometrics* 49, 677-689.

Fay, R. E. and Herriot, R. A. (1979), Estimates of income for small places: an application of James-Stein procedures to census data, *J. Amer. Statist. Assoc.* 74, 269-277.

Gallant, A. R. and Nychka, D. W. (1987), Semi-nonparametric maximum likelihood estimation, *Econometrica* 55, 363-390.

Ge, Z., Bickel, P. J. and Rice, J. A., (2004), A generic numerical integration algorithm via spline approximation for nonlinear mixed effects models, *Computational Statist. Data Anal.* 46,

Gelman, A., Bois, F. and Jiang, J. (1996), Physiological pharmacokinetic analysis using population modeling and informative prior distribution, *J. Amer. Statist. Assoc.* 91, 1400-1412.

Gianola, D. and Fernando, R. L. (1986), Bayesian methods in animal breeding theory, *J. Animal Sci.* 63, 217-244.

Ghosh, M., and Rao, J.N.K. (1994), Small area estimation: An appraisal (with discussion), *Statist. Sci.* 9, 55-93.

Ghosh, M., Natarajan, K., Stroud, T. W. F. and Carlin, B. P. (1998), Generalized linear models for small-area estimation, *J. Amer. Statist. Assoc.* 93, 273-282.

Hajek, J. (1971), Comment, in *Foundations of Statistical Inference*, V. P. Godambe and D. A. Sprott Eds., Toronto: Holt, Rinchart and Winston.

Hartley, H. O. and Rao, J. N. K. (1967), Maximum likelihood estimation for the mixed analysis of variance model, *Biometrika* 54, 93-108.

Harville, D. A. (1974), Bayesian inference for variance components using only error contrasts, *Biometrika* 61, 383-385.

Harville, D. A. (1977), maximum likelihood approaches to variance components estimation and related problems, *J. Amer. Statist. Assoc.* 72, 320-340.

Harville, D. A. (1990), BLUP (best linear unbiased prediction) and beyond, in *Advances in Statistical Methods for Genetic Improvement of Livestock* (D. Gianola and K. Hammond, eds.) 239-276, Springer, New York.

Henderson, C. R. (1948), Estimation of general, specific and maternal combining abilities in crosses among inbred lines of swine, Ph.D. Thesis, Iowa State Univ., Ames, Iowa.

Henderson, C. R. (1950), Estimation of genetic parameters (abstract), *Ann. Math. Statist.* 21, 309-310.

Henderson, C. R. (1953), Estimation of variance and covariance components, *Biometrics* 9, 226-252.

Henderson, C. R. (1973), Sire evaluation and genetic trends, in *Proceedings of the Animal Breeding and Genetics Symposium in Honor of Dr. Jay L.*

Lush, 10-41, Amer. Soc. Animal Sci. - Amer. Dairy Sci. Assoc. - Poultry Sci. Assoc., Champaign, IL.

Henderson, C. R., Kempthorne, O., Searle, S. R. and von Krosigk, C. N. (1959), Estimation of environmental and genetic trends from records subject to culling, *Biometrics* 15, 192-218.

Heyde, C. C. (1994), A quasi-likelihood approach to the REML estimating equations, *Statist. & Probab. Letters* 21, 381-384.

Heyde, C. C. (1997), *Quasi-Likelihood and Its Application*, Springer, New York.

Hill, B. M. (1965), Inference about variance components in the one-way model, *J. Amer. Statist. Assoc.* 60, 806-825.

Hinde, J. (1982), Compound Poisson regression models, in *GLIM 82: Proceedings of the International Conference on Generalized Linear Models*, R. Gilchrist ed., Springer, Berlin, 109-121.

Jeske, D. R. and Harville, D. A. (1988), Prediction-interval procedures and (fixed-effects) confidence-interval procedures for mixed linear models, *Commun. Statist. - Theory Meth.* 17, 1053-1087.

Jiang, J. (1996), REML estimation: Asymptotic behavior and related topics, *Ann. Statist.* 24, 255-286.

Jiang, J. (1997a), Wald consistency and the method of sieves in REML estimation, *Ann. Statist.* 25, 1781-1803.

Jiang, J. (1997b), A derivation of BLUP — best linear unbiased predictor, *Statist. Probab. Letters* 32, 321-324.

Jiang, J. (1998a), Consistent estimators in generalized linear mixed models, *J. Amer. Statist. Assoc.* 93, 720-729.

Jiang, J. (1998b), Asymptotic properties of the empirical BLUP and BLUE in mixed linear models, *Statistica Sinica* 8, 861-885.

Jiang, J. (1999), Conditional inference about generalized linear mixed models, *Ann. Statist.* 27, 1974-2007.

Jiang, J. (2000a), A matrix inequality and its statistical applications, *Linear Algebra Appl.* 307, 131-144.

Jiang, J. (2000b), A nonlinear Gauss-Seidel algorithm for inference about GLMM, *Computational Statistics* 15, 229-241.

Jiang, J. (2001a), Goodness-of-fit tests for mixed model diagnostics, *Ann. Statist.* 29, 1137-1164.

Jiang, J. (2001b), Mixed-effects models with random cluster sizes, *Statist. & Probab. Letters* 53, 201-206.

Jiang, J. (2003a), Empirical method of moments and its applications, *J. Statist. Plann. Inference* 115, 69-84.

Jiang, J. (2003b), Empirical best prediction for small area inference based on generalized linear mixed models, *J. Statist. Plann. Inference* 111, 117-127.

Jiang, J. (2005a), Partially observed information and inference about non-Gaussian mixed linear models, *Ann. Statist.*, in press.

Jiang, J. (2005b), Comment on Song, Fan and Kalbfleisch: Maximization by parts in likelihood inference, *J. Amer. Statist. Assoc.*, in press.

Jiang, J., Jia, H. and Chen, H. (2001), Maximum posterior estimation of random effects in generalized linear mixed models, *Statistica Sinica* 11, 97-120.

Jiang, J. and Lahiri, P. (2001), Empirical best prediction for small area inference with binary data, *Ann. Inst. Statist. Math.* 53, 217-243.

Jiang, J. and Lahiri, P. (2005), Estimation of finite population domain means - a model assisted empirical best prediction approach, *J. Amer. Statist. Assoc.*, to appear.

Jiang, J., Lahiri, P. and Wan, S. (2002), A unified jackknife theory for empirical best prediction with M-estimation, *Ann. Statist.* 30, 1782-1810.

Jiang, J., Lahiri, P. and Wu, C. H. (2001), A generalization of Pearson's χ^2 goodness-of-fit test with estimated cell frequencies, *Sankhyā* A 63, 260-276.

Jiang, J. and Rao, J. S. (2003), Consistent procedures for mixed linear model selection, *Sankhyā* 65, 23-42.

Jiang, J. and Zhang, W. (2001), Robust estimation in generalized linear mixed models, *Biometrika* 88, 753-765.

Jiang, J. and Zhang, W. (2002), Distribution-free prediction intervals in mixed linear models, *Statistica Sinica* 12, 537-553.

Karim, M. R. and Zeger, S. L. (1992), Generalized linear models with random effects: Salamander mating revisited, *Biometrics* 48, 631-644.

Kackar, R. N. and Harville, D. A. (1981), Unbiasedness of two-stage estimation and prediction procedures for mixed linear models, *Commun. Statist. - Theory Meth.* 10, 1249-1261.

Kackar, R. N. and Harville, D. A. (1984), Approximations for standard errors of estimators of fixed and random effects in mixed linear models, *J. Amer. Statist. Assoc.* 79, 853-862.

Khuri, A. I. and Sahai, H. (1985), Variance components analysis: a selective literature survey, *Internat. Statist. Rev.* 53, 279-300.

Khuri, A. I., Mathew, T. and Sinha, B. K. (1998), *Statistical Tests for Mixed Linear Models*, Wiley, New York.

Kuk, A. Y. C. (1995), Asymptotically unbiased estimation in generalized linear models with random effects, *J. Roy. Statist. Soc. B* 57, 395-407.

Laird, N. M. and Ware, J. M. (1982), Random effects models for longitudinal data, *Biometrics* 38, 963-974.

Lange, N., and Ryan, L. M. (1989), Assessing normality in random effects models, *Ann. Statist.* 17, 624-642.

Lee, Y. and Nelder, J. A. (1996), Hierarchical generalized linear models (with discussion), *J. Roy. Statist. Soc. B* 58, 619-678.

Liang, K. Y. and Zeger, S. L. (1986), Longitudinal data analysis using generalized linear models, *Biometrika* 73, 13-22.

Lin, X. (1997), Variance components testing in generalized linear models with random effects, *Biometrika* 84, 309-326.

Lin, X. and Breslow, N. E. (1996), Bias correction in generalized linear mixed models with multiple components of dispersion, *J. Amer. Statist. Assoc.* 91, 1007-1016.

Lin, X. and Carroll, R. J. (1999), SIMEX variance component tests in generalized linear mixed measurement error models, *Biometrics* 55, 613-619.

Lindstrom, M. J. and Bates, D. M. (1990), Nonlinear mixed effects models for repeated measures data, *Biometrics* 46, 673-687.

Malec, D., Sedransk, J., Moriarity, C. L. and LeClere, F. B. (1997), Small area inference for binary variables in the National Health Interview Survey, *J. Amer. Statist. Assoc.* 92, 815-826.

McCullagh, P. and Nelder, J. A. (1989). *Generalized Linear Models*, 2nd ed., London; Chapman and Hall.

McCulloch, C. E. (1994), Maximum likelihood variance components estimation for binary data, *J. Amer. Statist. Assoc.* 89, 330-335.

McCulloch, C. E. (1997), Maximum likelihood algorithms for generalized linear mixed models, *J. Amer. Statist. Assoc.* 92, 162-170.

McCulloch, C. E. (2003), *Generalized Linear Mixed Models*, CBMS/IMS Monograph Series, Vol. 7.

McCulloch, C. E. and Searle, S. R. (2000), *Generalized, Linear, and Mixed Models*, Wiley, New York.

McGilchrist, C. A. (1994), Estimation in generalized mixed models, *J. Roy. Statist. Soc. B* 56, 61-69.

Miller, J. J. (1977), Asymptotic properties of maximum likelihood estimates in the mixed model of analysis of variance, *Ann. Statist.* 5, 746-762.

Patterson, H. D. and Thompson, R. (1971), Recovery of interblock information when block sizes are unequal, *Biometrika* 58, 545-554.

Pinhero, J. C. and Bates, D. M. (1995), Approximations to the log-likelihood function in the nonlinear mixed-effects model, *J. Computational & Graphical Statist.* 4, 12-35.

Prasad, N. G. N. and Rao, J. N. K. (1990), The estimation of mean squared errors of small area estimators, *J. Amer. Statist. Assoc.* 85, 163-171.

Prentice, R. L. (1988), Correlated binary regression with covariates specific to each binary observation, *Biometrics* 44, 1033-1048.

Rao, J. N. K. (2003), *Small Area Estimation*, Wiley, New York.

Rao, C. R. (1972), Estimation of variance and covariance components in linear models, *J. Amer. Statist. Assoc.* 67, 112-115.

Rao, C. R. and Kleffe, J. (1988), *Estimation of Variance Components and Applications*, North-Holland, Amsterdam.

Richardson, A. M. and Welsh, A. H. (1994), Asymptotic properties of restricted maximum likelihood (REML) estimates for hierarchical mixed linear models, *Austral. J. Statist.* 36, 31-43.

Robinson, D. L. (1987), Estimation and use of variance components, *The Statistician* 36, 3-14.

Robinson, G. K. (1991), That BLUP is a good thing: The estimation of random effects (with discussion), *Statist. Sci.* 6, 15-51.

Schall, R. (1991), Estimation in generalized linear models with random effects, *Biometrika* 78, 719-727.

Searle, S. R., Casella, G. and McCulloch, C. E. (1992), *Variance Components*, Wiley, New York.

Shaw, R. G. (1987), Maximum-likelihood approaches applied to quantitative genetics of natural populations, *Evolution* 41, 812-826.

Song, P. X-K., Fan, Y. and Kalbfleisch, J. D. (2005), Maximization by parts in likelihood inference (with discussion), *J. Amer. Statist. Assoc.*, in press.

Song, P. X-K. and Jiang, W. (2000), Assessing conditional independence for log-linear Poisson models with random effects, *Commun. Statist. - Theory Meth.* 29, 1233-1245.

Speed, T. P. (1997), Restricted maximum likelihood (REML), *Encyclopedia of Statistical Sciences* 1, 472-481.

Thompson, W. A., Jr. (1962), The problem of negative estimates of variance components, *Ann. Math. Statist.* 33, 273-289.

Tiao, G. C. and Tan, W. Y. (1965), Bayesian analysis of random effects models in the analysis of variance I: posterior distribution of the variance components, *Biometrika* 52, 37-53.

Tiao, G. C. and Tan, W. Y. (1966), Bayesian analysis of random effects models in the analysis of variance II: effect of autocorrelated errors, *Biometrika* 53, 477-495.

Verbyla, A. P. (1990), A conditional derivation of residual maximum likelihood, *Austral. J. Statist.* 32, 227-230.

Vonesh, E. F. and Chinchilli, V. M. (1997), *Linear and Nonlinear Models for the Analysis of Repeated Measurements*, New York, NY: Marcel Dekker.

Walker, S. (1996), An EM algorithm for nonlinear random effects models, *Biometrics* 52, 934-944.

Wolfinger, R. (1993), Laplace's approximation for nonlinear mixed models, *Biometrika* 80, 791-795.

Wolfinger, R. and O'Connell, M. (1993), Generalized linear mixed models: a pseudo-likelihood approach, *J. Statist. Computn Simuln* 48, 233-vi243.

Yuh, L., Beal, S., Davidian, M., Harrison, F., Hester, A., Kowalski, K., Vonesh, E. F. and Wolfinger, R. (1994), Population pharmacokinetic/pharmacodynamic methodology and applications: A bibliography, *Biometrics* 50, 566- 575.

Zeger, S. L. and Karim, M. R. (1991), Generalized linear models with random effects: a Gibbs sampling approach, *J. Amer. Statist. Assoc.* 86, 79-86.

CHAPTER 21

Robust Location and Scatter Estimators in Multivariate Analysis

Yijun Zuo

Department of Statistics and Probability
Michigan State University
East Lansing, MI 48824, USA
zuo@msu.edu

The sample mean vector and the sample covariance matrix are the corner stone of the classical multivariate analysis. They are optimal when the underlying data are normal. They, however, are notorious for being extremely sensitive to outliers and heavy tailed noise data. This article surveys robust alternatives of these classical location and scatter estimators and discusses their applications to the multivariate data analysis.

21.1. Introduction

The sample mean and the sample covariance matrix are the building block of the classical multivariate analysis. They are essential to a number of multivariate data analysis techniques including multivariate analysis of variance, principal component analysis, factor analysis, canonical correlation analysis, discriminant analysis and classification, and clustering. They are optimal (most efficient) estimators of location and scatter parameters at any multivariate normal models. It is well-known, however, that these classical location and scatter estimators are extremely sensitive to unusual observations and susceptible to small perturbations in data. Classical illustrative examples showing their sensitivity are given in Devlin, Gnanadesikan and Kettenring (1981), Huber (1981), Rousseeuw and Leory (1987), and Maronna and Yohai (1998).

Bickel (1964) seems to be the first who considered the robust alternatives of the sample mean vector – the coordinate-wise median and the coordinate-wise Hodges-Lehmann estimator. Extending the univariate trimming and Winsorizing idea of Tukey (1949) and Tukey (1960) to higher dimensions,

Bickel (1965) proposed the metrically trimmed and Winsorized means in the multivariate setting. All these estimators indeed are much more robust (than the sample mean) against outliers and contaminated data (and some are very efficient as well). They, however, lack the desired *affine equivariance* (see Section 21.2.2) property of the sample mean vector.

Huber (1972) discussed a "peeling" procedure for location parameters which was first proposed by Tukey. A similar procedure based on iterative trimming was presented by Gnanadesikan and Kettenring (1972). The resulting location estimators become affine equivariant but little seems to be known about their properties. Hampel (1973) was the first to suggest an affine equivariant iterative procedure for a scatter matrix, which turns out to be a special *M*-estimator (see Section 21.3.1) of the scatter matrix.

Inspired by Huber (1964)'s seminal paper, Maronna (1976) first introduced and treated systematically general *M*-estimators of multivariate location and scatter parameters. Huber (1977) considered the robustness of the covariance matrix estimator with respective to two measures: *influence function* and *breakdown point* (defined in Section 21.2).

Multivariate *M*-estimators are not greatly influenced by small perturbations in a data set and have reasonably good efficiencies over a broad range of population models. Ironically, they, introduced as robust alternatives to the sample mean vector and the sample covariance matrix, were frequently mentioned in the robust statistics literature in the last two decades, not because of their robustness but because of their not being robust enough globally (in terms of their breakdown point). Indeed, *M*-estimators have a relatively very low breakdown point in high dimensions and are not very popular choices of robust estimators of location and scatter parameters in the multivariate setting. Developing *affine equivariant* robust alternatives to the sample mean and the sample covariance matrix that also have *high breakdown points* consequently was one of the fundamental goals of research in robust statistics in the last two decades.

This paper surveys some influential robust location and scatter estimators developed in the last two decades. The list here is by no means exhaustive. Section 21.2 presents some popular robustness measures. Robust location and scatter estimators are reviewed in Section 21.3. Applications of robust estimators are discussed in Section 21.4. Concluding remarks and future research topics are presented in Section 21.5 at the end of the paper.

21.2. Robustness Criteria

Often a statistic T_n can be regarded as a functional $T(\cdot)$ evaluated at an empirical distribution F_n, where F_n is the empirical version of a distribution F based on a random sample $X_1, \cdots, X_n$ from F, which assigns mass $1/n$ to each sample point X_i, $i = 1, \cdots, n$. In the following we describe three most popular robustness measures of functional $T(F)$ or statistic $T(F_n)$.

21.2.1. *Influence function*

One way to measure the robustness of the functional $T(F)$ at a given distribution F is to measure the effect on T when the true distribution slightly deviates from the assumed one F. In his Ph.D thesis, Hampel (1968) explored this robustness and introduced the influence function concept. For a fix point $x \in \mathbb{R}^d$, let δ_x be the point-mass probability measure that assigns mass 1 to the point x. Hampel (1968) and Hampel (1971) defined the *influence function* of the functional $T(\cdot)$ at a fixed point x and the given distribution F as

$$IF(x; T, F) = \lim_{0 < \epsilon \to 0} \frac{T((1 - \epsilon)F + \epsilon\delta_x) - T(F)}{\epsilon}, \qquad (21.1)$$

if the limit exists. That is, the influence function measures the relative effect (influence) on the functional T of an infinitesimal point mass contamination of the distribution F. Clearly, the relative effect (influence) on T is desired to be small or at least bounded. A functional $T(\cdot)$ with a bounded influence function is regarded as robust and desirable.

A straightforward calculation indicates that for the classical mean and covariance functionals $\mu(\cdot)$ and $\Sigma(\cdot)$ at a fixed point x and a given F in $\mathbb{R}^d$,

$$IF(x; \mu, F) = x - \mu(F), \qquad IF(x; \Sigma, F) = (x - \mu)(x - \mu)' - \Sigma(F).$$

Clearly, both influence functions are unbounded with respect to standard vector and matrix norms, respectively. That is, an infinitesimal point mass contamination can have an arbitrarily large influence (effect) on the classical mean and covariance functionals. Hence these functionals are not robust.

The model, $(1 - \epsilon)F + \epsilon\delta_x$, a distribution with a slight departure from the F, is also called the ϵ-*contamination model*. Since only a point-mass contamination is considered in the definition, the influence function measures the *local* robustness of the functional $T(\cdot)$. General discussions and treatments of influence functions of statistical functionals could be found in Serfling (1980), Huber (1981), and Hampel, Ronchetti, Rousseeuw, and Stahel (1986).

In addition to being a measure of local robustness of a functional $T(F)$, the influence function can also be very useful for the calculation of the asymptotic variance of $T(F_n)$. Indeed, if $T(F_n)$ is asymptotically normal, then the asymptotic variance of $T(F_n)$ is just $E(IF(X;T,F))^2$ in general. Furthermore, under some regularity conditions, the following asymptotic representation is obtained in terms of the influence function:

$$T(F_n) - T(F) = \int IF(x;T,F)d(F_n - F)(x) + o_p(n^{-1/2}),$$

which leads to the asymptotic normality of the statistic $T(F_n)$.

21.2.2. *Breakdown point*

The influence function captures the local robustness of a functional $T(\cdot)$. The breakdown point, on the other hand, depicts the *global* robustness of $T(F)$ or $T(F_n)$. Hampel (1968) and Hampel (1971) apparently are the first ones to consider the breakdown point of $T(F)$ in an asymptotic sense.

Donoho and Huber (1983) considered a finite sample version of the notion, which since then has become the most popular quantitative measure of global robustness of an estimator $T_n = T(F_n)$, largely due to its intuitive appeal, non-probabilistic nature of the definition, and easy calculation in many cases. Roughly speaking, the finite sample breakdown point of an estimator T_n is the minimum fraction of "bad" (or contaminated) data points in a data set $X^n = \{X_1, \cdots, X_n\}$ that can render the estimator useless. More precisely, Donoho and Huber (1983) defined the finite sample *breakdown point* of a *location* estimator $T(X^n) = T(F_n)$ as

$$BP(T; X^n) = \min\{\frac{m}{n} : \sup_{X_m^n} |T(X_m^n) - T(X^n)| = \infty\}, \qquad (21.2)$$

where X_m^n is a contaminated data set resulting from replacing (contaminating) m points of X^n with arbitrary m points in $\mathbb{R}^d$. The above notion sometimes is called *replacement* breakdown point. Donoho (1982) and Donoho and Huber (1983) also considered *addition* breakdown point. The two versions, however, are actually interconnected quantitatively; see Zuo (2001). Thus we focus on the replacement version throughout in this paper.

For a *scatter* (or covariance) estimator S of the matrix Σ in the probability density function $f((x-\mu)'\Sigma^{-1}(x-\mu))$, to define its breakdown point one can still use (21.2) but with T on the left side replaced by S and $T(\cdot)$ on the right side by the vector of the logarithms of the eigenvalues of $S(\cdot)$. Note that for a location estimator, it becomes useless if it approaches ∞.

On the other hand, for a scatter estimator, it becomes useless if one of its eigenvalues approaches 0 or ∞ (this is why we use the logarithm).

Clearly, the higher the breakdown point of an estimator, the more robust the estimator against outliers (or contaminated data points). It is not difficult to see that one bad (or contaminating one) point of a data set of size n is enough to ruin the sample mean or the sample covariance matrix. Thus, their breakdown point is $1/n$, the lowest possible value. That is, the sample mean vector and the sample covariance matrix are not robust globally (and locally as well due to the unbounded influence functions).

On the other hand, to have the sample median (Med) breakdown (unbounded), one has to move 50% of data points to the infinity. Precisely, the univariate median has a breakdown point $\lfloor (n+1)/2 \rfloor /n$ for any data set of size n, where $\lfloor x \rfloor$ is the largest integer no larger that x. Likewise, it can be seen that the median of the absolute deviations (from the median) (MAD) has a breakdown point $\lfloor n/2 \rfloor /n$ for a data set with no overlapping data points. These breakdown point results turn out to be the best for any *reasonable* location and covariance (or scale) estimators, respectively. Note that the breakdown point of a constant estimator is 1 but the estimator is not reasonable since it lacks some equivariance property.

Location and scatter estimators T and S are called *affine equivariant* if

$$T(AX^n + b) = A \cdot T(X^n) + b, \quad S(AX^n + b) = A \cdot S(X^n) \cdot A', \quad (21.3)$$

respectively, for any $d \times d$ non-singular matrix A and any vector $b \in \mathbb{R}^d$, where $AX^n + b = \{AX_1 + b, \cdots, AX_n + b\}$. They are called *rigid-body* or *translation* equivariant if (21.3) holds for any orthogonal A or any identity A $(A = I_d)$, respectively. When $b = 0$ and $A = sI_d$ for a scalar $s \neq 0$, T and S are called *scale* equivariant. The following breakdown point upper bound results are due to Donoho (1982). We provide here a much simpler proof.

Lemma 1: *For any translation (scale) equivariant location (scatter) estimator T (S) at any sample X^n in $\mathbb{R}^d$, $BP(T(S), X^n) \leq \lfloor (n+1)/2 \rfloor /n$.*

Proof It suffices to consider the location case. For $m = \lfloor (n+1)/2 \rfloor$ and $b \in \mathbb{R}^d$, let $Y_m^n = \{X_1 + b, \cdots, X_m + b, X_{m+1}, \cdots, X_n\}$. Both Y_m^n and $Y_m^n - b$ are data sets resulting from contaminating at most m points of X^n. Observe

$$\|b\| = \|T(Y_m^n) - T(Y_m^n - b)\| \leq \sup_{X_m^n} 2 \cdot \|T(X_m^n) - T(X^n)\| \to \infty \text{ as } \|b\| \to \infty.$$

Here (and hereafter) $\| \cdot \|$ is the Euclidean norm for a vector and $\|A\| = \sup_{\|u\|=1} \|Au\|$ for a matrix A.

The coordinate-wise and the L_1 (also called *spatial*) medians are two known location estimators that can attain the breakdown point upper bound in the lemma; see Lopuhaä and Rousseeuw (1991), for example. Both estimators, however, are not affine equivariant (the first is only translation and the second is just rigid-body equivariant). On the other hand, no scatter matrices constructed can reach the upper bound in the lemma. In fact, for affine equivariant scatter estimators and for data set X^n in a *general position* (that is, no more than d data points lie in the same $d-1$ dimensional hyperplane), Davies (1987) provided a negative answer and proved the following breakdown point upper bound result.

Lemma 2: *For any affine equivariant scatter estimator S and data set X^n in general position in $\mathbb{R}^d$, $BP(S, X^n) \leq \lfloor (n-d+1)/2 \rfloor /n$.*

MAD is a univariate affine equivariant scale estimator that attains the upper bound in this lemma. Higher dimensional affine equivariant *scatter* estimators that reach this upper bound have been proposed in the literature. The following questions about *location* estimators, however, remain open:

(1) Is there any affine equivariant location estimator in high dimensions that can attain the breakdown point upper bound in Lemma 1? If not,
(2) What is the breakdown point upper bound of an affine equivalent location estimator?

A partial answer to the first question is given in Zuo (2004a) where under a slightly narrow definition of the finite sample breakdown point a location estimator attaining the upper bound in Lemma 1 is introduced.

21.2.3. *Maximum bias*

The point-mass contamination in the definition of influence function is very special. In practice, a deviation from the assumed distribution can be due to the contamination of any distribution. The influence function consequently measures a special *local* robustness of a functional $T(\cdot)$ at F. A very broad measure of *global* robustness of $T(\cdot)$ at F is the so-called maximum bias; see Huber (1964) and Huber (1981). Here any possible contaminating distribution G and the contaminated model $(1-\epsilon)F + \epsilon G$ are considered for a fixed $\epsilon > 0$ and the *maximum bias* of $T(\cdot)$ at F is defined as

$$B(\epsilon; T, F) = \sup_G \|T((1-\epsilon)F + \epsilon G) - T(F)\|. \qquad (21.4)$$

$B(\epsilon; T, F)$ measures the worst case bias due to an ϵ amount contamination of the assumed distribution. $T(\cdot)$ is regarded as robust if it has a moderate maximum bias curve for small ϵ. It is seen that the standard mean and covariance functionals have an unbounded maximum bias for any $\epsilon > 0$ and hence are not robust in terms of this maximum bias measure.

The minimum contamination amount ϵ^* that can lead to an unbounded maximum bias is called the *asymptotic breakdown point* of T at F. Its finite sample version is exactly the one given by (21.2). On the other hand, if G is restricted to a point-mass contamination, then the rate of the change of $B(\epsilon; T, F)$ relative to ϵ, for ϵ arbitrarily small, is closely related to $IF(x; T, F)$. Indeed the "slope" of the maximum bias curve $B(\epsilon; T, F)$ at $\epsilon = 0$ is often the same as the supremum (over x) of $\|IF(x; T, F)\|$. Thus, the maximum bias really depicts the entire picture of the robustness of the functional T whereas the influence function and the breakdown point serve for two extreme cases. Though a very important robustness measure, the challenging derivation of $B(\epsilon; T, F)$ for a location or scatter functional T in high dimensions makes the maximum bias a less popular one than the influence function and the finite sample breakdown point in the literature.

To end this section, we remark that robustness is one of the most important performance criteria of a statistical procedure. There are, however, other important performance criteria. For example, efficiency is always a very important performance measure for any statistical procedure. In his seminal paper, Huber (1964) took into account both the robustness and the efficiency (in terms of the asymptotic variance) issues in the famous "minimax" (minimizing worst case asymptotic variance) approach. Robust estimators are commonly not very efficient. The univariate median serves as a perfect example. It is the most robust affine equivariant location estimator with the best breakdown point and the lowest maximum bias at symmetric distributions (see Huber (1964)). Yet for its best robustness, it has to pay the price of low efficiencies relative to the mean at normal and other light-tailed models. In our following discussion about the robustness of location and scatter estimators, we will also address the efficiency issue.

21.3. Robust Multivariate Location and Scatter Estimators

This section surveys important affine equivariant robust location and scatter estimators in high dimensions. The efficiency issue will be addressed.

21.3.1. *M-estimators and variants*

As pointed out in Section 21.1, affine equivariant M-estimators of location and scatter parameters were the early robust alternatives to the classical sample mean vector and sample covariance matrix. Extending Huber (1964)'s idea of the univariate M-estimators as minimizers of objective functions, Maronna (1976) defined multivariate M-estimators as the solutions T (in $\mathbb{R}^d$) and V (a positive definite symmetric matrix) of

$$\frac{1}{n}\sum_{i=1}^{n} u_1(((X_i - T)'V^{-1}(X_i - T))^{1/2})(X_i - T) = 0, \qquad (21.5)$$

$$\frac{1}{n}\sum_{i=1}^{n} u_2((X_i - T)'V^{-1}(X_i - T))(X_i - T)(X_i - T)' = V, \qquad (21.6)$$

where u_i, $i = 1, 2$, are weight functions satisfying some conditions. They are a generalization of the *maximum likelihood estimators* and can be regarded as weighted mean and covariance matrix as well. Maronna (1976) discussed the existence, uniqueness, consistency, asymptotic normality, influence function and breakdown point of estimators. Though possessing bounded influence functions for suitable u_i's, $i = 1, 2$, T and V have relatively low breakdown points ($\leq 1/(d+1)$) (see, Maronna (1976) and p. 226 of Huber (1981)) and hence are not robust globally in high dimensions. The latter makes the M-estimators less appealing choices in robust statistics, though they can be quite efficient at normal and other models.

Tyler (1994) considered some sufficient conditions for the existence and uniqueness of M-estimators with special redescending weight functions. Constrained M-estimators, which combine both good local and good global robustness properties, are considered in Kent and Tyler (1994).

21.3.2. *Stahel-Donoho estimators and variants*

Stahel (1981) and Donoho (1982) "outlyingness" weighted mean and covariance matrix appear to be the first location and scatter estimators in high dimensions that can integrate affine equivariance with high breakdown points. In $\mathbb{R}^1$, the outlyingness of a point x with respect to (w.r.t.) a data set $X^n = \{X_1, \cdots, X_n\}$ is simply $|x - \mu(X^n)|/\sigma(X^n)$, the absolute deviation of x to the center of X^n standardized by the scale of X^n. Here μ and σ are univariate location and scale estimators with typical choices including (mean, standard deviation), (median, median absolute deviation), and more generally, univariate M-estimators of location and scale (see Huber

(1964) and Huber (1981)). Mosteller and Tukey (1997) (p. 205) introduced an outlyingness weighted mean in $\mathbb{R}^1$. Stahel and Donoho (SD) considered a multivariate analog and defined the outlyingness of a point x w.r.t. X^n in $\mathbb{R}^d$ $(d \geq 1)$ as

$$O(x, X^n) = \sup_{\{u:\ u \in R^d,\ \|u\|=1\}} |u'x - \mu(u \cdot X^n)|/\sigma(u \cdot X^n) \qquad (21.7)$$

where $u'x = \sum_{i=1}^{d} u_i x_i$ and $u \cdot X^n = \{u'X_1, \cdots, u'X_n\}$. If $u'x - \mu(u \cdot X^n) = \sigma(u \cdot X^n) = 0$, we define $|u'x - \mu(u \cdot X^n)|/\sigma(u \cdot X^n) = 0$. Then

$$T_{\mathrm{SD}}(X^n) = \sum_{i=1}^{n} w_i X_i \Big/ \sum_{i=1}^{n} w_i, \qquad (21.8)$$

$$S_{\mathrm{SD}}(X^n) = \sum_{i=1}^{n} w_i (X_i - T_{\mathrm{SD}}(X^n))(X_i - T_{\mathrm{SD}}(X^n))' \Big/ \sum_{i=1}^{n} w_i \qquad (21.9)$$

are the SD outlyingness weighted mean and covariance matrix, where $w_i = w(O(X_i, X^n))$ and w is a weight function down-weighting outlying points.

Since μ and σ^2 are usually affine equivariant, $O(x, X^n)$ is then *affine invariant*: $O(x, X^n) = O(Ax+b, AX^n+b)$ for any non-singular $d \times d$ matrix A and vector $b \in \mathbb{R}^d$. It follows that T_{SD} and S_{SD} are affine equivariant.

Stahel (1981) considered the asymptotic breakdown point of the estimators. Donoho (1982) derived the finite sample breakdown point for (μ, σ) being median (Med) and median absolute deviation (MAD), for X in a general position, and for suitable weight function w. His result, expressed in terms of *addition* breakdown point, amounts to (see, e.g., Zuo (2001))

$$BP(T_{\mathrm{SD}}, X^n) = \frac{\lfloor (n - 2d + 2)/2 \rfloor}{n}, \quad BP(S_{\mathrm{SD}}, X^n) = \frac{\lfloor (n - 2d + 2)/2 \rfloor}{n}.$$

Clearly, BPs of the SD estimators depend essentially on the BP of MAD (since Med already provides the best possible BP). As a scale estimator, MAD breaks down (*explosively* or *implosively*) as it tends to ∞ or 0. Realizing that it is easier to implode MAD with a projected data set $u \cdot X^n$ for X^n in high dimension (since there will be d overlapping projected points along some projection directions), Tyler (1994), Gather and Hilker (1997), and Zuo (2000) all modified MAD to get a higher BP of the SD estimators:

$$BP(T_{\mathrm{SD}}^*, X^n) = \lfloor (n - d + 1)/2 \rfloor/n, \quad BP(S_{\mathrm{SD}}^*, X^n) = \lfloor (n - d + 1)/2 \rfloor/n.$$

Note that the latter is the best possible BP result for S_{SD} by Lemma 2.

The SD estimators stimulated tremendous researches in robust statistics. Seeking *affine equivariant* estimators with *high BPs* indeed was one primary goal in the field in the last two decades. The asymptotic behavior of the SD estimators, however, was a long-standing problem. This hindered the estimators from becoming more popular in practice. Maronna and Yohai (1995) first proved the $\sqrt{n}$-consistency. Establishing the limiting distributions, however, turned out to be extremely challenging. Indeed, there once were doubts in the literature about the existence or the normality of their limit distributions; see, e.g., Lopuhaä (1999) and Gervini (2003).

Zuo, Cui and He (2004) and Zuo and Cui (2005) studied general data depth weighted estimators, which include the SD estimators as special cases, and established a general asymptotic theory. The asymptotic normality of the SD estimators thus follows as a special case from the general results there. The robustness studies of the general data depth induced estimators carried out in Zuo, Cui and He (2004) and Zuo and Cui (2005) also show that the SD estimators have bounded influence functions and moderate maximum bias curves for suitable weight functions. Furthermore, with suitable weight functions, the SD estimators can outperform most leading competitors in the literature in terms of robustness and efficiency.

21.3.3. *MVE and MCD estimators and variants*

Rousseeuw (1985) introduced affine equivariant *minimum volume ellipsoid* (MVE) and *minimum covariance determinant* (MCD) estimators as follows. The MVE estimators of location and scatter are respectively the center and the ellipsoid of the minimum volume ellipsoid containing (at least) h data points of X^n. It turns out that the MVE estimators can possess a very high breakdown point with a suitable h $(= \lfloor (n + d + 1)/2 \rfloor)$ (Davies (1987)). They, however, are neither asymptotically normal nor $\sqrt{n}$ consistent (Davies (1992a)) and hence are not very appealing in practice. The MCD estimators are the mean and the covariance matrix of h data points of X^n for which the determinant of the covariance matrix is minimum. Again with $h = \lfloor (n + d + 1)/2 \rfloor$, the breakdown point of the estimators can be as high as $\lfloor (n - d + 1)/2 \rfloor / n$, the best possible BP result for any affine equivariant *scatter* estimator by Lemma 2; see Davies (1987) and Lopuhaä and Rousseeuw (1991). The MCD estimators have bounded influence functions that have jumps (Croux and Haesbroeck (1999)). The estimators are $\sqrt{n}$-consistent (Butler, Davies and Jhun (1993)) and the asymptotical normality is also established for the *location* part but not for the scatter part (Butler, Davies and Jhun (1993)). The estimators are not very efficient at normal models and this is especially true at the h selected in order for the estimators to have a high breakdown point; see Croux and Haesbroeck (1999). In spite of their low efficiency, the MCD estimators are quite popular in the literature, partly due to the availability of fast computing algorithms of the estimators (see, e.g., Hawkins (1994) and Rousseeuw and Van Driessen (1999)).

To overcome the low efficiency drawback of the MCD estimators, reweighted MCD estimators were introduced and studied; see Lopuhaä and Rousseeuw (1991), Lopuhaä (1999), and Croux and Haesbroeck (1999).

21.3.4. *S-estimators and variants*

Davies (1987) introduced and studied S-estimators for multivariate location and scatter parameters, extending an earlier idea of Rousseeuw and Yohai (1984) in regression context to the location and scatter setting. Employing a *smooth* ρ function, the S-estimators extend Rousseeuw's MVE estimators which are special S-estimators with a non-smooth ρ function. The estimators become $\sqrt{n}$-consistent and asymptotically normal. Furthermore they can have a very high breakdown point $\lfloor (n - d + 1)/2 \rfloor / n$, again the upper bound for any affine equivariant *scatter* estimator; see Davies (1987). The S-estimators of location and scatter are defined as the vector T_n and the positive definite symmetric (PDS) matrix C_n which minimize the determinant of C_n, $\det(C_n)$, subject to

$$\frac{1}{n} \sum_{i=1}^{n} \rho\Big(((X_i - T_n)C_n^{-1}(X_i - T_n))^{1/2} \Big) \leq b_0, \qquad (21.10)$$

where the non-negative function ρ is symmetric and continuously differentiable and strictly increasing on $[0, c_0]$ with $\rho(0) = 0$ and constant on $[c_0, \infty)$ for some $c_0 > 0$ and $b_0 < a_0 = \sup \rho$. As shown in Lopuhaä (1989), S-estimators have a close connection with M-estimators and have bounded influence functions. They can be highly efficient at normal models; see Lopuhaä (1989) and Rocke (1996). The latter author, however, pointed out that there can be problems with the breakdown point of the S-estimators in high dimensions and provided remedial measures. Another drawback is that the S-estimators can not simultaneously attain a high breakdown point and a given efficiency at the normal models. Modified estimators that can overcome the drawback were given in Lopuhaä (1991), Lopuhaä (1992) and Davies (1992b). The S-estimators can be computed with a fast algorithm such as the one given in Ruppert (1992).

21.3.5. *Depth weighted and maximum depth estimators*

Data depth has recently been increasingly pursued as a promising tool in multi-dimensional exploratory data analysis and inference. The key idea of data depth in the location setting is to provide a center-outward ordering of multi-dimensional observations. Points deep inside a data cloud receive high depth and those on the outskirts get lower depth. Multi-dimensional points then can be ordered based on their depth. Prevailing notions of data depth include Tukey (1975) halfspace depth, Liu (1990) simplicial depth and projection depth (Liu (1992), Zuo and Serfling (2000a) and Zuo (2003)). All these depth functions satisfy desirable properties for a general depth functions; see, e.g., Zuo and Serfling (2000b). Data depth has found applications to nonparametric and robust multivariate analysis. In the following we focus on the application to multivariate location and scatter estimators.

For a give sample X^n from a distribution F, let F_n be the empirical version of F based on X^n. For a general depth function $D(\cdot, \cdot)$ in $\mathbb{R}^d$, depth-weighted location and scatter estimators can be defined as

$$L(F_n) = \frac{\int x w_1(D(x, F_n)) dF_n(x)}{\int w_1(D(x, F_n)) dF_n(x)}, \tag{21.11}$$

$$S(F_n) = \frac{\int (x - L(F_n))(x - L(F_n))' w_2(D(x, F_n)) dF_n(x)}{\int w_2(D(x, F_n)) dF_n(x)}, \tag{21.12}$$

where w_1 and w_2 are suitable weight functions and can be different; see Zuo, Cui and He (2004) and Zuo and Cui (2005). These depth-weighted estimators can be regarded as generalizations of the univariate L-statistics. A similar idea is first discussed in Liu (1990) and Liu, Parelius and Singh (1999), where the depth-induced location estimators are called *DL-statistics*. Note that equations (21.11) and (21.12) include as special cases depth trimmed and Winsorized multivariate means and covariance matrices; see Zuo (2004b) for related discussions. With the projection depth (PD) as the underlying depth function, these equations lead to as special cases the Stahel-Donoho location and scatter estimators, where the projection depth is defined as

$$PD(x, F_n) = 1/(1 + O(x, F_n)), \tag{21.13}$$

where $O(x, F_n)$ is defined in (21.7). Replacing F_n with its population version F in (21.11), (21.12) and (21.13), we obtain population versions of above definitions.

Common depth functions are affine invariant. Hence $L(F_n)$ and $S(F_n)$ are affine equivariant. They are unbiased estimators of the center θ of symmetry of a symmetric F of X (i.e., $\pm(X - \theta)$ have the same distribution) and of the covariance matrix of an elliptically symmetric F, respectively; see Zuo, Cui and He (2004) and Zuo and Cui (2005). Under mild assumptions on w_1 and w_2 and for common depth functions, $L(F_n)$ and $S(F_n)$ are strongly consistent and asymptotically normal. They are locally robust with bounded influence functions and globally robust with moderate maximum biases and very high breakdown points. Furthermore, they can be extremely efficient at normal and other models. For details, see Zuo, Cui and He (2004) and Zuo and Cui (2005).

General depth weighted location and scatter estimators include as special cases the re-weighted estimators of Lopuhaä (1999) and Gervini (2003), where Mahalanobis type depth (see Liu (1992)) is utilized in the weight calculation of sample points. With appropriate choices of weight functions, the re-weighted estimators can possess desirable efficiency and robustness properties. Since Mahalanobis depth entails some initial location and scatter estimators, the performance of the re-weighted estimators depends crucially on the initial choices in both finite and large sample sense, though.

Another type of depth induced estimators is the maximum depth estimators, which could be regarded as an extension of the univariate median type estimators to the multivariate setting. For a given location depth function $D_L(\cdot, \cdot)$ and scatter depth function $D_S(\cdot, \cdot)$ and a sample X^n (or equivalently F_n), maximum depth estimators can be defined as

$$MDL\,(F_n) = \arg\sup_{x \in \mathbb{R}^d} D_L(x, F_n) \tag{21.14}$$

$$MDS\,(F_n) = \arg\sup_{\Sigma \in \mathcal{M}} D_S(\Sigma, F_n), \tag{21.15}$$

where $\mathcal{M}$ is the set of all positive definite $d \times d$ symmetric matrices. Aforementioned depth notions are all location depth functions. An example of the scatter depth function, given in Zuo (2004b), is defined as follows. For a given univariate scale measure σ, define the outlyingness of a matrix $\Sigma \in \mathcal{M}$ with respect to F_n (or sample X^n) as

$$O(\Sigma, F_n) = \sup_{u \in S^{d-1}} g\big(\sigma^2(u \cdot X^n)/u'\Sigma u\big), \tag{21.16}$$

where g is a nonnegative function on $[0, \infty)$ with $g(0) = \infty$ and $g(\infty) = \infty$; see, e.g., Maronna, Stahel and Yohai (1992) and Tyler (1994). The (projection) depth of a scatter matrix $\Sigma \in \mathcal{M}$ then can be defined as (Zuo

(2004b))

$$D_S(\Sigma, F_n) = 1/(1 + O(\Sigma, F_n)).\tag{21.17}$$

A scatter depth defined in the same spirit was first given in Zhang (2002).

The literature is replete with discussions on location depth D_L and its induced deepest estimator $MDL(F_n)$; see, e.g., Liu (1990), Liu, Parelius and Singh (1999),Zuo and Serfling (2000a), Arcones, Chen and Giné (1994), Bai and He (1999), Zuo (2003) and Zuo, Cui and He (2004). There are, however, very few discussions on scatter depth D_S and its induced deepest estimator $MDS(F_n)$ (exceptions are made in Maronna, Stahel and Yohai (1992), Tyler (1994), Zhang (2002), and Zuo (2004b) though). Further studies on D_S and MDS such as robustness, asymptotics, efficiency, and inference procedures are called for.

Maximum depth estimators tend to be highly robust locally and globally as well. Indeed, the maximum projection depth estimators of location have bounded influence functions and moderately maximum biases; see Zuo, Cui and Young (2004). Figure 21.1 clearly reveals the boundedness of the influence functions of the maximum projection depth estimator (PM) (and the projection depth weighted mean (PWM)) with Med and MAD for μ and σ.

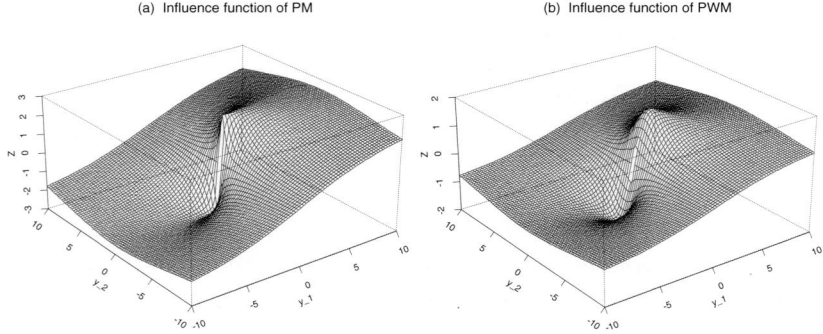

Figure 21.1. (a) The first coordinate of the influence function of maximum projection depth estimator of location (projection median (PM)). (b) The first coordinate of the influence function of the projection depth weighted mean (PWM).

Maximum depth estimators can also possess high breakdown points. For example, both the maximum projection depth estimators of location and scatter can possess the highest breakdown points among their competitors,

$$BP(MDL, X^n) = \frac{\lfloor (n-d+2)/2 \rfloor}{n}, \quad BP(MDS, X^n) = \frac{\lfloor (n-d+1)/2 \rfloor}{n},$$

where PD is the depth function with Med and a modified version of MAD as μ and σ in its definition; see Zuo (2003) and Tyler (1994). Maximum depth estimators can also be highly efficient. For example, with appropriate choices of μ and σ, the maximum projection depth estimator of location can be highly efficient; see Zuo (2003) for details.

21.4. Applications

Robust location and scatter estimators find numerous applications to multivariate data analysis and inference. In the following we survey some major applications including robust Hotelling's T^2 statistic, robust multivariate control charts, robust principal component analysis, robust factor analysis, robust canonical correlation analysis and robust discrimination and clustering. We skip the application to the multivariate regression (see, e.g., Croux and Dehon (2001), Croux, Aelst and Dehon (2003) and Rousseeuw, Van Aelst, Van Driessen and Agulló (2004) for related studies).

21.4.1. *Robust T^2 and control charts*

Hotelling's T^2 statistic: $n(\bar{X} - E(X))S^{-1}(\bar{X} - E(X))$ is the single most fundamental statistic in the classical inference about the multivariate mean vectors of populations as well as in the classical multivariate analysis of variance. It is also the statistic for the classical multivariate quality control charts. Built on the sample mean $\bar{X}$ and the sample covariance matrix S, T^2, unfortunately, is not robust. The T^2 based procedures also depend heavily on the normality assumption.

A simple and intuitive way to robustify the Hotelling's T^2 statistic is to replace $\bar{X}$ and S with robust location and scatter estimators, respectively. An example was given in Willems, Pison, Rousseeuw and Van Aelst (2002), where re-weighted MCD estimators were used instead of the mean and the covariance matrix. A major issue here is the (asymptotic) distribution of the robust version of T^2 statistic. Based on the multivariate sign and sign-rank tests of Randles (1989), Peters and Randles (1991) and Hettmansperger et al. (1994), robust control charts are constructed by Ajman and Vining (1998) and Ajman, Randles and Vining (1998).

Another approach to construct robust multivariate quality charts is via data depth. Here a quality index is introduced based on the depth of points and the multivariate processes are monitored based on the index. Representative studies include Liu (1995) and Liu and Singh (1993). Others include Ajman, Randles, Vining and Woodall (1997) and Stoumbos and Allison (2000).

Finally, the projection (depth) pursuit idea has also been employed to construct multivariate control charts; see, e.g., Ngai and Zhang (2001).

21.4.2. *Robust principal component analysis*

Classical principal component analysis (PCA) is carried out based on the eigenvectors (eigenvalues) of the sample covariance (or correlation) matrix. Such analysis is extremely sensitive to outlying observations and the conclusions drawn based on the principal components may be adversely affected by the outliers and misleading. A most simple and appealing way to robustify the classical PCA is to replace the matrix with a robust scatter estimator. Robust PCA studies started in as early as 1970s and include Maronna (1976), Campbell (1980) and Devlin, Gnanadesikan and Kettenring (1981), where M-estimators of location and scatter were utilized instead of the sample mean and covariance matrix. Some recent robust PCA studies focus on the investigation of the influence function of the eigenvectors and eigenvalues; see, e.g., Jaupi and Saporta (1993), Shi (1997) and Croux and Haesbroeck (2000).

A different approach to robust PCA uses projection pursuit (PP) techniques; see Li and Chen (1985), Croux and Ruiz-Gazen (1996) and Hubert Rousseeuw and Verboven (2002) . It seeks to maximize a robust measure of spread to obtain consecutive directions along which the data points are projected. This idea has been generalized to common principal components in Boente, Pires, and Rodrigues (2002).

Recently, Hubert Rousseeuw and Vanden Branden (2005) combined the advantages of the above two approaches and proposed a new method to robust PCA where the PP part is used for the initial dimension reduction and then the ideas of robust scatter estimators are applied to this lower-dimensional data space.

21.4.3. *Robust factor analysis*

The classical factor analysis (FA) starts with the usual sample covariance (or correlation) matrix and then the eigenvectors and eigenvalues of the matrix are employed for estimating the loading matrix (or the matrix is used in the likelihood equation to obtain the maximum likelihood estimates of the loading matrix and specific variances). The analysis, however, is not robust since outliers can have a large effect on the covariance (or correlation matrix) and the results obtained may be misleading or unreliable.

A straightforward approach to robustify the classical FA is to replace the sample covariance (or correlation) matrix with a robust one. One such example was given in Pison, Rousseeuw, Filzmoser and Croux (2003) where MCD estimators were employed. Further systematic studies on robust FA such as robustness, efficiency and performance, and inference procedures are yet to be conducted.

21.4.4. *Robust canonical correlation analysis*

The classical canonical correlation analysis (CCA) seeks to identify and quantify the associations between two sets of variables. It focuses on the correlation between a linear combination of the variables in one set and a linear combination of the variables in another set. The idea is to determine first the pair of linear combinations having the largest correlation, then the next pair of linear combinations having the largest correlation among all pairs uncorrelated with the previous selected pair, and so on. In practice, sample covariance (or correlation) matrix is utilized to achieve the goal. The result obtained, however, is not robust to outliers in the data since the sample covariance (or correlation) matrix is extremely sensitive to unusual observations. To robustify the classical approach, Karnel (1991) proposed to use M-estimators and Croux and Dehon (2002) the MCD estimators. The latter paper also studied the influence functions of canonical correlations and vectors. Robustness and asymptotics of robust CCA were discussed in Taskinen, Croux, Kankainen, Ollila and Oja (2005). More studies on robust CCA are yet to be seen.

21.4.5. *Robust discrimination, classification and clustering*

In the classical discriminant analysis and classification, the sample mean and the sample covariance matrix are often used to build discriminant rules which however are very sensitive to outliers in data. Robust rules can be obtained by inserting robust estimators of location and scatter into the classical procedures. Croux and Dehon (2001) employed S-estimators to carry out a robust linear discriminant analysis. A robustness issue related to the quadratic discriminant analysis is addressed by Croux and Joossens recently. He and Fung (2000) discussed the high breakdown estimation and applications in discriminant analysis. Huber and Van Driessen (2004) discussed fast and robust discriminant analysis based on MCD estimators.

In the classical clustering methods, the sample mean and the sample covariance matrix likewise are often employed to build clustering rules. Robust estimators of location and scatter could be used to replace the mean vector and the covariance matrix to obtain robust clustering rules. References on robust clustering methods include Kaufman and Rousseeuw (1990) Kaufman and Rousseeuw (1990). Robust clustering analysis is a very active research area of computer scientists; see, e.g., Drost and Klaassen (1997) and Fred and Jain (2003) and references therein. More studied on clustering analysis from statistical perspective with robust location and scatter estimators are needed.

21.5. Conclusions and Future Works

Simulation studied by Maronna and Yohai (1995), Gervini (2002), Zuo, Cui and He (2004), Zuo, Cui and Young (2004) and Zuo and Cui (2005) indicate that the projection depth weighted mean and covariance matrix (the Stahel-Donoho estimators) with suitable weight functions can outperform most of its competitors in terms of local and global robustness as well as efficiency at a number of distribution models. We thus recommend the Stahel-Donoho estimators and more generally projection depth weighted mean and covariance matrix as favorite choices of robust location and scatter estimators. Maximum depth estimators of location and scatter are strong competitors, especially from robustness view point. They (especially maximum depth scatter estimators) deserve further attention and investigations.

Computing high breakdown point robust affine equivariant location and scatter estimators is always a challenging task and there is no exception for the projection depth related estimators. Recent studies of this author, however, indicate that some of these estimators can be computed exactly in two and higher dimensions for robust μ and σ such as Med and MAD. Though fast approximate algorithms for computing these estimators already exist for moderately high dimensional data, issues involving the computing of these depth estimators such as how accurate and how robust are the approximate algorithms are yet to be addressed.

At this point, all applications of robust location and scatter estimators to multivariate data analysis are centered around the MCD based procedures. Since MCD estimators are not very efficient and can sometime have unstable behavior, we thus recommend replacing MCD estimators with the projection depth weighted estimators and expect that more reliable and efficient procedures are to be obtained. Asymptotic theory involving the robust multivariate analysis procedures is yet to be established.

Finally we comment that data depth is a natural tool for robust multivariate data analysis and more researches along this direction which can lead to very fast, robust, and efficient procedures are needed.

Acknowledgments

The author thanks Professors Hira Koul and Catherine Dehon for their insightful and constructive comments and suggestions. The work is partly supported by NSF grant DMS-0234078.

References

Ajmani, V. and Vining, G. (1998). A ronust multivariate exponentially weighted moving average chart. Preprint.

Ajmani, V., Randles, R. and Vining G. (1998). Robust multivariate control charts. Preprint.

Ajmani, V., Randles, R., Vining, G. and Woodall, W. (1997). Robustness of multivariate control charts. In *Proceedings of the Section on Quality and Productivity, 1997 Quality and Productivity Reserach Conference*, American Statistical Association, Alexandria, VA. 152-157.

Arcones, M. A., Chen, Z. and Giné, E. (1994). Estimators related to U-processes with applications to multivariate medians: asymptotic normality. *Ann. Statist.* **22** 1460-1477.

Bai, Z. D. and He, X. (1999). Asymptotic distributions of the maximum depth estimators for regression and multivariate location. *Ann. Statist.* **27** 1616-1637.

Bickel, P. J. (1964). On some alternative estimates for shift in the p-variate one sample problem. *Ann. Math. Statist.* **35** 1079-1090.

Bickel, P. J. (1965). On some robust estimates of location. *Ann. Math. Statist.* **36** 847-858.

Boente, G, Pires, A. M. and Rodrigues, I. (2002). Influence functions and outlier detection under the common principal components model: a robust approach. *Biometrika* **89** 861-875.

Butler, R. M, Davies, P. L. and Jhun, M. (1993). Asymptotics for the minimum covariance determinant estimator. *Ann. Statist.* **21** 1385-1400.

Campbell, N. A. (1980). Robust procedures in multivariate analysis: robust covariance estimation. *Appl. Statist.* **29** 231-237.

Croux, C. and Dehon, C. (2001). Robust linear discriminant analysis using S-estimators. *Canad. J. Statist.* **29** 473-492.

Croux, C. and Dehon, C. (2002). Analyse canonique Basée sur des estimateurs robustes de la matrice de covariance. *La Revue de Statistique Appliquee* **2** 5-26

Croux, C. and Dehon, C., Van Aelst, S.,and Rousseeuw, P. J. (2001). Robust Estimation of the Conditional Median Function at Elliptical Models. *Statist. Probab. Letters* **51** 361-368.

Croux, C. and Haesbroeck, G. (1999). Influence function and efficiency of the minimum covariance determinant scatter matrix estimator. *J. Multivariate Anal.* **71** 161-190.

Croux, C. and Haesbroeck, G. (2000). Principal compoment analysis based on robust estimators of the covariance or correlation matrix: influence functions and efficiencies. *Biometrika* **87** 603-618.

Croux, C. and Joossens, K. Influence of Observations on the Misclassification Probability in Quadratic Discriminant Analysis *J. Multivariate Anal.* (to appear).

Croux, C. and Ruiz-Gazen, A. (1996). A fast algorithm for robust principal components based on projection pursuit. In *COMPSTAT 1996, Proceedings in Computational Statistics* (Ed A. Prat). Heidelberg, Physica-Verlag. 211-217.

Croux, C., Van Aelst, S. and Dehon, C. (2003). Bounded Influence Regression using High Breakdown Scatter Matrices. *Ann. Instit. Statist. Math.* **55** 265-285.

Dav, R. N. and Krishnapuram, R. (1997). Robust clustering methods: A unified view. *IEEE Trans. Fuzzy Syst.* **5** 270-293.

Davies, P. L. (1987). Asymptotic behavior of S-estimators of multivariate location parameters and dispersion matrices. *Ann. Statist.* **15** 1269-1292.

Davies, P. L. (1992a). The asymptotics of Rousseeuw's minimum volume ellipsoid estimator. *Ann. Statist.* **20** 1828-1843.

Davies, P. L.. An efficient Fréchet-differentiable high breakdown multivariate location and dispersion estimator. *J. Multivariate Anal.* **40** 311-327.

Devlin, S. J., Gnanadesikan, R. and Kettenring, J. R. (1981). Robust estimation of dispersion matrics and principal components. *J. Amer. Statist. Assoc.* **76** 354-762.

Donoho, D. L. (1982). Breakdown properties of multiavraite location estiamtors. Ph.D. thesis, Harvard University.

Donoho, D. L. and Huber, P. J. (1983). The notion of breakdown point. In *A Festschrift for Erich Lehmann* (Eds. P. J. Bickel, K. Doksum, and J. L. Hodges, Jr.). Wadsworth, Belmont, CA. 157-184.

Fred, A. and Jain, A. K. (2003). Robust data clustering. In *Proc. of IEEE Computer Society Conference on Computer Vision and Pattern Recognition II*. 128-133.

Gather, U. and Hilker, T. (1997). A note on Tyler's modification of the MAD for the Stahel-Donoho estimator. *Ann. Statist.* **25** 2024-2026.

Gervini, D. (2002). The influence function of the Stahel-Donoho estimator of multivariate location and scatter. *Statist. Probab. Lett.* **60** 425-435.

Gervini, D. (2003). A robust and efficient adaptive reweighted estimator of multivariate location and scatter. *J. Multivariate Anal.* **84** 116-144.

Gnanadesikan, R. and Kettenring, J. R. (1972). Robust estimates, residuals, and outlier detection with multiresponse data. *Biometrics* **28** 81-124.

Hampel, F. R. (1968). Contributions to the theory of robust estimation. Ph.D. dissertion, University of California, Berkeley, CA.

Hampel, F. R. (1971). A general qualitative definition of robustness. *Ann. Math. Statis.* **42** 1887-1896.

Hampel, F. R. (1973). Robust estimation: A condensed partial survey. *Z. Wahrscheinlichkeitstheorie und Verw. Gebiete* **27** 87-104.

Hampel, F. R., Ronchetti, E. M., Rousseeuw, P. J. and Stahel, W. A. (1986). *Robust Statistics: The approach based on influence functions.* John Wiley & Sons, New York.

Hawkins, D. M. (1994). The feasible solution algorithm for the minimum covariance determinant estimator in multivariate data. *Comput. Statist. Data Anal.* **17** 197-210.

He, X. and Fung, W. K. (2000). High breakdown estimation for multiple populations with applications to discriminant analysis. *J. Multivariate Anal.* **72** 151-162.

Huber, P. J. (1964). Robust estimation of a location parameter. *Ann. Math. Statist.* **35** 73-101.

Huber, P. J. (1972). Robust Statistics: A review. *Ann. Math. Statist.* **35** 1041-1067.

Huber, P. J. (1977). Robust covariances. In *Statistical Decision Theory and Related Topics* (Eds. S. S. Gupta and D. S. Moore). Academic Press, New York. 165-191.

Huber, P. J. (1981). *Robust Statistics.* (John Wiley & Sons, New York.

Hubert, M., Rousseeuw, P. J. and Verboven, S. (2002). A fast method for robust principal components with applications to chemometrics. *Chemom. Intell. Lab. Sys.* **60** 101-111.

Hubert, M., Rousseeuw, P. J. and Vanden Branden, K. (2005). ROBPCA: a new approach to robust principal component analysis. *Technometrics* **47** 64-79.

Hubert, M. L. and Van Driessen, K. (2004). Fast and robust discrimant analysis. *Comput. Statist. Data Anal.* **45** 301-320.

Jaupi, L. and Saporta, G. (1993). Using the influence function in robust principal components analysis. In *New Directions in Statistical Data Analysis and Robustness* (Eds, S. Morgenthaler, E. Ronchetti and W. A. Stahel). Basel, Birkhäuser. 147-156.

Karnel, G. (1991). Robust canonical correlation and correspondance analysis. In *The Frontiers of Statistical Scientific and Industrial Applications*, Volume II of the Proceedings of ICOSCO-I. 335-354.

Kaufman, L. and Rousseeuw, P. J. (1990). *Finding Groups in Data: An Introduction to Cluster Analysis.* Wiley-Interscience, New York.

Li, G. and Chen, Z. (1985). Projection-Pursuit Approach to Robust Dispersion Matrices and Principal Components: Primary Theory and Monte Carlo. *J. Amer. Statist. Assoc.* **80** 759-766.

Liu, R. Y. (1990). On a notion of data depth based on random simplices. *Ann. Statist.* **18** 405-414.

Liu, R. Y. (1992). Data depth and multivariate rank tests. In L_1-*Statistical Analy-*

sis and Related Mathods (Ed. Y. Dodge). North-Holland, Amsterdam. 279-294.

Liu, R. Y. (1995). Control Charts for Multivariate Processes. *J. Amer. Statist. Assoc.* **90** 1380-1387.

Liu, R. Y., Parelius, J. M., and Singh, K. (1999). Multivariate analysis by data depth: Descriptive statistics, graphics and inference (with discussion). *Ann. Statist.* **17** 1662-1683.

Liu, R. Y. and Singh, K. (1993). A quality index based on data depth and multivariate rand tests. *J. Amer. Statist. Assoc.* **88** 252-260.

Lopuhaä, H. P. (1989). On the relation between S-estimators and M-estimators of multivariate location and covariance. *Ann. Statist.* **17** 1662-1683.

Lopuhaä, H. P. (1991). Multivarite τ-estimators for location ans scatter. *Can. J. Statist.* **19** 307-321.

Lopuhaä, H. P. (1992). Highly efficient estimators of multivariate location with high breakdown point. *Ann. Statist.* **20** 398-413.

Lopuhaä, H. P. (1999). Asymptotics of reweighted estimators of multivariate location and scatter. *Ann. Statist.* **27** 1638-1665.

Lopuhaä, H. P. and Rousseeuw, P. J. (1991). Breakdown points of affine equivariant estimators of multivariate location and covariance matrices. *Ann. Statist.* **19** 229-248.

Maronna, R. A. (1976). Robust *M*-estimators of multivariate location and scatter. *Ann. Statist.* **1** 51-67.

Maronna, R. A. and Yohai, V. J. (1995). The behavior of the Stahel-Donoho robust multivariate estimator. *J. Amer. Statist. Assoc.* **90** 330-341.

Maronna, R. A. and Yohai, V. J. (1998). Robust estimation of multivariate location and scatter. In *Encyclopedia of Statistical Sciences, Updated volume 2* (Eds. S. Kotz, C. Read and D. Banks). Wiley, New York. 589-596.

Maronna, R. A., Stahel, W. A. and Yohai, V. J. (1992). Bias-robust estimates of multivariate scatter based on projections. *J. Multivariate Anal.* **42** 141-161.

Mosteller, F. and Tukey, J. W. (1997). *Data Analysis and Regression.* Addison Wesley, Reading, MA.

Ngai, H. and Zhang, J. (2001). Multivariate cumulative sum control charts based on projection pursuit. *Statist. Sinica* **11** 747-766.

Pison, G., Rousseeuw, P. J., Filzmoser, P. and Croux, C. (2003). Robust Factor Analysis. *J. Multivariate Anal.* **84** 145-172.

Rocke, D. M. (1996). Robustness Properties of *S*-estimators of multivariate location and shape in high dimension. *Ann. Statist.* **24** 1327-1345.

Rousseeuw, P. J. (1985). Multivariate estimation with high breakdown point. In *Mathematical Statistics and Applications* (Eds. W. Grossmann, G. Pflug, I. Vincze and W. Wertz). Reidel. 283-297.

Rousseeuw, P. J. and Leroy, A. M. (1987). *Robust Regression and Outlier Detection.* Wiley, New York.

Rousseeuw, P. J., Van Aelst, S., Van Driessen, K. and Agulló, J. (2004). Robust multivariate regression. *Technometrics* **46** 293-305.

Rousseeuw, P. J. and Van Driessen, K. (1999). A fast Algorithm for the minimum covariance determinant estimator. *Technometrics* **41** 212-223.

Rousseeuw, P. J. and Yohai, V. J. (1984). Robust regression by means of *S*-estimators. In *Robust and Nonlinear Time Series Analysis. Lecture Notes in Statist.* Springer, New York. **26** 256-272.

Ruppert, D. (1992). Computing S-Estimators for Regression and Multivariate Location/Dispersion. *J. Comput. Graph. Statist.* **1** 253-270

Serfling, R. J. (1980). *Approximation Theorems of Mathematical Statistics.* John Wiley & Sons, New York.

Shi, L. (1997). Local influence in principal components analysis. *Biometrika* **84** 175-186.

Stahel, W. A. (1981). Robust estimation: Infinitesimal optimality and covariance matrix estimators. Ph.D. thesis, ETH, Zurich.

Taskinen, S., Croux, C., Kankainen, A., Ollila, E. and Oja, H. (2005). Influence functions and efficiencies of the canonical correlation and vector estimates based on scatter and shape matrices. *J. Multivariate Anal.* (to appear).

Tukey, J. W. (1949). Memorandum Reports 31-34, *Statist. Research Group.* Princeton Univ.(unpublished).

Tukey, J. W. (1960). A survey of sampling from contaminated distributions. In *Contributions to Probability and Statistics* (Ed. I. Olkin). Standford Univ. Press. 448-485.

Tukey, J. W. (1975). Mathematics and the picturing of data. In *Proc. International Congress of Mathematicians Vancouver 1974.* Canadian Mathematics Congress, Montreal. **2** 523-531.

Tyler, D. E. (1994). Finite sample breakdown points of projection based multivarite location and scatter statistics. *Ann. Statist.* **22** 1024-1044.

Willems, G., Pison, G., Rousseeuw, P. J. and Van Aelst, S. (2002). A Robust Hotelling Test. *Metrika* **55** 125-138.

Zhang, J. (2002). Some extensions of Tukey's depth function. *J. Multivariate Anal.* **82** 134-165.

Zuo, Y. (2000). A note of finite sample breakdown points of projection based multivariate location and scatter statistics. *Metrika* **51** 259-265.

Zuo, Y. (2001). Some quantitative relationships between two types of finite sample breakdown point. *Statit. Probab. Lett.* **51** 369-375.

Zuo, Y. (2003). Projection-based depth functions and associated medians. *Ann. Statist.* **31** 1460-1490.

Zuo, Y. (2004a). Projection Based Affine Equivariant Multivariate Location Estimators with the Best Possible Finite Sample Breakdown Point. *Statistica Sinica* **14** 1199-1208.

Zuo, Y. (2004b). Statistical depth functions and some applications. *Adv. Math.* (China) **33** 1-25.

Zuo, Y. and Cui, H. (2005). Depth weighted scatter estiamtors. *Ann. Statist.* **33** 381-413.

Zuo, Y., Cui, H. and He, X. (2004). On the Stahel-Donoho estimator and depth weighted means for multiavraite data. *Ann. Statist.* **32** 167-188.

Zuo, Y., Cui, H. and Young, D. (2004). Influence function and maximum bias of projection depth based estimators. *Ann. Statist.* **32** 189-218.

Zuo, Y. and Serfling, R. J. (2000a). General notions of statistical depth function. *Ann. Statist.* **28** 461-482.

Zuo, Y. and Serfling, R. J. (2000b). Structural properties and convergence results for contours of sample statistical depth functons. *Ann. Statist.* **28** 483-499.

CHAPTER 22

Estimation of the Loss of an Estimate

Wing Hung Wong

Department of Statistics
Stanford University
Stanford, CA 94305-4065
whwong@stanford.edu

22.1. Introduction

Suppose $y_1, \ldots, y_n$ are independent random variables. The density $p_{\theta_i}(\cdot)$ of y_i is supposed to be known up to a parameter θ_i. Let $\widehat{\boldsymbol{\theta}} = \left(\widehat{\theta}_1(\boldsymbol{y}), \ldots, \widehat{\theta}_n(\boldsymbol{y})\right)$ be an estimate of $\boldsymbol{\theta}$ constructed from the sample $\boldsymbol{y} = (y_1, \ldots, y_n)$. Note that the estimate of θ_i can depend on y_j, $j \neq i$. Let $\ell_i(\cdot, \cdot)$ be a loss function so that $\ell_i(\theta_i, \widehat{\theta}_i)$ represents the loss of using $\widehat{\theta}_i$ as the estimate of θ_i. The purpose of the present paper is to introduce some methods for the estimation of the average loss $L(\boldsymbol{\theta}, \widehat{\boldsymbol{\theta}}) = \frac{1}{n} \sum_1^n \ell_i(\theta_i; \widehat{\theta}_i)$.

Let $\boldsymbol{y}_{(-i)} = (y_1, \ldots, y_{i-1}, y_{i+1}, \ldots, y_n)$ be the sample with y_i deleted, and write

$$g_i(y_i) = \widehat{\theta}_i(y_i; \boldsymbol{y}_{(-i)}),$$

then

$$\ell_i(\theta_i, \widehat{\theta}_i) = \ell_i\big(\theta_i; g_i(y_i)\big).$$

Conditional on $\boldsymbol{y}_{(-i)}$, $g_i(\cdot)$ is a known function. This leads us to a one-dimensional estimation problem: Given the known functions $\ell_i(\cdot, \cdot)$ and $g_i(\cdot)$, and an observation y_i from $p_{\theta_i}(\cdot)$, find an estimate $\widehat{\ell}_i(y_i)$ of the quantity $\ell_i\big(\theta_i, g_i(y_i)\big)$. One obvious possibility is to estimate $\ell_i\big(\theta_i, g_i(y_i)\big)$ by a Bayes estimate, i.e. setting $\widehat{\ell}_i(y_i)$ to be

$$\frac{\int \ell(\theta_i, g_i(y_i)) p_{\theta_i}(y_i) \pi(\theta_i) d\theta_i}{\int p_{\theta_i}(y_i) \pi(\theta_i) d\theta_i}$$

where $\pi(\theta_i)$ is a prior for θ_i. This would be a reasonable procedure if n is small and each y_i is strongly informative on θ_i.

In this paper, however, we are mainly interested in the situation when n is large and each y_i by itself is not strongly informative on θ_i. In this case, it is desirable to require $\widehat{\ell}_i(y_i)$ to be an unbiased estimator of $\ell_i(\theta_i, g_i(y_i))$.

To cite a scenario where unbiasedness is clearly the appropriate requirement, suppose $\{\widehat{\theta}_i, i = 1, \ldots, n\}$ are weakly dependent random variables in the sense that each $\widehat{\theta}_i$ is approximately independent of most but a small fraction of the other $\widehat{\theta}_j$'s. Conditional on $\boldsymbol{y}_{(-i)}$, let $t_i = t(y_i; g_i(\cdot))$ be an estimator of $\ell_i(\theta_i, g_i(y_i))$, and

$$r_i = t_i - \ell_i(\theta_i, g_i(y_i)),$$

then $\{r_i, i = 1, \ldots, n\}$ are also weakly dependent variables. Hence $\mathrm{Var}\left(\frac{1}{n}\sum_1^n r_i\right) \to 0$, and

$$E\left(\frac{1}{n}\sum_1^n t_i - \frac{1}{n}\sum_1^n \ell_i(\theta_i, \widehat{\theta}_i)\right)^2$$

$$= \mathrm{Var}\left(\frac{1}{n}\sum_1^n r_i\right) + \left(\frac{1}{n}\sum_1^n E(r_i)\right)^2$$

will be determined by the average biases of the t_i's. In this case, unbiasedness is clearly desirable if it can be achieved. Otherwise, we should try to keep the bias of $\widehat{\ell}_i$ small over a reasonable range of values of θ_i. It is, however, important to note that the variance of $\frac{1}{n}\sum_1^n r_i$ can be small under much more general conditions than weak dependency of the $\widehat{\theta}_i$'s. This will be discussed in Section 22.6.

Thus, for each i, we need to find unbiased or approximately unbiased estimator of $\ell_i(\theta_i, g_i(y_i))$. If there is more than one unbiased estimator, we choose the one giving, in some sense, the smallest value for $E_{\theta_i}(\widehat{\ell}_i(y_i) - \ell_i(\theta_i, g_i(y_i)))^2$.

In many applications, we are interested in the loss of $\widehat{\theta}_i$ only to compare it to the loss of another estimator $\tilde{\theta}_i$. It is clear that in this case it suffices to compare $\ell_i(\theta_i, \widehat{\theta}_i) + k(\theta_i)$ to $\ell(\theta_i, \tilde{\theta}_i) + k(\theta_i)$ where $k(\cdot)$ is a constant function of θ_i. We call such a function $\ell_i(\theta_i, \widehat{\theta}_i) + k(\theta_i)$ a comparative loss function. For comparison among estimators, it is enough to find unbiased estimates of their comparative losses (corresponding to a common $k(\cdot)$). We then have the freedom of choosing $k(\cdot)$ to make it easy to construct unbiased estimates.

As a final remark, we note that if the family of densities $\{p_{\theta_i}(\cdot)\}$ is a complete family as θ_i vary in its range, then an unbiased estimator of the loss (or a comparative loss), must be unique if it exists.

22.2. Kullback-Leibler Loss and Exponential Families

Suppose y_i has density $p_{\theta_i}(\cdot)$ and $g_i(\cdot)$ is a given function of y_i as defined in the introduction. To simplify notations, we will suppress the subscript i in the rest of this section. The Kullback-Leibler pseudo-distance between two densities $p(\cdot)$ and $q(\cdot)$ are defined by

$$K(p, q) = \int p \log \frac{p}{q} \, dy.$$

The Kullback-Leibler loss (KL loss) of the estimator $g(y)$ is then defined by

$$\ell\big(\theta, g(y)\big) = K\big(p_\theta, p_{g(y)}\big).$$

Let $p_\theta(z)$ be an exponential family distribution, then

$$\log p_\theta(z) = \phi(\theta)t(z) + \alpha(\theta) + m(z)$$

for some functions $\phi(\cdot)$, $\alpha(\cdot)$ of θ and $t(\cdot)$, $m(\cdot)$ of z, and $\ell\big(\theta, g(y)\big) = \phi(\theta)\mu(\theta) + \alpha(\theta) - \phi\big(g(y)\big)\mu(\theta) - \alpha\big(g(y)\big)$, where $\mu(\theta) = E_\theta t(z)$. Even though the exponential family structure leads to a relatively simple form for the KL loss, in most cases it is still not possible to find exactly unbiased estimate of this loss. We will discuss the construction of approximately unbiased estimates in a later section. However, a corresponding comparative loss function $\ell\big(\theta, g(y)\big) - \phi(\theta)\mu(\theta) - \alpha(\theta) = -\phi\big(g(y)\big)\mu(\theta) - \alpha\big(g(y)\big)$ is particularly simple.

To estimate this comparative loss, we only need to solve the following problem:

(i) Find a function $h(y)$ so that

$$E_\theta h(y) = \mu(\theta) E_\theta \phi\big(g(y)\big) \qquad \text{for all } \theta. \tag{22.1}$$

(ii) If there exist more than one solution to (22.1), choose the one that minimizes

$$\int E_\theta \big(h(y) - \mu(\theta)\phi\big(g(y)\big)\big)^2 \pi(\theta) d\theta$$

where $\pi(\cdot)$ is an appropriate weight function.

If a function $h(\cdot)$ satisfying (22.1) can be found, then $-h(y) - \alpha\big(g(y)\big)$ will be an unbiased estimator of the comparative KL loss $\ell\big(\theta, g(y)\big) - \phi(\theta)\mu(\theta) - \alpha(\theta)$. Exact solutions to the above problem can be found in several important exponential family models. We list two examples.

Example 22.1: Poisson distribution

Suppose y has a Poisson distribution with mean θ, then $\log p_\theta(y) = y\log(\theta) - \theta - \log(y!)$.

Hence $\mu(\theta) = E_\theta(y) = \theta$, $\phi(\theta) = \log(\theta)$, $\alpha(\theta) = -\theta$. To obtain an unbiased estimate of the comparative KL loss, notice that

$$\mu(\theta)E_\theta\phi\big(g(y)\big) = \theta\sum_{y=0}^{\infty}\log\big(g(y)\big) \cdot e^{-\theta}\theta^y/y!$$

$$= \sum_{z=1}^{\infty} z\log\big(g(z-1)\big)\frac{e^{-\theta}\theta^z}{z!} \qquad (z = y+1)$$

$$= E_\theta y\log\big(g(y-1)\big).$$

It follows that $y\log\big(g(y-1)\big)$ is an unbiased estimate of $\mu(\theta)\phi\big(g(y)\big)$, and $g(y) - y\log\big(g(y-1)\big)$ is an unbiased estimator of the comparative KL loss $K(p_\theta, p_{g(y)}) - \theta\log(\theta) + \theta$. It is the unique unbiased estimator because the Poisson family is complete.

Example 22.2: Gamma scale family

Suppose y has a Gamma distribution with a known shape parameter k and an unknown scale parameter θ, then

$$\log p_\theta(y) = -\frac{y}{\theta} - k\log(\theta) + \log\big(y^{k-1}/\Gamma(k)\big).$$

Hence

$$\mu(\theta) = E_\theta(y) = k\theta, \quad \phi(\theta) = -\frac{1}{\theta}, \quad \alpha(\theta) = -k\log(\theta).$$

To obtain an unbiased estimate of $\mu(\theta)E_\theta\phi\big(g(y)\big)$, notice that

$$\mu(\theta)E_\theta\phi\big(g(y)\big) = -k\theta E_\theta\left(\frac{1}{g(y)}\right)$$

$$= -\frac{k}{\Gamma(k)\theta^{k-1}}\int_0^{\infty}\left(\frac{y^{k-1}}{g(y)}\right)(e^{-\frac{y}{\theta}})\,dy$$

$$= -\frac{k}{\Gamma(k)\theta^{k-1}}\left\{\Big[v(y)e^{-\frac{y}{\theta}}\Big]_0^{\infty} - \int_0^{\infty}v(y)\left(-\frac{1}{\theta}e^{-\frac{y}{\theta}}\right)dy\right\}$$

where $v(y) = \int_0^y \frac{z^{k-1}}{g(z)}\, dz$. We assume that $g(z)$ does not converge to zero faster than z^k as $z \to 0$, so that $v(y)$ exists and $\left[v(y)e^{-\frac{y}{\theta}}\right] \to 0$ as $z \to 0$ or $z \to \infty$. Then it follows that

$$E_\theta \mu(\theta)\phi\big(g(y)\big) = E_\theta\big(-ky^{-(k-1)}v(y)\big).$$

Hence $ky^{-(k-1)}v(y)+k\log\big(g(y)\big)$ is an unbiased estimate of the comparative KL loss

$$K(p_\theta, p_{g(y)}) + 1 + k\log(\theta).$$

It is the unique unbiased estimate because of the completeness of the Gamma scale family.

22.3. Mean Square Error Loss

The mean square error loss (MLE loss) of an estimator $\widehat{\boldsymbol{\theta}}(\boldsymbol{y})$ is defined by

$$L(\boldsymbol{\theta}, \widehat{\boldsymbol{\theta}}) = \frac{1}{n}\sum_1^n \big(\widehat{\theta}_i(\boldsymbol{y}) - \theta_i\big)^2.$$

Write $\widehat{\theta}_i(y) = g_i(y_i)$ where $g_i(\cdot)$ is the (random) function of y_i defined in the introduction, then each term in $L(\boldsymbol{\theta}, \widehat{\boldsymbol{\theta}})$ is of the form

$$\big(g(y) - \theta\big)^2 = g(y)^2 - 2\theta g(y) + \theta^2,$$

where, for simplicity, the subscript i has been suppressed from g_i and θ_i. Hence, $g(y)^2 - 2\theta g(y)$ is a comparative MSE loss for $g(\cdot)$ and an unbiased estimator of it is of the form $g(y)^2 - 2e(y)$ where $e(\cdot)$ satisfies the equation

$$E_\theta e(y) = \theta E_\theta g(y) \tag{22.2}$$

for all θ. If, furthermore, there is an unbiased estimator $f(y)$ of the term θ^2, then

$$g(y)^2 - 2e(y) + f(y)$$

is an unbiased estimate of the MSE.

Example 22.3: Suppose y has a Poisson distribution with mean θ. By the same argument as used in Example 1, it is seen that $e(y) = yg(y-1)$ is an unbiased estimator of $\theta E_\theta g(y)$. Furthermore, it is easy to check that $f(y) = y^2 - y$ is an unbiased estimator of θ^2. Hence

$$g(y)^2 - 2yg(y-1) + y^2 - y$$

is the unique unbiased estimator of $\big(g(y) - \theta\big)^2$.

Example 22.4: Suppose y has a Gamma(k, θ) distribution with a known shape parameter k. The arguments used in examples 2 and 3 can be used to calculate an unbiased estimate of $(g(y) - \theta)^2$. The resulting estimate is

$$g(y)^2 - 2y^{-(k-1)}v(y) + \frac{y^2}{k + k^2}$$

where

$$v(y) = \int_0^y z^{k-1}g(z)dz.$$

22.4. Location Families

Suppose y_i has density $p_i(y_i - \theta_i)$ where, for each i, $p_i(\cdot)$ is a known density on $\mathbb{R}$ with mean zero. Suppressing the subscript i, we write

$$y = \theta + \epsilon$$

where ϵ has density $p(\cdot)$ and satisfies $E(\epsilon) = 0$. It was observed in the last section that to estimate the MSE of an estimator $g(y)$, we need to construct unbiased estimates of $\theta g(y)$ and θ^2. Unbiased estimation of θ^2 is easy: we may use $f(y) = y^2 - \sigma^2$ where $\sigma^2 = \text{Var}(\epsilon) = \int \epsilon^2 p(\epsilon)d\epsilon$. To construct an unbiased estimator of $\theta g(y)$, observe that

$$E\theta g(y) = Eyg(y) - E\epsilon g(\theta + \epsilon).$$

Thus it suffices to find an unbiased estimator $h(y)$ of the term $E\epsilon g(\theta + \epsilon)$, i.e. to find a function $h(\cdot)$ to satisfy the following equation for all θ

$$\int h(\theta + \epsilon)\, p(\epsilon)d\epsilon = \int g(\theta + \epsilon)\, \epsilon p(\epsilon)d\epsilon. \tag{22.3}$$

The solution to this integral equation, if it exists, can be obtained in the following way.

Let $H(\cdot)$, $P(\cdot)$, $G(\cdot)$ be the Fourier transforms of $h(\cdot)$, $p(\cdot)$ and $g(\cdot)$ respectively. For example,

$$P(s) = (\mathcal{F}p)(s) = \int p(\epsilon)e^{-i\cdot 2\pi\epsilon s}d\epsilon.$$

Let $q(\epsilon) = p(-\epsilon)$, then

$$\int h(\theta + \epsilon)\, p(\epsilon)d\epsilon = h * q(\theta)$$

where $*$ denotes the convolution operation. The Fourier transform of $h*q(\theta)$ is $H(s) \cdot Q(s) = H(s)\bar{P}(s)$ where $\bar{P}(s)$ is the complex conjugate of $P(s)$.

Similarly transforming the right hand side of (22.3), and using the fact that the Fourier transform of $\epsilon p(\epsilon)$ is $(\mathbf{i}/2\pi)P'(s)$, it is seen that the integral equation (22.3) is equivalent to

$$H(s)\bar{P}(s) = -\frac{\mathbf{i}}{2\pi}G(s)\bar{P}'(s) \qquad (22.4)$$

In other words, a solution $h(\cdot)$ exists for (22.3) iff we can find a L_1 function $H(\cdot)$ which satisfies (22.4) for all s. In particular, if $P(s) \neq 0$ for all s then h is determined uniquely as

$$h = -\frac{1}{2\pi}\mathcal{F}^{-1}(\mathbf{i}\cdot G\bar{P}'/\bar{P}). \qquad (22.5)$$

In general, $p(s)$ may vanish for some values of s and (22.5) cannot be used. In this case, we can get approximate solutions of (22.3) by the following device: first approximate $p(\cdot)$ by another density $p_1(\cdot)$ which has a nonvanishing Fourier transform, then compute $h = -\frac{1}{2\pi}\mathcal{F}^{-1}(\mathbf{i}\cdot G\bar{P}'_1/\bar{P}_1)$ and regard it as an approximate solution of (22.3). One possible choice of p_1 is $p_1(\epsilon) = (1-\alpha)p(\epsilon) + \alpha\phi(\epsilon)$ where $\phi(\cdot)$ is a normal density. The above choice of h will then satisfy (22.3) with p replaced by p_1:

$$\int h(\theta + \epsilon)p_1(\epsilon)d\epsilon = \int g(\theta + \epsilon)\,\epsilon p_1(\epsilon)\,d\epsilon.$$

$$\left[\int h(\theta + \epsilon)p(\epsilon)d\epsilon - \int g(\theta + \epsilon)\epsilon p(\epsilon)d\epsilon\right] \qquad \text{i.e.}$$

$$= \frac{\alpha}{(1-\alpha)}\left[-\int h(\theta + \epsilon)\phi(\epsilon)d\epsilon + \int g(\theta + \epsilon)\epsilon\phi(\epsilon)d\epsilon\right].$$

Thus the bias of this choice of $h(\cdot)$ is of order α. Typically, if α is chosen too small then $h(\cdot)$ will have high variance. In practice, one needs to choose each α_i (in estimating $(g_i - \theta_i)^2$) carefully in order to achieve a good bias/variance trade-off in the estimation of $\frac{1}{n}\sum(\widehat{\theta}_i - \theta_i)^2$.

Example 22.5: Normal location model

Let ϵ be $N(0,1)$, then

$$p(\epsilon) = (2\pi)^{-\frac{1}{2}}e^{-\frac{1}{2}\epsilon^2} \qquad \text{and} \qquad P(s) = e^{-2\pi^2 s^2}.$$

Hence $p'(s) = -(2\pi)^2 sP(s)$ and

$$H(s) = \mathbf{i}\cdot(2\pi)sG(s) = (\mathcal{F}g')(s).$$

It follows that $h(x) = g'(x)$, which leads to the unique unbiased estimate (of the MSE)

$$g(y)^2 - 2\big[yg(y) - g'(y)\big] + (y^2 - 1) = \big(g(y) - y\big)^2 + \big(2g'(y) - 1\big).$$

If the variance of ϵ is σ^2, then the term $(2g' - 1)$ should be multiplied by σ^2. Returning to the whole vector $\widehat{\boldsymbol{\theta}} = (\widehat{\theta}_1(\boldsymbol{y}), \ldots, \widehat{\theta}_n(\boldsymbol{y})) = (g_1(y_1), \ldots, g_n(y_n))$, the above result then leads to

$$\frac{1}{n} \sum_1^n (\widehat{\theta}_i - y_i)^2 + \frac{2}{n} \sum_1^n \frac{\partial}{\partial y_i} \widehat{\theta}_i - \sigma^2$$

as an unbiased estimator of the MSE loss $\frac{1}{n} \sum_1^n (\widehat{\theta}_i - \theta_i)^2$. This estimate was first obtained in Stein (1981) as an unbiased estimate of the MSE risk $\frac{1}{n} \sum_1^n E(\widehat{\theta}_i - \theta_i)^2$.

Example 22.6: Symmetric stable distributions

Let $p(\cdot)$ be a symmetric stable density with a known scale parameter, then its Fourier transform is given by

$$P(s) = e^{-c|s|^\alpha},$$

where c depends on α and the scale parameter. We assume that the index α is known and $\alpha \in (1, 2]$. It follows that $P'(s) = -\mathrm{sign}(s)\alpha c|s|^{\alpha-1}P(s)$, and, by (22.4),

$$H(s) = \alpha c\big(|s|^{-(2-\alpha)}\big) \cdot \left(\frac{\mathbf{i}}{2\pi} \, \mathrm{sign}(s)|s|G(s)\right)$$

$$= \alpha c\big(|s|^{-(2-\alpha)}\big) \cdot \left(\frac{\mathbf{i}}{2\pi} \, sG(s)\right).$$

Hence $h(x) = \alpha c(g' * t)(x)$, where

$$t(x) = \mathcal{F}^{-1}\big(|s|^{-(2-\alpha)}\big)(x)$$
$$= \big[2^{2-\alpha}\pi^{-(\alpha-1)}\Gamma(\alpha-1)\sin\big((2-\alpha)\pi/2\big)\big] \cdot |x|^{-(\alpha-1)}.$$

It is interesting to note that if $\alpha < 2$ then θ^2 cannot be estimated by $f(y) = y^2 - \mathrm{Var}(\epsilon)$ because $\mathrm{Var}(\epsilon)$ is infinite. However, the comparative loss $\big(g(y) - \theta\big)^2 - \theta^2$ can still be estimated by

$$g(y)^2 - 2yg(y) + 2\alpha c \cdot g' * t(y).$$

22.5. Approximate Solutions

In general, exactly unbiased approximation to the loss may not exist. This typically happens when the family $\{p_\theta(\cdot), \theta \in \Theta\}$ is rich but the range $\mathcal{Y}$ of the random variable y is small. In such cases, one has to be satisfied with an estimator which is in some sense close to being unbiased for the loss $\ell_i(\theta, \widehat{\theta})$. For example, suppose we want to find a function $e(\cdot)$ to satisfy (22.2). A generally applicable method is as follows: Let $e(\cdot) = \sum_1^\infty c_i f_i(\cdot)$ where $\{f_i(\cdot), i = 1, 2, \dots\}$ is a set of basis functions in a certain space of functions on $\mathcal{Y}$. Let $\gamma(\theta) = \theta E_\theta g(y)$. We can determine the coefficients c_i's so that (22.2) is approximately satisfied in a certain sense. For example:

(a) Restricted unbiasedness: choose a subset $\Theta_0 \subset \Theta$ and requires $e(\cdot)$ to satisfy (22.2) for all $\theta \in \Theta_0$. In particular, if Θ_0 is finite, we may approximate $e(\cdot)$ by taking the first m terms in the expansion and attempt to solve for the coefficients $c_1, \dots, c_m$ in the linear system

$$\sum_1^m c_j \left(E_{\theta_i} f_j(y) \right) = \gamma(\theta_i), \qquad \theta_i \in \Theta_0 = \{\theta_1, \cdots, \theta_m\}.$$

(b) Least-squares solution: Suppose $\gamma(\theta)$ and $\phi_i(\theta) = E_\theta f_i(y)$ $i = 1, 2, \dots$ are all elements of a L_2 space with inner product $\langle \phi, \gamma \rangle = \int \phi(\theta) \gamma(\theta) \, d\mu(\theta)$ where $\mu(\cdot)$ is an appropriate measure on Θ. We may then determine $c_1, \dots, c_m$ by projecting $\gamma(\cdot)$ into the space spanned by $\phi_1(\cdot), \dots, \phi_m(\cdot)$.

Although this approach can be implemented numerically in almost any problem, the degree to which $e(\cdot)$ is "approximately unbiased" must be investigated in each application. We give two examples.

Example 22.7: Location family with bounded error

As discussed in Section 22.4, the key to finding an unbiased approximation to the MSE is the solution of the integral equation (22.3), or equivalently (22.4). Unfortunately, if the error density $p(\cdot)$ has bounded support, then its Fourier transform $P(\cdot)$ may have isolated zeros and (22.4) may not be satisfied by any $H(\cdot)$. For example, if $p(\cdot)$ is the triangular density, i.e. $p(x) = 1 - |x|$, $|x| \leq 1$, then $P(s) = \left[\sin(\pi s)/\pi \right]^2$, and for $s \neq 0$,

$$P'(s)/P(s) = 2\pi \left[\cot(\pi s) - 1/\pi s \right].$$

In this case, the "formal" solution (22.5) will have singularities at $s = \pm 1, \pm 2, \dots$.

Thus, in many applications the equation (22.3) cannot be satisfied for all θ. A method for constructing approximate solutions has already been given in Section 22.4. We now describe another method which can often be used to construct a function $h(\cdot)$ which satisfies (22.3) for all $\theta \in [-L, L]$ where L is a suitably large constant.

Suppose the support of $p(\cdot)$ is contained in an interval $[-\delta, \delta]$, and $T > L + 2\delta$. Let $\tilde{g}(\cdot)$ be a periodic function with period $2T$ such that $\tilde{g}(y) = g(y)$ for $y \in [-L - \delta, L + \delta]$. It is easy to check that $E_\theta \tilde{g}(y) = E_\theta g(y)$ for all $\theta \in [-L, L]$. Thus it suffices to consider the problem of finding a periodic function $h(\cdot)$ (with period $2T$) to satisfy the equation

$$\int h(\theta + \epsilon)\, p(\epsilon)\, d\epsilon = \int \tilde{g}(\theta + \epsilon)\, \epsilon p(\epsilon)\, d\epsilon$$

for all $\theta \in (-\infty, \infty)$. Since $h(\cdot)$ is periodic, we can expand it in Fourier series:

$$h(x) = \sum_{n=-\infty}^{\infty} H_n e^{\mathrm{i} \cdot 2\pi s_n x}$$

where $s_n = \frac{n}{T}$, $n = 0, \pm 1, \pm 2, \cdots$

and

$$H_n = \frac{1}{2T} \int_{-T}^{T} h(x) e^{-\mathrm{i} \cdot 2\pi s_n x}\, dx.$$

Similarly, $\tilde{g}(x) = \sum_n \tilde{G}_n e^{\mathrm{i} \cdot 2\pi s_n x}$. Putting these into the integral equation and equating coefficients, we have

$$H_n \cdot \bar{P}(s_n) = \tilde{G}_n \cdot \bar{R}(s_n) \tag{22.6}$$

where $P(s_n) = \int_{-\infty}^{\infty} p(\epsilon) e^{-\mathrm{i} \cdot 2\pi s_n \cdot \epsilon} d\epsilon$ and $R(s_n) = \int_{-\infty}^{\infty} \epsilon p(\epsilon) e^{-\mathrm{i} \cdot 2\pi s_n \cdot \epsilon} d\epsilon = \frac{\mathrm{i}}{2\pi} P'(s_n)$ are the Fourier transforms of $p(\epsilon)$ and $\epsilon p(\epsilon)$. To solve (22.6) for H_n, we must make sure that $P(s_n) \neq 0 \quad$ for all $\quad n$. This is usually achievable by choosing T appropriately. For example, if $p(\cdot)$ is the triangular density, then $P(s)$ vanishes only at $s = \pm 1, \pm 2, \ldots$. Thus, we need to make sure that, for all integers $n = \pm 1, \pm 2, \ldots$, $s_n = \frac{n}{T}$ is not a nonzero integer. Theoretically, any positive irrational T will do.

From the point of view of controlling bias, one would like to choose L, (and hence T) as large as possible. However, there is a price to be paid: if T is too large then some of the values of $s_n = \frac{n}{T}$ will be very close to the zeros of $P(s)$ at $s = \pm 1, \pm 2, \ldots$. Consequently, the corresponding values of H_n will be very large, leading to an estimator $h(\cdot)$ with very high conditional variance. The choice of T_i from the considerations of bias/variance trade-off in the estimation of the average MSE $n^{-1} \sum_{1}^{n} (\hat{\theta}_i - \theta_i)^2$ is an interesting question which, however, will not be discussed further in this paper.

Example 22.8: Binomial distribution

Suppose each y_i has a Binomial (m_i, θ_i) distribution. From Section 22.3, we know that the construction of unbiased approximation to the MSE depends on the solution of the following problem: Find $e(\cdot)$ such that

$$E_\theta e(y) = \theta E_\theta g(y) \qquad \text{for all} \qquad \theta \in [0, 1] \qquad (22.7)$$

where y is a Binomial (m, θ) variable.

We will see that for a large class of functions $g(\cdot)$, there are exact solutions to (22.7). Furthermore, even when (22.7) is not exactly solvable, we can often construct $e(\cdot)$'s which satisfy (22.7) to a high degree of accuracy.

Let us represent the functions $e(y)$ and $g(y)$ by the vectors $v_e = (e_0, \ldots, e_m)$ and $v_g = (g_0, \ldots, g_m)$. Choosing $e_0 = 0$ and dividing both sides of (22.7) by θ, we have

$$\sum_{i=0}^{m-1} e_{i+1} \binom{m}{i+1} \theta^i (1 - \theta)^{m-1-i} = \sum_{i=0}^{m} g_i \binom{m}{i} \theta^i (1 - \theta)^{m-i}. \qquad (22.8)$$

The first term in the right hand side can be expanded in the following manner:

$$\begin{aligned}
g_0 (1 - \theta)^m &= g_0 \left[(1-\theta)^{m-1} - \theta (1-\theta)^{m-1} \right] \\
&= g_0 \left[(1-\theta)^{m-1} - \theta (1-\theta)^{m-2} + \theta^2 (1-\theta)^{m-2} \right] = \cdots \\
&= g_0 \left[(1-\theta)^{m-1} - \theta (1-\theta)^{m-2} + \cdots + (-1)^{m-1} \theta^{m-1} + (-1)^m \theta^m \right].
\end{aligned}$$

Expanding the other terms similarly, we obtain after some calculation that

$$\sum_{i=0}^{m} g_i \binom{m}{i} \theta^i (1-\theta)^{m-i} = \left[\sum_{i=0}^{m-1} \langle c_i, v_g \rangle \theta^i (1-\theta)^{m-1-i} \right] + \langle c_m, v_g \rangle \theta^m, \qquad (22.9)$$

where $\langle \cdot, \cdot \rangle$ is the inner product in $\mathbb{R}^{m+1}$ and the vectors $c_i, i = 0, \ldots, m$ are defined by the relations

$$\langle c_i, v_g \rangle = \sum_{j=0}^{i} (-1)^j \binom{m}{i-j} g_{i-j}.$$

It can be checked that $\langle c_m, v \rangle = 0$ if v is a vector representing any of the monomials in y of degree $\leq m - 1$. If $g(y)$ is a polynomial in y of degree at most $m - 1$, then $\langle c_m, v_g \rangle = 0$ and it follows that (22.7) is satisfied if we set $e(0) = 0$ and

$$e(i + 1) = \frac{\langle c_i, v_g \rangle}{\binom{m}{i+1}} \qquad i = 0, \ldots, m - 1. \qquad (22.10)$$

Hence, we have the result that, if $g(\cdot)$ is a degree $m - 1$ polynomial, then an unbiased estimate of $E\big(g(y) - \theta\big)^2$ is $g(y)^2 - 2e(y) + (y^2 - y)/(m^2 - m)$.

In general, for an arbitrary $g(\cdot)$, we have $\langle c_m, g \rangle = \|u_g\|^2$ where u_g is the component of v_g perpendicular to the vectors representing polynomials of degree $\leq m - 1$. If $g(\cdot)$ is any reasonable estimator of θ, this component u_g should be very small. In this general case, the estimator (22.10) can be improved in the following way. Let $\nu(\cdot)$ be an appropriate measure on $[0, 1]$ such that θ^m and $\theta^i (1 - \theta)^{(m-1)-i}$, $i = 0, \ldots, m-1$, are all square integrable w.r.t. $\nu(\cdot)$. Let $\theta^m = s(\theta) + r(\theta)$ where $s(\theta) = \sum_{i=0}^{m-1} \alpha_i \theta^i (1 - \theta)^{(m-1)-i}$ is the projection of θ^m onto the space spanned by $\theta^i (1 - \theta)^{m-1-i}$, $i = 0, \ldots, m-1$. Then the expression (22.10) should be modified to

$$e(i + 1) = \big(\langle c_i, v_g \rangle + \langle c_m, v_g \rangle \alpha_i\big) \Big/ \binom{m}{i+1}. \qquad (22.11)$$

In this case, $e(y)$ is not exactly unbiased for $\theta E_\theta g(y)$, the bias is $\|u_g\|^2 \theta r(\theta)$. The L_2 norm (w.r.t. ν) of this bias is often very small. For example, with $m = 3$, $\nu(\cdot) = $ Lebesque measure, exact calculation shows that the L_2 norm of $r(\theta)$ is 0.0189. If m is larger, the norm of $r(\cdot)$ would be much smaller.

22.6. Convergence of the Loss Estimate

In the preceeding sections we have provided constructions of unbiased (or nearly unbiased) estimator of $\frac{1}{n} \sum_{1}^{n} \ell_i(\theta_i, \widehat{\theta}_i)$. The estimator is of the form $\frac{1}{n} \sum_{1}^{n} t_i(\boldsymbol{y})$ where $t_i(\boldsymbol{y})$ is determined by the form of $\widehat{\theta}_i(y_1, \ldots, y_n)$ as an univariate function of y_i. The error of this estimator of the loss is $\frac{1}{n} \sum_{1}^{n} (t_i - \ell_i)$ where, by construction, each term $(t_i - \ell_i)$ has zero expectation. We now argue that, under quite general conditions, the error $\frac{1}{n} \sum_{1}^{n} (t_i - \ell_i)$ is expected to converge to zero.

One condition for $\frac{1}{n} \sum_{1}^{n} (t_i - \ell_i)$ to converge to zero can be stated loosely as follows: The value of $\frac{1}{n} \sum_{1}^{n} (t_i - \ell_i)$ should not depend much on $(y_1, \ldots, y_m)$ if $n \gg m$. Basically, we want events concerning the limiting behavior of $\frac{1}{n} \sum_{1}^{n} (t_i - \ell_i)$ to belong to the tail σ-field generated by $y_1, y_2, \ldots$. If this is true, then the zero-one law implies that $\frac{1}{n} \sum_{1}^{n} (t_i - \ell_i)$ converges to a constant which must necessarily be zero. Unfortunately, it is not easy to formulate the technical conditions on $\widehat{\theta}_i(\cdot, \cdot)$ and $\ell_i(\cdot, \cdot)$ to ensure measurability of $\frac{1}{n} \sum_1^n (t_i - \ell_i)$ with respect to the tail σ-field. Furthermore, the zero-one law does not give any indication on the speed of the convergence. For these reasons we will instead investigate the convergence of the loss estimate by direct variance calculations. We will show that for a very large class of estimators $\widehat{\theta}(\boldsymbol{y})$, the variance of $\frac{1}{n} \sum_1^n (t_i - \ell_i)$ is of order n^{-1}. For simplicity, we will only consider the case of mean square error loss.

Recall that we are estimating $\ell_i = g_i(y_i)^2 - 2\theta_i g_i(y_i) + \theta_i^2$ by $t_i = g_i(y_i)^2 - 2e_i(y_i) + f_i(y_i)$ where $e_i(y_i)$ and $f_i(y_i)$ are constructed to be unbiased estimators of $\theta_i g_i(y_i)$ and θ_i^2 respectively. Hence

$$\frac{1}{n} \sum_1^n (t_i - \ell_i) = -\frac{2}{n} \sum_1^n \left[e_i(y_i) - \theta_i g_i(y_i)\right] + \frac{1}{n} \sum_1^n \left[f_i(y_i) - \theta_i^2\right].$$

Since the functions $f_i(\cdot)$ are (non-random) functions of y_i alone, we have

$$\mathrm{Var}\left(\frac{1}{n} \sum_1^n (f_i - \theta_i^2)\right) \leq \frac{c}{n}$$

provided each $\mathrm{Var}(f_i) = \int p_{\theta_i}(y) \left(f_i(y) - \theta_i^2\right)^2 dy \leq c$. The analysis of the variance of $\frac{1}{n} \sum (e_i - \theta_i g_i)$ is considerably more complicated. The reason is that both $e_i(\cdot)$ and $g_i(\cdot)$ are random functions, i.e. the values of $g_i(y_i)$ and $e_i(y_i)$ depends not only on y_i but also on $\boldsymbol{y}_{(-i)}$. As a result, all terms in the average are generally dependent on each other. To proceed further, let T_i be the operator which maps the function $g_i(\cdot)$ to the function $e_i(\cdot)$, i.e. T_i is constructed so that for any (non-random) function $g(\cdot)$ of y_i, $(T_i g)(y_i)$ is unbiased for $\theta_i g(y_i)$. T_i is assumed to have the following properties:

a) (Unbiasedness) For all $g \in \mathbb{G}_i$ where $\mathbb{G}_i$ is a large linear space of (non-random) functions of y_i (which is defined separately in each application) we have

$$E\left[(T_i g)(y_i) - \theta_i g(y_i)\right] = 0. \tag{22.12}$$

b) (Linearity) For any $f, g \in \mathbb{G}_i$, $a, b \in \mathbb{R}$,

$$T_i(af + bg) = a T_i f + b T_i g. \tag{22.13}$$

c) If $g(y) \equiv c$ where c is a constant, then

$$(T_i g)(y) = c u_i(y). \tag{22.14}$$

where $u_i(y)$ is an unbiased estimate for θ.

The forms of T_i in several examples have been obtained in the preceeding sections:

$$\begin{aligned}
(T_i g)(y_i) &= y_i g(y_i - 1) &&\text{(Poisson example)} \\
(T_i g)(y_i) &= y_i g(y_i) - g'(y_i) &&\text{(Normal)} \\
(T_i g)(y_i) &= \frac{1}{y_i^{k-1}} \int_0^{y_i} z^{k-1} g(z) dz &&\text{(Gamma)} \\
(T_i g)(y_i) &= y_i g(y_i) + 2\alpha c (t * g')(y_i) &&\text{(Stable laws)}
\end{aligned}$$

T_i can also be applied to a multivariate function $h(y_1, \ldots, y_n)$, in which case $(T_i h)(y_1, \ldots, y_n)$ is obtained by regarding $h(y_1, \ldots, y_n)$ as a univariate function of y_i, with $\boldsymbol{y}_{(-i)}$ fixed, and then applying T_i to this univariate function. For example, it follows from (22.14) that; if h is a function of $\boldsymbol{y}_{(-i)}$, then

$$(T_i h)(y_1, \ldots, y_n) = u_i(y_i) h(\boldsymbol{y}_{(-i)}). \tag{22.15}$$

A function $h(y_1, \ldots, y_n)$ is said to belong to the domain of T_i $\left(h \in \mathcal{D}(T_i)\right)$ if $E(h^2) < \infty$ and

$$E\big[(T_i - \theta_i)h\big]^2 < c_i E(h^2), \tag{22.16}$$

where c_i is a constant which is typically equal to the squared norm of T_i as an operator on the class $\mathbb{G}_i$ of univariate functions of y_i. Also, let R_j be the operator representing expectation over y_j conditional on $\boldsymbol{y}_{(-j)}$, i.e.

$$\begin{aligned}
(R_j h)(y_1, \ldots, y_n) &= E(h \mid \boldsymbol{y}_{(-j)}) \\
&= \int h(y_1, \ldots, y_n) p_{\theta_j}(y_i) dy_j.
\end{aligned}$$

Suppose that $\widehat{\theta}_i(y_1, \ldots, y_n)$ has an ANOVA decomposition

$$\widehat{\theta}_i(\boldsymbol{y}) = g_i(y_i) = \mu_i + \sum_{j=1}^n \alpha_j^i H_j^i(y_j) + \sum_{\{j_1, j_2\}} \beta_{\{j_1, j_2\}}^i H_{\{j_1, j_2\}}^i(y_{j_1}, y_{j_2}) + \cdots) \tag{22.17}$$

where H_j^i, H_{j_1, j_2}^i etc. are orthogonal random variables satisfying the conditions

$$R_j H_{\{j_1, \ldots, j_k\}}^i = 0 \qquad \text{if} \qquad j \in \{j_1, \ldots, j_k\}. \tag{22.18}$$

For the construction of such decompositions, see Efron and Stein (1981). We assume that each $\widehat{\theta}_i$ has an expansion (22.17) up to m terms where m is independent of n, and that H_j^i, $H_{\{j_1,j_2\}}^i$ etc. are in $\mathcal{D}(T_i)$ and all of them have variance $\leq M$. Finally, the coefficients α_j^i, $\beta_{\{j_1,j_2\}}^i$ etc. are non-negative constants such that

$$\sum_j \alpha_j^i = \sum_{\{j_1,j_2\}} \beta_{\{j_1,j_2\}}^i = \cdots = 1.$$

Theorem 22.1: Suppose T_i, $i = 1,\ldots,n$ satisfy (22.12)–(22.16) with $c_i \leq c_0 < \infty$, and T_i commutes with R_j whenever $j \neq i$. Suppose $\widehat{\theta}_i$ $i = 1,\ldots,n$ have ANOVA decompositions satisfying the assumptions of the preceeding paragraph, then there exists a constant $c > 0$ such that

$$\operatorname{Var}\left[n^{-1} \sum_1^n \left(e_i(y_i) - \theta_i g_i(y_i) \right) \right] \leq \frac{c}{n}.$$

Proof: Observe that

$$\sum_1^n (e_i - \theta_i g_i) = \sum_{i=1}^n (T_i - \theta_i)\widehat{\theta}_i$$

$$= \left[\sum_{i=1}^n (T_i - \theta_i)\mu_i \right] + \left[\sum_{i=1}^n \sum_j \alpha_j^i (T_i - \theta_i) H_j^i \right]$$

$$+ \left[\sum_{i=1}^n \sum_{\{j_1,j_2\}} \beta_{\{j_1,j_2\}}^i (T_i - \theta_i) H_{\{j_1,j_2\}}^i \right] + \cdots,$$

where there are m terms in this expansion. We will only demonstrate the bound for the variance of, say, the third order interaction term. By (22.15), this third order term can be written as $A + B$ where

$$A = \sum_{i=1}^n \sum_{i \notin \{j_1,j_2,j_3\}} \gamma_{\{j_1,j_2,j_3\}}^i \left(u_i(y_i) - \theta_i \right) H_{\{j_1,j_2,j_3\}}^i,$$

$$B = \sum_{i=1}^n \sum_{\{j_2,j_3\}} \gamma_{\{i,j_2,j_3\}}^i (T_i - \theta_i) H_{\{i,j_2,j_3\}}^i.$$

To bound the variance of A, consider

$$E\left[(u_i - \theta_i) H_{\{j_1,j_2,j_3\}}^i (u_k - \theta_k) H_{\{\ell_1,\ell_2,\ell_3\}}^k \right]. \tag{22.19}$$

If $k \notin \{i, j_1, j_2, j_3\}$ we can replace $(u_k - \theta_k)$ by $R_k(u_k - \theta_k) = 0$ in (22.19). Similarly, if $\ell_2 \notin \{i, j_1, j_2, j_3\}$, we can replace $H_{\{\ell_1,\ell_2,\ell_3\}}^k$

by $R_{\ell_2}(H^k_{\{\ell_1,\ell_2,\ell_3\}}) = 0$. Hence (22.19) is zero unless $\{k,\ell_1,\ell_2,\ell_3\} = \{i,j_1,j_2,j_3\}$. Also, by our assumptions on the norm of T_i and the variance of $H^i_{\{j_1,j_2,j_3\}}$, the absolute value of (22.19) is bounded by $c_0 M$. It follows that

$$
\begin{aligned}
\text{Var}(A) &\leq c_0 M \sum_{i=1}^{n} \sum_{i \notin \{j_1,j_2,j_3\}} \gamma^i_{\{j_1,i,j_3\}} \, [\gamma^i_{\{j_1,j_2,j_3\}} + \gamma^{j_1}_{\{i,j_2,j_3\}} + \gamma^{j_2}_{\{j_1,i,j_3\}} \\
&\quad + \gamma^{j_3}_{\{j_1,j_2,i\}}] \\
&\leq 4c_0 M \sum_{i=1}^{n} \left(\sum_{\{j_1,j_2,j_3\}} \gamma^i_{\{j_1,j_2,j_3\}} \right) \\
&\leq 4c_0 M n.
\end{aligned}
$$

To bound the variance of B, notice that by applying (22.12) with $g(y_i) = H^i_{\{i,j_2,j_3\}}(y_i, y_{j_2}, y_{j_3})$ where y_{j_2} and y_{j_3} are fixed, we have $R_i(T_i - \theta_i)H^i_{\{i,j_2,j_3\}} = 0$. Also, since $j_2 \neq i$, it follows that

$$
R_{j_2}(T_i - \theta_i)H^i_{\{i,j_2,j_3\}} = (T_i - \theta_i)R_{j_2}H^i_{\{i,j_2,j_3\}} = 0.
$$

Using these equalities and repeating the same type of arguments used to bound $\text{Var}(A)$, we have

$$
\text{Var}(B) \leq 3c_0 M n.
$$

This completes the derivation of the bound for the third order interaction term. The same argument can be applied to bound the variance of any other term. ∎

Acknowledgments

The research was supported in part by the NSF grants DMS 9204504 and 0505732.

References

Efron, B. and Stein, C. (1981). The jackknife estimate of variance. *Ann. Statist.* **9**, 586–596.

Stein, C. (1981). Estimation of the mean of a multivariate normal distribution. *Ann. Statist.* **9**, 1135–1151.

Subject Index

Author Index